Progress in Wall Turbulence 2

ERCOFTAC SERIES

VOLUME 23

Series Editor

Bernard Geurts
Faculty of Mathematical Sciences, University of Twente, Enschede,
The Netherlands

Aims and Scope of the Series

ERCOFTAC (European Research Community on Flow, Turbulence and Combustion) was founded as an international association with scientific objectives in 1988. ERCOFTAC strongly promotes joint efforts of European research institutes and industries that are active in the field of flow, turbulence and combustion, in order to enhance the exchange of technical and scientific information on fundamental and applied research and design. Each year, ERCOFTAC organizes several meetings in the form of workshops, conferences and summerschools, where ERCOFTAC members and other researchers meet and exchange information.

The ERCOFTAC Series will publish the proceedings of ERCOFTAC meetings, which cover all aspects of fluid mechanics. The series will comprise proceedings of conferences and workshops, and of textbooks presenting the material taught at summerschools.

The series covers the entire domain of fluid mechanics, which includes physical modelling, computational fluid dynamics including grid generation and turbulence modelling, measuring-techniques, flow visualization as applied to industrial flows, aerodynamics, combustion, geophysical and environmental flows, hydraulics, multiphase flows, non-Newtonian flows, astrophysical flows, laminar, turbulent and transitional flows.

More information about this series at http://www.springer.com/series/5934

Michel Stanislas · Javier Jimenez
Ivan Marusic
Editors

Progress in Wall Turbulence 2

Understanding and Modelling

Springer

Editors
Michel Stanislas
LML UMR 8107
Ecole Centrale de Lille
Villeneuve d'Ascq
France

Ivan Marusic
Mechanical Engineering
University of Melbourne
Parkville, VIC
Australia

Javier Jimenez
E.T.S. Ingenieros Aeronauticos
Polytechnic University of Madrid
Madrid
Spain

ISSN 1382-4309 ISSN 2215-1826 (electronic)
ERCOFTAC Series
ISBN 978-3-319-20387-4 ISBN 978-3-319-20388-1 (eBook)
DOI 10.1007/978-3-319-20388-1

Library of Congress Control Number: 2015945138

Springer Cham Heidelberg New York Dordrecht London

Printed on acid-free paper

Springer International Publishing AG Switzerland is part of Springer Science+Business Media
(www.springer.com)

Preface

This book brings together contributions from participants of the second WALLTURB workshop on “Understanding and modelling of wall turbulence” held in Lille (France) from June 18 to 20, 2014.

This workshop follows the inaugural workshop organized in 2009 by the WALLTURB EC project, and aimed to assess the progress made in the field of near-wall turbulence in the 5 years separating the two workshops.

The workshop assembled 60 participants from all over the world, with 6 invited lecturers and 39 contributions, and provided an opportunity to review the recent progress in theoretical, experimental and numerical approaches to wall turbulence.

This book gathers papers from most of the contributors to the workshop. It is aimed as being a milestone in the research field, thanks to the high level of invited speakers and the involvement of the contributors.

Lille
Madrid
Melbourne
August 2009

Michel Stanislas
Javier Jimenez
Ivan Marusic

Contents

Part I
Invited Lectures

On the Size of the Eddies in the Outer Turbulent Wall Layer: Evidence from Velocity Spectra

Sergio Pirozzoli

Abstract The scaling of size of the momentum-bearing eddies in wall-parallel planes in the outer part of turbulent wall layers is analyzed, by examining spectra of the fluctuating velocities taken from direct numerical simulations and experiments. For all flows under scrutiny the normalized spectra highlight growth of the eddies size with the wall distance. The results indicate the capability of a modified mixing length (Pirozzoli, J Fluid Mech 702:521–532, 2012 [1]) of accounting with greater precision for the wall-normal variation of the size of the eddies bearing streamwise momentum. This observation can be explained by assuming that outer layer momentum streaks (superstructures) spread under the collective action of the other eddies, which impart a (nearly) uniform eddy diffusivity throughout the outer wall layer.

Keywords Wall-bounded flows · Turbulence · Coherent structures

1 Introduction

This is the follow-up of a previous study [1] (hereafter referred to as P12), in which the issue of the proper scaling of the size of eddies (broadly defined as regions which retain some degree of coherence in turbulent flow) was addressed. The main results of that study were that, in the outer part of turbulent wall layers (including channels and boundary layers, also in the compressible regime) the spanwise size of the eddies, as measured through the integral correlation length scale, is controlled by a modified mixing length, defined as

$$\frac{\ell_{1/2}}{\delta} \sim \left(\frac{\partial u^+}{\partial \eta}\right)^{-1/2}, \tag{1}$$

where $\eta = y/\delta$, $u^+ = \tilde{u}/u_\tau^*$, y is the distance from the wall, $u_\tau^* = (\tau_w/\overline{\rho})^{1/2}$ is the effective friction velocity at a given off-wall location (at the wall, u_τ^* coincides with

S. Pirozzoli (✉)
Dipartimento di Ingegneria Meccanica E Aerospaziale, Università di Roma 'La Sapienza', Rome, Italy
e-mail: sergio.pirozzoli@uniroma1.it

M. Stanislas et al. (eds.), *Progress in Wall Turbulence 2*, ERCOFTAC Series 23, DOI 10.1007/978-3-319-20388-1_1 the conventional definition of the friction velocity, u_τ), τ_w is the wall shear stress, $\overline{\rho}$ is the mean local density, δ is the thickness of the wall layer, and $\tilde{u} = \overline{\rho u}/\overline{\rho}$ (which coincides with $\overline{u}$ in strictly incompressible flow). The scaling given in Eq. (1) (hereafter referred to as the modified mixing length) should be contrasted with Prandtl's classical mixing length, namely $\ell_m \sim y$, and with its proposed generalization [2]

$$\frac{\ell_m}{\delta} \sim \left(\frac{\partial u^+}{\partial \eta}\right)^{-1}, \tag{2}$$

which reduces to Prandtl's scaling in the presence of a sizeable log layer. As shown by Mizuno and Jiménez [2], Eq. (2) provides a better prediction for the growth of the eddies with the wall distance than the wall distance itself, and hence it will be hereafter used instead of the classical Prandtl's scaling.

In this paper, we address the issue of the proper scaling of the outer layer eddies in greater detail than in P12 by: (i) scrutinizing the spectral densities of the velocity components, to separately establish the growth of the various scales of motion with the wall distance; (ii) including additional numerical and experimental data, to further verify the robustness of the observations.

2 Numerical and Experimental Data

The numerical database used for the analysis includes direct numerical simulations (DNS) of compressible boundary layer flows and incompressible channel flows of the Couette–Poiseuille family. The boundary layer data cover the range of friction Reynolds numbers, $Re_\tau = 250 - 4000$ (with $Re_\tau = \delta/\delta_v$, where $\delta_v = \nu_w/u_\tau$ is the inner length scale, and δ is the boundary layer thickness), and free-stream Mach number $M_\infty \leq 4$. Details on the numerical implementation and mesh resolution are provided in earlier studies [3, 4], and will not be repeated here. The channel flow database includes DNS performed in a rectangular channel (whose half-height is h), and it includes simulations with stationary walls (corresponding to pure turbulent Poiseuille flow, cases P, PH, PHH), and simulations with with one stationary (S) and one moving (M) wall, the controlling parameters being the bulk Reynolds number and the ratio of the shear stress at the two walls $\gamma = \tau_M/\tau_S$ ($\gamma = -1$ for Poiseuille flow). Specifically, flow case P1 corresponds to a Poiseuille-like flow with reduced shear; flow cases C1 and C are Couette-like flows, the latter being very close to pure Couette flow. Details on the channel flow simulations can be retrieved in the original references [5, 6]. Two Poiseuille flow cases from the DNS database of del Álamo et al. [7] were also included (labeled as TO1, TO2). It is noteworthy that the M2HH and the PHH simulations represent the upper limit of Reynolds number currently reached in DNS of boundary layers and channel flows, and they both exhibit slightly less than a decade of quasi-logarithmic variation of the mean velocity [4, 6], thus satisfying the requirements [8, 9] for being representative of high-Reynolds-number wall turbulence.

Table 1 Flow properties for numerical turbulent boundary layers (left) and Couette–Poiseuille flows (right)

Flow case	M_∞	Re_τ	Re_θ	Flow case	γ	$Re_{\tau M}$	$Re_{\tau S}$
M03	0.3	432	1098	P	−1	284	284
M2L	2	251	1122	P1	−0.24	128	261
M2	2	508	2434	C1	0.27	130	250
M2H	2	1116	6046	C	0.98	242	245
M2HH	2	3940	18787	PH	−1	541	541
M3	3	504	3867	PHH	−1	4080	4080
M4	4	506	5824	TO1	−1	991	991
				TO2	−1	2000	2000

In the table on the left, $Re_\theta = \rho_\infty u_\infty \theta/\mu_\infty$, $Re_\tau = \delta/\delta_v$, where θ is the momentum thickness. In the table on the right, $\gamma = \tau_M/\tau_S$, $Re_{\tau M,S} = u_{\tau M,S} h/\nu$, $u_{\tau M,S} = (|\tau_{M,S}|/\rho)^{1/2}$, where the subscripts M and S refer to the moving and the stationary wall, respectively

In the following, δ is stipulated to be the 99 % velocity thickness for boundary layers, even though other choices are possible. In the channel flow cases the outer length scale is defined to be the distance from the stationary wall to the point where the mean shear becomes zero, if any (hence, $\delta = h$ for flow cases P, PH, PHH, and TOx). The main parameters for the DNS hereafter reported are shown in Table 1.

As will be made clearer in the analysis, verification of the scaling formulas (1) and (2) requires access to the spectral densities of the velocity components in the spanwise and/or the streamwise direction at several wall distances. For that purpose, we have considered the boundary layer experimental database of Hutchins et al. [10] which covers friction Reynolds numbers in the range $Re_\tau = 500-2300$, and which includes two-point correlations of streamwise velocity fluctuations in the spanwise direction at several off-wall stations. For the sake of preciseness, we point out that the values of the friction Reynolds number quoted here are slightly lower than those reported in the original reference, because of the different definition of the boundary layer thickness, which was estimated in experiments through a Coles law-of-the-wake fit of the mean velocity profile. We also consider the pipe flow experimental database by Bailey and Smits [11], which covers a single Reynolds number, $Re_\tau = 3400$, and which includes spanwise and streamwise correlations of u at five off-wall stations [11]. The reader is referred to the original references for details on the experimental setup.

3 Spectral Densities of Velocity Fluctuations

The power spectral densities associated with the u, v, w velocity fluctuations (respectively, the streamwise, wall-normal, and spanwise components) have been analyzed for all flows presented in Sect. 2, at various wall distances, limited to $y/\delta \geq 0.1$. Here and in the following '+' superscripts will be used to denote quantities made nondimensional with respect to u_τ and δ_v. The attention is mainly focused on the spanwise spectra, but information on the scaling of the streamwise spectra is also provided. For the sake of the analysis we define the power spectral densities of the generic property φ in the ith direction, E^i_φ, in such a way that

$$\overline{\varphi'^2} = \int_0^\infty E^i_\varphi(k_i)\,\mathrm{d}k_i, \tag{3}$$

where k_i is the Fourier wavenumber in the ith direction. To account for the effect of turbulence kinetic energy variation across the wall layer, we consider the normalized energy spectra, defined as

$$\hat{E}^i_\varphi(k_i) = E^i_\varphi(k_i)/\overline{\varphi'^2}. \tag{4}$$

As shown in standard textbooks, the integral length scale of φ in a given homogeneous direction, defined as the integral of the corresponding two-point correlation function

$$\Lambda^i_\varphi = \int_{-\infty}^\infty C_{\varphi\varphi}(\Delta x_i)\,\mathrm{d}\Delta x_i, \tag{5}$$

is linked with the corresponding normalized spectra through the relation

$$\Lambda^i_\varphi = \lim_{k_i \to 0} \pi \hat{E}^i_\varphi(k_i). \tag{6}$$

Equation (6) shows that the integral length scale is closely connected with the asymptotic behavior of the power spectrum at the largest scales.

The typical organization of the spanwise velocity spectra is shown in Fig. 1, for some boundary layer DNS at $y/\delta = 0.3$. Note that, for the sake of representation, wavelengths ($\lambda_z = 2\pi/k_z$) are shown on the horizontal axis, rather than the corresponding wavenumbers, here scaled by the boundary layer thickness. The spectra of u and v (those of w are similar to the former) are both bump shaped, and they exhibit

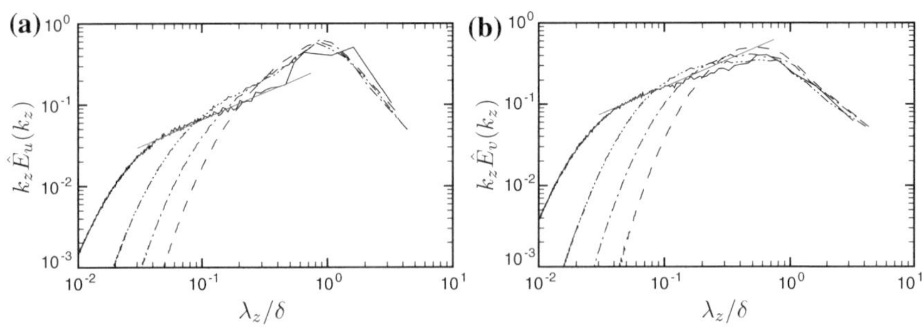

Fig. 1 Premultiplied spanwise spectral densities of u (**a**) and v (**b**) at $y/\delta = 0.3$ for flow cases M2L (*dashes*); M2 (*dash-dot*); M2H (*dash-dot-dot*); M2HH (*solid*). The straight diagonal lines mark a $k_z^{-5/3}$ behavior

a $k_z^{-5/3}$ power-law behavior in the small-scale range, which reflects the establishment of a locally isotropic energy cascade [12], and which becomes evident at the highest Re_τ.

It is the main goal of the forthcoming discussion to verify whether this variation can be compensated through a suitable scale transformation. For that purpose, in the following we will inspect the premultiplied normalized energy spectra as a function of the wavelength ($\lambda_i = 2\pi/k_i$), scaled by a suitable mixing length (either ℓ_m or $\ell_{1/2}$), in semi-log scale. This type of representation provides hints on the possible universality of the distributions, while retaining the illustrative advantage that equal areas underneath the graphs correspond to equal energies.

3.1 DNS Data

We first consider the spanwise velocity spectra extracted from DNS of boundary layer and channel flows. The data are shown at several off-wall stations, namely $y/\delta = 0.1\,i, i = 1, \ldots, 7$, thus covering the whole outer layer, well into the wake region. Only the spectra of u and v are shown in the following for space limitations, those of w being similar to those of u. The u spectra scaled in classical mixing length units, shown in the left column of Fig. 2 for selected flow cases, have a reasonable degree of universality. This observation was also made by Mizuno and Jiménez [2], and taken in support for the validity of the classical mixing length scaling, in the generalized form given in Eq. (2). However, careful inspection of the figures shows consistent shift of the scaled spectra to shorter wavelengths as the wall distance increases (from the dotted curves to the solid curves). This discrepancy is more evident at the large scales (i.e., the right end of the spectra), which explains the poor collapse of the integral length scales with ℓ_m observed in P12. The spectra of v scaled with ℓ_m, shown in the right column of Fig. 2, have a slightly different behavior. While the same shift of the normalized spectra of the small scales is found as for u, in this case the trend at the largest resolved scales appears to be nearly insensitive to y, and this is again consistent with the success of ℓ_m in parameterizing the integral length scales of v observed in P12.

The u and v spanwise spectra scaled with the modified mixing length are shown in Fig. 3. Of course, the qualitative nature of the graphs does not change from Fig. 2, since a y-dependent shift of the horizontal axis is involved. However, all the u-spectra exhibit approximate collapse with the wall distance in this case. In fact, the spectral energy density of both the small and the large scales becomes nearly y-independent, and scatter is mainly found in the peak amplitude associated with what we defined as energy-containing eddies. Greater scatter is observed in the DNS at the higher Reynolds number (M2HH and PHH), which is likely caused by slow convergence of the outer layer spectra. It is noteworthy that in boundary layers the spectral peak is placed at $\lambda_z \approx 3\ell_{1/2}$, and at $\lambda_z \approx 4\ell_{1/2}$ in channel flows. Hence, in agreement with previous observations [13], eddies in channels are found to be somewhat larger than

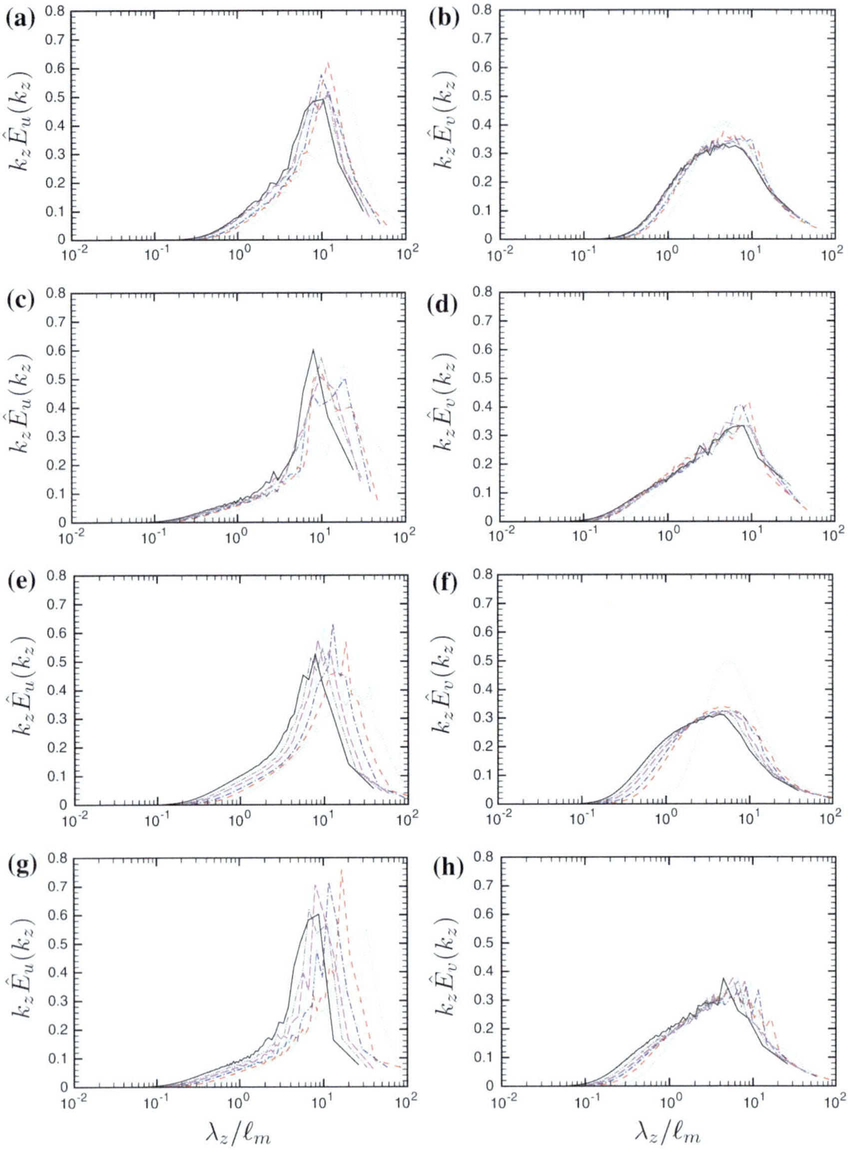

Fig. 2 DNS data: premultiplied spanwise spectral densities of u (*left column*) and v (*right column*) in classical mixing length scaling (ℓ_m, as defined in Eq. (2)), for flow cases M2H (**a**-**b**), M2HH (**c**-**d**), TO2 (**e**-**f**), PHH (**g**-**h**). The data are shown from $y/\delta = 0.1$ (*dots*) to $y/\delta = 0.7$ (*solid lines*), in intervals of 0.1δ

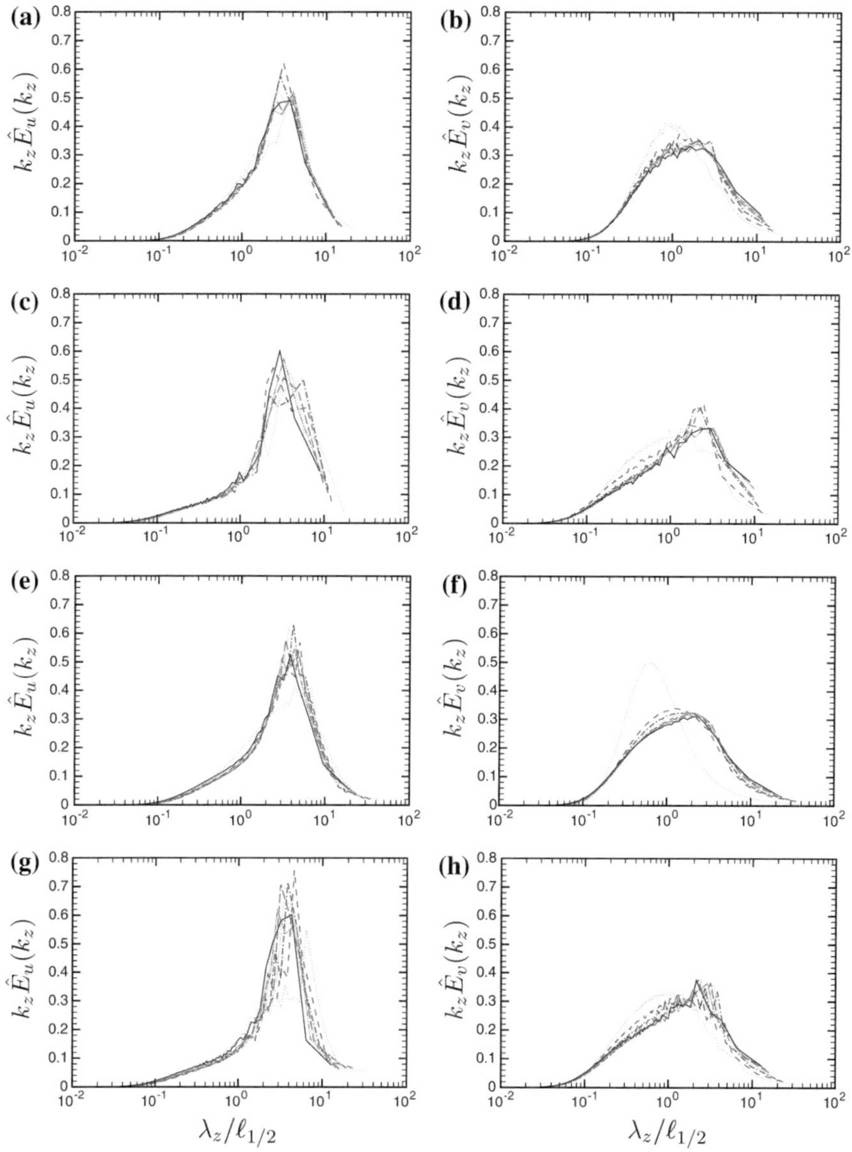

Fig. 3 DNS data: premultiplied spanwise spectral densities of *u* (*left column*) and v (*right column*) in modified mixing length scaling ($\ell_{1/2}$, as defined in Eq. (1)), for flow cases M2H (**a**-**b**), M2HH (**c**-**d**), TO2 (**e**-**f**), PHH (**g**-**h**). The data are shown from $y/\delta = 0.1$ (*dots*) to $y/\delta = 0.7$ (*solid lines*), in intervals of 0.1δ

in boundary layers. Comparing the v spectra in Fig. 3 with those in Fig. 2, one will see that both ℓ_m and $\ell_{1/2}$ perform similarly in removing the y dependence for all the flow cases. However, the $\ell_{1/2}$ scaling seems to be more effective in parameterizing the small eddies, whereas the large eddies are probably more accurately parameterized by ℓ_m. The position of the spectral peak is also less sensitive to y in the $\ell_{1/2}$ scaling, with $\lambda_z \approx 1 - 2\ell_{1/2}$ for boundary layers, and $\lambda_z \approx 2\ell_{1/2}$ for channels. Normalization of the spectra with respect to the wall distance was also attempted as a further possibility, but it was found to yield poorer data collapse than both $\ell_{1/2}$ and ℓ_m.

Streamwise spectra of u and v are shown in Fig. 4, limited to the TO2 case [7], scaled with respect to δ, ℓ_m, and $\ell_{1/2}$. In this case, δ scaling yields the best collapse of the u spectra across the whole range of scales, which implies that the longitudinal size of the u-bearing eddies does not depend on the wall distance to leading order,

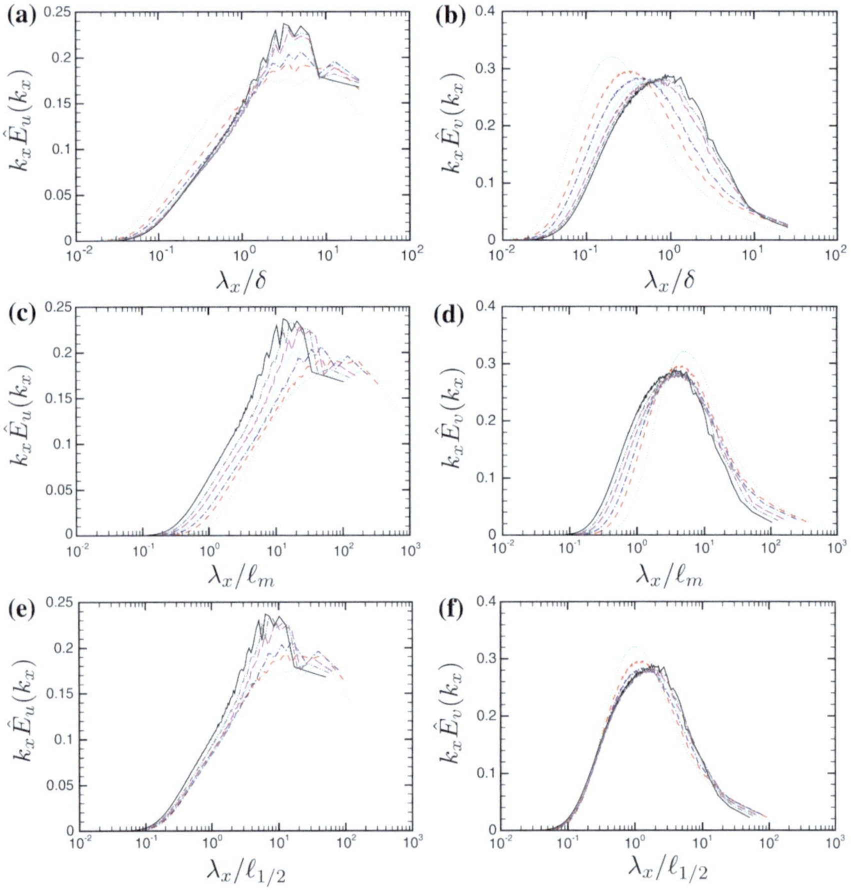

Fig. 4 Premultiplied streamwise spectral densities of u (*left column*) and v (*right column*) for flow case TO2. Wavelengths are scaled with respect to δ (**a-b**), ℓ_m (**c-d**), and $\ell_{1/2}$ (**e-f**). The data are shown from $y/\delta = 0.1$ (*dots*) to $y/\delta = 0.7$ (*solid lines*), in intervals of 0.1δ

and the energy-containing eddies are about $4-5\,\delta$ long. As pointed out by Hutchins and Marusic [9], the spectral peak of u in the outer layer is the imprint of large-scale momentum streaks, which have a typical meandering pattern in the streamwise direction. Thus, streamwise spectra do not necessarily convey exhaustive information on their actual length, and indeed visual inspection supports lengths in excess of 20δ [9]. As a consequence, the observed invariance of the δ-scaled u spectra with y is probably to be more correctly interpreted as evidence that the turns of the meanders have a spacing that does not depend on y. The streamwise spectra of v (right column of Fig. 4) appear again to be very accurately parameterized by $\ell_{1/2}$, both at the small and at the energy-containing scales. However, the large-scale end of the spectra seems to scale on δ, and the longitudinal integral scale of v is nearly y-independent.

3.2 Experimental Data

The boundary layer data from Hutchins et al. [10] are considered first. The spanwise spectra of u, which we estimated by Fourier transforming the corresponding two-point velocity correlations, shown in Fig. 5, further support the notion that the modified mixing length is significantly more accurate in accounting for the growth of

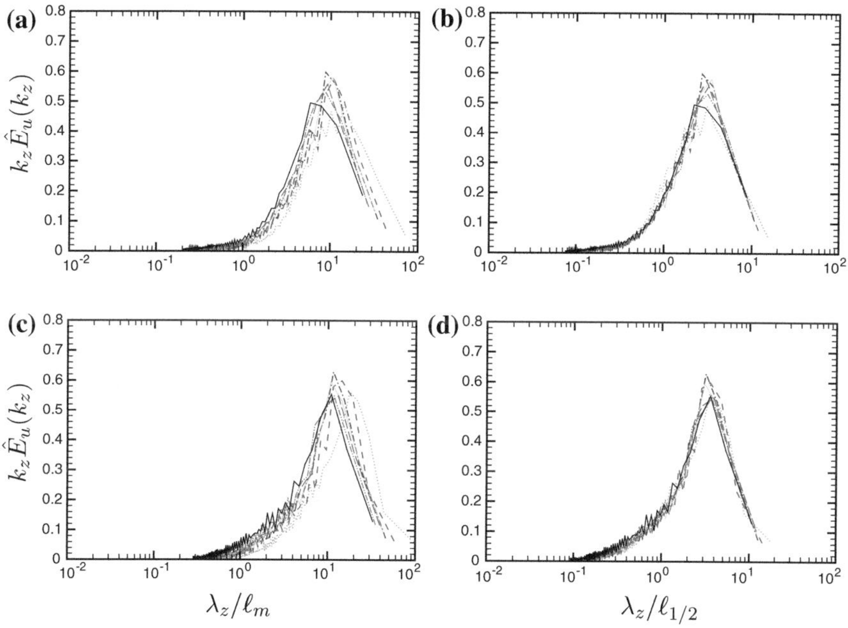

Fig. 5 Boundary layer experiments [10]: premultiplied spanwise spectral densities of u in classical mixing length scaling (*left column*) and in modified mixing length scaling (*right column*) at $Re_\tau = 546$ (**a-b**); 2287 (**c-d**). The data are shown from $y/\delta = 0.1$ (*dots*) to $y/\delta = 0.7$ (*solid lines*), in intervals of 0.1δ

the whole range of u-bearing eddies with y. Comparison with the numerical boundary layer spectra of u presented in Fig. 3 shows very close similarities, including the position of the spectral peak, which is very nearly constant ($\lambda_z \approx 3\ell_{1/2}$) across the range of wall distances and Reynolds numbers.

We next consider the pipe flow data from Bailey and Smits [11], shown in Fig. 6. The spanwise spectra (left column) exhibit some energy pileup at the smallest scales, caused by lack of smoothness of the spatial correlations. Nevertheless, we refrained from using any smoothing procedure, and we present the raw data. With this caveat, the spectra at the large-scale end appear to have little sensitivity on y, when scaled with $\ell_{1/2}$, with a peak at $\lambda_z \approx 3 - 4\ell_{1/2}$. On the other hand, the streamwise spectra

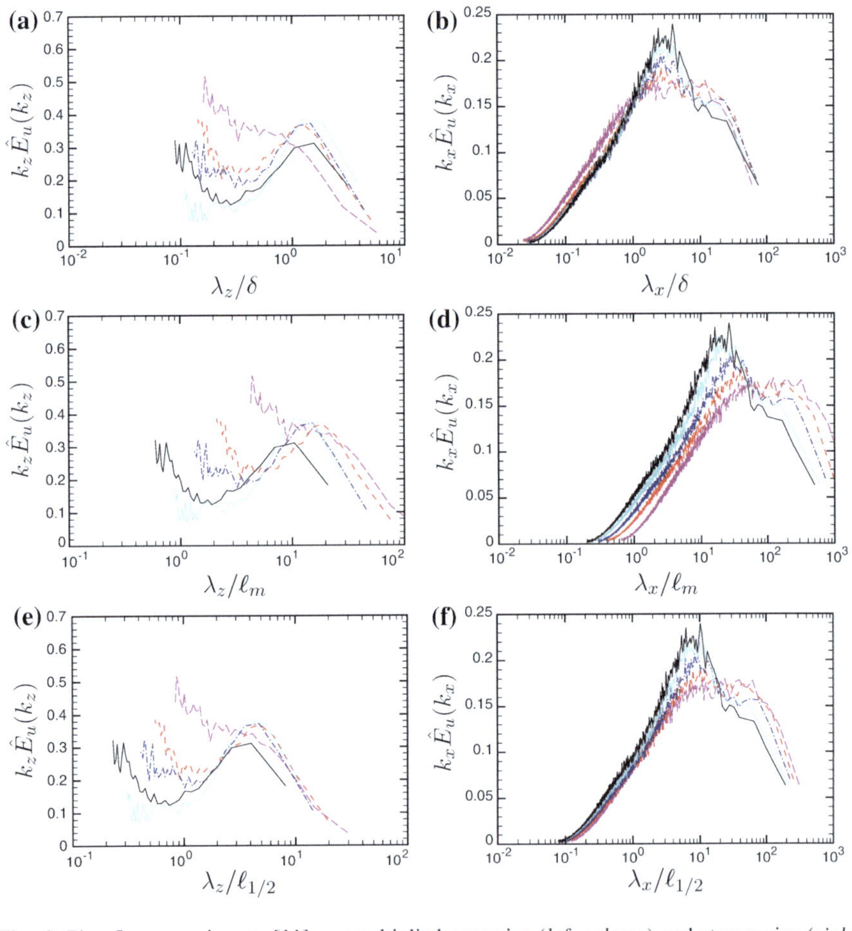

Fig. 6 Pipe flow experiments [11]: premultiplied spanwise (*left column*) and streamwise (*right column*) power spectral densities of u in δ scaling (**a**-**b**), classical mixing length scaling (**c**-**d**), and modified mixing length scaling (**e**-**f**). The data are shown from $y/\delta = 0.1$ (*dashed lines*), up to $y/\delta = 0.5$ (*solid lines*), in intervals of 0.1δ

(right column) in the large-scale range appear to be insensitive to y (and thus to scale on δ), whereas the small scales are perhaps equally well parameterized by $\ell_{1/2}$. The apparent insensitivity of the streamwise velocity spectra to y was also noticed by Bailey and Smits [11], who identified two separate peaks in the spectrum, corresponding to large-scale and very-large-scale motions, whose size varies weakly with y. They found that this dependence can be further minimized by using the same convection velocity at all off-wall stations when applying Taylor's hypothesis, and interpreted this fact with the argument that spectral peaks at different wall-normal positions correspond to the same coherent motions.

4 Conclusions

The key finding is certainly that the typical length scale of most eddies in wall-parallel planes is proportional to $\ell_{1/2}$, rather than to ℓ_m. Although the two alternatives are not vastly different, we believe that the consistent improvement in the universality of the scaled spectra in the $\ell_{1/2}$ normalization, observed for a variety of flow cases (only a part of which has been shown), is a sufficient proof of the validity of the new scaling. In this respect, as also observed by Chernyshenko and Baig [14], we argue that the growth of eddies in wall-parallel planes must be controlled by the interplay between the local mean shear and diffusion. Turbulent diffusion in wall-parallel planes in mainly associated with u and w velocity fluctuations. These "sloshing" motions (inactive motions, in Townsend's terminology, [15]) are not constrained by the impermeability condition at the wall, and hence their typical length scale is not bound to be proportional to the wall distance. This is not the case for turbulent transport in the vertical direction, which is controlled by wall-normal velocity fluctuations (active motions, according to Townsend), whose typical length scale is bound to be proportional (at least sufficiently close to the wall) to the wall distance. This observation points to the relevance of an eddy viscosity based on the friction velocity and on the total wall layer thickness in controlling turbulent diffusion processes in wall-parallel planes,

$$\nu_{tI} \sim u_\tau^* \delta, \tag{7}$$

where the subscript I denotes the contribution of inactive motions, to discriminate from the classical eddy viscosity. Note that the effective friction velocity (u_τ^*) may vary across the wall layer in compressible flow, owing to mean density variations. However, since density variations are concentrated in the wall proximity, ν_{tI} is approximately constant in the outer layer. Next, based on this observation, we assume that the growth of eddies in wall-parallel planes is controlled by turbulent diffusion. Borrowing arguments used by del Álamo et al. [7] we note that, on a time scale t, diffusion would yield eddies with size $\ell \sim (\nu_{tI} t)^{1/2}$. If the relevant eddy viscosity for diffusion is (nearly) constant across the wall layer, as given in Eq. (7), it appears that the size of the eddies should be the same at all off-wall locations. However, diffusion processes at a given wall distance are limited in time by the disruptive effect of the mean shear, whose typical time scale is

$$\tau_S = \left(\frac{\partial \tilde{u}}{\partial y}\right)^{-1}. \tag{8}$$

It is then reasonable to conclude that the typical size of eddies resulting from time-limited wall-parallel diffusion processes is

$$\ell \sim (\nu_{tI}\tau_S)^{1/2} \sim \left(u_\tau^* \delta\right)^{1/2} \left(\frac{\partial \tilde{u}}{\partial y}\right)^{-1/2}, \tag{9}$$

which is identical to the mixing length defined in Eq. (1).

It is important to state that by no means our arguments imply that the classical Prandtl's theory should be discarded. Indeed, our observations do not apply to turbulent transport in the wall-normal direction (hence, to the Reynolds shear stress), which is probably still controlled by the classical mixing length, as suggested from the analysis of the wall-normal velocity correlations [2]. Hence, it is expected that a signature of Prandtl's scaling is observed in the wall-parallel spectra of v. This is consistent with the observation that the largest v-bearing eddies scale on ℓ_m, rather than on $\ell_{1/2}$.

We expect that the present results, while not directly relevant for mean momentum balance in wall-bounded flows, may shed some light onto the mechanisms of growth of turbulent eddies in the strongly anisotropic wall layer, and perhaps to build deterministic models for the outer layer eddies, in the same line of thought as Townsend's conical eddies [15]. Also, our results might be relevant for the study of the dispersion of contaminants/pollutants in shear flow. Investigation of the two-dimensional spectra in wall-parallel planes and of the vertical velocity correlations, currently ongoing, might lead to fuller characterization of the shape and size of the typical outer layer coherent structures.

Acknowledgments I acknowledge that some results in this paper have been achieved using the PRACE Research Infrastructure resource JUGENE based at the Forschungszentrum Jülich (FZJ) in Jülich, Germany. Thanks are also due to Sean C.C. Bailey, Alexander J. Smits, Nicholas Hutchins, and Ivan Marusic, for providing experimental data and other useful information.

References

1. S. Pirozzoli, J. Fluid Mech. **702**, 521–532 (2012)
2. Y. Mizuno, J. Jiménez, Phys. Fluids **23**, 085112 (2011)
3. S. Pirozzoli, M. Bernardini, J. Fluid Mech. **688**, 120–168 (2011)
4. S. Pirozzoli, M. Bernardini, Phys. Fluids **25**, 021704 (2013)
5. S. Pirozzoli, M. Bernardini, P. Orlandi, J. Fluid Mech. **680**, 534–563 (2011)
6. M. Bernardini, S. Pirozzoli, P. Orlandi, J. Fluid Mech. (2013). Under review

7. J.C. del Álamo, J. Jiménez, P. Zandonade, R.D. Moser, J. Fluid Mech. **500**, 135 (2004)
8. J. Jiménez, R. Moser, Philos. Trans. R. Soc. Lond. A **365**, 715 (2007)
9. N. Hutchins, I. Marusic, J. Fluid Mech. **579**, 1 (2007)
10. N. Hutchins, W.T. Hambleton, I. Marusic, J. Fluid Mech. **541**, 21 (2005)
11. S. Bailey, A. Smits, J. Fluid Mech. **651**, 339 (2010)
12. J. Jiménez, Annu. Rev. Fluid Mech. **44**, 27 (2012)
13. J. Monty, J. Stewart, R. Williams, M. Chong, J. Fluid Mech. **589**, 147 (2007)
14. S.I. Chernyshenko, M.F. Baig, J. Fluid Mech. **544**, 99 (2005)
15. A. Townsend, *The Structure of Turbulent Shear Flow*, 2nd edn. (Cambridge University Press, Cambridge, 1976)

Sensitized-RANS Modelling of Turbulence: Resolving Turbulence Unsteadiness by a (Near-Wall) Reynolds Stress Model

Suad Jakirlić and Robert Maduta

Abstract A turbulence model designed and calibrated in the steady RANS (Reynolds-Averaged Navier-Stokes) framework has usually been straightforwardly applied to an unsteady calculation. It mostly ended up in a steady velocity field in the case of confined wall-bounded flows; a somewhat better outcome is to be expected in globally unstable flows, such as bluff body configurations. However, only a weakly unsteady mean flow can be returned with the level of unsteadiness being by far lower compared to a referent database. The latter outcome motivated the present work dealing with an appropriate extension of a near-wall Second-Moment Closure (SMC) RANS model towards an instability-sensitive formulation. Accordingly, a Sensitized-RANS (SRANS) model based on a differential, near-wall Reynolds stress model of turbulence, capable of resolving the turbulence fluctuations to an extent corresponding to the model's self-balancing between resolved and modelled (unresolved) contributions to the turbulence kinetic energy, is formulated and applied to several attached and separated wall-bounded configurations—channel and duct flows, external and internal flows separating from sharp-edged and continuous curved surfaces. In most cases considered the fluctuating velocity field was obtained started from the steady RANS results. The model proposed does not comprise any parameter depending explicitly on the grid spacing. An additional term in the corresponding length scale-determining equation providing a selective assessment of its production, modelled in terms of the von Karman length scale (formulated in terms of the second derivative of the velocity field) in line with the SAS (Scale-Adaptive Simulation) proposal (Menter and Egorov, Flow Turbul Combust 85:113–138, (2010) [14]), represents here the key parameter.

S. Jakirlić (✉) · R. Maduta
Institute of Fluid Mechanics and Aerodynamics / Center of Smart Interfaces, Technische Universität Darmstadt, Alarich-Weiss-Straße 10, 64287 Darmstadt, Germany
e-mail: s.jakirlic@sla.tu-darmstadt.de

R. Maduta
Outotec GmbH, Ludwig-Erhard-Strasse 21, 61440 Oberursel, Germany
e-mail: roberttibor@gmx.li

M. Stanislas et al. (eds.), *Progress in Wall Turbulence 2*,
ERCOFTAC Series 23, DOI 10.1007/978-3-319-20388-1_2

1 Introduction

The work on development of the hybrid RANS/LES (Large-Eddy Simulation) methods and novel Unsteady RANS (URANS) methods (RANS model plays here the role of a sub-scale model) has been greatly intensified in recent years. The relevant methods have been proposed by Spalart et al. ([20], DES—Detached Eddy Simulation; see Spalart, [19] for the DES method upgrades, namely Delayed DES and Improved Delayed DES), Menter and Egorov ([14]; SAS—Scale-Adaptive Simulations), Girimaji ([7]; PANS—Partially Averaged Navier Stokes; see also Basara et al. [1], for the PANS method extension to account for the near-wall effects), and Chaouat and Schiestel ([4]; PITM—Partially Integrated Transport Model). The common feature of all these models is an appropriate modification of the scale-determining equation providing a dissipation rate level which suppresses the turbulence intensity towards the subgrid (i.e. sub-scale) level in the regions where large coherent structures with a broader spectrum dominate the flow, allowing in such a way evolution of structural features of the associated turbulence. Whereas an appropriate dissipation level enhancement in the PANS method (similar is in the case of the PITM method) is achieved by reducing selectively (e.g. in the separated shear layer region) the destruction term in the model dissipation equation, i.e. its coefficient $C_{\varepsilon,2}$ (e.g. the grid-spacing-dependent model coefficient function in the PANS method provides appropriate decrease of the standard value $C_{\varepsilon,2} = 1.92$, prevailing in the near-wall region, towards a significantly lower value in the separated shear layer of the periodic 2D hill flow, see e.g. [3]), an additional production term was introduced into the ω equation ($\omega \propto \varepsilon/k$—inverse turbulent time scale) in the SAS framework. This term is modelled in terms of the von Karman length scale comprising the second derivative of the velocity field ($\nabla^2\mathbf{U}$), which is capable of capturing the vortex size variability, [14].

The work reported here aims at developing an instability sensitive, anisotropy-resolving Second-Moment Closure (SMC) model. This model scheme, functioning as a ‘sub-scale’ model in the Unsteady Sensitized-RANS (SRANS) framework, represents a differential near-wall Reynolds stress model formulated in conjunction with the scale-supplying equation governing the homogeneous part of the inverse turbulent time scale: $\omega_h = \varepsilon_h/k$. The model capability to account for the vortex length and time scales variability was enabled through a selective enhancement of the production of the dissipation rate in line with the SAS proposal (Scale-Adaptive Simulation, Menter and Egorov, [14]) pertinent particularly to the highly unsteady separated shear layer region. The predictive performances of the proposed model are checked by computing series of internal and external, two-dimensional and three-dimensional flows in channels, ducts and past bluff bodies including separation from sharp-edged and continuous curved surfaces in a range of Reynolds numbers.

2 Computational Method

The following section briefly outlines the computational model proposed. It is followed by description of the numerical method and associated details.

2.1 Computational Model

The equation governing the homogeneous part of the total viscous dissipation rate, $\varepsilon_h = \varepsilon - 0.5\nu\partial^2 k/(\partial x_j \partial x_j)$, modelled in term-by-term manner by Jakirlic and Hanjalic [8] represents the starting point for the present development. The RSM-based ω_h-equation following directly from the ε_h-equation (here, instead of originally used General-Gradient-Diffusion-Hypothesis (GGDH) for the turbulent diffusion modelling, the Simple GDH with diffusion coefficient modelled in terms of turbulence viscosity was applied; thereby, no difference between the Prandtl-Schmidt numbers corresponding to the quantities k and ε_h was made; $\sigma_k = \sigma_\varepsilon = \sigma_\omega = 1.1$ is adopted finally) by using well-known relationship

$$\frac{\mathbf{D}\omega_h}{\mathbf{D}t} = \frac{1}{k}\frac{\mathbf{D}\varepsilon_h}{\mathbf{D}t} - \frac{\varepsilon_h}{k^2}\frac{\mathbf{D}k}{\mathbf{D}t} \tag{1}$$

with

$$\frac{\mathbf{D}k}{\mathbf{D}t} = \frac{\partial}{\partial x_k}\left[\left(\frac{1}{2}\nu + \frac{\nu_t}{\sigma_k}\right)\frac{\partial k}{\partial x_k}\right] + P_k - \varepsilon_h \quad \text{and}$$

$$\frac{\mathbf{D}\varepsilon_h}{\mathbf{D}t} = \frac{\partial}{\partial x_k}\left[\left(\frac{1}{2}\nu + \frac{\nu_t}{\sigma_\varepsilon}\right)\frac{\partial \varepsilon_h}{\partial x_k}\right] + C_{\varepsilon,1}\frac{\varepsilon_h}{k}P_k - C_{\varepsilon,2}\frac{\varepsilon_h\varepsilon_h}{k} + 2C_{\varepsilon,3}\nu\nu_t\frac{\partial^2 U_i}{\partial x_j \partial x_l}\frac{\partial^2 U_i}{\partial x_j \partial x_l}$$

reads:

$$\begin{aligned}\frac{\mathbf{D}\omega_h}{\mathbf{D}t} = {} & \frac{\partial}{\partial x_k}\left[\left(\frac{1}{2}\nu + \frac{\nu_t}{\sigma_\omega}\right)\frac{\partial \omega_h}{\partial x_k}\right] + C_{\omega,1}\frac{\omega_h}{k}P_k - C_{\omega,2}\omega_h^2 \\ & + \frac{2}{k}\left(C_{cr,1}\frac{1}{2}\nu + C_{cr,2}\frac{\nu_t}{\sigma_\omega}\right)\frac{\partial \omega_h}{\partial x_k}\frac{\partial k}{\partial x_k} + \frac{2}{k}C_{\omega,3}\nu\nu_t\frac{\partial^2 U_i}{\partial x_j \partial x_l}\frac{\partial^2 U_i}{\partial x_j \partial x_l}\end{aligned} \tag{2}$$

with $P_k = -\overline{u_i u_j}\partial U_i/\partial x_j$ representing production of the kinetic energy of turbulence and coefficients $C_{\omega,1} = C_{\varepsilon,1} - 1 = 0.44$, $C_{\omega,2} = C_{\varepsilon,2} - 1 = 0.8$ and $C_{\omega,3} = 1.0$ taking their standard values. The introduction of the 'correction' coefficients $C_{cr,1} = 0.55$ and $C_{cr,2} = 0.275$ into the cross-derivative term copes with the correction of the near-wall behaviour of the ω_h-variable (see [9] for more details). Last term on the right-hand side represents the gradient production term; here, instead of the original formulation (modelled by using the vorticity transport theorem) comprising both the mean rate of strain and second derivative of the velocity field a simplified version (pertinent to an eddy-viscosity model) is applied in line with the request for a practical model usage. The model for turbulent viscosity ν_t, accounts for both Reynolds stress anisotropy (being beyond the reach of the eddy-viscosity model group) and viscosity effects, with characteristic length representing a switch between the Kolmogorov length scale and the turbulent length scale.

The latter equation is appropriately extended through the introduction of the SAS term [14] into the ω_h-equation:

$$\frac{\mathbf{D}\omega_{h,SAS}}{\mathbf{D}t} = \frac{\mathbf{D}\omega_h}{\mathbf{D}t} + P_{SAS}; \quad P_{SAS} = C_{RSM,1} \max\left[P^*_{SAS}, 0\right]$$

$$P^*_{SAS} = 2.3713\kappa S^2 \left(\frac{L}{L_{vk}}\right)^{1/2} - 3C_{RSM,2} k \max\left[\frac{(\nabla\omega_h)^2}{\omega_h^2}, \frac{(\nabla k)^2}{k^2}\right] \tag{3}$$

with $L = k^{1/2}/\omega_h$ being the turbulent length scale, $L_{vk} = \kappa S/|\nabla^2 U|$ (with $\nabla^2 U = [\partial^2 U_i/\partial x_j^2 \times \partial^2 U_i/\partial x_j^2]^{1/2}$) representing the 3D generalization of the classical boundary layer definition of the von Karman length scale and S the invariant of the mean strain tensor ($S = \sqrt{2S_{ij}S_{ij}}$; $S_{ij} = 0.5(\partial U_i/\partial x_j + \partial U_j/\partial x_i)$). It should be noted that the P_{SAS} term introduced in the ω_h-equation has almost identical form as the one being used in the eddy-viscosity-based $k-\omega$ SST-SAS model [14]. However, two coefficients, $C_{RSM,1} = 0.004$ and $C_{RSM,2} = 8$ and exponent of the length scales ratio ((1/2) instead of 2)are introduced adjusting its use in the framework of a Second-Moment Closure model. The natural decay of the homogeneous isotropic turbulence, fully developed channel flows in a range of Reynolds number (with underlying velocity field following the logarithmic law) and the non-equilibrium 2D hill flow at two different Reynolds numbers ($Re_H = 10{,}600$ and 37,000) have been interactively computed in the process of the coefficients calibration. Equation (3) represents a grid-spacing-free formulation. This explicit non-dependence on the grid-spacing represents certainly an advantage over some hybrid LES/RANS models, especially in the case of unstructured grids with arbitrary grid-cell topology. The contours of the P_{SAS} term in the flow over a periodical arrangement of 2D hills and past a tandem cylinder configuration depicted in Fig. 1 clearly shows that it is active only in the region of the separated shear layer. In the reminder of the flow domain, especially in the near-wall regions, its effect vanishes.

The proposed model is solved in conjunction with the Jakirlic and Hanjalic's [8] Reynolds stress model equation ($\varepsilon_h = \omega_h k$, $k = \overline{u_i u_i}/2$, $P_{ij} = -\overline{u_i u_k}\partial U_j/x_k - \overline{u_j u_k}\partial U_i/\partial x_k$); similar to the scale-supplying equation (Eq. 2) the GGDH turbulent diffusion model is replaced by the corresponding SGDH model formulation:

$$\frac{\partial \overline{u_i u_j}}{\partial t} + \frac{\partial U_k \overline{u_i u_j}}{\partial x_k} = \frac{\partial}{\partial x_k}\left[\left(\frac{1}{2}\nu + \frac{\nu_t}{\sigma_{\overline{u_i u_j}}}\right)\frac{\partial \overline{u_i u_j}}{\partial x_k}\right] + P_{ij} - \varepsilon^h_{ij} + \Phi_{ij} + \Phi^w_{ij} \tag{4}$$

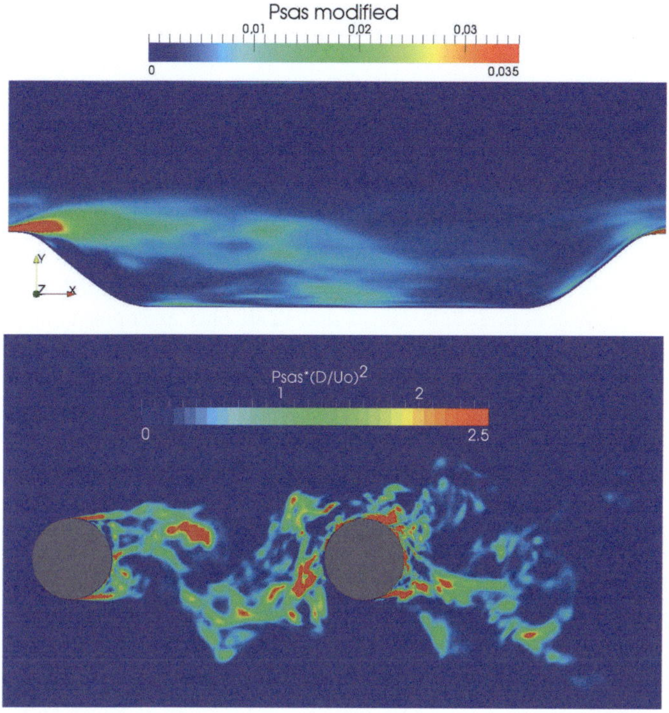

Fig. 1 Contours of the P_{SAS} term (Eq. 3) coloured by its magnitude in the 2D hill flow (*upper*) and flow past a tandem-cylinder (*lower*)

For more detailed insight into the modelling rationale interested readers are referred to [9].

2.2 Numerical Method

All computations were performed using the code Open-FOAM, an open source Computational Fluid Dynamics toolbox (www.opencfd.co.uk/openfoam), utilizing a cell-center-based finite volume method on an unstructured numerical grid and employing the solution procedure based on the implicit pressure algorithm with splitting of operators (PISO) for coupling between pressure and velocity fields. SIMPLE procedure was applied when computing the steady flows using the RANS-RSM model. The convective transport was discretized by a scheme blending between the second order central differencing (CDS) and first order upwind (UDS) schemes with $\gamma_{CDS} = 0.95$ and $\gamma_{UDS} = 0.05$ in most of the cases considered. For the time integration the second order three point backward scheme was used. The code is parallelized applying the Message Passing Interface (MPI) technique for communication between the processors.

3 Results and Discussion

The predictive performances of the proposed instability-sensitive Reynolds stress model (denoted by IS-RSM throughout the work) are intensively assessed in numerous aerodynamic-type flows of different complexity featured also by 2D and 3D separation along with available experimental, DNS (Direct Numerical Simulation) and LES reference results: fully developed flow in a plane channel, flow over a series of axisymmetric 2D hills, flow over a backward-facing step, flow over a wall-mounted fence, flow in a three-dimensional diffuser and flow past tandem cylinder configurations. Figures 2, 3, 4, 5, 6, 7, 8, 9, 10, 11 and 12 display exemplarily some selected results obtained by the consequent models application. For purpose of the mutual comparison the results of the 'background' RANS-RSM model are also depicted. For more extensive result presentation and more detailed discussion, also with respect to the computational issues, interested readers are referred to [9].

3.1 Fully Developed Flow in a Plane Channel

Fully developed turbulent flow in a plane channel represents most important representative of wall-bounded flow configurations for studying wall proximity effects on turbulence structure characterized by enhanced anisotropy of both the Reynolds stress tensor and stress dissipation tensor. Channel flow represents a globally stable, unidirectional (in mean) flow ($\partial/\partial x_1 = 0$, $\partial/\partial x_3 = 0$) characterized by a strong mean shear ($\partial U_1/\partial x_2$) but with a low level of inherent forcing. In such a flow, the employment of conventional RANS models, especially those on the second-moment closure level, leads traditionally to correctly predicted distribution of the Reynolds stress components and mean velocity. Accordingly, capturing the turbulence instabilities is here not of decisive importance (as e.g. in the flow over a 2D hill). However, as the consequence of the enhanced sensitivity to turbulence unsteadiness, appropriate eddy-structure resolving is enabled also in such a globally stable flow configuration.

Figure 2-left illustrates the instantaneous flow field obtained by the present instability-sensitive model starting from the mean flow and turbulence fields obtained by the RSM model within the steady RANS framework (periodic inlet/outlet boundary conditions have been applied with the streamwise pressure gradient imposed in accordance with the relevant Reynolds number). Presently, friction-velocity-based Reynolds number $Re_\tau = 395$ is considered; reference DNS database is from [15]. The solution domain adopted ($L_x \times L_y \times L_z = 4h \times 2h \times 2h$); with h representing the half channel width) was meshed by a grid comprising 462,000 ($N_x \times N_y \times N_z = 70 \times 110 \times 60$) grid cells, implying the near-wall resolution in terms of the height of the wall-next grid cell corresponding to $\Delta y^+ = 1.6$. Figure 2-right shows the modelled and resolved fractions of the streamwise and shear Reynolds stress components obtained using the present IS-RSM model. The maximum ratio of the modelled to total kinetic energy related to both turbulent stress components corresponding approximately to 25 % is found in the near-wall region.

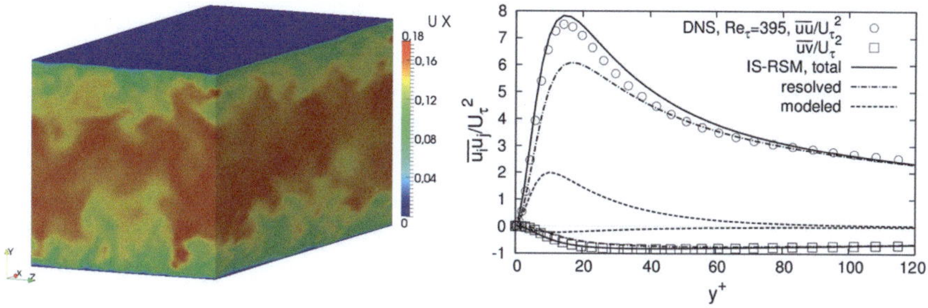

Fig. 2 Fully developed flow in a plane channel at $Re_\tau = 395$—instantaneous axial velocity field obtained by the present IS-RSM (*left*) and the streamwise ($\overline{u^2}$) and shear ($\overline{uv}$) Reynolds stress components (*right*); DNS from [15]

3.2 Periodic Flow Over a 2D Hill

Flow over a series of the hill-shaped constrictions (reference LES results have been made available by Fröhlich et al. [6] and Breuer et al. [2]; complementary experimental investigations have been performed by Rapp and Manhart, [18]) exhibits a number of features typically associated with a separating flow: boundary layer separation from a continuous curved surface, reattachment, highly unsteady shear layer that separates the main stream from the recirculation flow, relaxation in the post-reattachment region, alternating adverse (flow deceleration) and favourable (flow acceleration) pressure gradient effects (globally along the flow but even across the same streamwise location), strong departure from the equilibrium conditions, streamline curvature effects, wall proximity effects, Reynolds stress anisotropy, etc. This flow configuration is characterized by high level of natural instability, originating primarily from the highly intermittent separation region oscillating over a wider wall area. Consequently, a highly unsteady separated shear layer spread over a larger portion of the flow domain was generated. Accordingly, it could be concluded without going into greater details that the correct capturing of the present 2D-hill flow configuration is beyond the reach of the conventional, inherently steady RANS closures, almost independent of the modelling level (the complementary RANS-RSM results are also illustrated). The incapability of accounting for any spectral dynamics makes RANS closures limited for capturing correctly such flows dominated by large-scale dynamics. A direct consequence is inadequate (low) level of turbulence activity (controlling the reattachment process) in the separated shear layer and correspondingly longer recirculation zone (see e.g., Fig. 4). The solution domain (with dimensions $(L_x, L_y, L_z) = (9H, 3.03H, 4.5H)$, see Fig. 3-left, is in accordance with the reference LES simulation. The mesh consisting of $N_x \times N_y \times N_z = 160 \times 160 \times 60$ grid cells was designed by an appropriate coarsening of the 13 Mio. cells fine grid made available by Breuer et al. [2]. Both the $Re_H = 10{,}600$ case and the $Re_H = 37{,}000$ case (the mean velocity and Reynolds stress results of the former case are not shown here due to sake of brevity—these are of the similar quality) were computed using the same mesh (lower Re-number case was computed also by using a substantially coarser grid—$N_x \times N_y \times N_z = 80 \times 100 \times 30$—no important difference in results was obtained). Similar to the fully developed channel flow (previous section), periodic inlet/outlet boundary conditions have been utilized with the streamwise pressure gradient corresponding to the prescribed Reynolds number. It is interesting to report that no initial turbulence fluctuations were necessary in this periodical flow configuration. The mean flow and turbulent quantities obtained by the steady RANS computations using the background $\overline{u_i u_j} - \omega_h$ model served for the initialization of the computations with the present IS-RSM formulation.

The results obtained by computing the 2D hill configuration, displayed in Fig. 3-left, document appropriate vortex structure reproduction—visualized here by the instantaneous velocity field—being beyond the reach of any RANS model. The introduction of the P_{SAS}-term (Eq. 3) within the instability sensitive second-moment closure in the Unsteady RANS framework contributed strongly to the turbulence activity intensification (originating from the resolved motion) in the region around the separation point (see the P_{SAS}-field in Fig. 1-upper). The model capability to account for the large-scale structures and bulk unsteadiness led consequently to the increased magnitude of the turbulent shear stress component (Fig. 4-lower; typical result pertinent to any RANS model is a significantly lower turbulence intensity in the separated shear layer; for comparison, the results of the RANS computations by the $\overline{u_i u_j} - \omega_h$ model denoted by RSM are also displayed), improved shape of the mean velocity profiles (Fig. 4-upper) and correctly predicted reattachment length, Figs. 3-right and 5. The latter figures illustrate appropriate recirculation zone shortening, from $(x/H)_{RP} = 4.62$ (pertinent to the lower Reynolds number $Re_H = 10{,}600$) towards $(x/H)_{RP} = 3.72$ (pertinent to the higher Reynolds number $Re_H = 37{,}000$) in good agreement with the LES ($(x/H)_{RP} = 4.62$) and experimental ($(x/H)_{RP} = 3.76$) reference results.

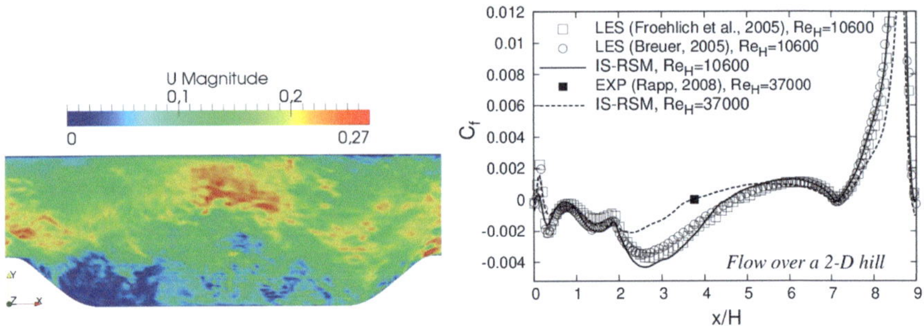

Fig. 3 Periodic flow over a 2D hill: instantaneous velocity field obtained by the present IS-RSM at $Re_H = 10{,}600$ (*left*) and friction coefficient development at the lower wall for both Reynolds numbers considered (*right*)

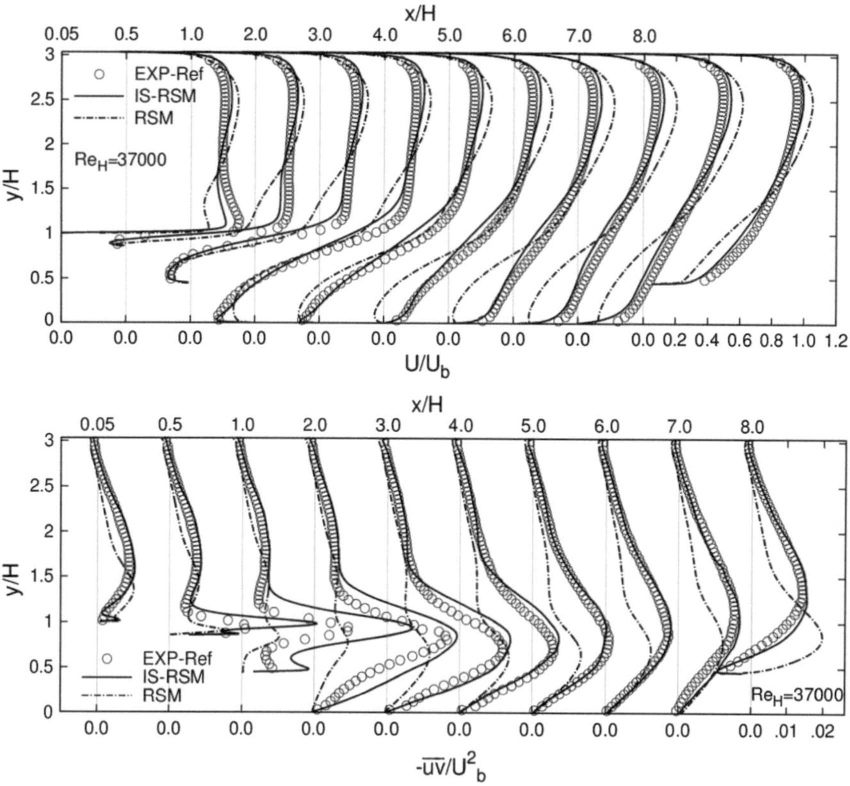

Fig. 4 Periodic flow over a 2D hill at $Re_H = 37{,}000$—mean velocity (*upper*) and shear stress component (*lower*) profile developments obtained by the present IS-RSM; Exp. from [18]

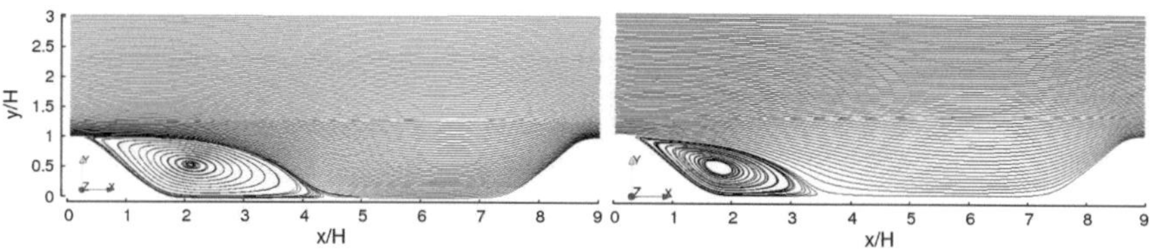

Fig. 5 Periodic flow over a 2D hill at $Re_H = 10{,}600$ (*left*) and $Re_H = 37{,}000$ (*right*)—mean streamlines obtained by the present IS-RSM

3.3 Turbulent Flow over a Backward-Facing Step

The low Reynolds number configuration ($Re_H = 5100$ based on the step height H; the Reynolds number based on the expanding channel height is $Re_{(10H)} = 51{,}000$) investigated experimentally by Jovic and Driver [10] and by means of DNS by Le et al. [13] was chosen as the next test case. 1.59 Million grid cells in total were used (30 cells are distributed uniformly over the spanwise extension of $4H$). The inflow plane was located at the step wall at $-3.5H$; the solution domain is extended up to $20H$. The fluctuating field was generated by applying the method of Kornev and Hassel [11] onto the flow field obtained by the background RSM model. This configuration possesses all the features typical for a separating flow, as described in the previous section. However, as the flat plate boundary layer separates at the sharp edge (fixed separation point with the time-averaged mean dividing streamline running parallel to the step wall for a certain distance) its subsequent transformation into a shear layer is characterized by a much less intensive oscillations compared to the separation from a curved surface. Consequently, the capturing of the unsteady character of the flow is not of decisive importance for correct representation of the mean flow and time-averaged turbulence quantities. Good results could be obtained even if computing the flow in a steady manner by applying an advanced RANS model (the present Reynolds stress model is certainly such a model). The only important departure from the reference database is pertinent to a slight underprediction of the turbulence level immediately after separation similar to the separation at a curved surface (not shown here); however, unlike in the latter flow it recovers by itself leading consequently to a correct prediction of the mean reattachment length ($(x/H)_{RP} = 6.28$; see the friction coefficient development in Fig. 6-lower-left). A certain weakening of the flow reversal

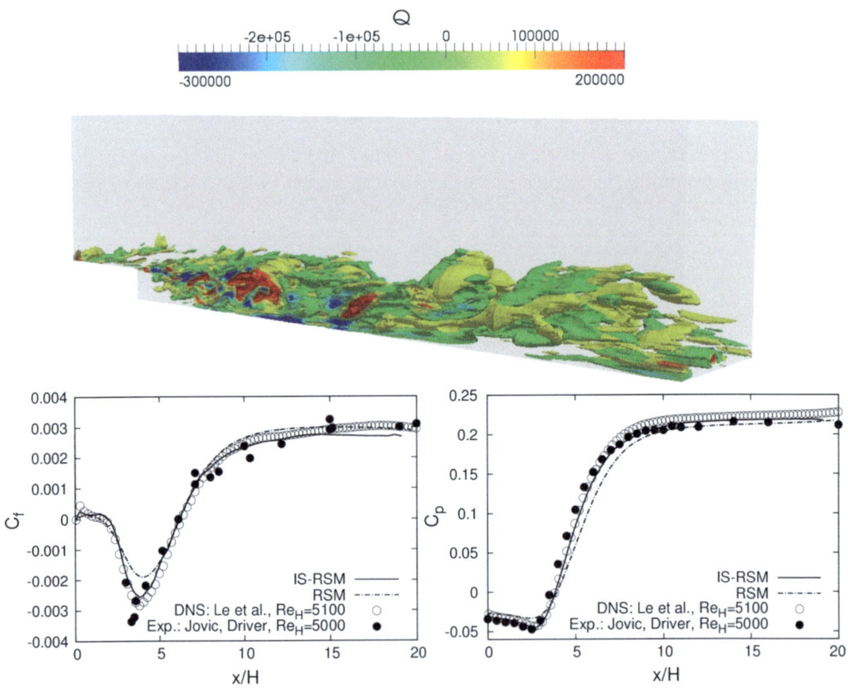

Fig. 6 Flow over a backward-facing step—vortex structure illustrated by the Q-criteria (*upper*), friction and pressure coefficients (*lower*)

intensity leads to an important underprediction of the friction coefficient magnitude whose (negative) peak coincides with the location $x/H = 4$. It is a typical RANS result pertinent especially to such a low Reynolds number. Its correct prediction (this is valid also for the pressure coefficient, Fig. 6-lower-right) requires the employment of a model being capable of capturing the instantaneous flow field. The application of the present IS-RSM model led to the vortex structure capturing (Fig. 6-upper) leading consequently to an appropriate intensification of the back-flow (not shown here) and the magnitude enhancement (negative peak) of the friction coefficient. The quality of the latter result is especially dependent on the correct capturing of the inherently unsteady impact of the near-wall streams corresponding to the mean recirculation zone and the corner bubble at the secondary reattachment point (C_f curve crosses zero value at $x/H \approx 2$).

3.4 Turbulent Flow over a Wall-Mounted Fence

The structure of the flow separated at a fence-shaped, sharp-edged obstacle mounted at the bottom wall of a plane channel is extremely complex despite the fact that the separation occurs at a fixed point coinciding with the fence tip. An impression about the flow structure complexity could be gained from Fig. 7-upper illustrating instantaneous flow field. The flow conditions upstream of the fence comply with the fluid impingement onto the fence and, consequently, with a strong upward skewing. The flow separating from the fence tip is characterized by a strongly curved, highly unsteady separated shear layer oscillating and spreading over an expanded flow region. The strong shear layer oscillations occur in a broader frequency range; the entire flow domain is dominated by large coherent structures (unlike in the case of the flat boundary layer separation from the backward-facing step, see previous section)—accordingly, the mean recirculation zone is much longer (almost two times) compared to a relevant backward-facing step configuration with $(x/H)_{RP} \approx 6-7$ (see the previously computed backward-facing step flow). Expectedly, these features are beyond the reach of RANS equations independent of the modeling level. The outcome is significant underprediction of the turbulence activity in the separated shear layer (see Fig. 8-lower), causing a too long recirculation zone, up to $(x/H)_{RP} \approx 14.5$ (see C_f-coefficient evolution in Fig. 7-lower).

The reference database is provided experimentally by Larsen et al. [12]. The fence-height-based ($H = 40$ mm) Reynolds number corresponds to $Re_H = 3000$ and the Reynolds number based on the channel height ($7.5H$) equals to $Re_{(7.5H)} = 22{,}000$. The inlet plane of the solution domain adopted is situated $8.25H$ upstream of the fence and the outlet plane is positioned at $32.5H$. The spanwise extent corresponds to $10H$. Numerical mesh comprising 1.0 Million cells in total is squeezed towards the channel and fence walls providing the dimensionless distance of the wall-adjacent grid node corresponding to $y^+ \approx 1$; the uniform distribution of 45 grid cells is adopted

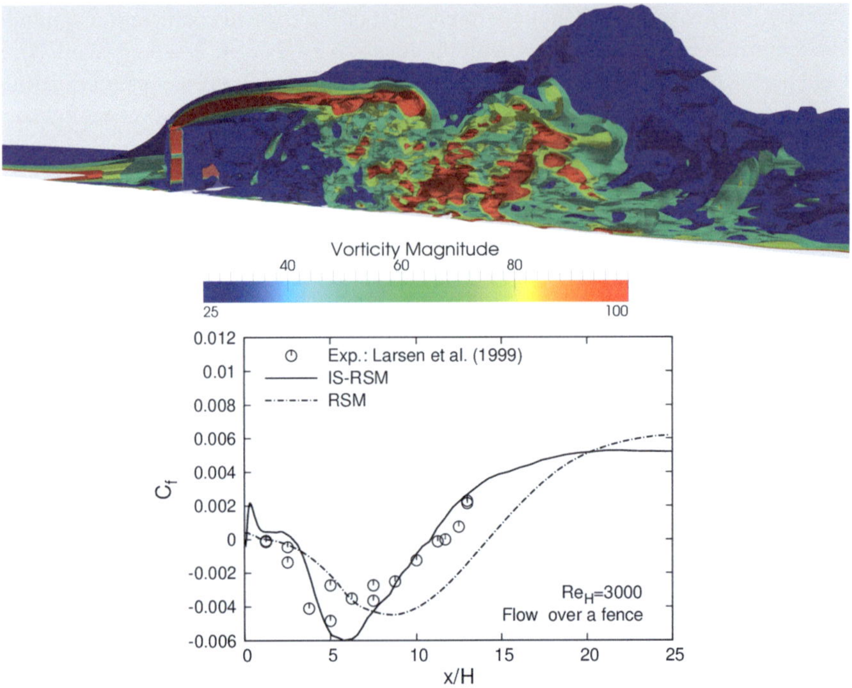

Fig. 7 Flow over a 2D fence—vortex structure illustrated by the instantaneous vorticity field (*upper*) and friction coefficient evolution at the bottom wall (*lower*)

in the spanwise direction. The RANS-RSM computations have been performed by prescribing the experimentally available velocity profile at the inflow plane. The fluctuating inflow at $-8.25H$ for the simulations using the instability-sensitive RSM model is generated by a precursor simulation of the corresponding, fully developed channel flow at $Re_{(7.5H)} = 22{,}000$ (with bulk velocity $U_b = 1.17\,\text{m/s}$) using the same turbulence model (see Fig. 8; however, the difference in the inflow conditions is not regarded to be of decisive importance concerning the objectives of the present work dealing with the turbulence intensity enhancement being appropriately correlated with the mean velocity field).

Similarly to the case of the flow over a 2D hill important improvement is obtained by applying the IS-RSM model with the scale-supplying equation extended appropriately to account for the turbulence level enhancement in the separated shear layer region. Figure 8-lower displays the profile development of turbulent shear stress component. The improvement in the results compared to the initial Reynolds stress model is obvious. The intensified turbulence activity in the region of separation led subsequently to the separated shear layer reattachment at a distance corresponding to $(x/H)_{RP} \approx 11$ (the experimentally determined reattachment point position is at $(x/H)_{RP} = 11.7$; see also the friction coefficient development, Fig. 7-lower). This contributed strongly to the correct reproduction of the $\overline{uv}$ profile development in its

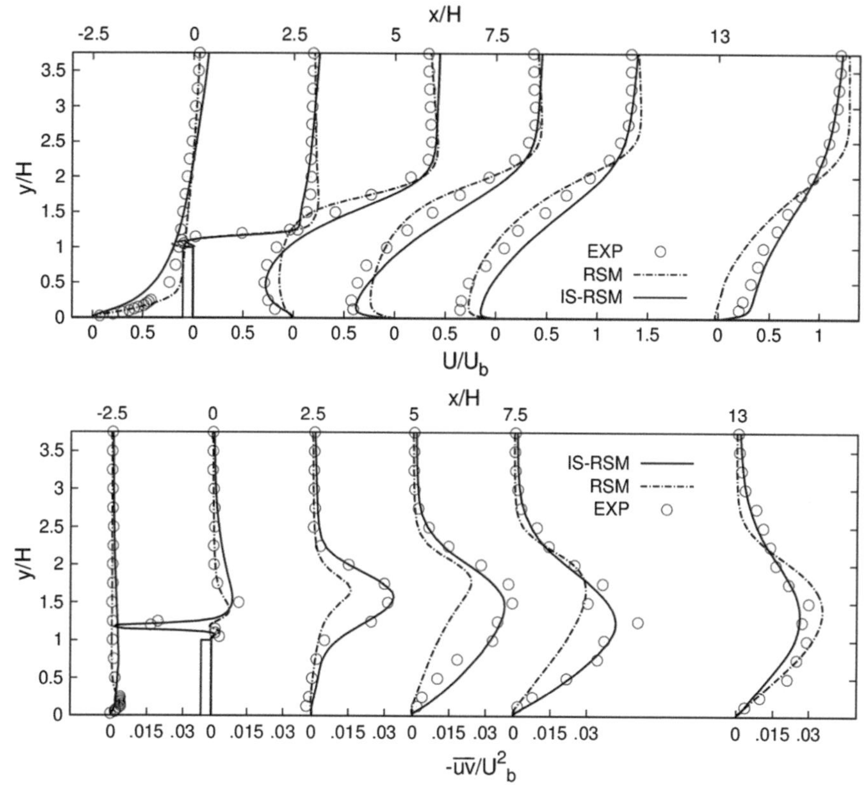

Fig. 8 Flow over a 2D fence—mean velocity (*upper*) and shear stress component (*lower*) profile developments obtained by the present IS-RSM; Exp. from [12]

entirety. The improved prediction is especially reflected in the correct capturing of the specific sign change of the shear stress component at the fence tip. The direct consequence of the correctly returned turbulent stress level increase is the greatly improved predictions of the mean velocity development, Fig. 8-upper.

3.5 Turbulent Flow in a 3D Diffuser

The flow in a presently considered three-dimensional diffuser is featured by an incompressible fully developed duct flow (height $h = 1$ cm; width $B = 3.33$ cm) discharging into a diffuser (of the length $L = 15h$), whose upper-wall and one-side wall are appropriately inclined with the expansion angles of 11.3° and 2.56°, respectively, Fig. 9-upper. The bulk velocity in the inflow duct is 1 m/s resulting in the Reynolds number based on the duct height of $Re_h = 10{,}000$. The reference experimental and DNS databases were provided by Cherry et al. [5] and Ohlsson et al. [17].

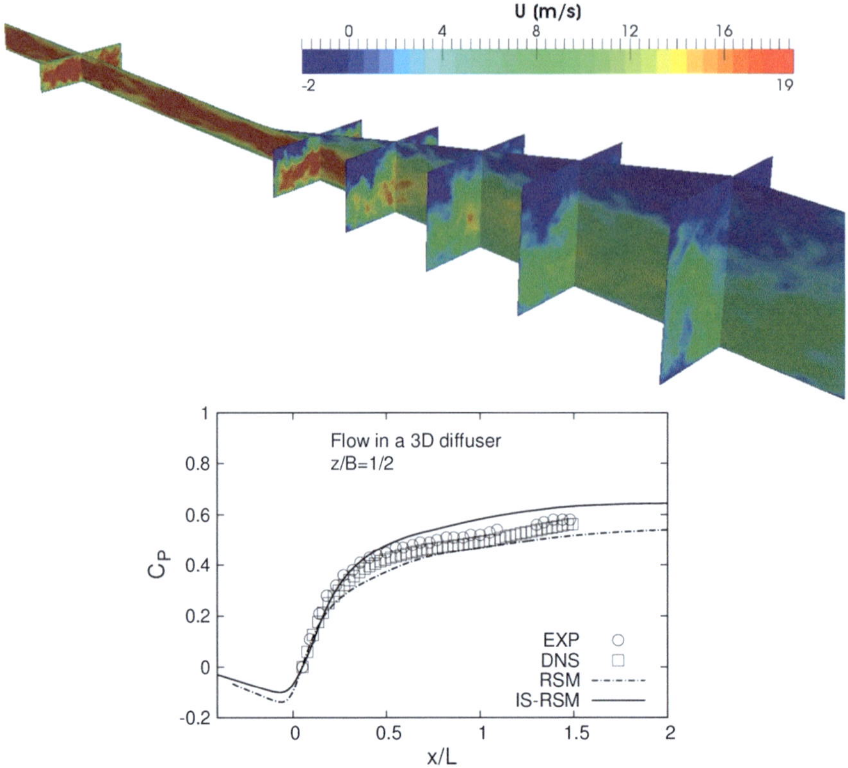

Fig. 9 Flow in a three-dimensional diffuser—instantaneous velocity field obtained by the present IS-RSM model (*upper*) and pressure coefficient development at the lower flat diffuser wall (*lower*)

The solution domain whose inlet plane is located at $x/h = -15$ (origin of the coordinate system coincides with the duct expansion onset—$x/L = 0$—and the non-expanded side wall—$z/B = 0$) in the inflow duct and exit plane at $x/h = 45$ in the straight outflow duct was meshed with the grid consisting of 3.75 Mio. cells in total; it corresponds to the position of the wall-adjacent computational node at $y^+ \approx 1.5$. Similar to the previous 2D fence case, the fluctuating inflow was generated by performing the simulation of the fully developed flow in the 3D duct by applying the IS-RSM model.

This is a fairly complex flow characterized by the boundary layer separation starting in the corner built by two sloped walls (corner separation) and spreading over the entire upper wall (Figs. 9-upper and 10) due to an adverse pressure gradient imposed on the duct flow by expanding the cross-section area. The results obtained by the IS-RSM model are in a good qualitative agreement with the experimental findings despite a somewhat thinner recirculation zone, unlike the RANS-RSM model resulting in a growth of the corner bubble without occupying the upper wall in its entirety (not shown here). Figures 9-lower and 11 display a quantitative comparison

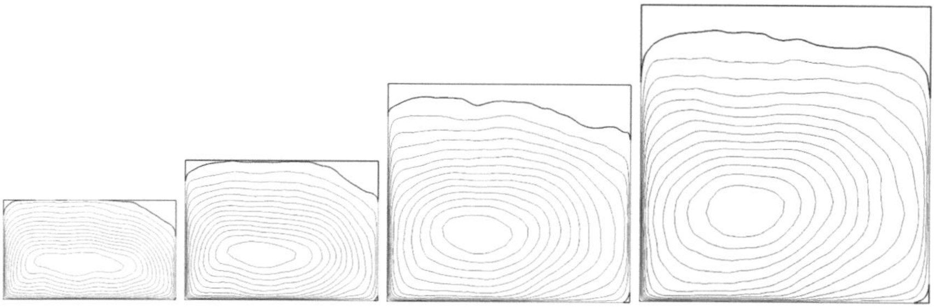

Fig. 10 Flow in a three-dimensional diffuser—iso-contours of the axial velocity field in the cross planes $y-z$ at selected streamwise locations (x/h = 5, 8, 12 and 15, respectively) within the diffuser section (*thick lines* denote the zero-velocity lines) obtained by the present IS-RSM model

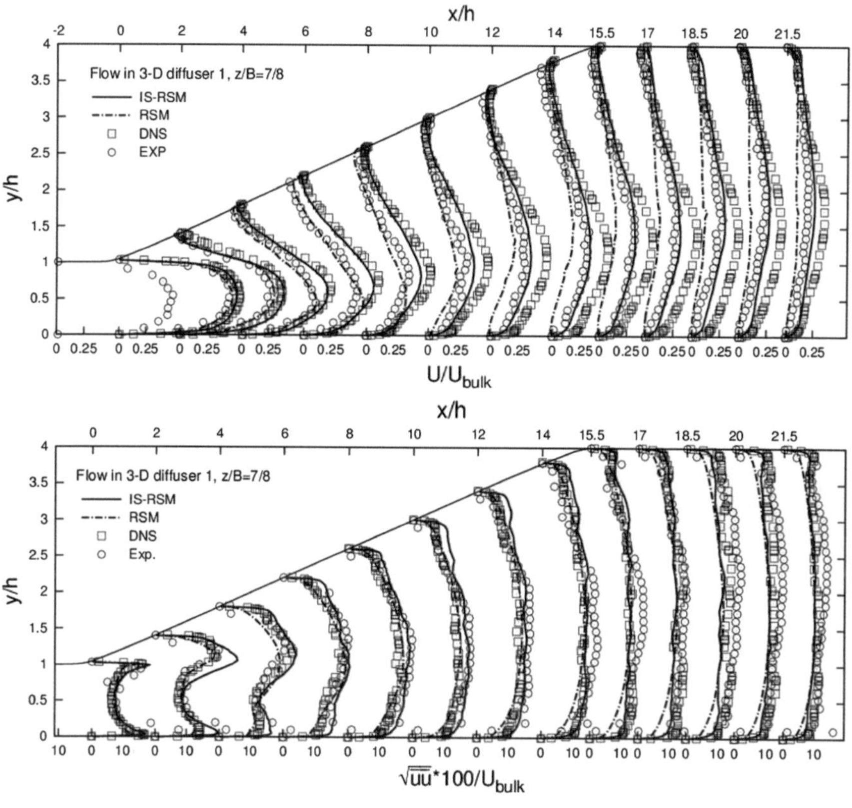

Fig. 11 Flow in a three-dimensional diffuser—evolution of the axial velocity and streamwise turbulence intensity profiles in the vertical x–y plane positioned at $z = 7B/8$. Exp. from [5, 17]

between the presently computed pressure coefficients, axial velocity and streamwise turbulence intensity profile developments and the reference experimental and DNS databases. The results obtained exhibit reasonable agreement with both data sets, especially with respect to the pressure recovery.

3.6 Flow Past Tandem-Cylinder Configurations

The presently considered pair of circular cylinders (with diameter D) in tandem (Fig. 12) represents a simplified landing gear configuration of an aircraft. The interaction between the flow fields past cylinder components was in focus of relevant experimental investigations by Neuhart et al. [16] providing details about the mean velocity, mean surface pressure, root-mean-square of the fluctuation pressure and time-averaged turbulence intensity. Two distinct tandem-cylinder configurations characterized by different in-between spacing were considered: $L = 3.7D$ (long-distance case) and $L = 1.435D$ (short-distance case). The corresponding Reynolds and Mach numbers based on the cylinder diameter D and velocity of the oncoming flow amount 166,000 ($U_{inlet} = 1.66$ m/s) and 0.1285 respectively. The grid adopted consists of the 60,000 cells in the x–y plane for both cases. This two-dimensional grid was extended in the spanwise direction by two cylinder diameters for the IS-RSM simulations. 80 uniformly distributed grid cells were placed in the spanwise direction resulting in 4.8 Million cells in total.

According to Zdravkovich [21] the distance $L/D = 3.7$ relates to the so-called bistable case corresponding to a configuration in which the flow structure at/behind the first cylinder switches from the continuous shedding resembling the well-known von Karman vortex street to a continuously separated shear layer reattaching temporarily at the front side of the rear cylinder; behind the second cylinder a continuous vortex street develops, Fig. 12-upper-left. Unlike the long-distance case, in the short-distance configuration ($L/D = 1.435$) the cylinders are that close to each other resembling one long obstacle for the oncoming flow, Fig. 12-lower-left. Accordingly, the wall boundary layer separating quasi-stationary from the front cylinder transforms into a shear layer which reattaches at the rear cylinder. The flow in the wake behind the downstream cylinder exhibits a continuous shedding behaviour. These descriptions reveal the flow structure past the large-distance case being more challenging for turbulence models due to the bistable (intermittent) behaviour. Capturing of turbulence unsteadiness by applying the present IS-RSM model is illustrated in Fig. 1-lower displaying the field of the P_{SAS} production term. As mentioned previously, the tandem cylinder configuration can be regarded as a simplified version of a landing gear and can therefore serve as the first step in testing turbulence models for predicting the airframe noise. The unsteady pressure field is the most important flow variable acting as the noise-source representative. Conventional RANS models fail traditionally in predicting it because of their time-averaging rationale. Only unsteady interactions involving large scales can be reasonably captured. The unsteady feature of the pressure field is represented through the root-mean-square of the fluctuating

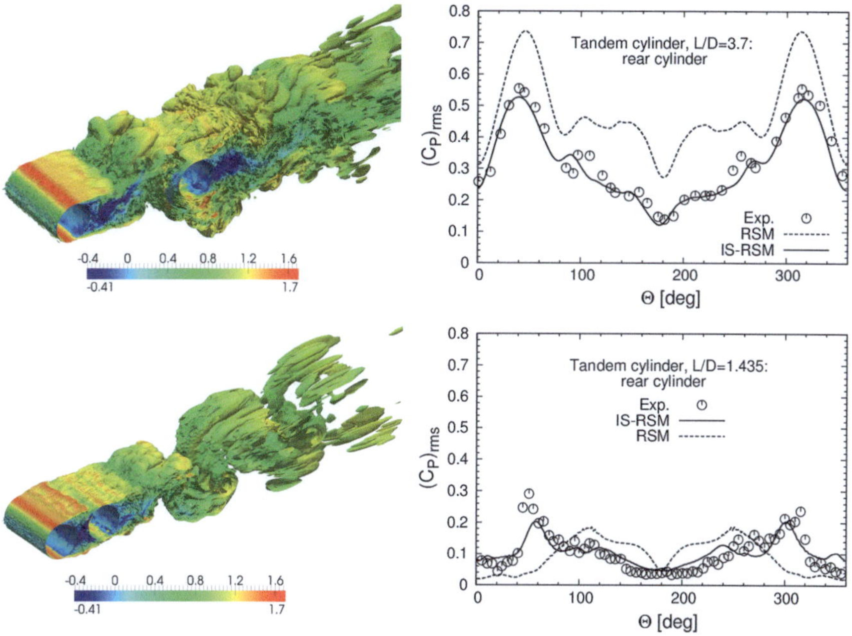

Fig. 12 Flow past tandem cylinder configurations—large ($L/D = 3.7$; *upper*) and small in-between spacing ($L/D = 1.435$; *lower*); vorticity magnitude coloured by the normalized axial velocity (U_x/U_{inlet}) obtained by the present IS-RSM (*left*) and root-mean-square (rms) of the fluctuating pressure on the downstream cylinder (*right*); Exp. from [16]

pressure on downstream cylinder for both cylinder separations, Fig. 12-right. It was experimentally found that the second cylinder is the main source of noise as the relevant $(C_p^{'})_{rms}$ values are four to five times higher than those measured on the upstream cylinder. Therefore, the model results compared to the experiment relate only to this downstream cylinder. The IS-RSM model results exhibit reasonable agreement in regard to both peak values and $(C_p^{'})_{rms}$ distribution over the most of the cylinder surface indicating high potential for being used as a tool for the noise prediction.

4 Conclusion

Potential of the presently formulated near-wall differential Reynolds stress model (RSM) extended appropriately to account for turbulence instabilities within 'Sensitized RANS' framework (SRANS) was illustrated by computing a series of 2D and 3D wall-bounded flow configurations featured by separation and reattachment in a broad range of Reynolds numbers. The key element in the present model is an additional SAS-based production term (in line with Menter and Egorov, [14]) introduced into the scale-determining equation governing the inverse time scale

ω_h ($\propto \varepsilon_h/k$). This term, formulated in terms of the ratio of the turbulent length scale to the von Karman length scale (comprising the second derivative of the velocity field), enables appropriate model receptivity to the turbulence unsteadiness promoting a selective enhancement of the turbulent dissipation rate production influencing consequently an adequate suppression of the modelled turbulence intensity towards the respective sub-scale level (related mostly to the separated shear layer region). Significantly improved predictions, compared to the baseline RSM model, with respect to the structural characteristics of the instantaneous flow field, some most important integral characteristics (e.g. friction and surface pressure coefficients), the mean velocity field and turbulence quantities demonstrate the model's potential in solving the complex flows separated from continuous curved surfaces and obstacles exhibiting broader frequency range. In addition, the capability of the model to operate in scale-resolving mode in some globally stable flows, such as flow in a plane channel and flow over a backward-facing step, is also illustrated.

Acknowledgments The work of R. Maduta has been funded by the EU project ATAAC (ACP8-GA-2009-233710).

References

1. B. Basara, S. Krajnovic, S. Girimaji, Z. Pavlovic, Partially Averaged Navier Stokes (PANS) method for turbulence simulations: near-wall eddy-viscosity transport model implementation. AIAA J. **49**(12), 2627–2636 (2011)
2. M. Breuer, N. Peller, Ch. Rapp, M. Manhart, Flow over periodic hills—numerical and experimental study in a wide range of Reynolds numbers. Comput. Fluids **38**, 433–457 (2009)
3. C.-Y. Chang, S. Jakirlic, B. Basara, C. Tropea, Predictive capability assessment of the PANS-ζ-f model of turbulence. Part I: physical rationale by reference to wall-bounded flows including separation, in *Advances in Hybrid RANS-LES Modelling 5. Notes on Numerical Fluid Mechanics and Multidisciplinary Design*, ed. by S. Girimaji, W. Haase, S.-H. Peng, D. Schwamborn (Springer, Berlin, 2015), pp. 371–383
4. B. Chaouat, R. Schiestel, A new partially integrated transport model for subgrid-scale stresses and dissipation rate for turbulent developing flows. Phys. Fluids **17**(065106), 1–19 (2005)
5. E.M. Cherry, C.J. Elkins, J.K. Eaton, Geometric sensitivity of three-dimensional separated flows. Int. J. Heat Fluid Flow **29**, 803–811 (2008)
6. J. Fröhlich, C.P. Mellen, W. Rodi, L. Temmerman, M.A. Leschziner, Highly resolved large-eddy simulation of separated flow in a channel with streamwise periodic constrictions. J. Fluid Mech. **526**, 19–66 (2005)
7. S.S. Girimaji, Partially-averaged Navier-Stokes model for turbulence: a Reynolds-averaged Navier-Stokes to direct numerical simulation bridging method. J. Appl. Mech. **73**, 413–421 (2006)
8. S. Jakirlic, K. Hanjalic, A new approach to modelling near-wall turbulence energy and stress dissipation. J. Fluid Mech. **439**, 139–166 (2002)
9. S. Jakirlic, R. Maduta, Extending the bounds of 'steady' RANS closures: towards an instability-sensitive Reynolds stress model. Int. J. Heat Fluid Flow **51**, 175–194 (2015). doi:10.1016/j.ijheatfluidflow.2014.09.003
10. S. Jovic, D. Driver, Reynolds number effect on the skin friction in separated flows behind a backward-facing step. Exp. Fluids **18**, 464–467 (1995)

11. N. Kornev, E. Hassel, Synthesis of homogeneous anisotropic divergence-free turbulent fields with prescribed second-order statistics by vortex dipoles. Phys. Fluids **19**(6), 067101 (2007)
12. P.S. Larsen, J.J. Schmidt, U. Ullum, Experimental study of temporal and spatial structures in fence-on-wall test case, in *IUTAM Symposium on Simulation and Identification of Organized Structures in Flows*. Fluid Mechanics and Its Applications, vol. 52 (Kluwer Academic Publisher, 1999), pp. 25–37
13. H. Le, P. Moin, J. Kim, Direct numerical simulation of turbulent flow over a backward-facing step. J. Fluid Mech. **330**, 349–374 (1997)
14. F.R. Menter, Y. Egorov, The scale-adaptive simulation method for unsteady turbulent flow predictions. Part 1: theory and model description. Flow Turbul. Combust. **85**, 113–138 (2010)
15. R.D. Moser, J. Kim, N.N. Mansour, Direct numerical simulation of turbulent channel flow up to $Re_\tau = 590$. Phys. Fluids **11**(4), 943–945 (1999)
16. D.H. Neuhart, L.N. Jenkins, M.M. Choudhari, M.R. Khorrami, Measurements of the flowfield interaction between tandem cylinders. AIAA-2009-3275 (2009)
17. J. Ohlsson, P. Schlatter, P.F. Fischer, D.S. Henningson, DNS of separated flow in a three-dimensional diffuser by the spectral-element method. J. Fluid Mech. **650**, 307–318 (2010)
18. Ch. Rapp, M. Manhart, Flow over periodic hills: an experimental study. Exp. Fluids **51**, 247–269 (2011)
19. P.R. Spalart, Detached-eddy simulation. Annu. Rev. Fluid Mech. **41**, 181–202 (2009)
20. P.R. Spalart, W.-H. Jou, M. Strelets, S. Allmaras, Comments on the feasibility of LES for wings and on a hybrid RANS/LES approach, 1st AFOSR Int. Conf. on DNS and LES, in *Advances in DNS/LES*, ed. by C. Liu, Z. Liu (Greyden Press, Columbus, 1997), pp. 137–147
21. M.M. Zdravkovich, Flow induced oscillations of two interfering circular cylinders. J. Sound Vib. **101**(4), 511–521 (1985)

Coherent Structures in Wall-Bounded Turbulence

Javier Jiménez and Adrián Lozano-Durán

Abstract The current knowledge about some particular kinds of coherent structures in the logarithmic and outer layers of wall-bounded turbulent flows is briefly reviewed. It is shown that a lot has been learned about their geometry, flow properties and temporal behaviour. It is also shown that, although the wall-attached structures carry the largest fraction of most flow properties, they are only extreme cases of smaller wall-detached eddies, and that the latter connect with the more classical behaviour of homogeneous turbulence away from walls. Nevertheless, it is argued that little is known about the dynamical origin of these structures, and that a concerned effort is required to quantitatively identify which one (or ones) of the plausible available dynamical models is a better representation of the observed behaviour.

1 Introduction

It is 'a-priori' unclear whether there are coherent structures in turbulence, or how they should be defined. Their most compelling support derives from free-shear flows, where visualisations reveal a wave-like organisation of advected scalars [1], which can be traced for relatively long times, and that can be linked to the Kelvin–Helmholtz instability of the mean velocity profile. Besides that visual impression, which did a lot to crystallise a structural view of turbulent flows with which to complement earlier stochastic descriptions, the identification of those waves in terms of a known dynamical process was important. It opened the way to the prediction of their properties, and later to effective control strategies [2, 3].

The situation is not as satisfactory for wall-bounded turbulence, in which the mean velocity profile does not sustain a linear modal instability. Long-lived structures were identified in wall turbulence even earlier than in free-shear flows [4], mainly

J. Jiménez (✉) · A. Lozano-Durán
School of Aeronautics, U. Politécnica, 28040 Madrid, Spain
e-mail: jimenez@torroja.dmt.upm.es

A. Lozano-Durán
e-mail: adrian@torroja.dmt.upm.es

M. Stanislas et al. (eds.), *Progress in Wall Turbulence 2*,
ERCOFTAC Series 23, DOI 10.1007/978-3-319-20388-1_3

in the form of long streaky structures of the streamwise velocity, but it was soon realised that the streaks could not survive by themselves, and that other structures were needed to complete a self-sustaining regeneration cycle [5–7]. In the past 20 years, numerous models for the dynamics of such flows have been proposed, such as linearised approximations, both modal and non-modal [8–10]; specific structures such as hairpin vortices [11–13]; instabilities of the velocity streaks [14, 15] and equilibrium and time-periodic exact solutions of the Navier–Stokes equations [16, 17].

In most cases, structures that approximately answer to the description in those models can be identified in turbulent flows, and there are physical grounds to believe that they really exist and may even be common. Moreover, all of them share some degree of intellectual appeal that is at the root of the original proposal. However, the question of how often the structures and the evolution that they predict actually occur in the flow, and how relevant are they to its dynamics, has generally been addressed at most qualitatively or partially (e.g. hairpin 'heads', or wall-normal velocity 'bursts').

There is little doubt that exact structures of any kind are unlikely to be found in turbulence. All of the examples given above either exist only at Reynolds numbers well below those of fully developed turbulence, or are unstable or transient. On the other hand, it has been persuasively argued that equilibrium solutions and other simple trajectories in phase space, even if unstable or transient, are approached by the flow more often than other random non-equilibrium states, and are therefore more relevant to the statistics than other kinematically possible flow fields [17, 18].

In recent years, temporally and spatially resolved numerical databases have begun to allow the quantitative analysis of how closely a particular solution is approached by the flow, and how often that happens, although the analysis often involves isolating a particular subset of the flow, a restricted range of wavenumbers, or both. It is usually, but not always, also necessary to inhibit or discount the effect of interactions between different structures. We will discuss some examples in which it has recently become possible to test particular simplified turbulence 'cartoons' in essentially natural flows, examine their shortcomings, and point to possible avenues for future tests of other popular (or less popular) ones.

The organisation of the paper is as follows. Section 2 describes in some detail what has been found about two particular kinds of long-lived structures in wall-bounded flows, about how they evolve in time, and about how they are related to similar structures in shear flows far from the wall. It will be seen that a lot is known, but that optimistic impression is tempered in Sect. 3, which discussed how little we really know about the dynamical origin of what is observed, and how far are we from conclusively identifying the observations with any of the conceptual models mentioned above. Section 4 summarises and concludes.

2 We Know Everything About Turbulent Structures

Channels are often used as archetypes for the theoretical characterisation of wall-bounded turbulence because they are relatively simple to simulate (although apparently hard to realise experimentally), and because numerical simulations offer the best chance of studying the flow in detail. There are channel simulations available at relatively high Reynolds numbers [19–23], and time-resolved databases of their evolution have recently become available for analysis [24].

Two kinds of structures have been studied in detail for this flow: Vortex 'clusters' are connected objects of particularly strong discriminant of the velocity gradient [25], and 'Qs' [26] are connected regions of strong Reynolds stresses that are the three-dimensional analogues of the classical 'quadrant' events studied by experimentalists from single-point signals [27]. The most important Qs are those in the second 'quadrant' (Q2 or ejections) and in the fourth quadrant (Q4 or sweeps), i.e. those for which the wall-normal velocity fluctuations, v, have opposite sign to the streamwise velocity fluctuations, u, so that they carry a Reynolds stress consistent with the mean shear ($uv < 0$ for $\partial_y U > 0$). Both the clusters and the Qs can be classified into wall-attached and wall-detached families, depending on whether or not their roots reach the neighbourhood of the wall. The wall-attached structures have been studied in most detail. They are larger than the local Corrsin scale [28], interact directly with the ambient mean shear, and presumably draw their energy from it [29]. The wall-attached sweeps and ejections carry over 60 % of the total tangential Reynolds stress.

Above the viscous layer near the wall, Qs and clusters are complicated objects with fractal dimensions of the order of $D_F = 2 - 2.5$. Two examples are given in Fig. 1, both of which are wall-attached and large enough to extend into the logarithmic layer ($L_y/h = 0.15 - 0.20$), where L_y is the wall-normal dimension, and h is the channel half-width. It is known that the large attached eddies of both kinds ($L_y^+ \gtrsim 100$) have self-similar aspect ratios, $L_z \approx L_y$ and $L_x \approx 3L_y$ [25, 26], and lifetimes that also scale linearly with their height, $t^+ \approx L_y^+$ [30].

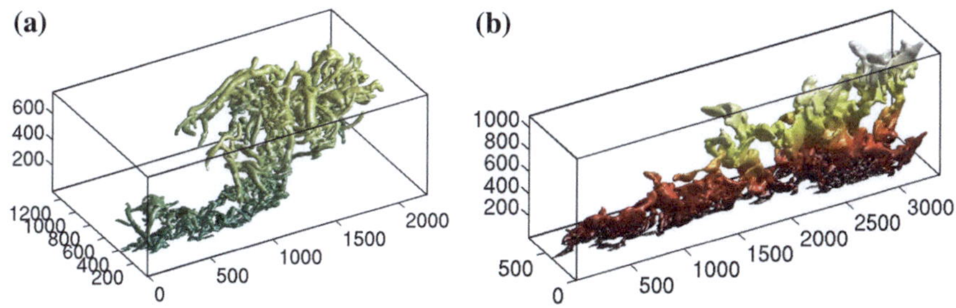

Fig. 1 **a** Wall-attached vortex cluster. **b** Ejection ('second quadrant' Reynolds stress event). Turbulent channel at $Re_\tau = 4200$ [22]. Flow is from *left* to *right*, and axes are in wall units

Attached sweeps and ejections are almost always found in side-by-side pairs, with a vortex cluster near their base. The mean flow conditioned to one of these pairs is a quasi-streamwise roller sitting at the boundary between a high- and a low-velocity streak of the streamwise velocity, in an arrangement reminiscent of the better-known associations of streaks and vortices in the viscous layer [15, 31]. Above the buffer region, they are typically much larger than those closer to the wall, and involve disorganised turbulent objects similar to those in Fig. 1. In those cases, the association of Qs with rollers can only be recognised in a conditional statistical sense. For an example of an individual pair, the reader is referred to Fig. 12 in [26], or to the three-dimensional version of that figure in the supplementary material to that paper.

It is interesting that, even if attached eddies play a dominant role in wall-bounded flows, the proximity of the wall does not appear to be required for their formation. A recent simulation of homogeneous shear flow [32] was found to contain vortex clusters, sweeps and ejections with statistical properties very similar to those of the large detached eddies in channels (see Fig. 2a for an example), and these large Qs are also responsible for most of the tangential Reynolds stress. It is also known that channels with rough-like walls [33] or even with no wall at all [34] have vortex clusters indistinguishable from those of normal channels.

In fact, attached eddies appear to be particular cases of detached ones that have become large enough to collide with the wall. Figure 2b shows the probability distribution of the vertical dimensions of sweeps and ejections in a channel and in homogeneous shear. Those in the channel have been separated in bands of the position of their farthest point from the wall. The probability of their vertical dimension decreases

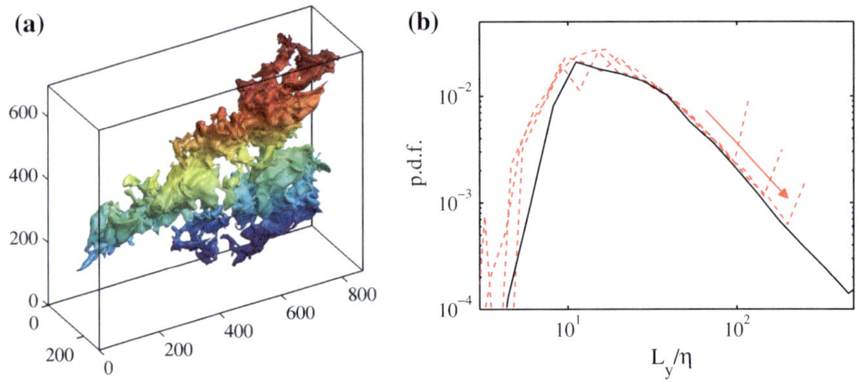

Fig. 2 **a** Large ejection in a homogeneous shear flow with microscale Reynolds number $Re_\lambda = 100$ [32]. Flow is from *left* to *right*. Compare with Fig. 1b. **b** Probability density function of the logarithm of the wall-normal dimension of sweeps and ejections (Q2+Q4). ‒ ‒ ‒ ‒ , Attached or detached eddies with a given maximum height, $y_{max}/h = 0.3, 0.5, 0.7$, increasing in the direction of the *arrow*. Turbulent channel at $Re_\tau = 2000$ [22]. ——— , Homogeneous shear, $Re_\lambda = 100$ [32]. Axes in both panels are in Kolmogorov units

for larger eddies, roughly as $\text{pdf}(L_y) \sim L_y^{-5/2}$, until it reaches $L_y = y_{max}$. The eddies then hit the wall, and the p.d.f. accumulates into the distribution of attached eddies. Choosing a higher band simply extends the probability tail to taller eddies before they reach the wall. The figure also contains data from a homogeneous shear, where the eddies follow the same probability distribution, but never accumulate at the (non-existent) wall.

Even more revealing is the temporal evolution of these structures, which requires tracking them in time and unravelling the numerous interactions in which they split or merge with one another along their lifetimes [24, 30]. It was in this way that the previously mentioned lifetimes were computed, although they refer to 'primary' eddies that are not born from a split or disappear into a merger. When interactions are taken into account, it turns out that most eddies larger than a few Kolmogorov scales (η) lose or receive some fragments of comparable size at some point in their lives [30], and that such direct or inverse 'cascade' events are responsible for the largest part of their growth and decay. Numerically, however, most of the splits and mergers take place between a larger eddy and a smaller fragment of the order of the Kolmogorov scale.

The connection of these events with a cascade can be made more quantitative. Centring on the direct cascade, the interactions between eddies can be best studied by ordering the different branches, each of which represents the evolution of a particular eddy, into a graph in which the nodes are the splits. In each split, one 'main' branch survives and one is created anew. Each branch can be assigned a 'split index' as the number of splits that separate it from its farthest descendent. A sketch is Fig. 3a. It can be expected that eddies with a larger index are also bigger, and that their size decreases as they approach the extremal branches in which they are dissipated by viscosity. This is demonstrated in Fig. 3b which shows the p.d.f. of the eddy size (the cubic root of its volume) for eddies with a given index [35]. The characteristics size of the smallest eddies (index $= 0$) is $L \approx 0.05h \approx 10\eta$, comparable to the

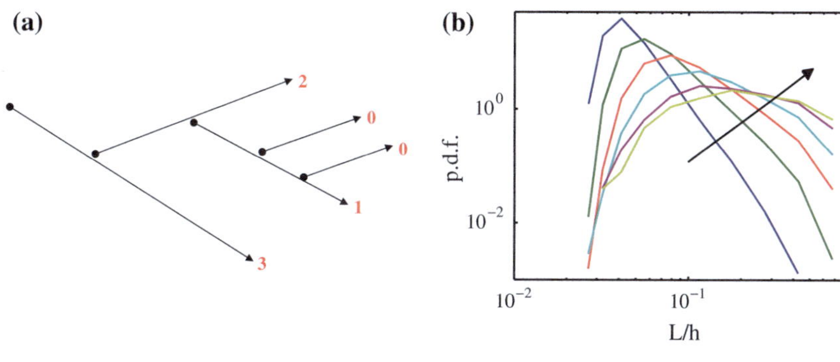

Fig. 3 **a** Sketch of the splitting tree of a large eddy into smaller fragments. The index of each branch is the number of splits that separates it from its farthest descendent. **b** Probability density function of the (cube root of) the volume of Q-structures, separated by their splitting index, ranges from 0 to 6 in the direction of the arrow [35]. Turbulent channel at $Re_\tau = 950$ [22]

diameter of the small-scale vortices [36], and those with the highest index are of the order of the integral scale, $L \approx 0.2h$. It is interesting that the ratio between the mean volume of consecutive indices changes from about 0.2 in the viscous range ($L < 30\eta$) to about 0.5 later on. There is a complicated relation between this ratio and the average volume fraction typically lost in a split, because a given eddy can break into fragments with very different future histories, and therefore with very different indices. But, if we take the volume ratio between consecutive indices as an estimate of the ratio between the volume of successive eddy 'generations', the first value can be interpreted as being dominated by the loss of Kolmogorov scale fragments, and the second one as a more classical equilateral inertial cascade. A similar analysis can be done for the merging of structures, giving information on the inverse component of the cascade.

3 We Know Nothing About Turbulent Structures

We have seen in the previous section that what amounts to a complete characterization of the behaviour of two kinds of coherent structures in channel flows. This has been an important advancement of the last few years that would have been difficult to predict a decade ago. Although most of it has been based on information from simulations, some experimental time-resolved datasets, generally limited to two-dimensional sections, have contributed substantially [37–40]. All these information have given us a fresh outlook on what is going on inside the turbulent wall-bounded flows, but it can be argued that our understanding of these flows is still well below that of the free-shear turbulence.

We lack a theoretical model for the behaviour that we observe. As mentioned in the introduction, the problem is not as much a lack of models, but a superabundance. It is probably true that several of those models are equivalent, although that remains to be proved. For example, it is not inconceivable that the structures described in the previous section can be described as packets of approximate hairpin vortices. There have been numerous attempts to show that particular theories are 'compatible' with statistical data, e.g. [41]. They are often successful, but that should not be considered as a sufficient proof. We know from RANS modelling that suitably tuned semiempirical models with little or no structural information can predict turbulence statistics very accurately.

That is true even when fairly detailed statistics are used. For example, it was shown in [42] that the temporal evolution of minimal channels was consistent with the linearised Orr's mechanism of inviscid transient growth. The similarities include such high-order quantities as temporal correlation functions and the detailed exchange of energy among velocity components. Similarly, it was shown in [9] that optimally amplified linearised perturbations agree well with the highest POD modes of full channels. However, it has proved difficult in both cases to identify in the flow the three-dimensional structures implied by the linear models, or their temporal evolution.

There may be several reasons for these failures. In the first place, as mentioned in the introduction, it is highly unlikely that any structure predicted by an essentially laminar model would be found in the chaotic environment of real turbulence at even moderate Reynolds numbers. The most that can probably be expected is some approximation, such as is evident, for example in the steady retreat from the identification of hairpin packets into 'heads', 'legs' or 'incomplete' hairpins. This is not necessarily bad, but it may easily become meaningless without a clear definition of the metric used to define the 'presence' of a structure.

The second problem is that many of these models actually predict transient phenomena. This is probably the best that can be expected from a system without modal instabilities, but it means that we have to identify something that only lives for a fraction of the system evolution. For example, the Orr model in [42] describes a relatively short burst in the wall-normal velocity that quickly decays after creating a streamwise velocity streak that lasts much longer. This is a fairly accurate description of the evolution of the flow in minimal channels [5–7, 43], but it means that we are trying to identify something that is not there most of the time. Again, we need to define a metric to describe not only how close is the system to a given solution, but also how often it happens. Note that this may be more complicated than just tracking time intervals or the contribution to the statistics. In the previous example, the bursts are probably not large contributors to the statistics of u or v [44], but they are necessary 'catalysts' for the injection of energy into the turbulent fluctuations.

A newer development has been the attempt to identify turbulent structures with exact solutions of the Navier–Stokes equations, whether permanent waves or more complex invariant sets. These efforts have generally centred on transitional flows, but some of them extend into the incipient turbulent regime. Early attempts to represent the statistics of pipes in terms of permanent waves were only moderately successful [45], and the focus quickly moved to recurrent solutions [46, 47], and to homoclinic and heteroclinic orbits as models for bursting [48–51]. All these studies share the usual ambiguity about the norm used as a measure of proximity. A promising technique that partially bypasses this limitation is to use an arbitrary norm to identify approximately a recurrent flow state, which is then used as an initial condition for the computation of exact recurrent solutions [46, 47]. However, most of these techniques have only been used on flows at low Reynolds numbers, or even on two-dimensional ones, and it is unclear whether they can be extrapolated to more complex situations.

4 Discussion and Conclusions

We have tried to show how we are both very near and very far from understanding the dynamics of shear-dominated wall-bounded turbulent flows. Very near in the sense that we have collected in the past decades what is probably a reasonably complete catalogue of coherent structures in these flows, of their properties, and of their temporal evolution. But very far in the sense that we lack a consensus theoretical model for their behaviour. Two questions suggest themselves. The first one is whether coherent structures are more than an observational construct, and whether we should really expect a theory for them. The second one is whether such a theory would be of any use.

Tackling the second question, first, our best guide could be the experience with free-shear flows mentioned at the beginning of this paper. In that case, a linearised theory for the nature of the structures naturally led to control strategies. The same has been true of our partial theoretical understanding of wall-bounded flows, even if the applications are as yet limited to simulations at moderate Reynolds numbers. For example, the early realisation of the role of sublayer streaks in determining friction drag, and of the buffer-layer vortices in sustaining the streaks, led to successful drag reduction strategies based on damping the vortices [52]. Further removed from intuition, [53] was able to laminarise turbulent Couette flow by acting on the system as it spontaneously approaches a fixed point in the 'edge' of the basin of attraction of turbulence. Although it is impossible to predict whether a better theoretical understanding would lead to better control strategies at realistic Reynolds numbers, the success rate of past 'intuitive' approaches has been at most moderate, and the previous examples offer some hope that better methods lay hidden within deeper theories.

The question about the relation between theory and structure is more complex, and cannot probably be answered conclusively at this point. Many of the theoretical models mentioned above were motivated by attempts to explain ('postdict') structures observed in highly constrained simulations. They were usually successful, but we have seen that the inverse question of identifying theoretical solutions in real flows has been more problematic. It is also clear that the idea of a coherent blob of strong vorticity is a very different concept from an invariant set of the Navier–Stokes equations. On the other hand, structures stay coherent for some reason, and turbulent flows are nothing but solutions of the Navier–Stokes equations. Consider one of the problems of describing any statistically stationary solution of a dissipative system such as turbulence, which is to understand how the energy is injected into the system. This is a basic question as much in fully turbulent flows, in permanent waves, or in any invariant set. In shear-dominated flows, it is known that energy injection is mediated by the production term, $-\langle uv\rangle \partial_y U$, and it is reasonable to expect that the Q-structures described in Sect. 2, which are the dominant carriers of $\langle uv\rangle$, play a role in that process. Similarly, vortex clusters, which label regions of high energy dissipation can be expected to be controlled by, and to modulate, the energy balance.

The basic message of this paper should be that, in spite of all the new observational information on the behaviour of turbulent wall-bounded flows, a lot remains to be understood about the reasons for that behaviour. It is the opinion of the present authors that, in the same way as a dominant thread of turbulence research during the past 50 years has been the reconciliation of the structural and statistical views of the flow, an important task for the next years will be to relate theoretical models with structural observations.

Acknowledgments This work was supported in part by the European Research Council, under grant ERC-2010.AdG-20100224, and by CICYT, under grant TRA2009-11498. A. Lozano–Durán was supported by an FPI fellowship from the Spanish Ministry of Education and Science. We are grateful to S. Dong for Fig. 2a and for some of the data in Fig. 2b, and to C. Huertas-Cerdeira for Fig. 3b.

References

1. G.L. Brown, A. Roshko, J. Fluid Mech. **64**, 775 (1974)
2. C.H. Ho, P. Huerre, Annu. Rev. Fluid Mech. **16**, 365 (1984)
3. M. Gaster, E. Kit, I. Wygnanski, J. Fluid Mech. **150**, 23 (1985)
4. S.J. Kline, W.C. Reynolds, F.A. Schraub, P.W. Runstadler, J. Fluid Mech. **30**, 741 (1967)
5. J. Jiménez, P. Moin, J. Fluid Mech. **225**, 221 (1991)
6. J.M. Hamilton, J. Kim, F. Waleffe, J. Fluid Mech. **287**, 317 (1995)
7. J. Jiménez, A. Pinelli, J. Fluid Mech. **389**, 335 (1999)
8. K.M. Butler, B.F. Farrell, Phys. Fluids A **5**, 774 (1993)
9. J.C. del Álamo, J. Jiménez, J. Fluid Mech. **559**, 205 (2006)
10. D.F. Gayme, B.J. McKeon, B. Bamieh, A. Papachristodoulou, J.C. Doyle, Phys. Fluids **23**, 065108 (2011)
11. A.E. Perry, M.S. Chong, J. Fluid Mech. **119**, 173 (1982)
12. J. Zhou, R.J. Adrian, S. Balachandar, T.M. Kendall, J. Fluid Mech. **387**, 353 (1999)
13. R.J. Adrian, Phys. Fluids **19**, 041301 (2007)
14. H.T. Kim, S.J. Kline, W.C. Reynolds, J. Fluid Mech. **50**, 133 (1971)
15. W. Schoppa, F. Hussain, J. Fluid Mech. **453**, 57 (2002)
16. M. Nagata, J. Fluid Mech. **217**, 519 (1990)
17. G. Kawahara, M. Uhlmann, L. van Veen, Annu. Rev. Fluid Mech. **44**, 203 (2012)
18. P. Cvitanović, Phys. Rev. Lett. **61**, 2729 (1988)
19. J. Kim, P. Moin, R.D. Moser, J. Fluid Mech. **177**, 133 (1987)
20. J.C. del Álamo, J. Jiménez, P. Zandonade, R.D. Moser, J. Fluid Mech. **500**, 135 (2004)
21. S. Hoyas, J. Jiménez, Phys. Fluids **18**, 011702 (2006)
22. A. Lozano-Durán, J. Jiménez, Phys. Fluids **26**, 011702 (2014)
23. M. Bernardini, S. Pirozzoli, P. Orlandi, J. Fluid Mech. **742**, 171 (2014)
24. A. Lozano-Durán, J. Jiménez, J. Phys.: Conf. Ser. **318**, 062016 (2011)
25. J.C. del Álamo, J. Jiménez, P. Zandonade, R.D. Moser, J. Fluid Mech. **561**, 329 (2006)
26. A. Lozano-Durán, O. Flores, J. Jiménez, J. Fluid Mech. **694**, 100 (2012)
27. S.S. Lu, W.W. Willmarth, J. Fluid Mech. **60**, 481 (1973)
28. S. Corrsin, Local isotropy in turbulent shear flow. Res. Memo 58B11, NACA (1958)
29. J. Jiménez, Phys. Fluids **25**, 101302 (2013)
30. A. Lozano-Durán, J. Jiménez, J. Fluid Mech. **759**, 432–471 (2014)
31. S.K. Robinson, Annu. Rev. Fluid Mech. **23**, 601 (1991)
32. S. Dong, A. Sekimoto, J. Jiménez, Bull. Am. Phys. Soc. **57**(17), D21.7 (2012)
33. O. Flores, J. Jiménez, J. Fluid Mech. **566**, 357 (2006)
34. Y. Mizuno, J. Jiménez, J. Fluid Mech. **23**, 085112 (2013)
35. C. Huertas–Cerdeira (private communication) (2014)
36. J. Jiménez, A.A. Wray, P.G. Saffman, R.S. Rogallo, J. Fluid Mech. **255**, 65 (1993)
37. D.J.C. Dennis, T.B. Nickels, J. Fluid Mech. **673**, 180 (2011)
38. D.J.C. Dennis, T.B. Nickels, J. Fluid Mech. **673**, 218 (2011)
39. G.E. Elsinga, C. Poelma, A. Schroder, R. Geisler, F. Scarano, J. Westerweel, J. Fluid Mech. **697**, 273 (2012)
40. J.A. LeHew, M. Guala, B.J. McKeon, Exp. Fluids **54**, 1508 (2013)
41. K.T. Christensen, R.J. Adrian, J. Fluid Mech. **431**, 433 (2001)

42. J. Jiménez, Phys. Fluids **25**, 110814 (2013)
43. J. Jiménez, S. Hoyas, M.P. Simens, Y. Mizuno, J. Fluid Mech. **657**, 335 (2010)
44. J. Jiménez, G. Kawahara, M.P. Simens, M. Nagata, M. Shiba, Phys. Fluids **17**, 015105 (2005)
45. R.R. Kerswell, O.R. Tutty, J. Fluid Mech. **584**, 69 (2007)
46. D. Viswanath, J. Fluid Mech. **580**, 339 (2007)
47. G.J. Chandler, R.R. Kerswell, J. Fluid Mech. **722**, 554 (2013)
48. N. Aubry, P. Holmes, J.L. Lumley, E. Stone, J. Fluid Mech. **192**, 115 (1988)
49. G. Kawahara, S. Kida, J. Fluid Mech. **449**, 291 (2001)
50. L. Van Veen, G. Kawahara, Phys. Rev. Lett. **107**, 114501 (2011)
51. J. Halcrow, J.F. Gibson, P. Cvitanović, D. Viswanath, J. Fluid Mech. **621**, 365 (2009)
52. H. Choi, P. Moin, J. Kim, J. Fluid Mech. **262**, 75 (1994)
53. G. Kawahara, Phys. Fluids **17**, 041702 (2005)

Attached Eddies and High-Order Statistics

Ivan Marusic and James D. Woodcock

Abstract The attached eddy hypothesis of Townsend [16] is the basis of a model of the logarithmic region in wall-bounded turbulent flows, in which the inertially dominated part of the flow is described by a hierarchy of self-similar eddying motions that extend to the wall. The hypothesis has gained considerable support from high Reynolds number experiments and recently from DNS Sillero et al., Phys. Fluids 25:105102, 2013, [14]. Meneveau and Marusic, J. Fluid Mech., 719:R1, 2013, [9] also recently used the attached eddy hypothesis, together with the central limit theorem, to deduce that all even-ordered moments of the streamwise velocity will exhibit a logarithmic dependence on the distance from the wall. This was also further supported by experimental evidence.

In this paper, we consider a more rigorous physical and mathematical basis for the attached eddy hypothesis than those considered in the past, and extend the proof of Campbells theorem to apply to the velocity field corresponding to a forest of variously sized eddies that are randomly placed on the wall. Thus, by modelling the flow as a hierarchy of self-similar eddies that are perfectly randomly and independently placed on the wall, we have been able to derive functional forms for the average velocity, as well as all higher order moments of the velocity fluctuations, and cross-correlations between velocity fluctuations in different directions (such as the Reynolds shear stress). This leads to a derivation of the Meneveau and Marusic result for high-order moments without invoking the central limit theorem and indicates that such logarithmic behaviours will only be realisable in the large Reynolds number limit.

The model predicts that the streamwise and spanwise fluctuations will be super-Gaussian, while the wall-normal fluctuations will soon approach a constant skewness as the Reynolds number increases. (This is notably different behaviour from that which would be predicted by the use of the central limit theorem.) In the spanwise and wall-normal cases, this agrees with experimental results. Conversely, however, experiments have also shown the streamwise fluctuations to be sub-Gaussian.

I. Marusic (✉) · J.D. Woodcock
University of Melbourne, Melbourne, VIC 3010, Australia
e-mail: imarusic@unimelb.edu.au

J.D. Woodcock
e-mail: jwoodcock@unimelb.edu.au

M. Stanislas et al. (eds.), *Progress in Wall Turbulence 2*,
ERCOFTAC Series 23, DOI 10.1007/978-3-319-20388-1_4

This indicates that the present model will need to be refined in order to capture the sub-Gaussian behaviour of the streamwise fluctuations. One obvious potential refinement can be seen in the interdependent placement of the eddies.

1 Introduction

The great complexity of turbulence can be seen in the scarcity of phenomenological models capable of reproducing its various features. Much of the early modelling of turbulent flows simply regarded the flow as consisting of perfectly random fluctuations, the average behaviours of which could be experimentally determined.

Taking a different approach, Townsend [15] proposed a remarkably simple model for turbulent flow in the near-wall log-region. This model, now known as the *attached eddy hypothesis*, states that the flow in this region will be dominated by geometrically self-similar eddies, whose corresponding flow fields extend to the wall.

Using this attached eddy hypothesis, Townsend [16] was able to derive simple relationships between the second-order velocity fluctuations and the distance from the wall, z. If u, v and w represent the fluctuations in the streamwise, spanwise and wall-normal fluctuations respectively, Townsend found that

$$\left\langle u^2\right\rangle^+ = B_1 - A_1 \log\left(\frac{z}{\delta}\right), \tag{1a}$$

$$\left\langle v^2\right\rangle^+ = B_{1,v} - A_{1,v} \log\left(\frac{z}{\delta}\right), \tag{1b}$$

$$\left\langle w^2\right\rangle^+ = B_{1,w}, \tag{1c}$$

$$-\left\langle uw\right\rangle^+ = 1, \tag{1d}$$

where the angled brackets represent ensemble averages. The superscript '+' indicates that the quantities have been scaled according to viscous wall-units, and δ denotes the maximum distance from the wall at which the flow is dominated by the presence of the attached eddies. All of the As and Bs above are constants. Recent experiments at high Reynolds number [2, 4, 6, 8, 10, 17], as well as numerical simulations [14], have supported such logarithmic behaviours for the profiles of the velocity fluctuations.

At the time, however, existing flow visualisation technology did not enable Townsend to search real turbulent flows for coherent structures or repeating patterns. Furthermore, scientific computing, then in its infancy, was not up to the challenge of simulating high (or even moderate) Reynolds number turbulent flows. Nonetheless, the attached eddy hypothesis was further refined and its use extended over the following decades.

The hypothesis has been used to derive the mean velocity of wall-flows by [11, 12]. The streamwise flow in the log-region obeys the well known equation

$$\langle U\rangle^+ = \frac{1}{\kappa}\log\left(z^+\right) + C, \tag{2}$$

where κ is von Kármán's constant, and C depends on the roughness of the surface, but is otherwise constant.

More recently, Meneveau and Marusic [9] proposed a method for extending the use of the attached eddy hypothesis to higher order moments of the velocity. For this, they assumed that the eddies would display perfectly Gaussian behaviour, which allowed the use of the central limit theorem. The even-ordered moments derived in this way are

$$\left\langle (u^+)^{2p} \right\rangle^{1/p} = B_p - A_p \log\left(\frac{z}{\delta}\right), \tag{3}$$

where B_p and A_p are constants, and the A_ps relate to each other via

$$A_p = (2p-1)!!\, A_1, \qquad \text{where } n!! \equiv n(n-2)(n-4)\ldots 1.$$

In this work, we do not assume Gaussian behaviour. In fact, we will subsequently show that the assumption of Gaussianity produces incorrect results for higher order moments.

To this end, we revisit the basic theory and physical assumptions that underpin the attached eddy hypothesis. This has allowed us to place the model on a firmer basis, and to expand the application of the model to higher order moments. We utilise Campbell's theorem, which was originally applied to the attached eddy hypothesis by Marusic [7], following a suggestion by R.J. Adrian (private communication). Following the analysis of Townsend and Perry and Chong, we model the flow as a random distribution of eddies that are geometrically self-similar and whose locations are entirely independent.

The model is formulated and the results are briefly discussed in Sect. 2. A proof of Campbell's theorem is given in Appendix.

2 Velocity Statistics

The velocity field is modelled as the superposition of the individual velocity fields corresponding each of many eddies. The eddies all have the same shape and relative dimensions, differing from each other only in height. One sketch of a possible eddy shape is given in Fig. 1.

Each eddy can therefore be considered as a separate entity, defined only by its own properties. Its only defining properties being its location on the wall, $\mathbf{x}_e$, and its height, h. It is clear therefore that the defining length scale of the eddy will be h, while its velocity scale will be the *friction velocity*.

It follows that if $\mathbf{Q}$ is the velocity field at $\mathbf{x}$ corresponding to an individual eddy, then its spatial and height dependence will be

$$\mathbf{Q} = \mathbf{Q}\left(\frac{\mathbf{x} - \mathbf{x}_e}{h}\right). \tag{4}$$

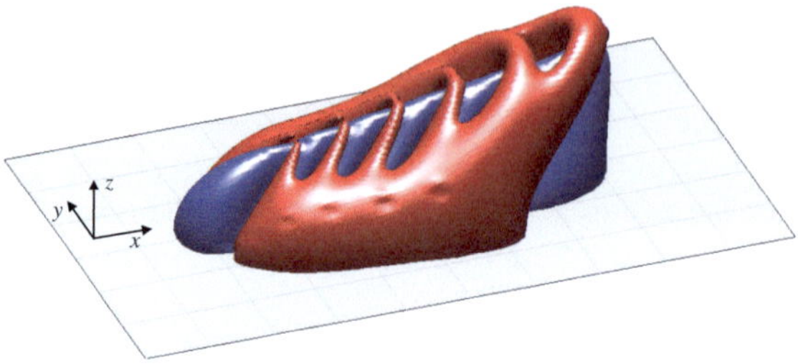

Fig. 1 Sketch of a typical representative eddying motion. The *red* and *blue* regions represent, respectively, the portion of the flow with streamwise velocity in excess of and below the mean. This shape is typical of the motions that have been observed to appear within turbulent flows [5]

The total velocity, $\mathbf{U}(\mathbf{x})$, will therefore simply be a sum over $\mathbf{Q}$ for all eddies. However, since we could never postulate the simultaneous locations of all eddies present, we must therefore consider instead the statistical properties of a large number of eddies.

The observation that h is the system's only natural length scale can be used to determine the size distribution of the eddies: If ρ_h denotes the density of eddies of size h, then it follows from simple dimensional analysis that

$$\rho_h \propto \frac{1}{h^3}. \tag{5}$$

The probability of an eddy having size h, which we denote by $P(h)$, will be proportional to ρ_h. It can therefore be derived via normalisation. If all eddies have sizes between h_{min} and h_{max}, then their probability distributions will be

$$P(h) = 2\left(h_{min}^{-2} - h_{max}^{-2}\right)^{-1}\frac{1}{h^3}. \tag{6}$$

The derivations of the first- and second-order moments of the velocity are given in Appendix. The extension to higher order moments will not be given here due to limited space.

To simplify the equations, particularly at higher orders, we introduce a new function, $\lambda_{k,l,m}$, known as a cumulant. It is defined by

$$\lambda_{k,l,m}(z) \stackrel{\text{def}}{=} \beta \int_{h_{min}}^{h_{max}} I_{k,l,m}(Z) h^2 P(h)\, \mathrm{d}h, \tag{7}$$

where β represents the average density (per area) of eddies on the wall. The function $I_{k,l,m}$ is known as the eddy contribution function, and is given by

$$I_{k,l,m}(Z) \stackrel{\text{def}}{=} \iint_{-\infty}^{\infty} Q_x^k\,(\mathbf{X})\, Q_y^l\,(\mathbf{X})\, Q_z^m\,(\mathbf{X})\ \mathrm{d}X\,\mathrm{d}Y. \tag{8}$$

The capital **X**, and its components X, Y and Z, represent the location scaled by h. Using cumulants, the mean velocity can be expressed as

$$\langle U\rangle = \lambda_1, \qquad \langle V\rangle = \lambda_{0,1,0}, \qquad \langle W\rangle = \lambda_{0,0,1}, \tag{9}$$

where we have used the shorthand

$$\lambda_n \equiv \lambda_{n,0,0} \tag{10}$$

for purely streamwise quantities. If we denote velocity fluctuations by **u**, so that

$$\mathbf{u}(\mathbf{x}) = \mathbf{U}(\mathbf{x}) - \langle \mathbf{U}(\mathbf{x})\rangle, \tag{11}$$

then

$$\begin{gathered}\langle u^2\rangle = \lambda_2, \qquad \langle v^2\rangle = \lambda_{0,2,0}, \qquad \langle w^2\rangle = \lambda_{0,0,2}, \\ \langle uv\rangle = \lambda_{1,1,0}, \qquad \langle uw\rangle = \lambda_{1,0,1},\end{gathered} \tag{12}$$

and the moments of the streamwise velocity fluctuations are given by

$$\langle u^3\rangle = \lambda_3, \tag{13a}$$

$$\langle u^4\rangle = \lambda_4 + 3\lambda_2^2, \tag{13b}$$

$$\langle u^6\rangle = \lambda_6 + 15\lambda_2\lambda_4 + 10\lambda_3^2 + 15\lambda_2^3, \qquad \text{etc.} \tag{13c}$$

and similarly for $\langle v^{2p}\rangle$ and $\langle w^{2p}\rangle$. It is instructive to compare these results to those obtained by Meneveau and Marusic [9], who assumed the distribution of eddies would be perfectly Gaussian. Their resulting even order moments were given in Eq. (3).

It can be seen in Eq. (13) that our results agree with the Gaussian behaviour in only the second-order statistics, but begin to deviate at higher order moments. We will now look more closely at the third- and fourth-order statistics.

2.1 Skewness Statistics

The skewness is denoted by S_u, S_v and S_w. They are defined by

$$S_u = \frac{\langle u^3\rangle}{\langle u^2\rangle^{3/2}}, \qquad S_v = \frac{\langle v^3\rangle}{\langle v^2\rangle^{3/2}}, \qquad S_w = \frac{\langle w^3\rangle}{\langle w^2\rangle^{3/2}}. \tag{14}$$

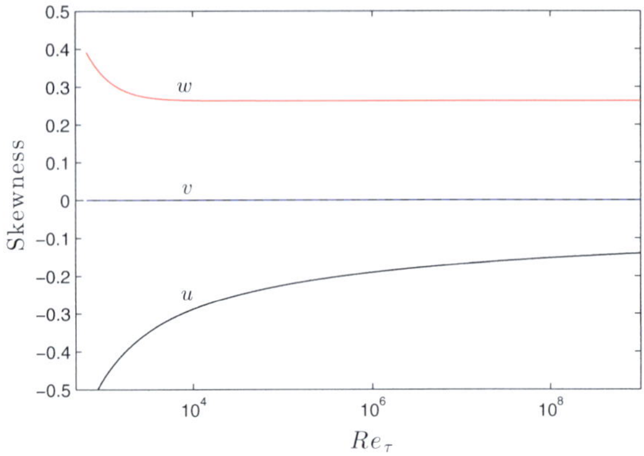

Fig. 2 Plots showing the skewness of the three velocity components calculated using the typical attached eddy depicted in Fig. 1 at $z^+ = 100$, where it is assumed that $Re_\tau = 100h_{max}/h_{min}$

By inspection of (13), we can see that the skewness is

$$S_u = \frac{\lambda_3}{\lambda_2^{3/2}}, \qquad S_v = \frac{\lambda_{0,3,0}}{\lambda_{0,2,0}^{3/2}}, \qquad S_w = \frac{\lambda_{0,0,3}}{\lambda_{0,0,2}^{3/2}}. \tag{15}$$

To investigate these and higher order moments of the velocity, we must have some conception of the behaviour of $\mathbf{Q}(\mathbf{x})$. To this end, we introduce an assumed eddy shape and derive a velocity field from it via the Biot–Savart equation. The procedure for doing so has been described by Marusic [7]. The eddy shape chosen is as given in Fig. 1, and consists of a train of six arch-shaped rods that are aligned in the streamwise direction in ascending order of height. This is intended to produce a purely qualitative appreciation for the output of the model.

The computed skewness is shown in Fig. 2. As expected, the spanwise skewness is zero. The streamwise skewness approaches zero gradually as the Reynolds number increases. The wall-normal skewness, on the other hand, rapidly approaches a constant.

2.2 Flatness Statistics

The flatness is of particular interest, since it allows us to compare the results of the attached eddy model to experimental results and to the Gaussian results reported by Meneveau and Marusic [9]. The flatness is denoted by F_u, F_v and F_w, and is defined by

$$F_u = \frac{\langle u^4 \rangle}{\langle u^2 \rangle^2}, \qquad F_v = \frac{\langle v^4 \rangle}{\langle v^2 \rangle^2}, \qquad F_w = \frac{\langle w^4 \rangle}{\langle w^2 \rangle^2}. \tag{16}$$

Following the example set by the derivation of the skewness, it can be shown that the flatness will be given by

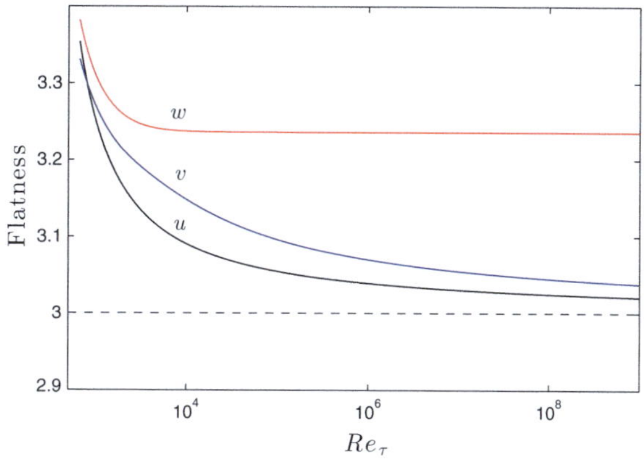

Fig. 3 Plots showing the flatness of the three velocity components calculated using the typical attached eddy depicted in Fig. 1 at $z^+ = 100$, where it is assumed that $Re_\tau = 100h_{max}/h_{min}$

$$F_u = 3 + \frac{\lambda_4}{\lambda_2^2}, \qquad F_v = 3 + \frac{\lambda_{0,4,0}}{\lambda_{0,2,0}^2}, \qquad F_w = 3 + \frac{\lambda_{0,0,4}}{\lambda_{0,0,2}^2}. \tag{17}$$

If the flow were perfectly Gaussian, the flatness would be 3, and it is clear from the above equation that it is here that the results of the attached eddy model differ from Gaussian behaviour.

The computed flatness is shown in Fig. 3. There it can be seen that all three components of the flatness will be super-Gaussian at finite Reynolds numbers, according to this attached eddy model (and in fact, a cursory glance at (7) and (8) will reveal that all of the above flatnesses must be super-Gaussian under the present model). However, experimental results compiled by Fernholz and Finley [3] up to $Re_\tau = 22,000$ show that in the log-region, $F_u \approx 2.8$, $F_v \approx 3.4$ and $F_w \approx 3.4$.

Therefore, while our results agree qualitatively with experimentation in the spanwise and wall-normal cases, it clearly disagrees in the streamwise case. This limitation has previously been recognised by Meneveau and Marusic [9], and will require some modification to the model.

3 Conclusions

Townsend's attached eddy hypothesis states that the flow in the log-region will be dominated by a forest of eddies, whose corresponding velocity fields extend to the wall. Previous studies have used the hypothesis to derive the first- and second-order moments of the velocity. In this work, we extend this analysis to higher order moments.

In order to do so, we have extended the use of Campbell's theorem to apply to the placement of eddies on a wall. As with earlier applications of the attached eddy hypothesis, we have modelled the flow as a random distribution of geometrically self-similar eddies.

Using Campbell's theorem, we have been able to derive the higher order moments of the velocity without resorting to the use of the central limit theorem (and the assumptions it carries with it). Importantly, we have found that the higher order moments we have derived in this way differ noticeably from those derived via the central limit theorem.

The model will not be complete until we also have an accurate function representing the flow field corresponding to the average eddy. However, the assumed eddy shape used in this work has been sufficient to demonstrate the qualitative behaviour of the model.

With this, we have shown that the model produces qualitatively correct skewnesses, as well as qualitatively correct flatness in the spanwise and wall-normal directions. The streamwise flatness, however, differs from experimentally reported the sub-Gaussian behaviour.

This indicates that the model will need to be modified in order to capture the true nature of real turbulent flows. The most obvious refinement is in the placement of the eddies, since they are reported to align in packets of various lengths [1], whereas the assumed eddies used in these calculations are all of exactly six arch-shaped vortex rods in length.

The authors wish to gratefully acknowledge the Australian Research Council for the financial support of this research, and Jason Monty and Charles Meneveau for valuable discussions.

Appendix

Since Campbell's theorem is usually applied to systems in which events are randomly placed in time, rather than space as the eddies are here, it is worthwhile to extend the existing proofs to cover spacial rather than temporal averages. This is an extension of a proof of Campbell's theorem by Rice [13], and applies to the mean velocity and the Reynolds stresses.

Central to the following proof are two physical assumptions we have made about the eddies. The first is the inherently reasonable assumption that there is a limited region of space over which the velocity field corresponding to a single eddy is non-negligible. Mathematically, this manifests as the fact that any integral over the x-y plane of the velocity field corresponding to a single eddy will be equivalent.

We illustrate here in one dimension: For any function $f(x)$,

$$\int_{-\infty}^{\infty} f(x-a)\,\mathrm{d}a = \int_{-\infty}^{\infty} f(x)\,\mathrm{d}x. \tag{18}$$

In fact, it is not necessary for the integral to be taken over the entire real line. It is sufficient that the value of $f(x)$ should be zero (or negligible) outside the bounds of integration. In this work, we will always take integrals over the entire x-y plane when using the above equation.

The second assumption we make use of is more controversial, namely that the locations of each eddy are independent of each other. As has been stated in Sect. 1, the angled brackets refer to ensemble averages in this work. Again we demonstrate with a one-dimensional example: If $F(x)$ represents the sum of K copies of the function $f(x)$, each of which is randomly located, so that

$$F(x) = \sum_{k=1}^{K} f(x - a_k), \tag{19}$$

then the ensemble average of $F(x)$ will be given by

$$\langle F(x) \rangle = \int_{-\infty}^{\infty} p_1(a_1) \int_{-\infty}^{\infty} p_2(a_2) \ldots \int_{-\infty}^{\infty} p_K(a_K) \sum_{k=1}^{K} f(x - a_k)\, \mathrm{d}a_1\, \mathrm{d}a_2 \ldots \mathrm{d}a_K, \tag{20}$$

where the $p_k(a_k)$ represent the probability that the kth variable has the value a_k. For each element k of the sum above, it is only the integral over a_k that will be non-zero. This is because, as we have assumed, the location of each eddy is independent of the location of every other eddy. The implication is that the above equation can be simplified to

$$\langle F(x) \rangle = \sum_{k=1}^{K} \int_{-\infty}^{\infty} p_k(a_k) f(x - a_k)\, \mathrm{d}a_k. \tag{21}$$

We will make use of (18) and (21) subsequently. We now explicitly consider eddies in three-dimensional space. Imagine there are N eddies randomly placed on a two-dimensional space of area L^2. (We will ultimately consider the limit as $L \to \infty$.) The velocity field that corresponds to these N eddies is given by

$$\mathbf{U}_N(\mathbf{x}) = \sum_{k=1}^{N} \mathbf{Q}\left(\frac{\mathbf{x} - \mathbf{x}_{e_k}}{h_k}\right). \tag{22}$$

When we take the ensemble average of $\mathbf{U}_N(\mathbf{x})$, we are averaging over infinitely many realisations in which N eddies are randomly placed on a plane of area L^2.

Because the eddies are perfectly randomly placed, $p(\mathbf{x}_e)$, the probability density function for the location of an eddy is given by

$$p(\mathbf{x}_e) = \frac{1}{L^2}. \tag{23}$$

The difference between the various realisations of this system will simply be the different placements of each of the eddies. Averaging over the ensemble therefore entails averaging over the possible locations of each of the eddies. Therefore,

$$\langle \mathbf{U}_N(\mathbf{x}) \rangle = \iint_{-L/2}^{L/2} p(\mathbf{x}_{e_1}) \ldots \iint_{-L/2}^{L/2} p(\mathbf{x}_{e_N}) \sum_{k=1}^{N} \int_{h_{min}}^{h_{max}} \mathbf{Q}\left(\frac{\mathbf{x} - \mathbf{x}_{e_k}}{h}\right) P(h)\, \mathrm{d}h\, \mathrm{d}\mathbf{x}_{e_N} \ldots \mathrm{d}\mathbf{x}_{e_1}. \tag{24}$$

(Note that because the wall-normal component of $\mathbf{x}_e$ is universally zero, it has been taken to be a two-dimensional vector in the above integrals. Throughout this work, any integral over $\mathbf{x}_e$ should be assumed to be over the x and y planes only.) By substituting (23) into the above, we get

$$\langle \mathbf{U}_N(\mathbf{x}) \rangle = \iint_{-L/2}^{L/2} \frac{1}{L^2} \cdots \iint_{-L/2}^{L/2} \frac{1}{L^2} \sum_{k=1}^{N} \int_{h_{min}}^{h_{max}} \mathbf{Q}\left(\frac{\mathbf{x} - \mathbf{x}_{e_k}}{h}\right) P(h)\, \mathrm{d}h\, \mathrm{d}\mathbf{x}_{e_N} \ldots \mathrm{d}\mathbf{x}_{e_1}. \tag{25}$$

We can now make use of the fact that the locations of each eddy are independent, in the same manner as Eq. (21), to simplify this to

$$\langle \mathbf{U}_N(\mathbf{x}) \rangle = \frac{1}{L^2} \sum_{k=1}^{N} \int_{h_{min}}^{h_{max}} P(h) \iint_{-L/2}^{L/2} \mathbf{Q}\left(\frac{\mathbf{x} - \mathbf{x}_{e_k}}{h}\right) \mathrm{d}\mathbf{x}_{e_k}\, \mathrm{d}h. \tag{26}$$

It is here that we make use of (18) and the fact that any integrals of a certain function over the entire x-y plane will be equivalent. The above equation therefore simplifies to

$$\langle \mathbf{U}_N(\mathbf{x}) \rangle = \frac{N}{L^2} \int_{h_{min}}^{h_{max}} P(h) \iint_{-L/2}^{L/2} \mathbf{Q}\left(\frac{\mathbf{x}}{h}\right) \mathrm{d}x\, \mathrm{d}y\, \mathrm{d}h. \tag{27}$$

We can now average $\mathbf{U}$ over all possible values of N. The overall velocity will therefore be expected to be

$$\langle \mathbf{U}(\mathbf{x}) \rangle = \sum_{N=0}^{\infty} \mathscr{P}(N) \langle \mathbf{U}_N(\mathbf{x}) \rangle, \tag{28}$$

where $\mathscr{P}(N)$ represents the probability that there are exactly N eddies. According to Poisson's law of small probabilities, the probability that there will be exactly N eddies on a plane of area L^2 will be

$$\mathscr{P}(N) = \frac{(\beta L^2)^N}{N!} e^{-\beta L^2}. \tag{29}$$

Using this value of $\mathscr{P}(N)$, it is easy to verify that

$$\sum_{n=0}^{\infty} N \mathscr{P}(N) = L^2 \beta. \tag{30}$$

Using (27) and (30) we can evaluate the sum in (28). In doing so we find that

$$\langle \mathbf{U}(\mathbf{x}) \rangle = \beta \int_{h_{min}}^{h_{max}} P(h) \iint_{-L/2}^{L/2} \mathbf{Q}\left(\frac{\mathbf{x}}{h}\right) \mathrm{d}x\, \mathrm{d}y\, \mathrm{d}h. \tag{31}$$

The mean velocity can now be related to the eddy contribution functions, $I_{k,l,m}(Z)$, which have been defined in (8). For the streamwise velocity in the limit as $L \to \infty$,

$$\langle U \rangle = \beta \int_{h_{min}}^{h_{max}} I_{1,0,0}(Z) h^2 P(h)\, \mathrm{d}h, \tag{32}$$

and similarly for $\langle V \rangle$ and $\langle W \rangle$. This result could be achieved through other means, since it essentially states that the velocity field corresponding to many geometrically identical eddies will simply be the sum of the velocity fields corresponding to each individual eddy. However, by extending this methodology to the averages of higher powers of the velocity, we see its utility.

We demonstrate this now by deriving the Reynolds shear stress. It would be trivial to modify the following derivation so that it instead derives $\langle u^2 \rangle$, $\langle v^2 \rangle$ or $\langle w^2 \rangle$.

$$\begin{aligned} \langle uw \rangle &= \langle (U - \langle U \rangle)(W - \langle W \rangle) \rangle \\ &= \langle UW \rangle - \langle U \rangle \langle W \rangle. \end{aligned} \tag{33}$$

The definition of $\mathbf{U}_N$ given in (22) can be extended to

$$UW_N(\mathbf{x}) = \sum_{k=1}^{N} \sum_{m=1}^{N} Q_x\left(\frac{\mathbf{x} - \mathbf{x}_{e_k}}{h_k}\right) Q_z\left(\frac{\mathbf{x} - \mathbf{x}_{e_m}}{h_m}\right). \tag{34}$$

Once again, the ensemble average is found by averaging over all of the possible heights and locations of the eddies, as it was in (24). This leads to

$$\langle UW_N\rangle = \sum_{k=1}^{N}\sum_{m=1}^{N}\iint_{h_{min}}^{h_{max}} P(h_k)P(h_m) \times \iint_{-L/2}^{L/2} p(x_{e_1})\ldots \iint_{-L/2}^{L/2} p(x_{e_N}) Q_x\left(\frac{\mathbf{x}-\mathbf{x}_{e_k}}{h_k}\right) Q_z\left(\frac{\mathbf{x}-\mathbf{x}_{e_m}}{h_m}\right) \mathrm{d}h_k\,\mathrm{d}h_m\,\mathrm{d}\mathbf{x}_{e_N}\ldots\mathrm{d}\mathbf{x}_{e_1}. \quad (35)$$

Naturally, there will be N cases in which Q_x and Q_z refer to the same eddy (i.e. where $k = m$). In these cases, the multiple integrals over h_k, h_m and the various $\mathbf{x}_e$ simplify to

$$\int_{h_{min}}^{h_{max}} P(h_k) \iint_{-L/2}^{L/2} p(x_{e_k}) Q_x\left(\frac{\mathbf{x}-\mathbf{x}_{e_k}}{h_k}\right) Q_z\left(\frac{\mathbf{x}-\mathbf{x}_{e_k}}{h_k}\right) \mathrm{d}\mathbf{x}_{e_k}\,\mathrm{d}h_k = \int_{h_{min}}^{h_{max}} P(h) \iint_{-L/2}^{L/2} Q_x\left(\frac{\mathbf{x}}{h}\right) Q_z\left(\frac{\mathbf{x}}{h}\right) \frac{\mathrm{d}\mathbf{x}}{L^2}\,\mathrm{d}h. \quad (36)$$

We have again made use of (18), and so the right-hand side above reflects the fact that we are taking the limit as $L \to \infty$, and in this limit, any integral over the entire x-y plane will be equivalent. It has also been made use of (23) on the right-hand side above.

This leaves the $N^2 - N$ cases in which Q_x and Q_z refer to different eddies (i.e. $k \neq m$). In these cases, the multiple integral in (35) becomes

$$\int_{h_{min}}^{h_{max}} P(h) \iint_{-L/2}^{L/2} Q_x\left(\frac{\mathbf{x}}{h_k}\right) \frac{\mathrm{d}\mathbf{x}}{L^2}\,\mathrm{d}h \int_{h_{min}}^{h_{max}} P(h) \iint_{-L/2}^{L/2} Q_z\left(\frac{\mathbf{x}}{h_m}\right) \frac{\mathrm{d}\mathbf{x}}{L^2}\,\mathrm{d}h. \quad (37)$$

By substituting these two into (35), we arrive at

$$\langle UW_N\rangle = \frac{N}{L^2}\int_{h_{min}}^{h_{max}} P(h) \iint_{-L/2}^{L/2} Q_x\left(\frac{\mathbf{x}}{h}\right) Q_z\left(\frac{\mathbf{x}}{h}\right) \mathrm{d}x\,\mathrm{d}y\,\mathrm{d}h + \frac{N^2-N}{L^4}\int_{h_{min}}^{h_{max}} P(h) \iint_{-L/2}^{L/2} Q_x(\mathbf{x})\,\mathrm{d}x\,\mathrm{d}y\,\mathrm{d}h \int_{h_{min}}^{h_{max}} P(h) \iint_{-L/2}^{L/2} Q_z(\mathbf{x})\,\mathrm{d}x\,\mathrm{d}y\,\mathrm{d}h. \quad (38)$$

To determine $\langle UW_N \rangle$, we must now sum over all possible values of N, as we did in (31). This gives

$$\langle UW \rangle = \sum_{N=0}^{\infty} \mathscr{P}(N) \langle UW_N \rangle. \tag{39}$$

By using (29) we can easily show that

$$\sum_{n=0}^{\infty} (N^2 - N) \mathscr{P}(N) = L^4 \beta^2. \tag{40}$$

After we substitute (30), (38), (40) and the above into (39), we find that the sum over N reduces to

$$\begin{aligned} \langle UW \rangle = \beta \int_{h_{min}}^{h_{max}} P(h) \iint_{-L/2}^{L/2} Q_x \left(\frac{\mathbf{x}}{h} \right) Q_z \left(\frac{\mathbf{x}}{h} \right) \mathrm{d}x \, \mathrm{d}y \, \mathrm{d}h \\ + \beta \int_{h_{min}}^{h_{max}} P(h) \iint_{-L/2}^{L/2} Q_x(\mathbf{x}) \, \mathrm{d}x \, \mathrm{d}y \, \mathrm{d}h \, . \, \beta \int_{h_{min}}^{h_{max}} P(h) \iint_{-L/2}^{L/2} Q_z(\mathbf{x}) \, \mathrm{d}x \, \mathrm{d}y \, \mathrm{d}h. \end{aligned} \tag{41}$$

If we take the limit as $L \to \infty$ then, after taking (9) into account, the above becomes

$$\langle UW \rangle = \beta \int_{h_{min}}^{h_{max}} P(h) \iint_{-\infty}^{\infty} Q_x \left(\frac{\mathbf{x}}{h} \right) Q_z \left(\frac{\mathbf{x}}{h} \right) \mathrm{d}x \, \mathrm{d}y \, \mathrm{d}h + \langle U \rangle \langle W \rangle. \tag{42}$$

By substituting the above into (33), we obtain

$$\langle uw \rangle = \beta \int_{h_{min}}^{h_{max}} P(h) \iint_{-\infty}^{\infty} Q_x \left(\frac{\mathbf{x}}{h} \right) Q_z \left(\frac{\mathbf{x}}{h} \right) \mathrm{d}x \, \mathrm{d}y \, \mathrm{d}h. \tag{43}$$

By comparing the above to (8), we can relate the Reynolds shear stress above to the eddy contribution function. This results in

$$\langle uw \rangle = \beta \int_{h_{min}}^{h_{max}} I_{1,0,1}(Z) h^2 P(h) \, \mathrm{d}h. \tag{44}$$

References

1. R.J. Adrian, C.D. Meinhart, C.D. Tomkins, Vortex organization in the outer region of the turbulent boundary layer. J. Fluid Mech. **422**, 1–53 (2000)
2. R.J. Etter, J.M. Cutbirth, S.L. Ceccio, D.R. Dowling, M. Perlin, High Reynolds number experimentation in the U.S. Navy's William B. Morgan large cavitation channel. Meas. Sci. Technol. **16**, 1701–1709 (2005)
3. H.H. Fernholz, P.J. Finley, The incompressible zero-pressure-gradient turbulent boundary layer: an assessment of the data. Prog. Aerosp. Sci. **32**, 245–267 (1996)
4. M. Hultmark, M. Vallikivi, S.C.C. Bailey, A.J. Smits, Turbulent pipe flow at extreme Reynolds numbers. Phys. Rev. Lett. **108**, 094501 (2012)
5. N. Hutchins, K. Chauhan, I. Marusic, J. Monty, J. Klewicki, Towards reconciling the large-scale structure of turbulent boundary layers in the atmosphere and laboratory. Bound.-Layer Meteorol. **145**, 273–306 (2012)
6. G.J. Kunkel, I. Marusic, Study of the near-wall-turbulent region of the high-Reynolds-number boundary layer using an atmospheric flow. J. Fluid Mech. **548**, 375–402 (2006)
7. I. Marusic, On the role of large-scale structures in wall turbulence. Phys. Fluids **13**(3), 735–743 (2001)
8. I. Marusic, J.P. Monty, M. Hultmark, A.J. Smits, On the logarithmic region in wall turbulence. J. Fluid Mech. **716**, R3 (2013)
9. C. Meneveau, I. Marusic, Generalized logarithmic law for high-order moments in turbulent boundary layers. J. Fluid Mech. **719**, R1 (2013)
10. T.B. Nickels, I. Marusic, S. Hafez, N. Hutchins, M.S. Chong, Some predictions of the attached eddy model for a high Reynolds number boundary layer. Philos. Trans. R. Soc. A **365**, 807–822 (2007)
11. A.E. Perry, M.S. Chong, On the mechanism of wall turbulence. J. Fluid Mech. **119**, 173–217 (1982)
12. A.E. Perry, S. Henbest, M.S. Chong, A theoretical and experimental study of wall turbulence. J. Fluid Mech. **165**, 163–199 (1986)
13. S.O. Rice, Mathematical analysis of random noise. Bell Syst. Tech. J. **23**, 282–332 (1944)
14. J.A. Sillero, J. Jiménez, R.D. Moser, One-point statistics for turbulent wall-bounded flows at Reynolds numbers up to $\delta^+ \approx 2000$. Phys. Fluids **25**, 105102 (2013)
15. A.A. Townsend, Equilibrium layers and wall turbulence. J. Fluid Mech. **11**, 97–120 (1961)
16. A.A. Townsend, *The Structure of Turbulent Shear Flow* (Cambridge University Press, Cambridge, 1976)
17. P. Vincenti, J.C. Klewicki, C. Morrill-Winter, C.M. White, M. Wosnik, Streamwise velocity statistics in turbulent boundary layers that spatially develop to high Reynolds number. Exp. Fluids **54**(1629), 1–13 (2013)

Part II
Papers

DNS of Turbulent Boundary Layers in the Quasi-Laminarization Process

Guillermo Araya, Luciano Castillo and Fazle Hussain

Abstract In this investigation, Direct Numerical Simulations (DNS) of turbulent spatially developing boundary layers (SDBL) with prescribed Very Strong Favorable Pressure Gradients (VSFPG) are performed by means of the Dynamic Multi-scale Approach (DMA) developed by Araya et al. JFM, 670:518–605, 2011 [1]. Although the prescription of an external VSFPG significantly reduces turbulence production, the flow never becomes completely laminar due to the finite value of the streamwise Reynolds normal stress, and thus the flow is quasi-laminar. In this sense, the mean flow carries the footprint of turbulence, particularly in the streamwise direction of the Reynolds stresses. In addition, the vertical transports toward the wall of $\overline{v'^{2+}}$ and $\overline{uv'^{+}}$ practically disappear in the inner region and significantly decrease in the outer region of the boundary layer during the quasi-laminarization stage. As a consequence, the "communication" between inner and outer regions is seriously restricted.

1 Introduction

Turbulent boundary layers subjected to severe acceleration or strong Favorable Pressure Gradients (FPG) are of great industrial interest, particularly, due to the drag reduction characteristics. If the external FPG is strong enough, the flow might experience a quasi-laminarization or a reversion process characterized by a meaningful depression of turbulent production and reduction of skin friction; however, there is a residual turbulence given by the flow history. Narasimha and Sreenivasan [2] concluded that the reversion process in highly accelerated flows was attributed to a dominance of pressure forces over almost unchanging (or frozen) Reynolds stresses in the outer region with the generation of a new laminar boundary in the inner region

G. Araya (✉) · L. Castillo · F. Hussain
Texas Tech University, Lubbock, TX 79409, USA
e-mail: araya@mailaps.org

L. Castillo
e-mail: luciano.castillo@ttu.edu

F. Hussain
e-mail: fazle.hussain@ttu.edu

M. Stanislas et al. (eds.), *Progress in Wall Turbulence 2*,
ERCOFTAC Series 23, DOI 10.1007/978-3-319-20388-1_5

stabilized by the favorable pressure gradient. Also, Narasimha and Sreenivasan [2] stated that the quasi-laminarization process was not associated to absorption or dissipation despite the fact that it could contribute, particularly in the near-wall region. Several studies in the past have recommended different methods to identify the reversion or laminarization process in highly accelerated low-speed boundary layers as well as various criteria, as summarized in [3, 4]. Traditionally, FPG flows have been characterized by the acceleration parameter $K = \frac{\nu}{U_\infty^2}\frac{dU_\infty}{dx}$, where ν is the kinematic viscosity and U_∞ is the freestream velocity [5]. For $K > 3.0 \times 10^{-6}$, it is expected that laminarization of the turbulent flow will take place. Spalart [6] performed DNS of sink flows and found that laminarization occurred at $K = 3.0 \times 10^{-6}$ by suddenly increasing the viscosity from simulations at $K = 2.5 \times 10^{-6}$ and $K = 2.75 \times 10^{-6}$. Kline et al. [7] proposed that relaminarization in highly accelerated flows could be associated with the ending of turbulent bursting. They measured the occurrence of bursts in a boundary layer subjected to strong favorable pressure gradient and inferred by extrapolation a critical acceleration parameter, K_c, of 3.5×10^{-6} at which the bursting process almost disappeared. Moretti and Kays [8] proposed a critical value for K ($K_c = 3.5 \times 10^{-6}$) based on the significant decrease of the heat transfer rate in the near-wall region during laminarization. On the other hand, Patel [9] suggested a critical value of -0.025 for Δp ($= \frac{\nu\, dP/dx}{u_\tau^3}$, where P is the mean pressure and u_τ is the friction velocity). Additionally, the Δp parameter can be expressed as $K(2/C_f)^{3/2}$. This clearly indicates that Δp carries information not only from the outer region, but also from the near-wall region, which is necessary to evaluate if the onset of laminarization is associated with the breakdown of the log law for the mean streamwise velocity [9]. Therefore, the parameter Δp might be considered a more complete indicator of the laminarization outbreak. More recently, Cal and Castillo [10] showed that a quasi-laminar state could be characterized by the pressure parameter Λ_δ ($= -\frac{\delta}{U_\infty d\delta/dx}\frac{dU_\infty}{dx}$). As a consequence, flows with $\Lambda_\delta > 0$ are in the quasi-laminar state, while for $\Lambda_\delta < 0$ a turbulent FPG quadrant can be defined. Consequently, understanding the mechanisms responsible for quasi-laminarization due to an external strong FPG is a challenging problem and not yet quite understood at the fundamental level. Most of the previous studies on quasi-laminar flows have focused on the analysis of the mean flow. The fact that in VSFPG the boundary layer thickness becomes thinner even at low Reynolds numbers, makes experimental measurements very challenging even to measure a single component of the velocity field very close to the wall. On the contrary, this does not represent a severe limitation for DNS. However, the computational cost (CPU time) is greatly penalized due to the necessary mesh resolution, which coupled with the need for long computational domains in the simulations of turbulent SDBL, make these simulations extremely difficult. In this article, the purpose is to shed some light on VSFPG turbulent flows with quasi-laminarization. Specifically, we seek to evaluate the vertical transport of the Reynolds stresses during the quasi-laminarization process. The uniqueness of the present study rely on two main aspects: (i) the acceleration parameter, $K = \frac{\nu dU_\infty/dx}{U_\infty^2}$, has been kept constant for a long distance ($\approx 40\delta_{inlet}$), while most of previous works have considered varying values of K, (ii) extensive DNS have been performed in a long computational domain with independent and realistic turbulent inflow conditions.

Finally, in this investigation we are performing extensive DNS of Very Strong Favorable Pressure Gradient (FPG) flows. The strength of the FPG is strong enough (acceleration) to provoke flow reversion or quasi-laminarization. Focus is given to the behavior and transport of the wall-normal component of the Reynolds normal stresses and the Reynolds shear stresses (the two more affected components during the quasi-laminarization process) under a constant value of the acceleration parameter K.

2 Numerical Details

The prescription of accurate turbulent inflow boundary conditions in SDBL is not a trivial task. For that reason, the DMA by Araya et al. [1] is employed in this investigation. The proposed approach is based on the rescaling-recycling method by Lund et al. [11]. The main philosophy of the rescaling-recycling method is to prescribe time-dependent turbulent information at the "inlet" plane based on the transformed flow solution downstream by using scaling laws from a downstream plane called "recycle". The principal differences of the DMA with respect to the original rescaling-recycling method are twofold: (i) the consideration of different transforming or scaling functions in the inner and outer regions of the boundary layer (i.e., Multi-scale) which may absorb more effectively the effects of external conditions such as the Reynolds number dependence; (ii) there is no need to use an empirical correlation in order to compute the inlet friction velocity, such information is deduced dynamically by involving an additional plane, the so-called "test plane" located between the inlet and recycle stations. More specifically, the scaling functions utilized in this investigation are based on the similarity analysis performed by George and Castillo [12]. Furthermore, the principal purpose of the upstream ZPG zone is to prescribe an equilibrium region to generate and feed turbulent information to the VSFPG zone. Additionally, the method has been tested and validated in a suite of DNS for zero (ZPG) and pressure gradient flows including in thermal boundary layers (Araya and Castillo [13, 14]) as well as in rough surfaces, Cardillo et al. [15].

Figure 1 shows the computational domain consisting of a region for ZPG with a length of $20\delta_{inlet}$ and the VSFPG region with a length of $40\delta_{inlet}$, where δ_{inlet} is the 99 % boundary layer thickness at the domain inlet. The dimensions of the composite computational box (L_x, L_y and L_z) are $60\delta_{inl}$, $3\delta_{inl}$ and $3\delta_{inl}$ along the streamwise, wall-normal, and spanwise directions, respectively. Notice that L_y is not constant in the FPG region but linearly decreases as in sink flows. The mesh configuration is: $600 \times 80 \times 80$, which represents the numbers of points along the streamwise, wall-normal, and spanwise directions, respectively. The mesh resolution is: $\Delta x^+ = 15$, $\Delta y^+_{min} = 0.2$, $\Delta y^+_{max} = 13$ and $\Delta z^+ = 8$. The Courant, Friedrichs, Levy (CFL) parameter remains constant during the simulation (~ 0.24) and the time step is $\Delta t^+ \approx 0.19$.

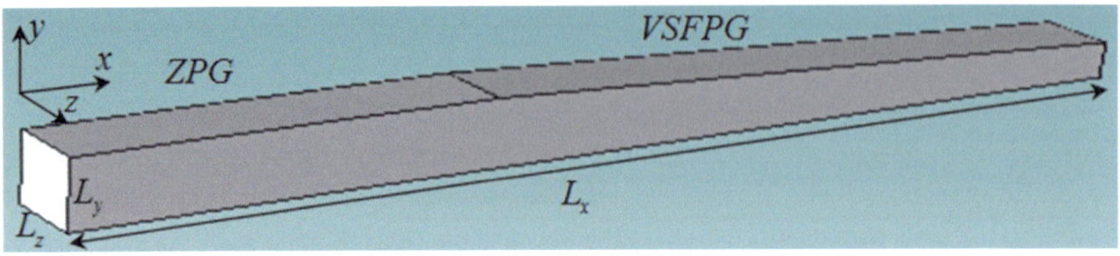

Fig. 1 Computational domain including the ZPG region and Very Strong FPG with eventual quasi-laminarization

Direct simulations of the Navier-Stokes equations for incompressible flows are performed using the PHASTA code (Parallel Hierarchic Adaptive Stabilized Transient Analysis). PHASTA is an open-source code and based on the Finite Element method with a Streamline Upwind Petrov-Galerkin (SUPG) stabilization [16]. The classical no-slip condition is prescribed for the three components of the velocity at the bottom wall. The top surface is prescribed as shear-less, the composite mesh possesses a constant cross-sectional area in the ZPG zone (where the freestream velocity U_∞ remains constant) and a zero-value for the normal component of velocity is imposed (streamline) in the VSFPG region. Moreover, periodic boundary conditions were prescribed in the spanwise direction. At the outlet plane, the pressure is assigned a value (Dirichlet condition).

3 Discussion of Results

Figure 2a shows the streamwise development of the momentum thickness Reynolds number Re_θ $(= U_\infty \theta/\nu$, where θ is the momentum thickness), the acceleration parameter K, and the shape factor H $(= \delta^*/\theta$, where δ^* is the displacement thickness). The streamwise coordinate is normalized in terms of δ_{inlet}, i.e., x/δ_{inlet}. The Reynolds number exhibits almost a linear behavior not only in the ZPG zone (increasing) but also in the VSFPG region (decreasing). In addition, the acceleration parameter, K, shows an abrupt rise in the ZPF-VSFPG intersection to finally tend asymptotically toward the value of 4.0×10^{-6} from around $x/\delta_{inlet} \approx 30$. At around $x/\delta_{inlet} \approx 32$, the shape factor H reaches a minimum, which can be attributed to the onset of quasi-laminarization. Moreover, the freestream velocity versus x/δ_{inlet} is depicted in Fig. 2b and a good agreement with the analytical solution (by assuming a value of $K = 4.0 \times 10^{-6}$) can be observed in the VSFPG region. This indicates that the design and boundary condition of the top surface are quite appropriate.

In Fig. 3 the skin-friction coefficient C_f $(= 2(u_\tau/U_\infty)^2)$ is shown. A short developing section of about $3\delta_{inlet}$ (defined as that inlet region where the turbulent structures behave in a nonphysical sense) can be observed for the C_f parameter. Downstream, the skin-friction coefficient exhibits the typical monotonic decay of ZPG flows and starts to sharply increases at the ZPG-VSFPG intersection as the

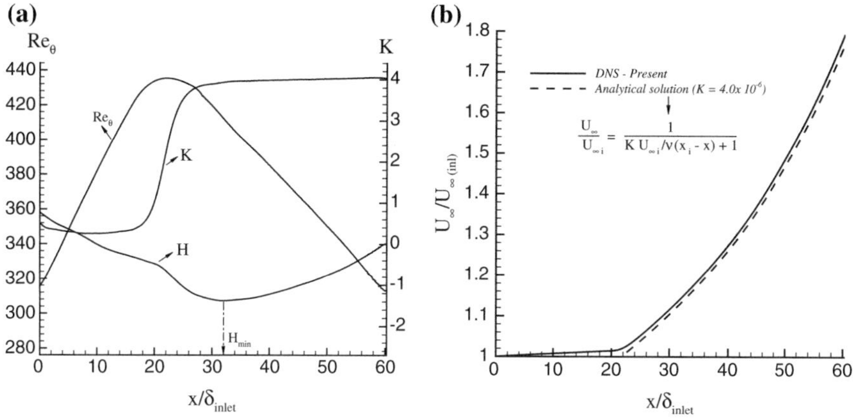

Fig. 2 Streamwise variation of **a** Re_θ, K and H; and, **b** the freestream velocity (numerical vs. theoretical)

Fig. 3 Streamwise variation of the skin-friction coefficient, C_f

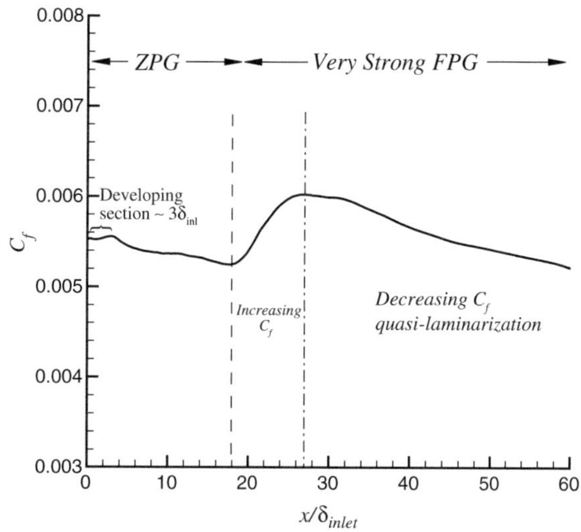

freestream accelerates. Nevertheless, the C_f profile reaches a maximum and quickly drops, in agreement with the quasi-laminar theory in Narasimha and Sreenivasan [2].

Streamwise developments of the local maxima of the turbulence intensities (u'^{+}_{rms}, v'^{+}_{rms} and w'^{+}_{rms}) as well as of Reynolds shear stresses ($\overline{uv'}^{+}$) in local wall units are shown in Fig. 4. It can be observed the sudden reduction of velocity fluctuations in the ZPG-FPG intersection around $x/\delta_{inlet} \approx 20$. Downstream, local maxima of v'^{+}_{rms}, w'^{+}_{rms} and $|\overline{uv'}^{+}|$ keep diminishing steadily, with similar slopes and with decreases between 50–60 % with respect to the ZPG values. It is interesting to highlight the evident change of slope (from steep to mild) in the $|\overline{uv'}^{+}|_{max}$ profile at $x/\delta_{inlet} \approx 32$.

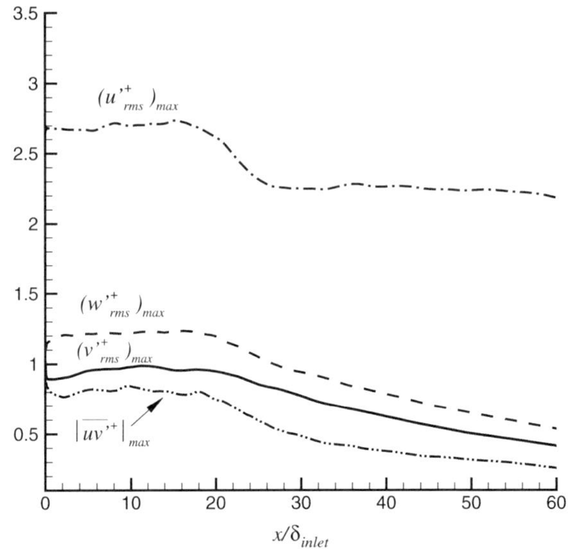

Fig. 4 Streamwise variation of local maxima of turbulence intensities and Reynolds shear stresses

This is consistent with the location of H_{min} (in Fig. 2a) that may be attributed to the onset of the quasi-laminarization process. On the other hand, local values of $(u'^{+}_{rms})_{max}$ exhibit reductions in the order of 15 %. Notice the almost constant values of $(u'^{+}_{rms})_{max}$ in the VSFPG region. Thus, the fact that $(u'^{+}_{rms})_{max}$ is finite and nonzero is the main reason why the FPG flow never settles to a laminar flow and the rationale for which we refer to quasi-laminar flow.

Triple correlations $\overline{v'v'^{2+}}$ and $\overline{v'uv'^{+}}$ are shown in Fig. 5 for the ZPG region (at $x/\delta_{inlet} = 15$) and for the VSFPG region (at $x/\delta_{inlet} = 58$). The vertical transport toward the wall of the Reynolds normal stresses $\overline{v'^{2+}}$ significantly decreases around $y^{+} \approx 10$ in the VSFPG zone. Similarly, the vertical transport away from the wall of the Reynolds normal stresses $\overline{v'^{2+}}$ exhibits a prominent reduction (up to 9 times) in the outer region ($y^{+} \approx 100$) during the quasi-laminarization stage. Analogous conclusions can be drawn with respect to the vertical transport of Reynolds shear stresses in Fig. 5b. In fact, this supports the proposed model by Sreenivasan [4] during quasi-laminarization, in which the inner and outer layers develop independently due to the restricted communication between layers.

The production (P_{12}) and velocity-pressure gradient (Π_{12}) terms of the Reynolds shear stress budget are depicted in Fig. 6 normalized by local values of u_{τ}^{4}/ν and in outer units y/δ, also for the ZPG and VSFPG regions. Both terms experience meaningful reductions in the entire boundary layer during the quasi-laminarization process; particularly, in peak values. Additionally, peak values of P_{12} and Π_{12} approximately remain at the same y/δ locations ($0.07 < y/\delta < 0.09$) even during quasi-laminarization. The observed residual values of the velocity-pressure gradient term at $x/\delta_{inlet} = 58$ can be attributed mainly to the contribution from $u'\partial p'/\partial y$ as

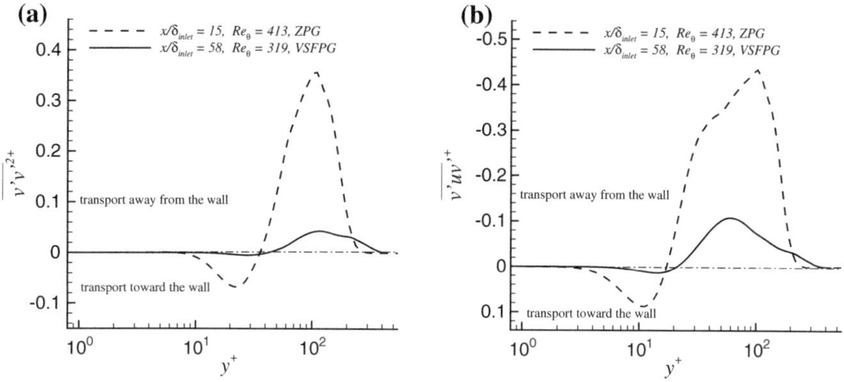

Fig. 5 Vertical transport of **a** $\overline{v'^{2+}}$; and, **b** $\overline{uv'^{+}}$

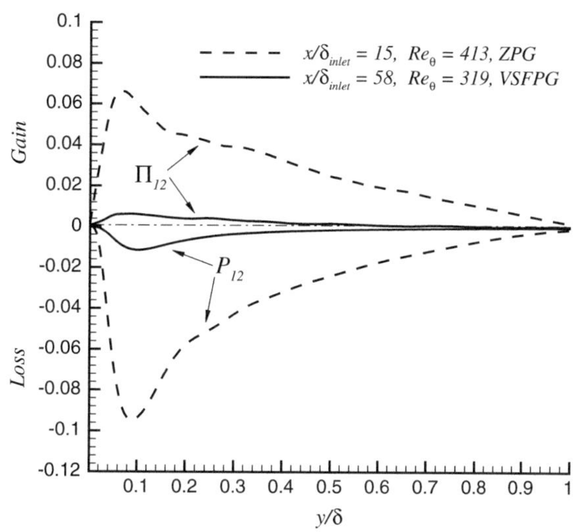

Fig. 6 Principal terms of the $\overline{uv'^{+}}$ budget: production rate and velocity-pressure gradient

a consequence of "frozen" streamwise Reynolds normal stresses, $\overline{u'^{2+}}$, as the flow laminarizes.

Figure 7 depicts iso-surfaces of instantaneous wall-normal velocity fluctuations v'/U_∞ in the composite domain. It is observed a prominent reduction of v' as the flow moves downstream in the Very Strong FPG region. As a consequence, the "messenger" (wall-normal component of velocity fluctuations) between the inner and outer regions is practically annihilated. This might be consistent with the proposed model by Narasimha and Sreenivasan [3]: inner and outer layers are quasi-independent in the quasi-laminarization process. However, further studies should be carried out to verify the proposed two-layer model by Narasimha and Sreenivasan [3], which is beyond the scope of the present article. Furthermore, the presence of highly stretched streaks

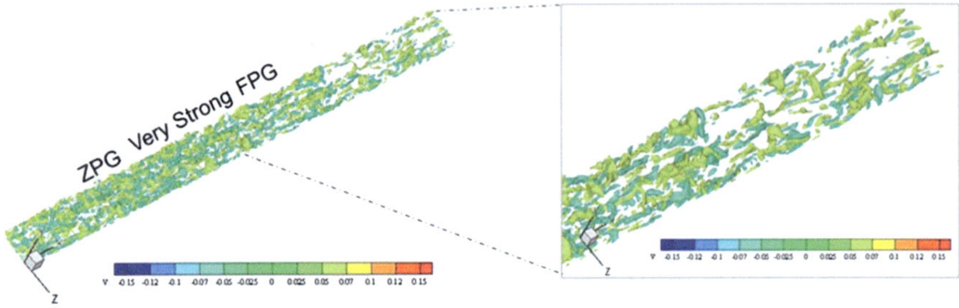

Fig. 7 Iso-surfaces of instantaneous wall-normal velocity fluctuations v'/U_∞

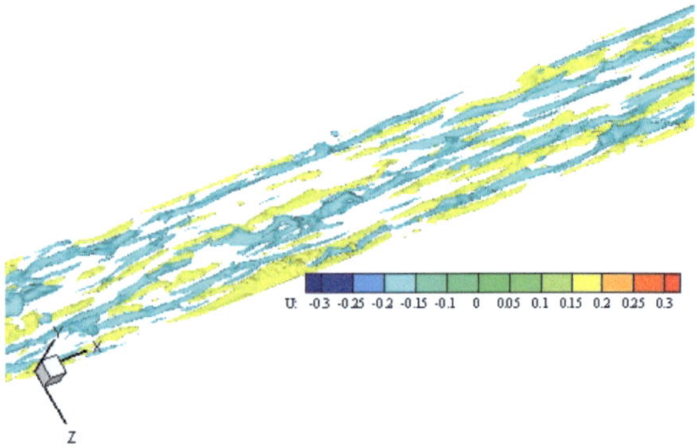

Fig. 8 Iso-surfaces of instantaneous streamwise velocity fluctuations u'/U_∞

are observed in the Very Strong FPG zone of Fig. 8. Also, notice that not only the low speed streaks experience the stretching process but also the high speed streaks.

4 Conclusions

The Dynamic Multi-scale Approach (DMA) by Araya et al. [1] is employed as turbulent inflow generator for Very Strong FPG flows in the quasi-laminarization process. It has been observed that the flow never becomes completely laminar due to the finite value of the streamwise Reynolds normal stresses. Furthermore, the vertical transports toward the wall of $\overline{v'^{2+}}$ and $\overline{uv'^{+}}$ practically disappear in the inner region and significantly decrease in the outer region of the boundary layer during the quasi-laminarization stage. Therefore, the "communication" between inner and outer regions is seriously suppressed.

Acknowledgments GA acknowledges XSEDE (Project Number: TG-CTS120046) and SCOREC (Scientific Computation Research Center, Rensselaer Polytechnic Institute) for supplying computational resources.

References

1. G. Araya, L. Castillo, C. Meneveau, K. Jansen, A dynamic multi-scale approach for turbulent inflow boundary conditions in spatially evolving flows. JFM **670**, 518–605 (2011)
2. R. Narasimha, K.R. Sreenivasan, Relaminarization in highly accelerated turbulent boundary layers. JFM **61**, 417–447 (1973)
3. R. Narasimha, K.R. Sreenivasan, Relaminarization of fluid flows, *Advances in Applied Mechanics*, vol. 19 (Academic Press Professional Inc., New York, 1979), pp. 221–309
4. K.R. Sreenivasan, Laminarescent, relaminarizing and retransitional flows. Acta Mech. **44**, 1–48 (1982)
5. B.E. Launder, Laminarization of the turbulent boundary layer in a severe acceleration. J. Appl. Mech. **31**(4), 707–708 (1964)
6. P. Spalart, Numerical study of sink-flow boundary layers. J. Fluids Mech. **172**, 307–328 (1986)
7. S.J. Kline, W.C. Reynolds, F.A. Schraub, P.W. Runstadler, The structure of turbulent boundary layers. JFM **30**, 741–773 (1967)
8. K.H. Moretti, W.M. Kays, Heat transfer to a turbulent boundary layer with varying free-stream velocity and varying surface temperature an experimental study. Int. J. Heat Mass Transf. **8**, 1181–1260 (1965)
9. V.C. Patel, Calibration of the Preston tube and limitation on its use in pressure gradients. J. Fluids Mech. **23**, 185–208 (1965)
10. R.B. Cal, L. Castillo, Similarity analysis of favorable pressure gradient turbulent boundary layers with eventual quasilaminarization. Phys. Fluids **20**, 105106 (2008)
11. T.S. Lund, X. Wu, K.D. Squires, Generation of turbulent inflow data for spatially-developing boundary layer simulations. J. Comput. Phys. **140**, 233–258 (1998)
12. W.K. George, L. Castillo, Zero-pressure-gradient turbulent boundary layer. Appl. Mech. Rev. **50**, 689–729 (1997)
13. G. Araya, L. Castillo, DNS of turbulent thermal boundary layers up to $Re_\theta = 2300$. Int. J. Heat Mass Transf. **55**(15), 4003–4019 (2012)
14. G. Araya, L. Castillo, DNS of turbulent thermal boundary layers subjected to adverse pressure gradients Phys. Fluids **25**, 095107 (2013)
15. J. Cardillo, Y. Chen, G. Araya, J. Newman, K. Jansen, L. Castillo, DNS of turbulent boundary layers with surface roughness. J. Fluids Mech. **729**, 603–637 (2013)
16. C.H. Whiting, K.E. Jansen, A stabilized finite element method for the incompressible Navier-Stokes equations using a hierarchical basis. Int. J. Numer. Methods Fluids **35**, 93–116 (2001)

Numerical ABL Wind Tunnel Simulations with Direct Modeling of Roughness Elements Through Immersed Boundary Condition Method

Bruno Lopez, Gabriel Usera, Gabriel Narancio, Mariana Mendina, Maritn Draper and Jose Cataldo

Abstract Reproduction of atmospheric boundary layer wind tunnel experiments by numerical simulation is achieved in this work by directly modeling, with immersed boundary method, the geometrical elements placed in the wind tunnel's floor to induce the desired characteristics to the boundary layer. The numerical model is implemented on the basis of the open-source flow solver caffa3d.MBRi, which uses a finite volume method over block structured grids, coupled with various LES approaches for turbulence modeling and parallelization through domain decomposition with MPI. The Immersed boundary method approach followed the work of Liao et al. (Simulating flows with moving rigid boundary using immersed-boundary method. Comput. Fluids 39, 152–167, 2010). Numerical simulation results are compared to wind tunnel measurements for the mean velocity profiles, rms profiles, and spectrums, providing good overall agreement. We conclude that the Immersed Boundary Condition method is a promising approach to numerically reproduce ABL Boundary Layer development methods used in physical modeling.

B. Lopez (✉) · G. Usera · G. Narancio · M. Mendina · M. Draper · J. Cataldo
IMFIA, J. Herrera Y Reissig 565, Montevideo, Uruguay
e-mail: brunolop@fing.edu.uy

G. Usera
e-mail: gusera@fing.edu.uy

G. Narancio
e-mail: gnaranci@fing.edu.uy

M. Mendina
e-mail: mmendina@fing.edu.uy

M. Draper
e-mail: mdraper@fing.edu.uy

J. Cataldo
e-mail: jcataldo@fing.edu.uy

M. Stanislas et al. (eds.), *Progress in Wall Turbulence 2*,
ERCOFTAC Series 23, DOI 10.1007/978-3-319-20388-1_6

1 Introduction

The global trend toward urbanization explains the growing interest in recent decades in studying the Atmospheric Boundary Layer (ABL), comprising the first layer of the atmosphere that extends about 1–2 km from ground and hosts a large part of human activities. The effect of winds on buildings and other structures, as well as on pedestrian in urban environments, the transport of pollutants in air, or wind power generation are a few examples of human activities that develop within the ABL and require its study [2, 3, 9]. The characteristics of the ABL flow are shaped up mainly by the interaction with the ground and so will vary depending on the characteristics of the terrain and its roughness which might correspond for example to different urban environments, sea, or vegetated fields.

Physically modeling ABL processes in a Wind Tunnel its traditional well established technique. Special Wind Tunnels are built for this purpose, named Atmospheric Boundary Layer Wind Tunnels, in which the test sections is preceded by a relatively long working zone in which the modeled ABL flow is developed over selected roughness elements.

While this technique is traditionally used for the study of the ABL, numerical modeling has emerged as a complementary contribution to it in the last decades.The development of modern parallel computers and computational fluid dynamics numerical methods presently allows to numerically simulate turbulent flows with different approaches, ranging from Reynolds Averaged Navier–Stokes methods (RANS), through Large Eddy Simulation (LES), up to Direct Numerical Simulation (DNS) [12].

The Numerical Wind Tunnel methodology targets the numerical simulation of flows which are physically modeled in Wind Tunnels. In the case of the ABL a primary concern in this methodology is to appropriately reproduce the boundary layer characteristics, which in turn are mainly driven by the roughness characteristics of the terrain. Different approaches have been proposed for this matter, including roughness wall functions and drag-based representation of vegetation [1].

This paper aims at a preliminar analysis of viability of using immersed boundary method to explicitly represent roughness elements in numerical simulations of ABL Wind Tunnel experiments. Section 2 presents the reference Wind Tunnel experiments used for comparision with the numerical simulations developed in this work. Section 3 presents the numerical method, describing the base open-source solver used, caffa3d.MBRi, as well as the immersed boundary approach followed to represent roughness elements. Section 4 presents the results and conclusions are derived in Sect. 5.

2 Wind Tunnel Experiments

The experimental data used in this work were obtained in the Atmospheric Boundary Layer Wind Tunnel of the Facultad de Ingenieria in Uruguay, which is an open-circuit wind tunnel with a 2.20 × 2.25 m square test section, and a 14 m long inlet region for

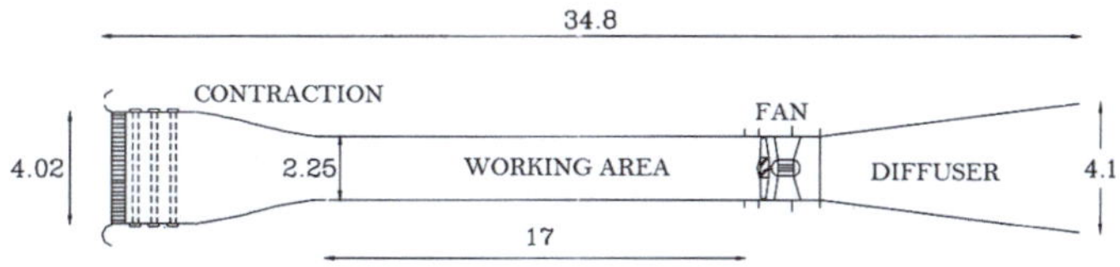

Fig. 1 Layout of the atmospheric boundary layer wind tunnel at IMFIA. Lengths in (m)

Fig. 2 General (*left*) and detailed (*right*) views of the wind tunnel test section and boundary layer preparation section, with roughness element cubes and vortex induction Standen spires type elements

the development of the ABL, capable of achieving maximum wind speeds of 30 m/s. In Fig. 1 the general layout of the Wind Tunnel is presented.

Boundary layer development is shaped up with a series of cubical elements, 3 cm in side, placed in a regular staggered arrangement with a 15 cm spacement, 3 Standen spires type vortex generators [14], of 134 cm height, and a 31.5 cm height wall placed at the inlet. This arrangement is used to reproduce an urban boundary layer with a length scale 1:200. Figure 2 gives general and detailed views of this arrangement.

Velocity was measured with a TSI IFA 100 hot wire anemometer. SN hot film probes were used for this purpose. The sampling rate was 4000 Hz and a low-pass filtering of the signal at 2000 Hz was used to avoid aliasing. 65,536 samples were taken in each location accounting to a sampling time of 16,384 s. The positioning of the hot film probe was done using a robotic arm designed to perform this task. Vertical profiles of mean velocity, root mean square fluctuations are obtained as well as power spectrums at selected locations. Experimental results are reported and compared with numerical simulation results in Sect. 4.

3 Numerical Method

The methodology developed to numerically simulate atmospheric boundary layer wind tunnel tests is based on the open-source solver caffa3d.MBRi [10], coupled with a specific immersed boundary conditions module, following the work of [7], to explicitly represent the geometry of roughness elements used in the Wind Tunnel.

The open source incompressible flow solver caffa3d.MBRi is a Fortran90 implementation of the finite volume method, evolved from the work of Ferziger and Peric [4]. It features a block structured framework to accommodate both a flexible approach to geometry representation and a straightforward implementation of parallel capabilities through the MPI library. Representation of complex geometries can be handled semi automatically through a combination of body fitted blocks of grids and the immersed boundary condition strategy over both Cartesian and body fitted grid blocks. The parallelization strategy is based on the same block structured framework, by means of encapsulated MPI calls performing a set of conceptually defined high level communication tasks.

3.1 Mathematical Model

The mathematical model comprises the mass balance equation (1) and momentum balance equation (2) for a viscous incompressible fluid, together with generic non-reacting scalar transport equation (3) for scalar field ϕ with diffusion coefficient Γ. Note that (2) has been written only for the first Cartesian direction here.

$$\int_S \left(\boldsymbol{v} \cdot \hat{\boldsymbol{n}}_S\right) dS = 0 \tag{1}$$

$$\int_\Omega \rho \frac{\partial u}{\partial t} d\Omega + \int_S \rho u \left(\boldsymbol{v} \cdot \hat{\boldsymbol{n}}_S\right) dS =$$

$$\int_\Omega \rho\beta \left(T - T_{ref}\right) \boldsymbol{g} \cdot \hat{\boldsymbol{e}}_1 d\Omega + \int_S -p\hat{\boldsymbol{n}}_S \cdot \hat{\boldsymbol{e}}_1 dS + \tag{2}$$

$$\int_S \left(2\mu \boldsymbol{D} \cdot \hat{\boldsymbol{n}}_S\right) \cdot \hat{\boldsymbol{e}}_1 dS$$

$$\int_\Omega \rho \frac{\partial \phi}{\partial t} d\Omega + \int_S \rho\phi \left(\boldsymbol{v} \cdot \hat{\boldsymbol{n}}_S\right) dS =$$

$$\int_S \Gamma \left(\nabla\phi \cdot \hat{\boldsymbol{n}}_S\right) dS \tag{3}$$

In these equations, $\boldsymbol{v} = (u, v, w)$ is the fluid velocity, ρ is the density, β is the thermal expansion factor, T the fluid temperature and T_{ref} a reference temperature, g is the gravity, p the pressure, μ the dynamic viscosity of the fluid and $\boldsymbol{D}$ the strain tensor. The balance equations are written for a region Ω, limited by a closed surface S, with unit outward pointing normal $\hat{\boldsymbol{n}}_S$. Finally, $\hat{\boldsymbol{e}}_1$ is the first Cartesian direction.

The generic transport equation (3) for non-reacting scalars can be used to implement in a straightforward manner further physical models like heat transport required for the temperature field, both Reynolds Averaged and Large Eddy Simulation turbulence closures, wet air processes which include evaporation and condensation, etc. An arbitrary number of scalar fields can be solved simultaneously, with coupling between them as for the case of temperature field influencing both momentum equations through buoyancy and wet air process equations through condensation and evaporation conditions. The use of equations in their global balance form together with the finite volume method, as opposed to the differential form, favors enforcing conservation laws for fundamental quantities such as mass and momentum into the solving procedure [4]. For the present simulations a standard Smagorinsky large eddy turbulence model was attached to the solver.

3.2 Discretization and Solving Procedure

Complete details for discretization of each term will not be given here but can be found in [16], together with various validations of the solver [10, 15, 16]. Second order central differencing scheme for diffusive terms is used, while convective terms are discretized blending first order upwind approximations and second order central differences.

Further, the SIMPLE [13] method for pressure-velocity coupling is used to obtain a discretized equation for the pressure, from the mass balance equation (1). Refined methods for pressure-velocity coupling can also be incorporated [6], together with improved linear interpolations [5]. Also different implicit time stepping schemes can be combined for the momentum equations, like first order backwards Euler or second order Cranck-Nicholson.

The usual lexicographical order in 3D implies that the resulting segregated linear equation system is hepta-diagonal in each grid block, and thus globally block hepta-diagonal. Either a block structured variant of the Stone-SIP solver [8] or a block structured Algebraic Multigrid (AMG) solver with SIP as a smoother [11, 16] are used for iterative solution of each linear system. The SIP solver algorithm accommodates well the block structure inherited from the grid, allowing efficient parallelization.

3.3 Immersed Boundary Conditions Method

Following the work of [7], a specific immersed boundary conditions module was included in the solver. In this approach the geometry of roughness elements used in the Wind Tunnel is explicitly represented over the structured grid, by means of a triangulated surface as shown in Fig. 3, and the distance from each grid node

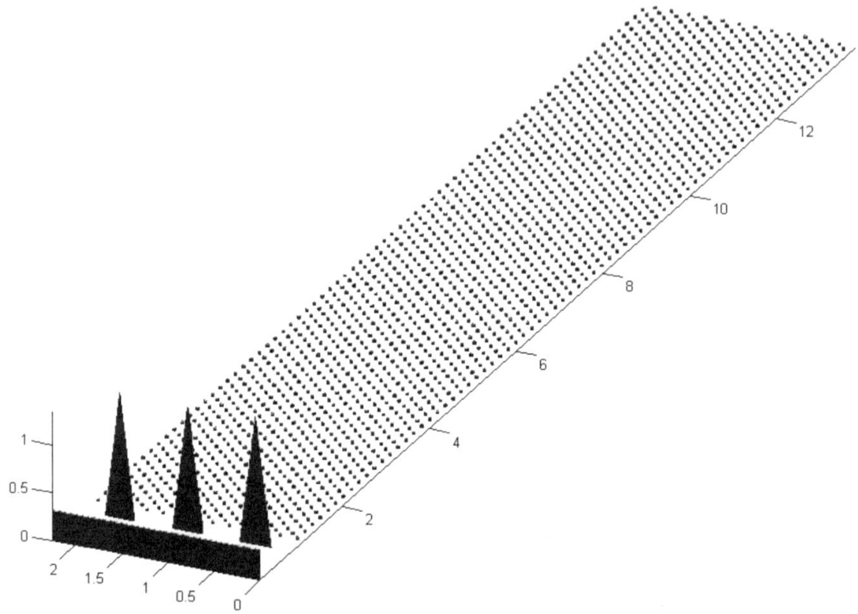

Fig. 3 Immersed boundary condition representation of roughness element cubes and vortex induction Standen spires type elements

to the closest roughness element wall is computed. For grid nodes that fall inside roughness elements an additional body force term is computed to enforce null velocity at that node. For grid nodes falling outside roughness elements, but close enough, an interpolation procedure is applied to estimate the target velocity at the node and the additional body force is applied based on that estimate. This computation is embedded into the overall implicit outer iteration procedure, so that the additional body force value at each node is adjusted within each time step until convergence is reached.

This procedure leads to an almost automatic meshing strategy for a geometry in which developing a body fitted block structured grid would be seldom feasible. Unstructured grids would be better suited for a body fitted approach, but still would require considerable meshing effort, especially considering the intricate global geometry of the roughness elements set.

While the grid resolution does not warrants that the boundary layer over each individual roughness element will be adequately captured, it is expected that the overall contribution to the development of the ABL like flow, comprised of the superposition of each elements wakes, will be indeed well represented.

3.4 Numerical Simulations Setup

The computational domain corresponds to the working area of the wind tunnel, 13.72 m long up to the test section, within which the boundary layer is developed. Buffer regions for inlet and outlet boundary conditions add up to the total length of 16.25 m for the computational domain, with a cross section 2.25 m wide and 2 m high. This domain was split into 26 identical regions or grid blocks, each 1.25 m long, 1.125 m wide and 2 m high. Two grids were setup with different spatial resolution. For the coarse grid, a uniform horizontal grid spacing of 1.56 cm was used. A vertical nonuniform distribution of cells was selected, with minimum vertical spacing of 1 mm at the floor. Each grid block then required 80 cells in the streamwise direction, 72 cells spanwise and 104 cells in the vertical, for a total of almost 600.000 cells per grid block. For the fine grid, the horizontal resolution was set to 1.04 cm while the vertical resolution was kept at 1 mm at the floor. The total number of cells per grid block for the fine grid was about 2 million cells, 120 cells streamwise, 108 cells spanwise and 156 in the vertical direction.

Wall boundary conditions with non slip condition were applied at the floor, while slip conditions were applied at the roof and side walls of the tunnel. The inlet was set to a uniform velocity of $Uo = 13.5$ m/s and null gradient boundary condition normal to the outlet was applied.

For the coarse grid a time step of 0.5 s was used, while for the fine grid computations were performed for time steps of 0.5 and 0.01 s. Computations were distributed in 26 cores on the Cluster-FING infrastructure [www.fing.edu.uy/cluster].

4 Results

Mean velocity profiles from Wind Tunnel experimental data and from the three numerical simulation runs are presented in Fig. 4, showing good agreement between numerical and experimental data, as well as almost grid and time-step independance. In Fig. 5, urms profiles are given for the same set of experimental and numerical data, showing a good match as well. Finally, power spectrums for longitudinal velocity are given at 100 mm from floor for experimental data in Fig. 6 and for numerical results from the fine grid in Fig. 7.

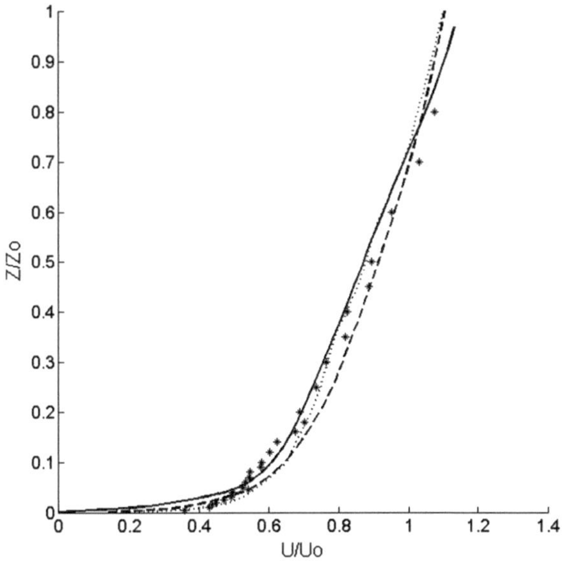

Fig. 4 Mean velocity profiles form experiments and numerical simulations. *symbols* experimental data; *solid line* coarse grid dt = 0.5 s; *dashed line* fine grid dt = 0.5 s; *dotted line* fine grid dt = 0.01 s

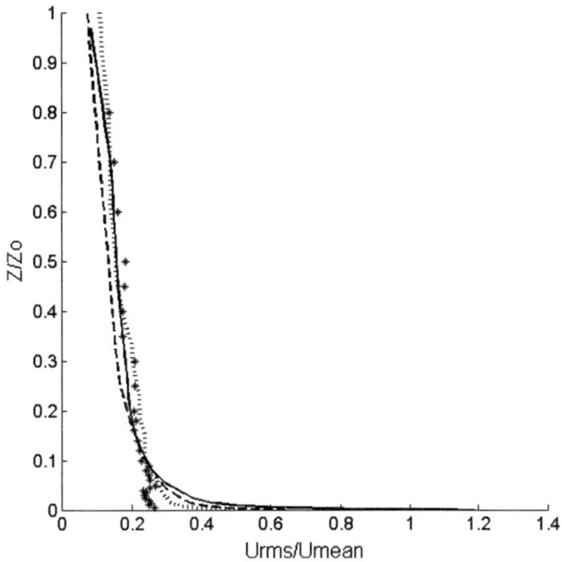

Fig. 5 Urms profiles form experiments and numerical simulations. symbols: experimental data; *solid line* coarse grid dt = 0.5 s; *dashed line* fine grid dt = 0.5 s; *dotted line* fine grid dt = 0.01 s

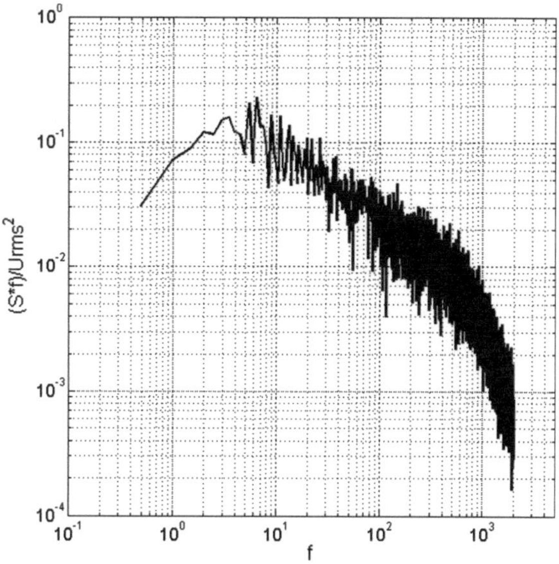

Fig. 6 Power spectrum for longitudinal velocity at 100 mm height, from experimental data

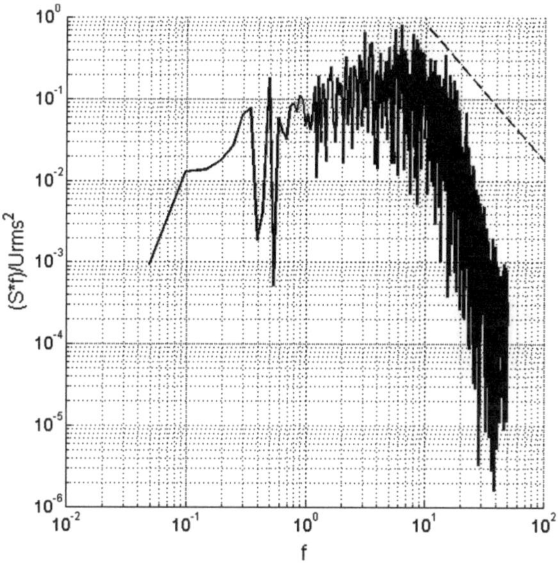

Fig. 7 Power spectrum for longitudinal velocity at 100 mm height, from fine grid numercial simulations

5 Conclusions

Preliminary results showing good overall agreement between experimental data and numerical simulations suggest that representation of roughness elements by immersed boundary condition method is an effective way of numerically modeling ABL Wind tunnel tests. Both mean velocity profile and urms profile were adequately captured at the applied grid resolutions, which must be fine enough to geometrically resolve the roughness elements. While computational intensive the proposed method has the advantage of requiring almost no calibration to reproduce wind tunnel test conditions.

Acknowledgments This work was supported by grant FSE-2011-6015 from ANII.

References

1. G.T. Bitsuamlak, T. Stathopoulos, C. Bedard, Numerical evaluation of wind flow over complex terrain: review (2004)
2. B. Blocken, J. Carmeliet, A review of wind-driven rain research in building science. J. Wind Eng. Ind. Aerodyn. **92**(13), 1079–1130 (2004)
3. E. Dayan, Wind energy in buildings: power generation from wind in the urban environment-where it is needed most. Refocus **7**(2), 338 (2006)
4. J. Ferziger, M. Peric, *Computational Methods for Fluid Dynamics* (Springer, New York, 2002)
5. T. Lehnhauser, M. Schäfer, Improved linear interpolation practice for finite-volume schemes on complex grids. Int. J. Numer. Methods Fluids **38**, 625–645 (2002)
6. T. Lehnhauser, M. Schäfer, Efficient discretization of pressure-correction equations on non-orthogonal grids. Int. J. Numer. Methods Fluids **42**, 211–231 (2003)
7. C. Liao, Y. Chang, C. Lin, J.M. McDonough, Simulating flows with moving rigid boundary using immersed-boundary method. Comput. Fluids **39**, 152–167 (2010)
8. Z. Lilek, S. Muzaferija, M. Peric, V. Seidl, An implicit finite-volume method using nonmatching blocks of structured grid. Numer. Heat Transf., Part B **32**, 385–401 (1997)
9. R.N. Meroney, Wind tunnel numerical simulations of pollution dispersion: a hybrid approach, Working paper, Croucher Advanced Study Institute on Wind tunnel Modelling. Hong Kong University of Science and Technology. 6–10 December (2004), pp. 60
10. M. Mendina, M. Draper, A.P. Kelm Soares, G. Narancio, G. Usera, A general purpose parallel block structured open source incompressible flow solver. Clust. Comput. **17**(2), 231–241 (2014)
11. J. Mora Acosta, Numerical algorithms for three dimensional computational fluid dynamic problems. Ph.D. thesis, UPC (2001)
12. S.B. Pope, *Turbulent Flow* (Cambridge University Press, Cambridge, 2001)
13. C.M. Rhie, W.L. Chow, A numerical study of the turbulent flow past an isolated airfoil with trailing edge separation. AIAA J. **21**, 1525–1532 (1983)
14. N.M. Standen, A spire array for generating thick turbulent shear layers for natural wind simulation in wind tunnels. National Research Council Canada, Aeronautical Establishment report no. LTR-LA-94 (1972)
15. G. Usera, A. Vernet, J.A. Ferré, Use of time resolved PIV for validating LES/DNS of the turbulent flow within a PCB enclosure model. Flow Turbul. Combust. **77**, 77–95 (2006)
16. G. Usera, A. Vernet, J.A. Ferré, A parallel block-structured finite volume method for flows in complex geometry with sliding interfaces. Flow Turbul. Combust. **77**, 471–495 (2008)

Three-Dimensional Nature of 2D Hairpin Packet Signatures in a DNS of a Turbulent Boundary Layer

S. Rahgozar and Y. Maciel

Abstract The existence of the outer region "hairpins" and "hairpin packets" is visually assessed in a well-resolved DNS of a zero-pressure-gradient turbulent boundary layer at moderately high Reynolds number. For this purpose, 50 independent 2D streamwise–wall-normal slices at $Re_\theta = 4300$ are extracted. The slices are then used to mimic the coarser resolution PIV velocity fields of Adrian et al. J. Fluid Mech, 422:1–54, 2000 [2] using the mimicking procedure of Rahgozar et al. J. Turbul, 14(10):37–66, 2013 [5] based on Gaussian filtering and linear interpolation. Afterwards, in the same manner as Adrian et al. J. Fluid Mech, 422:1–54, 2000 [2], the mimicked fields are inspected in order to discover the signatures of hairpin and hairpin packets. The vortices that are identified as hairpins are then isolated and visualized in three dimensions using the fully resolved DNS data. In agreement with Adrian et al. J. Fluid Mech, 422:1–54, 2000 [2], signatures associated by them to hairpin and hairpin packets are observed frequently in the mimicked planes. However, the 3D character of the 2D signatures is found to be more convoluted than the proposed hairpin packet model.

1 Introduction

Coherent vortices are believed to play an important role in the dynamics of turbulent flows. Among them, in wall turbulence, hairpins and hairpin packets have received great attention for more than a decade [1, 4]. By performing planar PIV measurements, Adrian et al. [2] (hereafter referred to as AMT) found that the cross sections of detected swirling regions in the streamwise–wall-normal plane are consistent with the idea that wall turbulence is populated by hairpin vortices. They also suggested

S. Rahgozar
Department of Mechanical and Materials Engineering, Queen's University, Kingston, ON K7L 3N6, Canada
e-mail: saeed.rahgozar.1@ulaval.ca

Y. Maciel (✉)
Department of Mechanical Engineering, Laval University, Quebec, Canada
e-mail: Yvan.Maciel@gmc.ulaval.ca

M. Stanislas et al. (eds.), *Progress in Wall Turbulence 2*,
ERCOFTAC Series 23, DOI 10.1007/978-3-319-20388-1_7

that the hairpins are often coherently aligned in the streamwise direction and tend to occur in packets. Since the study of AMT, the hairpin and hairpin packet paradigms have gained considerable support (see Marusic and Adrian [4]). The hairpin and hairpin packet paradigms are appealing concepts for numerous researchers in the turbulence community since these relatively simple structures can explain, or at least are consistent with, many observed flow events such as near-wall low-speed streaks, multiple ejections near the wall, scale variation observed throughout the logarithmic layer and bulges. However, the majority of evidence to support the hairpin and hairpin packet concepts comes from relatively coarse-resolution planar PIV measurements or low Reynolds number DNS data and moreover, three-dimensional visualizations of hairpins and hairpin packets have been so far restricted to the cases of near-wall generation and simplified artificial flows. Only recently it has become possible to fully resolve wall-bounded flows in both space and time at relatively high Reynolds numbers comparable to the ones in experimental investigations.

Throughout the paper, u, v and w are, respectively, the fluctuating streamwise, wall-normal and spanwise velocities and U, V and W denote their instantaneous values. The streamwise, wall-normal and spanwise coordinates are, respectively, x, y and z. Second quadrant motions ($u < 0$, $v > 0$) and fourth quadrant motions ($u > 0$, $v < 0$) are represented by $(uv)_2$ or Q_2 and $(uv)_4$ or Q_4, respectively.

1.1 Database

The high-resolution DNS data employed is that of a ZPG TBL of Schlatter and Örlü [6]. The simulations were done using a full spectral method to solve the time-dependent, incompressible Navier–Stokes equations. The DNS is finely resolved and the turbulence statistics are in very good agreement with experiments. The considered subdomains of the DNS data are two time-independent volumes of $5\delta \times 3\delta \times 1\delta$ size in the streamwise, spanwise and wall-normal directions, respectively. The Reynolds number Re_θ varies from 4139 to 4300 in the streamwise direction. The spatial resolution in outer units in the subdomains is approximately: $\Delta x/\delta = 0.005$ (streamwise), $\Delta z/\delta = 0.003$ (spanwise) and $\Delta y/\delta = 0.006$ at $y/\delta = 0.5$ (wall normal) which ranges roughly from 1η to 2η.

For the purpose of the present study, a total of 50 streamwise–wall-normal slices are extracted from the aforementioned sub-volumes. Furthermore, the DNS data is interpolated into a fine uniform mesh grid ($\Delta x/\delta = \Delta y/\delta = 0.0037$) to be able to mimic the AMT database and to identify vortices in the same manner as in AMT.

1.2 Signatures of Hairpins and Hairpin Packets According to AMT

AMT characterized the hairpin vortex signatures in a two-dimensional streamwise–wall-normal velocity vector field as follows: (i) a compact region of spanwise vorticity in the head; (ii) a Q_2 flow located beneath the head along a roughly 45° locus with a maximum somewhere below the vortex head; (iii) a region of low momentum fluid located below and upstream of the vortex head (iv) a shear layer caused by stagnation-point flow resulting from the Q_2/Q_4 interaction. The last signature was observed frequently but not always. AMT noticed that the hairpin vortex signatures very frequently occur in groups; the groups of hairpins, convecting with nearly the same streamwise velocity, form packets.

1.3 Approach

Since the packet structures were first conjectured in 2D velocity fields (see AMT), 2D streamwise–wall-normal slices of the DNS data are first extracted. Subsequently, the spatial resolution of the 2D well-resolved velocity fields is degraded using the technique described in [5] in order to have the same spatial resolution as in the PIV fields of AMT. Rahgozar et al. [5] showed that their technique can properly mimic the PIV database of AMT in terms of vortex characteristics (i.e. average swirling rate (λ_{ci}), diameter, circulation and vortex population density). As in AMT, the vortices are identified with swirling rate λ_{ci} in the PIV-mimicking 2D fields. These fields are then inspected, in the same manner as AMT, in order to discover the signatures of hairpin and hairpin packets. A zone of intense swirling rate is identified as the head of a hairpin if a region of Q_2 motion is located beneath the vortex with an inclination of roughly 45°. A train of hairpins are considered to form a packet candidate if the number of hairpins is at least five, their locus is coincident with a contour of constant velocity and a low momentum zone exists beneath the hairpin heads. In the last step, three-dimensional λ_2 [3], u and uv objects which are located in the vicinity of the detected 2D packets are identified and visualized in the fully resolved DNS data. For each parameter, different threshold values are also examined in order to determine the effect of thresholds on the observed structures. The features of "hairpins" and "hairpin packets" are then examined to investigate if the three-dimensional reality is consistent with the AMT proposed features.

2 Results and Discussion

2.1 Effects of a Coarse Spatial Resolution on the Cross Section of Vortices

Figure 1a shows a streamwise–wall-normal slice of the DNS data in which vortex cross sections are identified by the two-dimensional λ_{ci}. The same slice is shown in Fig. 1e, but this time, the velocity field was processed in order to have a spatial resolution corresponding to that of the PIV data of AMT, which was obtained with

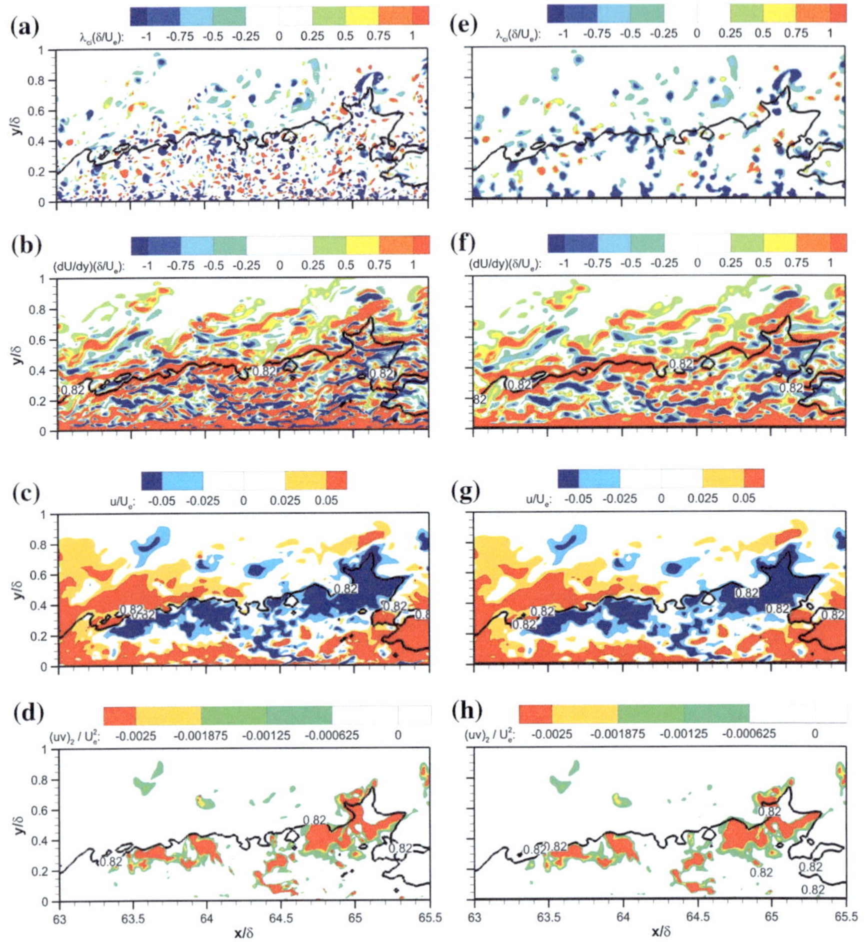

Fig. 1 **a**, **b**, **c** and **d** respectively, contours of two-dimensional λ_{ci}, dU/dy, u and $(uv)_2$ in a streamwise–wall-normal slice of the DNS data. *Righthand* side figures are same as those on the *left* but for the mimicked field. *Black lines* represent iso-velocity line of $U/U_e = 0.82$

an interrogation window size of 0.027δ (square windows) and a mesh size of 0.014δ. For both velocity fields, the only threshold applied on λ_{ci} is the minimum number of grid points necessary to consider a patch of non-zero λ_{ci} as a vortex and this number is adjusted according to the type of velocity field. Hence, in the original velocity field (Fig. 1a), a patch of non-zero λ_{ci} with at least four points is considered as a vortex. However, in the PIV-mimicking velocity field (Fig. 1e), the threshold size has to correspond to the size of the interrogation window that was used by AMT. Since an overlap of 50 % of the interrogation windows is used by AMT, a square composed of nine mesh points is the equivalent of the interrogation window size and therefore the threshold used is nine points.

The effects of spatial resolution of planar PIV on vortex size, swirling strength, circulation and population density were extensively analysed by [5] using a series of experimental and numerical databases. They found that typical values of mesh size and interrogation window size ($0.01 - 0.03\delta$) and typical levels of measurement uncertainties have significant effects on the vortex parameters. In agreement with [5], Fig. 1e shows a significant decrease in the vortex population density due to the lack of spatial resolution. This is caused partly by the filtering out of large and especially small vortices and partly by artificial merging of vortices. By comparing the AMT database and the well-resolved DNS data, Rahgozar et al. [5] reported that the ensemble average of the vortex diameter is double in the case of the AMT database, while the vortex population density is four times smaller. Moreover, as shown in Fig. 1b, f, the velocity gradient dU/dy is also significantly affected by the spatial resolution.

Nevertheless, the iso-velocity lines ($U/U_e = 0.82$, black lines) and the contours of u and $(uv)_2$ in Fig. 1c, d, g, h show that the spatial resolution has no important effect on the shape and size of large-scale u- and uv-structures.

2.2 Hairpin and Hairpin Packet Signatures in 2D Slices

In agreement with AMT, the conjectured hairpin and hairpin packet signatures are frequently observed in all slices of the resolution-degraded DNS data. Figure 1e epitomizes an "ideal packet" as described by AMT: an almost linear ramp with a growth angle of approximately 12°, regrouping the four aforementioned signatures of the hairpin and hairpin packets. The linear ramp is an iso-velocity line with $U/U_e = 0.82$ closely aligned with a train of vortices. The pattern is very much similar to the ramp-shaped patterns shown in Figs. 19 and 20 of AMT. Instantaneous velocity vectors in Fig. 2 clearly demonstrate rotational motions around the patches of λ_{ci} located on the iso-velocity line (candidates of hairpin head), inclined regions of Q_2 vectors and stagnation points between Q_2 and Q_4 motions. Note that the velocity vector map is viewed in a convective frame of reference $U_c = 0.82U_e$. Figure 1f presents contours of velocity gradients in the same velocity field. Consistent with

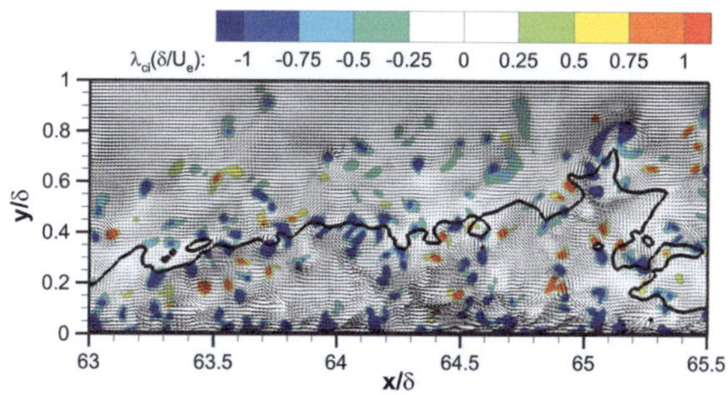

Fig. 2 Velocity vector map viewed in a convective frame of reference $U_c = 0.82U_e$ for the same slice shown in Fig. 1

the hairpin packet model, the contour of constant velocity (solid line) coincides closely with the regions of intense shear. The contours of u and $(uv)_2$ are shown in Fig. 1g and h. In agreement with the hairpin packet model, Fig. 1g, h show large-scale regions of low-speed flow and $(uv)_2$ motions beneath the constant velocity line. In brief, the patterns observed in Fig. 1 are in every respect consistent with the signatures conjectured to be those of hairpins and hairpin packets by AMT. Although the velocity field shown in this figure is a quintessential example of hairpin and hairpin packet signatures, broadly similar patterns are frequently observed in other slices.

As mentioned earlier, although the lack of spatial resolution has no important effect on the iso-velocity lines and the contours of u and $(uv)_2$, a large number of small vortices are filtered out or merged into larger vortices by the lack of spatial resolution. Hence, the lack of spatial resolution may reinforce the idea that the outer region of the boundary layer is mostly and densely populated by large-scale vortices which are sufficiently large and intense to form large-scale u- and uv-structures. However, a comparison between Fig. 1a, e rather suggests that these vortices only form a small subset of all vortices present and that many vortices are embedded within the low momentum zones. This means that the lack of spatial resolution indeed promotes the appearance of the hairpin packet signatures in 2D velocity fields.

2.3 Three-Dimensional Reality

Figure 3 shows the well-resolved, three-dimensional volume of the same velocity field shown in Figs. 1 and 2. The black line is the same iso-velocity line as in Fig. 1, which is situated in the streamwise–wall-normal slice at the mid-span of the volume. The colour contours are iso-surfaces of λ_2 with different thresholds, namely $-2, -1$ and $-0.5\sigma_{\lambda_2}$ (σ_{λ_2} is the standard deviation of λ_2) shown in Fig. 3a, b, c, respectively. In order to produce these iso-surfaces the following steps are taken: (1) Rotational regions are identified using λ_2 criteria. (2) Regions with λ_2 lower than the chosen thresholds are discarded. (3) Only sets of connected points comprising more than 500 points are kept. Note that voxels must share a face (and not only an edge or a vertex) to be considered as connected. (4) Among remaining λ_2 regions, only those located on or very near the iso-velocity line with $U/U_e = 0.82$ are shown.

A cursory glance at the field shown in Fig. 3a reveals that no hairpin-like vortices are observable. The vortex tubes have an irregular shape and are not necessarily spanwise oriented near the mid-span plane. The streamwise–wall-normal slice cuts of these vortices identified in Fig. 1 do not correspond in general to "hairpin heads". Figure 3b, c presents the same field as in Fig. 3a but with lower thresholds on λ_2. These figures confirm that the irregular shape of the vortices is not caused by the threshold since the lower threshold leads to a more complicated picture in which the vortices tend to merge into a few complex and intertwined vortex clusters.

Moreover, as it is shown in Fig. 3b, c, by lowering the threshold a hairpin-like vortex (marked with an arrow) appears above the iso-velocity line. However, this vortex is not a member of the conjectured hairpin heads (see Fig. 1) and it appears

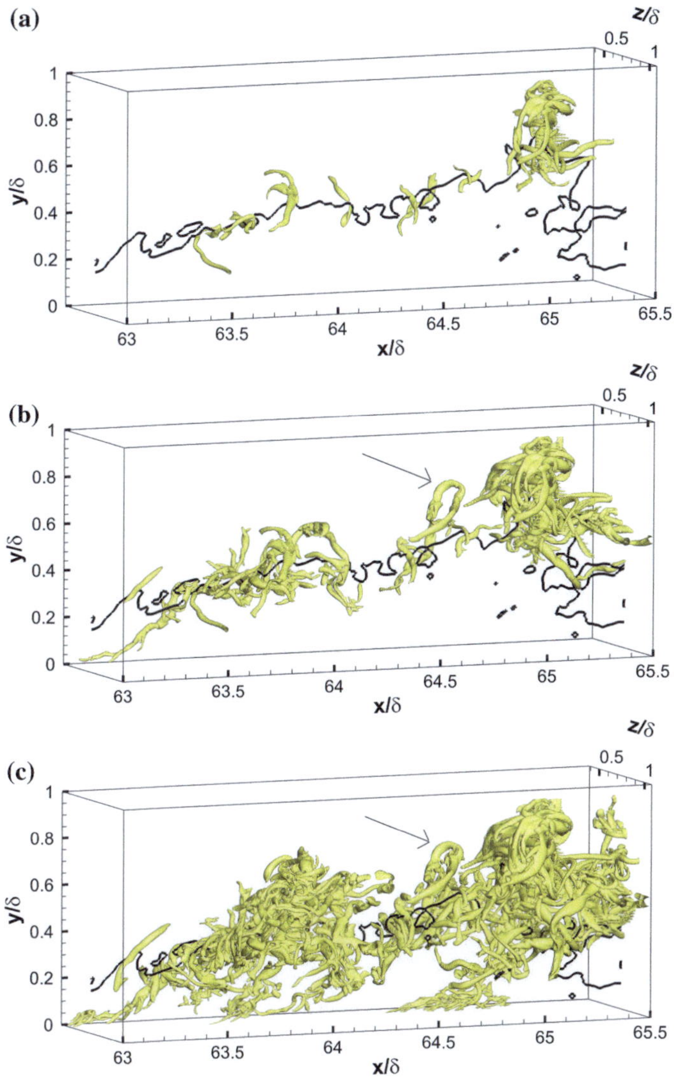

Fig. 3 Iso-surfaces of λ_2 structures identified according to the steps described in Sect. 2.3. **a** $\lambda_2/\sigma_{\lambda_2} = -2$, **b** $\lambda_2/\sigma_{\lambda_2} = -1$ and **c** $\lambda_2/\sigma_{\lambda_2} = -0.5$

here because, as the threshold is lowered, it becomes connected to another vortex which is located near the iso-velocity line. Furthermore, the head of this hairpin-like vortex is located far away from the iso-velocity line in both spanwise and wall-normal directions and is not aligned in the streamwise direction but rather crooked in a spanwise path.

In order to examine how comparable the extracted vortices in Fig. 3 are with other neighbouring vortices, all vortices satisfying the threshold of $\lambda_2 \leq -\sigma_{\lambda_2}$ and

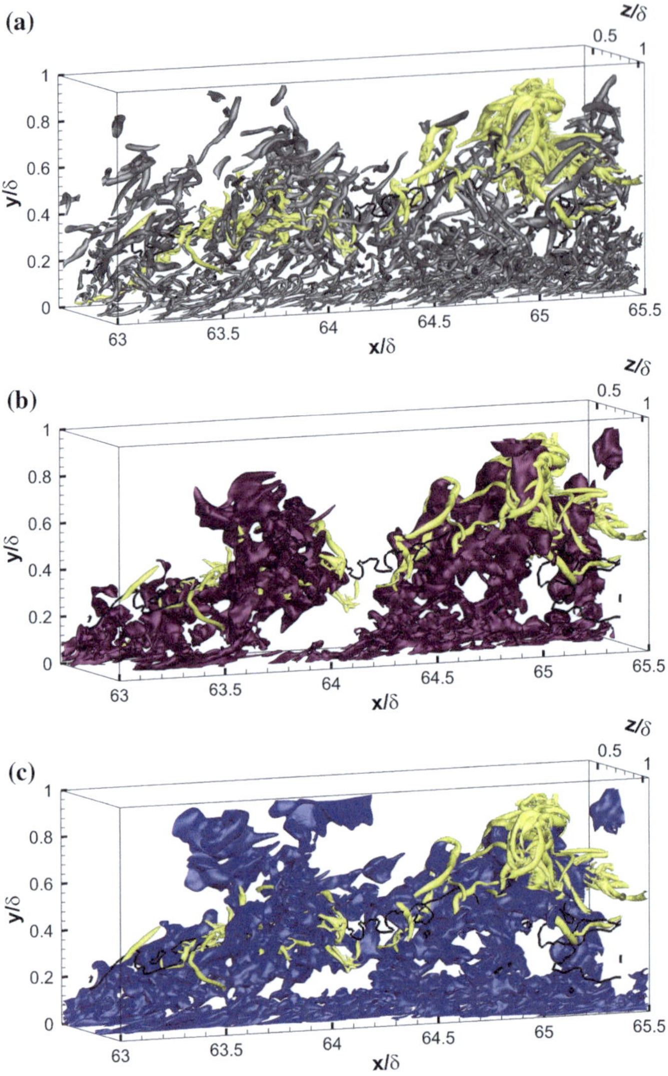

Fig. 4 Iso-surfaces of λ_2 structures shown in Fig. 3b (*yellow*) with: **a** other vortices in the volume satisfying the threshold of $\lambda_2 \leq -\sigma_{\lambda_2}$ and corresponding to more than 500 connected grid points (*grey*); **b** iso-surfaces of $\frac{Q_2}{\sigma_u \sigma_v} = -1$; **c** iso-surfaces of $\frac{u}{\sigma_u} = -1$

corresponding to more than 500 connected grid points are shown in Fig. 4a. The vortices belonging to the conjectured hairpin packet are shown in yellow, while all others are in grey. This figure shows no significant disparity between these two sets of vortices in terms of shape, size and intensity.

The iso-surfaces of Q_2 ($u < 0, v > 0$) and low-speed ($u < 0$) motions are shown in Fig. 4b, c. We use a similar approach as for λ_2 for detecting and visualizing the

Q_2 and low-speed regions. However, for these structures bigger than vortices, only regions comprising more than 5000 connected grid points are kept. These figures clearly show large-scale, three-dimensional structures which are closely intertwined with the vortex tubes. Similar to λ_2, the threshold has significant effects on the large-scale u- and uv-structures. Contrary to Q_2 structures, large-scale Q_4 ($u > 0, v < 0$) structures are mostly located around the vortex patches (not shown). Note that in the absence of a statistical analysis, the relation between the vortex patches and u-uv-structures cannot be generalized here. Nevertheless, we can confirm the frequent existence of large-scale u- and uv-structures and a possible association between these structures.

3 Conclusion

The outer region of a ZPG TBL is analysed from the perspective of AMT by applying their structure identification scheme on 2D streamwise–wall-normal slices of DNS data that mimic their coarser resolution planar PIV velocity fields. The 2D motions that are identified as signatures of "hairpins" and "hairpin packets" are compared to those found by AMT. These identified motions (vortices, low-momentum regions and quadrant motions) are then visualized in three dimensions using the fully resolved DNS data.

It is found that the 2D filtered fields of swirling strength significantly differ from the original well-resolved ones. Numerous vortices are filtered out by coarser spatial resolution and some vortices merge artificially into bigger vortices. Nevertheless, the shape and size of large-scale u- and uv-structures are not importantly altered by the mimicking approach. In agreement with AMT, signatures associated by them to hairpin and hairpin packets are observed frequently in the mimicked planes. This shows that the hairpin and hairpin packet signatures described by AMT are present in the perfect flow conditions of DNS and are not caused by imperfections in real flow fields such as wall roughness. However, the 3D character of the 2D signatures is found to be more convoluted than the proposed hairpin model. For instance, the vortex cross sections in 2D planes that were candidates for heads of hairpins are usually cross sections of vortex tubes with irregular and complex shapes, and locally not spanwise oriented. Moreover, a large vortex cross section in 2D low spatial resolution view is sometimes the imprint of intertwined smaller vortices. The 3D view of 2D packet signatures reveals a series of irregular vortex tubes which are not necessarily different from other nearby vortex tubes in terms of their shape, size and intensity. The large bunches of intertwined vortices generally cohabit with large-scale u- and uv-structures, but it seems unlikely that the fine and sparse irregular vortex tubes, detected from 2D packet signatures, can produce these large structures. Furthermore, although the threshold on the swirling strength has an important effect on the visual aspect of the vortices, the above conclusions are found to be not affected by the threshold.

Acknowledgments Financial support from NSERC of Canada is gratefully acknowledged by the authors. They also wish to thank P. Schlatter for sharing his DNS data as well as R.J. Adrian, C.D. Meinhart and C.D. Tomkins for sharing their experimental data.

References

1. R.J. Adrian, Hairpin vortex organization in wall turbulence. Phys. Fluids **19**(4), 041301 (2007)
2. R.J. Adrian, C.D. Meinhart, C.D. Tomkins, Vortex organization in the outer region of the turbulent boundary layer. J. Fluid Mech. **422**, 1–54 (2000)
3. J. Jeong, F. Hussain, On the identification of a vortex. J. Fluid Mech. **285**(1), 69–94 (1995)
4. I. Marusic, R.J. Adrian, *Ten Chapters in Turbulence, Chap. The Eddies and Scales of Wall Turbulence* (Cambridge University Press, Cambridge, 2013), pp. 176–220
5. S. Rahgozar, Y. Maciel, P. Schlatter, Spatial resolution analysis of planar PIV measurements to characterise vortices in turbulent flows. J. Turbul. **14**(10), 37–66 (2013)
6. P. Schlatter, R. Örlü, Assessment of direct numerical simulation data of turbulent boundary layers. J. Fluid Mech. **659**(1), 116–126 (2010)

Wall Pressure Signature in Compressible Turbulent Boundary Layers

N.A. Buchmann, Y.C. Kücükosman, K. Ehrenfried and C.J. Kähler

Abstract Large-scale turbulent flow structures associated with positive and negative wall pressure fluctuations in a compressible turbulent boundary layer are investigated. Experiments are conducted in a closed-loop transonic wind tunnel at Ma $= 0.5$–0.8, $\mathrm{Re}_\tau = 5{,}100$–$9{,}500$ with combined velocity field and wall pressure measurements. Both, velocity and pressure statistics are analysed and compare well with existing low Mach number data. Spatial two-point correlation is applied to determine the size and orientation of the large-scale flow structures, which depending on the wall height have an averaged length scales of 4–6 δ and a maximum inclination angel of $\approx 12°$–$13°$. The wall pressure fluctuations are associated with shear layer structures and it is shown that positive pressure fluctuations are correlated with low speed large-scale flow structures over streamwise extents of 4–5 δ.

1 Introduction

Understanding the relationship between coherent flow structures and associated wall pressure fluctuations in turbulent boundary layers is fundamental in the prediction and control of wall-bounded turbulent flows. In incompressible flows the turbulent velocity field and the fluctuating pressure field are connected by the solution of the Poisson's equation [10].

$$\nabla^2 p = -\rho\, \tilde{u}_{i,j} \tilde{u}_{j,i}, \tag{1}$$

where ρ is the fluid density and $\tilde{u}_{i,j}$ the partial derivative of the velocity $\tilde{u}_i = U_i + u_i$. Assuming the boundary layer flow is two-dimensional and homogenous in spanwise and streamwise direction Eq. 1 simplifies to

N.A. Buchmann (✉) · Y.C. Kücükosman · C.J. Kähler
Institute of Fluid Mechanics and Aerodynamics, Bundeswehr University Munich, 85577 Neubiberg, Germany
e-mail: nicolas.buchmann@unibw.de

K. Ehrenfried
Institut Für Aerodynamics and Fluid Mechanics, German Aerospace Center (DLR), 37073 Göttingen, Germany

M. Stanislas et al. (eds.), *Progress in Wall Turbulence 2*,
ERCOFTAC Series 23, DOI 10.1007/978-3-319-20388-1_8

$$\frac{1}{\rho}\nabla^2 p = -2\frac{dU}{dy}\frac{u_2}{dx} - u_{i,j}u_{j,i}. \qquad (2)$$

The two terms on the right hand side represents the pressure sources. The first term is frequently referred to as the fast, or linear *mean–turbulent* pressure source and the second slow, nonlinear $turbulent–turbulent$ interaction.

According to Farabee and Casarella [4] wall pressure fluctuations can be classified as low and high frequency fluctuations, which originate in the inner and outer region of the boundary layer. Significant contribution to the high wavenumber fluctuations comes from the viscous sublayer, while the logarithmic layer largely contributes to low wavenumber fluctuations [2]. Of particular interest are high-amplitude pressure peaks (HAPPs) with strong fluctuations 2–3 times above the r.m.s. pressure fluctuations. These pressure events have a low occurrence probability, but large contribution to the total r.m.s. pressure [11]. It is believed that positive pressure fluctuations are associated with shear layer structures and sweep–ejection events [13], while large negative pressure fluctuations are related to spanwise and quasi-streamwise vortices [6].

The wall pressure is affected by the entire velocity field as given by (2) and the role of large-scale flow structures, originating in the logarithmic and outer layer (i.e. super-structures,VLSM), in this relationship is yet nuclear. Large-scale flow structures are characterised by long meandering regions of positive and negative velocity fluctuations with significant contribution to the overall Reynolds stress [8]. More importantly, the large-scale flow structures have a direct influence on the near-wall cycle via amplitude modulation of the small-scale velocity fluctuations [9].

By logic extension one can hypothesis a similar connection between the large-scale structures and the fluctuating wall pressure field. Therefore, the present study applies simultaneous velocity and wall pressure measurements to first identify the large-scale flow structures and second to link them to the fluctuation wall pressure field.

2 Experimental Procedure

2.1 Turbulent Boundary Layer Experiment

Experiments are conducted in the Transonic-Wind-Tunnel at DLR Göttingen (TWG). The TWG can be operated continuously and has a 4.5 m long adaptive test section with $1 \times 1\ m^2$ cross-sectional area. The wind tunnel and plenum chamber are pressurised such that the Mach number Ma and stagnation pressure p_0 can be adjusted independently. A rigid plate of 2.75 m length and 1 m span is mounted symmetrically in the test section (Fig. 1). The plate has an elliptical nose and the flow is tripped 180 mm downstream of the leading edge. Adaption of the upper and lower test section wall ensures a nearly nearly zero pressure gradient, which decreases slightly towards

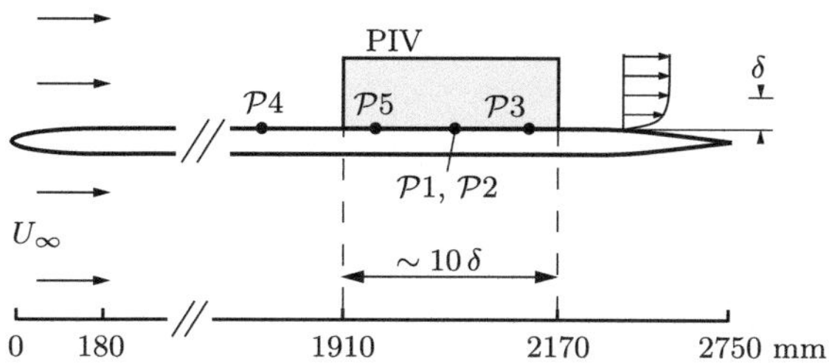

Fig. 1 Side view of the experimental setup showing with the boundary layer plate, pressure transducers $\mathcal{P}1-5$ and PIV measurement region

Table 1 Characteristics of the turbulent boundary layer for the different operating conditions (Ma, p_0) considered in the present study

Ma	p_0 (kPa)	U (m/s)	u_τ (m/s)	δ (mm)	θ (mm)	ν/u_τ (μm)	Re_τ	Re_θ
0.5	50	173.9	6.00	26.8	4.0	5.3	5,100	16,700
0.8	50	274.8	9.32	26.5	3.7	3.4	7,800	24,800
0.5	100	178.4	5.90	26.8	3.7	2.8	9,500	31,200

the end of the plate. The wind tunnel is operated at free-stream Mach numbers Ma = 0.5 and 0.8 and stagnation pressures $p_0 = 50$ and 100 kPa. Under these conditions the Reynolds number is $\mathrm{Re}_\tau = 5{,}100$–9,500 and the boundary layer thickness is $\delta = 26$–27 mm (see Table 1).

2.2 Velocity Field and Wall Pressure Measurements

In order to resolve the large and very large scale flow structures simultaneous planar PIV measurements with three sCMOS cameras are conducted to cover a large streamwise field-of-view of $1.5 \times 10\ \delta$. The cameras have a sensor size of 2560×2160 pixel^2 and 6.5 μm pixel pitch and are equipped with 100 mm focal length Zeiss macro lenses with an optical magnification of $\mathcal{M} = 0.17$. Illumination in the $x-y-$plane is provided by a Spectra Physics Quanta-Ray PIV 400 Nd:YAG dual-cavity laser. The light-sheet with a thickness of 0.5–1 mm thickness is introduced from the trailing edge of the flat plate model. The flow is seeded with DEHS droplets with 1 μm mean diameter. A total of 20,000 images are recorded and analysed with an iterative multi pass window-correlation routine. The final interrogation window size is 16×16 pixel^2 with 50 % overlap. This corresponds to a spatial resolution of 0.62 mm or $\Delta x^+ = 116 - 489$ depending on Re_τ.

The measurement section is equipped with 9 static pressure taps and an array of 5 dynamic piezo-resistive pressure transducers (EPE-S449-0.35B, Entran) (see [3] for details). Transducers $\mathcal{P}3-\mathcal{P}5$ are aligned with the light-sheet in streamwise direction and have a mean separation of approximately 7.5δ (see Fig. 1). Transducer $\mathcal{P}1-\mathcal{P}2$

are located of-axis at identical streamwise positions and spanwise separation of 0.23 δ and are used to assess the spanwise coherence of the pressure fluctuations across the light-sheet width. The pressure transducers have a sensing diameter of 2.4 mm and a 35 kPa pressure range. In order to improve the spatial resolution the sensors are mounted in an adaptor and connected to the surface pressure via a 0.5 mm orifice. Thus, depending on Re_τ the effective spatial resolution of the dynamic pressure measurements is $d^+ = du_\tau/\nu = 94$–147. Note that in order to fully resolve the wavenumber spectrum of the pressure fluctuations a resolution of $d^+ \leq 10$–12 is required [7, 11]. Consequently, the present measurements are restricted to large-scale pressure fluctuations only. A static calibration is performed for each transducer and the analog pressure signals are pre-amplified via a Endevco signal conditioner. The AC component of the signals are filtered and digitised by a Dewetron A/D converter at a rate of 20 kHz ($\Delta t u_\tau^2/\nu = 137$) over a time period of 500 s ($TU_\infty/\delta > 10^6$). In addition the camera and Q-switch trigger signal are also digitised such that the velocity and dynamic wall pressure measurements can be synchronised off-line.

3 Characterisation of the Turbulent Boundary Layer

3.1 Velocity Field Statistics

An example of the instantaneous velocity field at Ma $= 0.8$, $\mathrm{Re}_\tau = 7{,}800$ is given in Fig. 2. The velocity field consists of a hierarchy of coherent δ-scale structures that appear to be inclined in the streamwise direction by approximately 14°. The intermittency of the turbulent/non-turbulent interface is also clearly visible. The turbulent boundary layer develops in streamwise direction and δ and Re_τ increase by approximately 8 % over a distance of 10 δ. Mean flow statistic sampled in the center of the domain are summarised in Table 1 for the two Mach numbers investigated in this study. Mean velocity and Reynolds normal stress profiles are illustrated in Fig. 3 and are in good agreement with the data of Fernholz and Finley [5]. Since the focus of the current study is on the large-scale flow structure the measurements only resolve the flow in the logarithmic region and the outer part of the turbulent boundary layer (i.e. $800 \leq y^+ \leq 1.5\delta^+$).

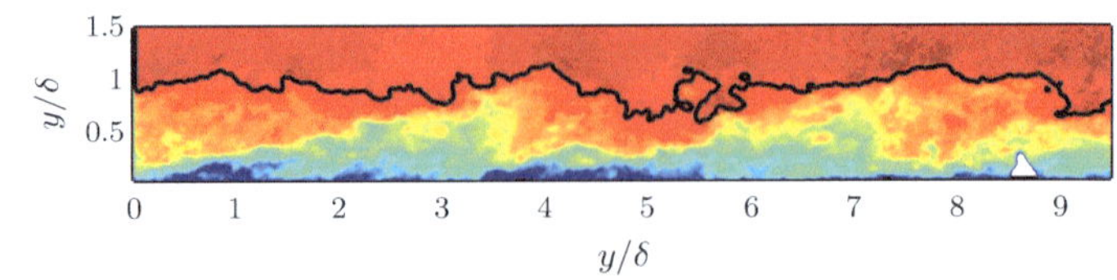

Fig. 2 Instantaneous velocity field U/U_∞ for $\mathrm{Re}_\tau = 7{,}800$ (Ma $= 0.8$). *Black line* indicates the boundary layer thickness δ and colour contours are from 0.5 to 1.0

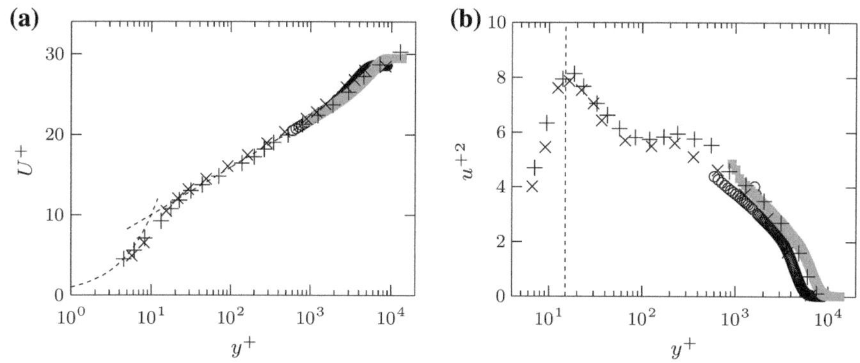

Fig. 3 Mean velocity profile U^+ (**a**) and Reynolds normal stress u^{+2} (**b**) in inner scaling. Present data: ∘, $\mathrm{Re}_\tau = 5{,}100$ (Ma = 0.5); ▪, $\mathrm{Re}_\tau = 7{,}800$ (Ma = 0.8); ×,+, Fernholz and Finley [5]

3.2 Statistical Properties of Wall Pressure Fluctuations

Typical power density spectra of the wall pressure fluctuations Φ_{pp} for sensor $\mathcal{P}1$–$\mathcal{P}4$ are shown in Fig. 4. The spectra show distinct peaks at the low frequency range, which are associated with tones produces by the wind tunnel turbine and structural vibrations of the test section. These acoustic fluctuations decay rapidly at higher frequencies and the power spectrum becomes dominated by the convective pressure fluctuations for frequencies above 1.5–2 kHz [3]. In the range from 2 to 10 kHz the spectra are flat and do not show a maximum typically seen in low Mach number boundary layers. The cut-off frequency of the Helmholz resonator is approximately 11 kHz above which the pressure fluctuations are attenuated by more than −3 dB (see [3] for details). In order to remove the acoustic pressure fluctuations the

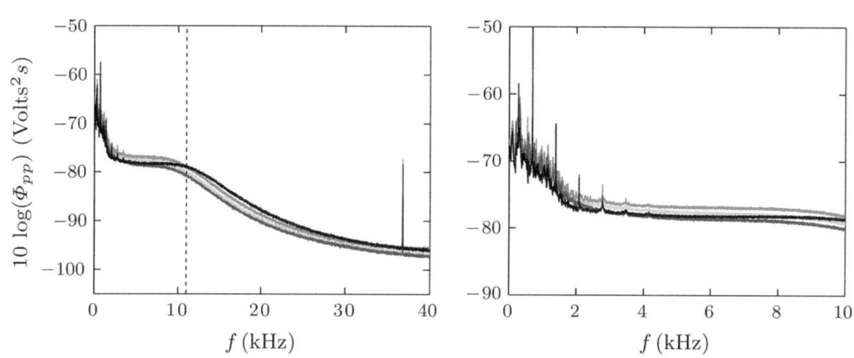

Fig. 4 Power spectral density of the wall pressure fluctuations Φ_{pp} for $\mathrm{Re}_\tau = 7{,}800$ (Ma = 0.8) and sensors: —— $\mathcal{P}1$; —— $\mathcal{P}2$; —— $\mathcal{P}3$; —— $\mathcal{P}4$. The *dashed line* indicates the −3 dB cut-off frequency of the Helmholz resonator

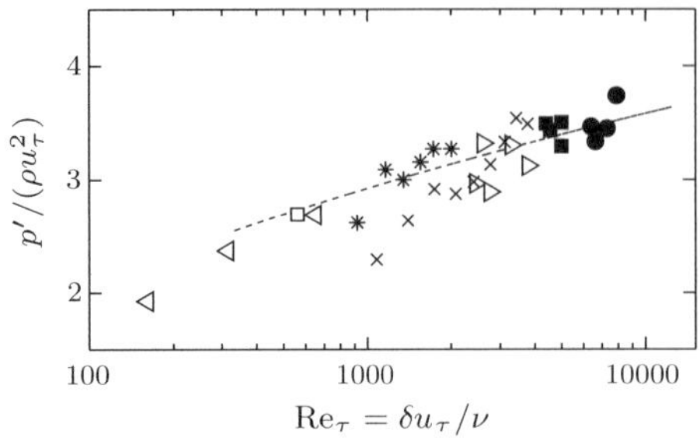

Fig. 5 Root-mean-square wall pressure p' in inner scaling as a function of the Reynolds number. Present results: ■, $\mathrm{Re}_\tau = 5{,}100$ (Ma = 0.5); •, $\mathrm{Re}_\tau = 7{,}800$ (Ma = 0.8). ▷, Bull and Thomas [1]; ∗, Farabee and Casarella [4]; □, Schewe [11]; ×, Tsuji et al. [14]; ◁, Spalart [12]; ‐ ‐ ‐ ‐ $(p'/\rho u_\tau^2)^2 = 6.5 + 1.86\ln(\mathrm{Re}_\tau/333)$ [4]

signals are filtered at 1.5 kHz using a phase preserving high-pass filter. The filter frequency is optimised by comparing the pressure auto-correlations, which remain largely unchanged for filter frequencies above 1.5 kHz.

The r.m.s. values of the filtered wall pressure fluctuations p' normalised by inner scaling are plotted in Fig. 5 together with previous results. The r.m.s. wall pressure increases slowly with Re_τ and follows the prediction $(p'/\rho u_\tau^2)^2 = 6.5 + 1.86\ln(\mathrm{Re}_\tau/333)$ of Farabee and Casarella [4]. The current result for Ma = 0.5 and Ma = 0.8 also follow this trend and extend the currently available data to larger Re_τ.

The probability density function (PDF) of the wall pressure fluctuations is shown in Fig. 6. The PDFs are normalised by p' and displayed in linear and logarithmic scaling. The PDFs show little variation with Mach number, respectively Reynolds number

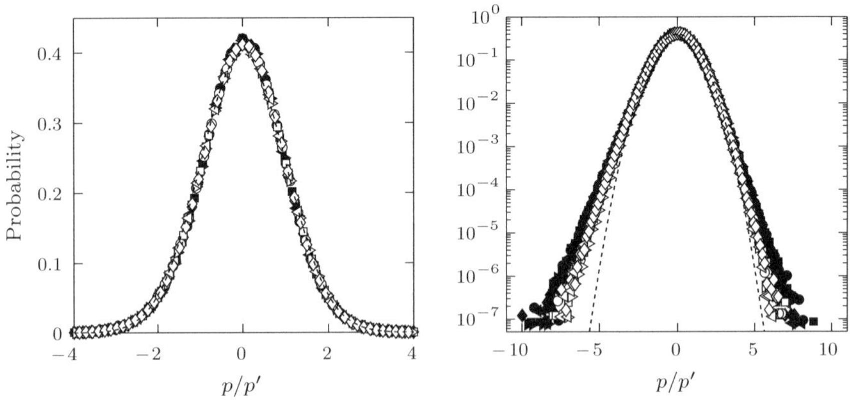

Fig. 6 Probability density function of wall pressure fluctuations. Symbols indicate the different sensors $\mathcal{P}1-\mathcal{P}5$. *Solid symbols*, $\mathrm{Re}_\tau = 5{,}100$ (Ma = 0.5); *open symbols*, $\mathrm{Re}_\tau = 7{,}800$ (Ma = 0.8); ⋯⋯⋯, Gaussian distribution

and obey a near Gaussian distribution. However, the tails of the PDFs approach an exponential distribution and show some slight dependency on the Reynolds/Mach number, which agrees well with observations in [14]. The skewness is approximately −0.09 for both cases and consistent with the low Mach number data of Gravante et al. [7]. The flatness decreases from 3.6 to 3.3 with increasing Re_τ and also follows the trends in [7] albeit at different spatial resolutions. It is important to note that spectral quantities and higher order statistics are dependent on the spatial resolution as pointed out in [7, 11] and that some caution needs to be taken when interpreting data from different sources.

4 Organisation of the Large-Scale Flow Structures

Two-point velocity correlations of the form $R_{uu}(\xi, y) = \langle\, u(x-\xi, y_p)\, u(x, y)\,\rangle$ for the Ma = 0.8 (Re_τ = 7,800) case and different wall heights of the conditioning point y_p are shown in Fig. 7. The averaged coherent flow structures extend over large streamwise domains (5–6 δ) and have large wall-normal extends (0.75–1 δ). The structures maintain their coherence for different wall height, but their streamwise length scale decreases with wall height (Fig. 8a). Different Mach/Reynolds numbers exhibits a good collapse when scaled on outer variables similar to the low Mach number data in Hutchins and Marusic (2007) [8]. Although the present data indicate averaged streamwise length-scales that are approximately 25–50 % larger than those given in [8] (see Fig. 9a). The contours of R_{uu} show a clear inclination in streamwise direction, which increase from approximately 3° near the wall ($y/\delta = 0.1$) to a maximum angle of 12°–13° at $y/\delta - 0.5$. The regions of positive correlations reach far into the boundary layer and persist over large streamwise extents and give an indication of the average size of the superstructures or VLSM-type events in the outer region of the turbulent boundary layer.

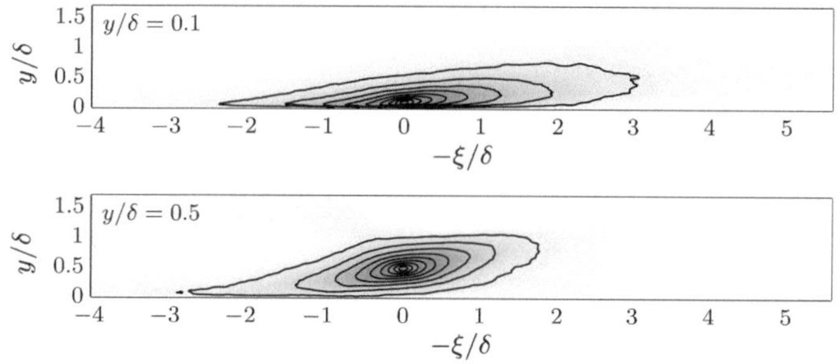

Fig. 7 Two-point correlation of the streamwise velocity fluctuation $R_{uu}(\xi, y)$ for Re_τ = 7,800 (Ma = 0.8) calculated at $y/\delta = 0.1$ (*top*) and $y/\delta = 0.5$ (*bottom*). Contour levels are from $R_{uu}/u'^2 = 0.1$ to 1.0 in increments of 0.1

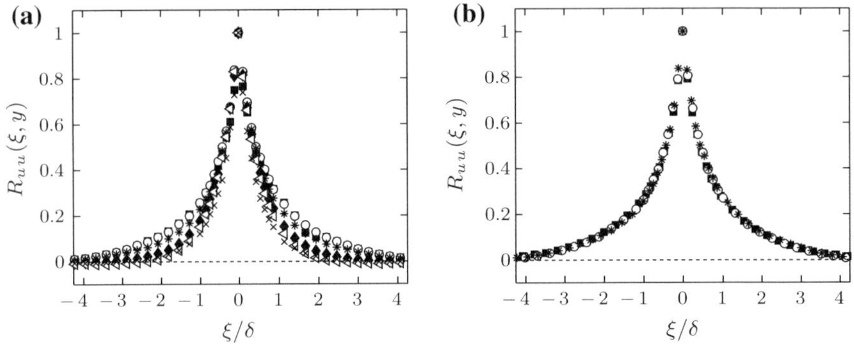

Fig. 8 Two-point velocity correlation $R_{uu}(\xi, y)$: (**a**) Ma = 0.8, Re_τ = 7,300 for different wall heights y/δ (■ = 0.15; ○ = 0.2; ∗ = 0.3; ◆ = 0.5; ◁ = 0.7; × = 1.0). (**b**) Reynolds numbers scaling at $y/\delta = 0.2$ (■, Re_τ = 5,100 (Ma = 0.5); ○, Re_τ = 9,500 (Ma = 0.5); ∗, Re_τ = 7,800 (Ma = 0.8))

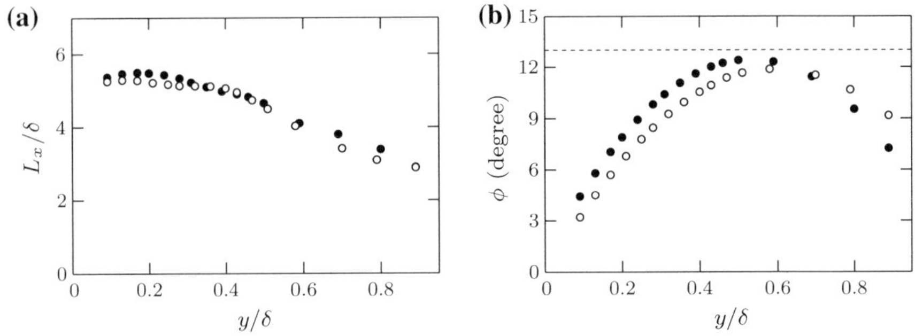

Fig. 9 Streamwise length-scale (**a**) and inclination angle (**b**) of the coherent large-scale structures for different wall heights y/δ: •, Re_τ = 5,100 (Ma = 0.5); ○, Re_τ = 7,300 (Ma = 0.8)

5 Space-Time PressureVelocity Correlations

Possible links between the large-scale flow structures and wall pressure fluctuations are assessed by means of the space-time correlation R_{pu} defined as:

$$R_{pu}(x, y, \tau) = \langle\, p(x_0, 0, t - \tau)\, u(x, y, t)\,\rangle \,/\, (\rho\, u_\tau^3), \tag{3}$$

where $(x_0, 0)$ is the position of the transducer $\mathcal{P}5$, t the time of the velocity field recording and $\langle\rangle$ the ensemble average over all velocity field realisations. Additionally, the space-time correlation is conditioned on the occurrence of positive and negative pressure fluctuation to obtain $R_{pu\,|\,p>0}$ and $R_{pu\,|\,p<0}$.

Figure 10a shows the space-time correlation $R_{pu}(x, y, 0)$ for Re_τ = 5,100 (Ma = 0.5). Positive correlations exists upstream of the reference point, while a large region (4–5 δ) of negative correlation extends downstream of the reference

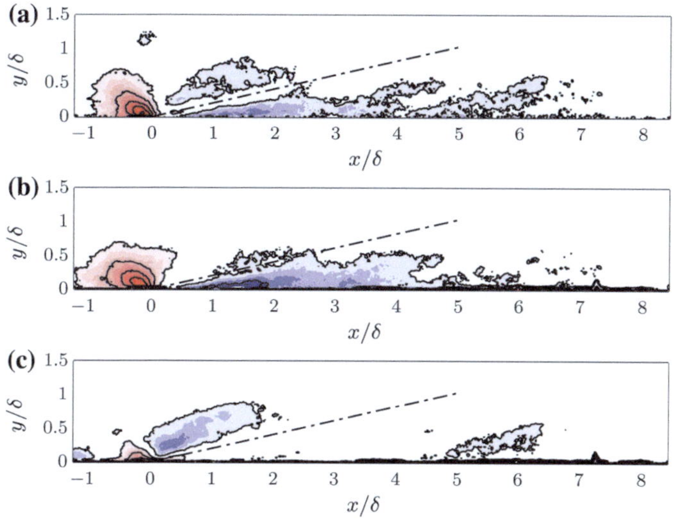

Fig. 10 Space-time correlations $R_{pu}(x, y, 0)$ at $\mathrm{Re}_\tau = 5{,}100$ (Ma = 0.5): (**a**) R_{pu}; (**b**) $R_{pu\,|\,p>0}$; (**c**) $R_{pu\,|\,p<0}$. The *dashed line* indicates an inclination angle of 12°. Contour levels are $R_{pu}/(\rho u_\tau^3) = -20$ to 75 in intervals of 15

point. According to Thomas and Bull [13], wall pressure fluctuations are associated with sweep–ejection events and the formation of the inclined shear layer upstream of the coherent structures. Indeed the negative pressure velocity fluctuations show an inclination angle of approximately 12°, which is consistent with the inclination of the coherent large-scale structures. Additionally, the conditioned space-time correlations reveal a strong link between the large-scale shear layer structures and positive pressure fluctuations (Fig. 10b). On the other hand, negative pressure fluctuations only exhibit a weak link with the large-scale streamwise flow organisation. The significant streamwise (6–7 δ) and wall-normal (0.75 δ) extend of the pressure-velocity correlation R_{pu} and $R_{pu\,|\,p>0}$ suggest that large-scale flow structures play an important role in determining the near wall velocity and wall pressure field.

6 Concluding Remarks

Relationships between large-scale streamwise velocity fluctuations and wall pressure fluctuations in a turbulent boundary layer are investigated. Velocity and pressure statistics at Ma = 0.5–0.8, $\mathrm{Re}_\tau = 5{,}100$–9,500 are analysed and compare well with existing low Mach number data. Analysis of the streamwise two-point velocity correlation reveals the size (4–6 δ) and orientation of the large-scale flow structures. When scaled on outer units, two-point correlations, averaged structure length and inclination angle collapse for the current range of Mach and Reynolds numbers.

Space-time correlations between wall pressure and streamwise velocity fluctuations reveal structures of large streamwise extent (5–6 δ). Positive pressure fluctuations are strongly correlated with negative velocity fluctuations, which provides evidence of a possible link between the large-scale flow structures in the buffer and outer layer and the near wall pressure field.

References

1. M.K. Bull, A.S.W. Thomas, High frequency wall pressure fluctuations in a turbulent boundary layer. Phys. Fluids **19**, 597–599 (1976)
2. P.A. Chang, U. Piomelli, W.K. Blake, Relationship between wall pressure and velocity-field sources. Phys. Fluids **11**, 3434–3448 (1999)
3. K. Ehrenfried, L. Koop, Experimental study of pressure fluctuations beneath a compressible turbulent boundary layer. In *29th AIAA Aeroacoustics Conference*, Vancouver, British Columbia Canada (2008)
4. T.M. Farabee, M.J. Casarella, Spectral features of wall pressure fluctuations beneath turbulent boundary layer. Phys. Fluids A **3**, 2419–2420 (1991)
5. H.H. Fernholz, P.J. Finley, The incompressible zero-pressure-gradient turbulent boundary layer: an assessment of the data. Prog. Aerosp. Sci. **32**, 245–311 (1996)
6. S. Ghaemi, F. Scarano, Turbulent structure of high-amplitude pressure peaks within the turbulent boundary layer. J. Fluid Mech. **735**, 381–426 (2013)
7. S.P. Gravante, A.M. Naguib, C.E. Wark, H.M. Nagib, Characterization of the pressure fluctuations under a fully developed turbulent boundary layer. AIAA J. **36**, 1808–1816 (1998)
8. N. Hutchins, I. Marusic, Evidence of very long meandering features in the logarithmic region of turbulent boundary layers. J. Fluid Mech. **579**, 1–28 (2007)
9. R. Mathis, N. Hutchins, I. Marusic, Large-scale amplitude modulation of the small-scale structures in turbulent boundary layers. J. Fluid Mech. **628**, 311—337 (2009)
10. A.M. Naguib, C.E. Wark, O. Juckenhöfel, Stochastic estimation and flow sources associated with surface pressure events in a turbulent boundary layer. Phys. Fluids **13**, 2611–2626 (2001)
11. G. Schewe, On the structure and resolution of wall-pressure fluctuations associated with turbulent boundary-layer flow. J. Fluid Mech. **134**, 311–328 (1983)
12. P.R. Spalart, Direct simulation if a turbulent boundary layer up to $R_\theta = 1410$. J. Fluid Mech. **187**, 61–98 (1988)
13. A.S.W. Thomas, M.K. Bull, On the role of wall-pressure fluctuations in deterministic motion in the turbulent boundary layer. J. Fluids Eng. **128**, 283–322 (1983)
14. Y. Tsuji, J.H.M. Fransson, P.H. Alfredsson, A.V. Johansson, Pressure statistics and their scaling in high-Reynolds-number turbulent boundary layers. J. Fluid Mech. **585**, 1–40 (2007)

Three-Dimensional Structure of Pressure–Velocity Correlations in a Turbulent Boundary Layer

Yoshitsugu Naka, Michel Stanislas, Jean-Marc Foucaut, Sebastien Coudert and Jean-Philippe Laval

Abstract Three-dimensional pressure–velocity correlations in a turbulent boundary layer have been investigated to understand the relationship between the pressure fluctuations and the coherent structures. Simultaneous measurements of the fluctuating pressure and velocity fields have been performed by the point pressure measurement technique and stereo PIV. The space–time three-dimensional structures of the pressure–velocity correlations, R_{pu}, R_{pv} and R_{pw}, are evaluated. The wall pressure fluctuations are closely coupled with large-scale coherent structures, i.e., large-scale sweep and ejection. For the pressure fluctuations in the field, the pressure–velocity correlations R_{pu} and R_{pw} exhibit a meaningful correlation in a region very extended in time in addition to the structures observed with the wall pressure. The Reynolds number effect is quantified from the data at $Re_\theta = 7\,300$, $10\,000$, and $18\,000$, it is mostly evidenced on the size and intensity of the correlations. Such 3D structures of the pressure–velocity correlations can be consistent with the evidence of large-scale and very-large-scale motions reported in the literature.

1 Introduction

The pressure fluctuations in incompressible flows are closely linked with the vortex structures and play a significant role in the transport of the turbulent kinetic energy and the Reynolds stress. In wall-bounded turbulent shear flows, the wall pressure pattern is observed as a footprint of the turbulent events occurring above it. The relationship

Y. Naka (✉)
Tokyo Institute of Technology, 2-12-1 Ookayama,
Tokyo, Meguro-ku 152-8550, Japan
e-mail: ynaka@navier.mes.titech.ac.jp

M. Stanislas · J. Foucaut
Laboratoire de Mécanique de Lille, Ecole Centrale de Lille, Université Lille Nord de France,
Boulevard Paul Langevin, 59655 Villeneuve d'ascq, France

S. Coudert · J. Laval
CNRS Laboratoire de Mécanique de Lille, Université Lille Nord de France,
Boulevard Paul Langevin, 59655 Villeneuve d'ascq, France

M. Stanislas et al. (eds.), *Progress in Wall Turbulence 2*,
ERCOFTAC Series 23, DOI 10.1007/978-3-319-20388-1_9

between the wall pressure fluctuation and the turbulent structure passing above the wall has extensively been studied theoretically, experimentally, and numerically ([1, 2] among others). Kim [2] investigated the pressure fluctuations from the database of a direct numerical simulation (DNS) of channel flow at low Reynolds number, and reported that the contributions to the pressure–strain correlations are local near the wall, but global away from the wall.

The large-scale correlated motions have been investigated by many authors, and it is known that structures typically larger than 3δ carry a significant fraction of the turbulent kinetic energy and Reynolds stress [3]. Adrian et al. [4] and Christensen and Adrian [5] performed PIV studies of such large-scale motions, called *hairpin packets*, in the boundary layer. Furthermore, very long structures called "superstructures" of the streamwise velocity component (more than $20\,\delta$) were also reported from atmospheric boundary layer measurements [6]. Such *hairpin packets* and *superstructures* are therefore important features of near-wall turbulence, but their relation to the pressure fluctuations, especially at the wall still remains unclear.

In the statistical sense, the large coherent structures can be observed in the two-point correlations. Foucaut et al. [7] visualized the three-dimensional structures of the two-point velocity correlations in a turbulent boundary layer from the stereo PIV measurements with two planes positioned orthogonally. The correlation of streamwise velocity shows an elongated ellipsoidal shape in the streamwise direction, inclined to the wall at an angle of approximately 10°. Tutkun et al. [8] evaluated the space–time correlations of the streamwise velocity component from the data of a rake of 143 single hot-wire probes, and found that the correlation spreads approximately $7\text{–}8\delta/U_e$ in time. Such an elongated shape of two-point streamwise velocity correlations is considered to be associated with the large-scale structures. Further, the wall-normal velocity correlation is much more localized and even close to isotropy as soon as the fixed point is far enough away from the wall.

Since an appropriate measurement technique is not yet available for the fluctuating pressure at a point in the flow, the pressure fluctuation in turbulent near-wall flows has been out of reach of experimental studies. The difficulty is mainly caused by the fact that the turbulent pressure fluctuations are weak, and are easily distorted by the ambient noise and probe intrusion, especially very close to the wall. Tsuji et al. [9] made a first attempt of measurements of the pressure fluctuations in a turbulent boundary layer by a small static pressure probe, which was originally developed by Toyoda et al. [10]. They reported successful results in the fundamental statistical quantities such as the mean, r.m.s., and power spectra of pressure fluctuations and investigated their scaling law.

In the present study, the simultaneous measurements of the fluctuating pressure and the three velocity components have been performed to investigate the three-dimensional structures of the space–time pressure–velocity correlations in a turbulent boundary layer. The pressure fluctuations at the wall and in the field are captured together with stereo PIV planes which are arranged perpendicular to the wall and to the mean flow direction. The stereo PIV system is designed to capture the large-scale motion which spreads over the whole boundary layer thickness and the small scales at the same time. Our target is to clarify how the pressure–velocity correlation is related to the large-scale coherent structure, and to evaluate their extension quantitatively.

2 Experiment

Simultaneous measurements of the fluctuating pressure and velocity field were performed in a closed-loop turbulent boundary layer wind tunnel at Laboratoire de Mécanique de Lille. The wind tunnel has a cross-section of 1 m × 2 m in transverse and spanwise directions, and a 21.6 m long development section. The maximum operating velocity is 10.5 m/s, and the boundary layer thickness δ reaches approximately 0.3 m at the measurement position.

An (x, y, z) Cartesian coordinate system is defined for the streamwise, wall-normal, and spanwise directions respectively. The origin is set at the center of the wall pressure hole which is located at the center in span on the lower wall of the wind tunnel section and at 18 m from the contraction outlet. The free stream velocity U_e was regulated at 3, 5 and 10 m/s with a stability of 0.5 % giving Reynolds numbers based on the momentum thickness θ and U_e, of $Re_\theta = 7\,300$, $10\,000$, and $18\,000$.

A schematic of the experiment is presented in Fig. 1. The fluctuating pressure signals at two points (at the wall and at the pressure probe) were captured, and the field of three velocity components close to the pressure holes was measured. The pressure probe consists of the tip, the pipe, and the connecting part to the microphone. The outer diameter of the pipe is 1.0 mm and its thickness is 0.05 mm. Two 0.4 mm diameter holes are opened with a separation angle of 180° at 19.5 mm from the tip. Therefore, the present pressure probe has a measurement volume of 0.4 mm × 0.4 mm × 1.0 mm in x-y-z directions. The design of the probe is mostly the same as in the previous study [11] except for the number of pressure holes. The wall pressure

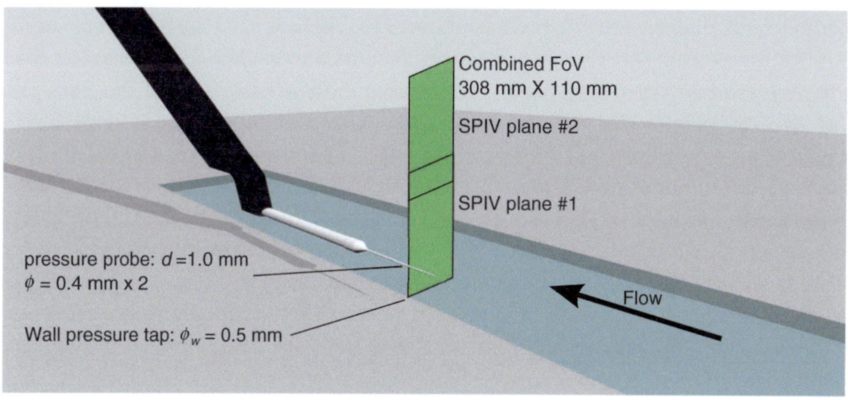

Fig. 1 Schematic of the simultaneous measurements of the fluctuating pressure at the wall tap and at the pressure probe and the velocity field by stereo PIV

tap is a 0.5 mm diameter hole, and is located next to the stereo PIV measurement plane. A 1/4-inch condenser microphone (B&K microphone 4938 and pre-amplifier 2670) was used to capture the pressure fluctuation. The signals were recorded by a 16-bit A/D converter board (Measurement Computing PCI-DAS6034) installed on a PC. The sampling rate was set to 40 kHz, and the Q-switch signals of the laser was simultaneously recorded for synchronization. The acoustic or electric frequency response of each component in the pressure measurement system is quantified, and the overall transfer function is taken into account in the post processing. The background noise in the pressure signal is reduced by the auxiliary microphone which is fixed at the center of the wind tunnel section ($y = 0.5$ m).

As depicted in Fig. 1, two stereo PIV planes were arranged in $y - z$ plane and were placed side by side in the wall-normal direction to cover the whole boundary layer thickness with a good spatial resolution. A 250 mJ/pulse Nd:YAG laser (BMI 5 000) was used for illumination. The scattered light from particles was captured by 4 CCD cameras with 2048 $\times$ 2048 pixels (Hamamatsu C9300-024) through Nikon 105 mm lenses. The aperture f#8 gave a diffraction spot of approximately 2 pixels. These cameras were mounted in the Scheimpflug condition [12] and the viewing angle and distance between the two cameras were 45° and 1.37 m, respectively. For seeding, polyethylene glycol particles whose diameter was about 1 μm were generated by a smoke generator. Each stereo PIV plane had a 16 $\times$ 11 cm field of view, and the combined field of 31 $\times$ 11 cm was obtained with a small overlap. The streamwise position of the light sheet was set to the holes of the pressure probe and the wall pressure tap. The two light sheets are slightly separated in the x direction (approximately 0.75 mm) to obtain better correlation. Note that the light sheet thickness is about 1.85 mm. The time separation between two exposures was set to $\Delta t = 250$, 150, and 75 μs for 3, 5, and 10 m/s to optimize maximum displacements to 10 pixels. The image signals from the four cameras were acquired by two-frame grabbers (X64 Xcelera-CL PX4), and the digitized images were recorded on PC hard drives. The sampling rate of stereo PIV was 4 Hz.

The PIV analysis was performed by a multi-pass FFT-based cross-correlation method with integer shift of both windows [13, 14]. A 1D Gaussian peak-fitting algorithm was used for the subpixel displacement determination. Three passes with different window sizes were used. The interrogation window size of the final pass was 26 $\times$ 39 pixels. The scale relationship between the physical and image spaces was 11.6 pixel/mm in y and 17.6 pixel/mm in z. Therefore, the physical size of the interrogation window was 2.24 $\times$ 2.21 mm in y and z. The PIV overlap ratio was 77 %. The Soloff method [15] was used to reconstruct the three velocity components. From a set of calibration and stereo PIV particle images, the misalignment between the light sheet and the calibration plane was compensated [16]. The characteristics of the boundary layer obtained by the present stereo PIV measurements are presented in Table 1 and coincide well with the previous hot-wire data obtained in the same facility [17].

Table 1 Experimental conditions

U_e (m/s)	δ_{99} (m)	δ^+	Re_θ
3.03	0.33	2 465	7 324
5.11	0.28	3 509	9 971
10.2	0.285	6 390	17 972

3 Space–Time Pressure–Velocity Correlations

The space–time correlation of the fluctuating pressure and velocity, R_{pu_i} is defined here as

$$R_{pu_i}(y_p, \Delta t, \Delta y, \Delta z) = \sum_{j=1}^{N} \left\{ p(t_j + \Delta t, y_p, 0) u_i(t_j, y_p + \Delta y, \Delta z) \right\} / N, \quad (1)$$

where t_j is the time of stereo PIV recording, y_p is the wall-normal position of the pressure probe, Δy and Δz are the separation of the moving point along y and z, respectively, (here $\Delta z = z$ as $z_p = 0$), Δt is the time separation of the pressure measurement with respect to t_j and N is the number of samples. Since the time-resolved pressure signal is available, the correlation of the velocity field at time t_j and pressure signal around t_j is evaluated. Each run of simultaneous measurement was repeated for nine different wall-normal positions of the pressure probe within $y_p = 3.8$ mm and 307 mm. The number of valid velocity field is 10 000 for each run. Note that the wall pressure–velocity correlation is evaluated from 90 000 samples, and is much better converged compared to the field pressure–velocity correlation.

3.1 3D Structure of the Wall Pressure–Velocity Correlations

First, we will look at the 3D structure of the correlation between the wall pressure fluctuations and the three velocity components. Figure 2 shows the 3D structures of the pressure–velocity correlations, R_{pu}, R_{pv} and R_{pw}, at Re_θ= 10 000. For R_{pu}, the positive correlation extends in the upstream ($\Delta t \leq 0$) on a significant part of the boundary layer thickness. The wall-normal extension at $\Delta t = 0$ is 0.3δ at Re_θ= 10 000 and exhibits the Reynolds number dependence. Taking into account the symmetry with respect to $z/\delta = 0$, the spanwise extent reaches to approximately 0.8δ ($\pm\, 0.4\delta$) in the upstream region. As this significant correlation region is positive just above the fixed point, positive pressure fluctuations at the wall are mostly correlated with positive streamwise velocity fluctuations and vice-versa. This is opposite to the idea that wall pressure fluctuations are generated directly by streamwise velocity fluctuations (which would lead to the opposite sign of the correlation). On the positive Δt side, negative correlation regions are evident but they are much weaker. These negative regions seem to be composed of two distinct subregions: an inclined

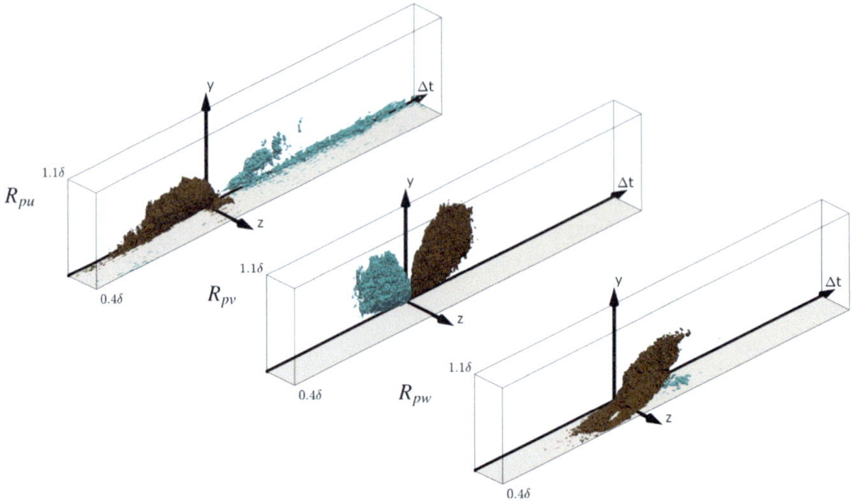

Fig. 2 3D structures of the pressure–velocity correlations, R_{pu}, R_{pv} and R_{pw} at $Re_\theta = 10000$. Isocontours of *dark orange* and *blue gray* indicate positive and negative values, respectively. Box size of each plot is $5\delta/U_e \times 1.1\delta \times 0.4\delta$

structure extending away from the wall and a near-wall thin layer spreading toward outside of the field of view downstream. The spanwise extent of the first one is comparable to the positive upstream region while the second one is limited near $z = 0$. The pressure–wall-normal velocity correlation R_{pv} is relatively localized in time but, as a difference to R_{pu}, it is nearly equally significant on both positive and negative Δt and it shows both positive and negative correlation regions of comparable size.

The positive and negative lobes are at a strong angle to the wall (approximately 45°). It is noted that the Reynolds number seems to have little effect on R_{pv}. The last and most complex one is the R_{pw} correlation. For homogeneity reasons, this correlation is zero in the $z/\delta = 0$ plane and antisymmetric with respect to this plane. Around the fixed point, it is mostly positive on the positive z side. The correlation has a spanwise extension larger than the field of view (which is 0.4δ) and is strongly inclined to the wall along z (20° to 25°). It is quite extensive in the wall-normal direction (up to $y/\delta = 0.6$), and also in time, on both positive and negative Δt sides. Weak negative correlation regions exist both downstream of the fixed point (positive Δt) close to the wall and upstream (negative Δt) further away from the wall. This leads globally to motions involved in the wall pressure fluctuations.

To facilitate understanding the physics of the pressure–velocity correlation, the wall pressure fluctuations are split into two parts: the positive ones which can be associated to either a decrease of the wall parallel (streamwise or spanwise) velocity component (Bernoulli static pressure effect) or an increase of the wall-normal component (stagnation point effect) with respect to the mean. The conditional averaging by the sign of the pressure fluctuation is introduced, and conditionally averaged R_{pu},

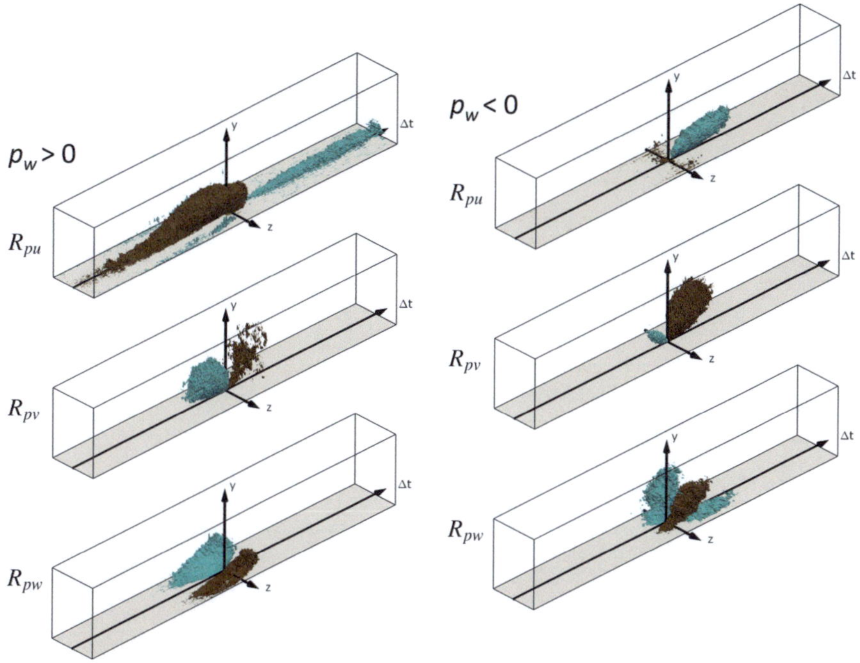

Fig. 3 3D structures of R_{pu}, R_{pv} and R_{pw} conditionally averaged based on the sign of the wall pressure signal. Box size of each plot is $5\delta/U_e \times 1.1\delta \times 0.8\delta$ (the distribution in negative z is synthesized with symmetry)

R_{pv} and R_{pw} at $Re_\theta = 10\,000$ are presented in Fig. 3. It is clear that both positive and negative pressure fluctuations at the wall are strongly correlated with coherent structures above them. These structures appearing at large scales are very different for the two signs of p. It is located predominantly upstream (negative Δt) for $p_w > 0$, while it is more on the downstream side for $p_w < 0$. These two signs are clearly associated to sweep, and the positive pressure fluctuations at the wall appear more as a consequence of the interaction of these large-scale motions with the wall.

3.2 3D Structure of the Field Pressure–Velocity Correlations

Having looked at the wall pressure–velocity correlations, the correlations of the pressure in the field and the velocity are discussed in this section. Figure 4 presents 3D structures of R_{pu} at $Re_\theta = 7\,300$, $10\,000$, and $18\,000$. The pressure probe position is closest to the wall ($y_p = 3.8$ mm) corresponding to the wall units of 28^+, 48^+, and 85^+. A Reynolds number effect on R_{pu} is visible on its size while the overall characteristics of the correlation are similar for the three different Reynolds numbers. As a difference to the pressure point at the wall, R_{pu} exhibits an inclined elongated

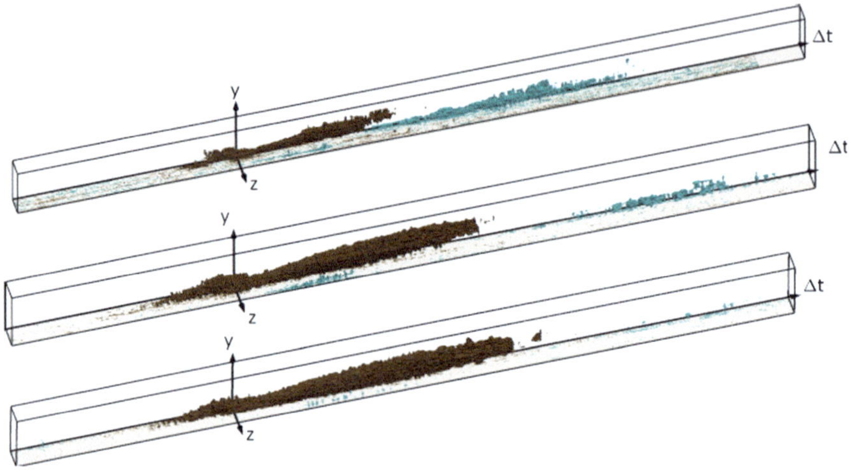

Fig. 4 3D structures of R_{pu} at Re_θ= 7300, 10000, and 18000 and at y_p = 3.8 mm (from *top* to *bottom*). Box size of each plot is $14\delta/U_e$ in time and the whole measurement area in y and z directions

shape in positive Δt. At Re_θ= 10000, it extends approximately down to $\Delta t U_e/\delta = 6$, which is similar to the observation in the velocity–velocity correlations [8]. Beside this dominant positive region, a second elongated region of correlation also appears clearly downstream of the previous one (larger Δt), but this time with a negative sign and closer to the wall. Both of these positive and negative correlation regions are inclined at a small angle (5° at Re_θ= 10000) with respect to the Δt axis. Another negative correlation region is visible on the spanwise side of the field, between $\Delta t U_e/\delta \simeq 1$ to 3. In Fig. 5, the distributions of R_{pu} at Re_θ= 10000 for different pressure measurement positions y_p^+ = 477, 959, 1 921, and 2 883 are presented. As y_p increases, the two positive correlation regions merge and form one large correlation pattern extending on both positive and negative Δt. This inclined structure shifts then progressively toward negative Δt when moving the pressure point away from the wall. The shape of this large positive correlation region changes significantly at $y_p^+ = 1\,921$, where it is detached from the wall.

Figure 6 illustrates the 3D distributions of R_{pv} and R_{pw} at the pressure measurement positions y_p^+ = 48, 477, 959, 1 921, and 2 883. In the distributions of R_{pv}, the correlation is much weaker than R_{pu} (isocontour value is approximately half of R_{pu}). This is not surprising as wall-normal velocity fluctuations are known to be smaller than streamwise ones. For the pressure reference position close to the wall ($y_p^+ = 48$), the R_{pv} correlation has a shape which is very similar to that of the pressure point at the wall. Unlike the wall point, an elongated region of negative correlation develops close to the wall on the positive Δt side. In the outer part, the positive correlation fades rapidly away with increasing y_p. The negative lobe becomes more isotropic and separates from the wall; the intensity decreases toward the edge of the boundary layer, and finally it disappears. The last correlation to be analyzed is

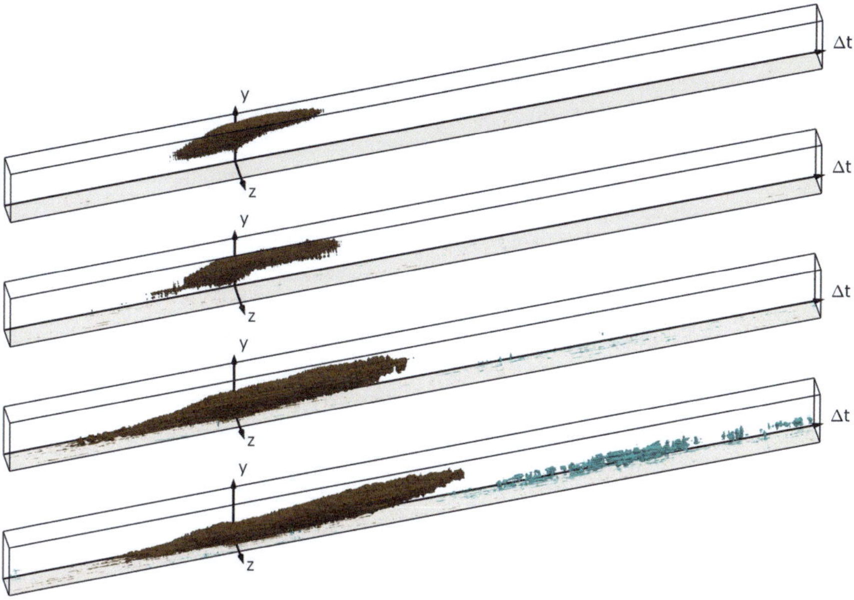

Fig. 5 3D structures of R_{pu} at $Re_\theta = 10\,000$ and at different pressure measurement positions, $y_p^+ = 477, 959, 1\,921$, and $2\,883$ (from *bottom* to *top*). Box size of each plot is $14\delta/U_e \times 1.1\delta \times 0.4\delta$

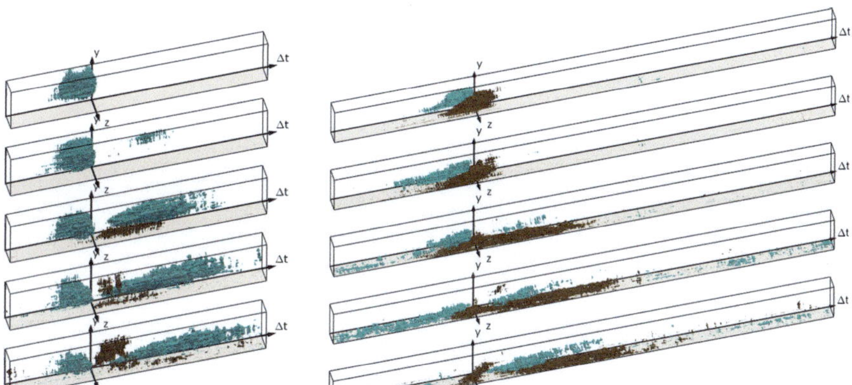

Fig. 6 3D structures of R_{pv} (*left*) and R_{pw} (*right*) at $Re_\theta = 10\,000$ and at different pressure measurement positions, $y_p^+ = 48, 477, 959, 1\,921$, and $2\,883$ (from *bottom* to *top*). Box sizes of R_{pv} and R_{pw} in time are $6U_e/\delta$ and $14U_e/\delta$, respectively

R_{pw}. Based on the previous analysis of R_{pu} and R_{pv}, it is possible to distinguish two regions in the correlation in the inner part. Close to the origin, there is a positive correlation region which is similar to the correlation with the fixed point at the wall although less converged due to the lower sampling. In addition, a group of two elongated structures, one positive and one negative, appear on the downstream side of the fixed point ($\Delta t > 0$). They obviously work together and are not independent from the elongated structures observed in the other correlations. In the outer part, the wall structure disappears first at $y_p^+ = 477$. The elongated structures are still visible in these wall-normal positions but become very weak. A new structure starts to be visible at $y_p^+ = 477$ with an elongated negative correlation region extending toward negative Δt and a positive correlation around the fixed point, which is obviously coupled to the negative one and laying below it. These two regions grow in wall-normal size and shrink in time when moving the fixed point further from the wall. It is strongly evoking an elongated streamwise-oriented large-scale vortical structure with a size of the order of the boundary layer thickness.

4 Conclusion

The 3D structures of pressure–velocity correlations, R_{pu}, R_{pv} and R_{pw}, have been visualized in a turbulent boundary layer at $Re_\theta = 7\,300$, $10\,000$, and $18\,000$.

The shape of the correlations indicates that the wall pressure fluctuations are strongly coupled with coherent structures which occupy a large region of the boundary layer. All correlations have their maximum close to the fixed point and decrease progressively away from it in a nonisotropic way. Wall pressure fluctuations are strongly linked to interfaces of sweep and ejection.

Field pressure–velocity correlations, R_{pu} and R_{pw} exhibit very elongated structures which are Reynolds number dependent. The correlations are found to be self-similar for the positions of the fixed pressure point in the inner layer, and they change in shape significantly in the outer layer. The shape of correlation R_{pu} is quite similar to those evidenced by velocity-velocity correlations.

Acknowledgments We are thankful to the "Conseil Régional Nord Pas de Calais" for providing a fellowship to Y. Naka which allowed to perform this experiment. The present research work was also supported by International Campus on Safety and Intermodality in Transportation, the European Community, the Regional Delegation for Research and Technology, the Ministry of Higher Education and Research, and the National Center for Scientific Research.

References

1. A.S.W. Thomas, M.K. Bull, J. Fluid Mech. **128**, 283 (1983)
2. J. Kim, J. Fluid Mech. **205**, 421 (1989)
3. R.J. Adrian, Phys. Fluids **19**(4), 041301 (2007)

4. R.J. Adrian, C.D. Meinhart, C.D. Tomkins, J. Fluid Mech. **422**, 1 (2000)
5. K.T. Christensen, R.J. Adrian, J. Fluid Mech. **431**, 433 (2001)
6. N. Hutchins, I. Marusic, J. Fluid Mech. **579**, 1 (2007)
7. J.M. Foucaut, S. Coudert, M. Stanislas, J. Delville, Exp. Fluids **50**, 839 (2011)
8. M. Tutkun, W.K. George, J. Delville, M. Stanislas, P.B.V. Johansson, J.M. Foucaut, S. Coudert, J. Turbul. **10** (2009)
9. Y. Tsuji, J.H.M. Fransson, P.H. Alfredsson, A.V. Johansson, J. Fluid Mech. **585**, 1 (2007)
10. K. Toyoda, T. Okamoto, Y. Shirahama, Appl. Sci. Res. **53**, 237 (1994)
11. Y. Naka, T. Omori, S. Obi, S. Masuda, Int. J. Heat Fluid Flow **737–746**, 27 (2006)
12. C. Willert, Meas. Sci. Technol. **8**, 1465 (1997)
13. J. Lin, J.M. Foucaut, J.P. Laval, N. Pérenne, M. Stanislas, *In Particle Image Velocimetry, Topics in Applied Physics*, vol. 112 (Springer, Berlin, 2008), pp. 191-221
14. S. Herpin, M. Stanislas, J.M. Foucaut, S. Coudert, J. Fluid Mech. **716**, 5 (2013)
15. S. Soloff, R.J. Adrian, Z.C. Liu, Meas. Sci. Technol. **8**, 1441 (1997)
16. S. Coudert, J.P. Schon, Meas. Sci. Technol. **12**, 1371 (2001)
17. J. Carlier, M. Stanislas, J. Fluid Mech. **535**, 143 (2005)

Computation of Complex Terrain Turbulent Flows Using Hybrid Algebraic Structure-Based Models (ASBM) and LES

C. Panagiotou, S.C. Kassinos and D. Grigoriadis

Abstract In this work, we revisit the coupling of the Algebraic Structure-Based Model with popular two-equation eddy viscosity models (EVM). We consider both the $v^2 - f$ model and variants of the κ-ω model. Our aim is to explore the role of the EVM in these couplings. Computations of turbulent boundary layer over a flat plate and a fully developed channel flow are initially performed for validation purposes. Then, the case of a 2D steep, smooth hill is considered, for which additional LES computations were performed in order to ascertain the validity of the experimental data. The coupling of the ASBM with the κ-ω-BSL model (hereafter called ASBM-BSL) showed superior robustness when compared to the ASBM-v^2-f hybrid model. Moreover, ASBM-BSL captures the size of the recirculation bubble more accurately, and overall yields a noticeable improvement in the prediction of the turbulent statistics in the recirculation region. All models fail to capture the correct shear stress profile at the top of the hill, exhibiting positive, non-realizable values near the wall. The present comparisons reveal a sensitivity of the hybrid closures to the choice of carrier model.

1 Introduction

In engineering practice, CFD computations rely heavily on two-equation eddy viscosity models (EVM). Undeniably, EVM offer certain computational advantages relative to more complex approaches, such as Reynolds Stress Transport (RST) models. Nevertheless, EVM would stand to benefit if a consistent method could be devised to augment the physics contained in them. Along this direction, a number of attempts have been made, such as nonlinear stress-strain relationships, etc.

Traditional stress models take as input the mean deformation field, the turbulence scales, and information about the componentality of the turbulence in order to predict the Reynolds stresses. Componentality is information about the relative strength of the fluctuating velocity components. It enters the model through the Reynolds

C. Panagiotou · S.C. Kassinos (✉) · D. Grigoriadis
Computational Science Laboratory UCY-CompSci, University of Cyprus,
Kallipoleos 75, 1678 Nicosia, Cyprus
e-mail: kassinos@ucy.ac.cy

M. Stanislas et al. (eds.), *Progress in Wall Turbulence 2*,
ERCOFTAC Series 23, DOI 10.1007/978-3-319-20388-1_10

stress tensor itself, or at least the representation of the tensor available in the model. Structure-Based Models (SBM) depart from this approach in that they additionally use information about the structure of the turbulence in order to predict the Reynolds stresses [5, 7]. This information is derived, directly or indirectly, from a set of exact transport equations for the one-point turbulence structure tensors. The model that is at the focus of the present work, i.e., the Algebraic Structure-Based Model (ASBM) [6, 8, 12], is an engineering simplification within the SBM family of models.

In recent years, there has been some interest in exploring couplings of ASBM with various two-equation EVM [1, 2, 6]. In such hybrid models, the turbulence scales are provided by solving the transport equations of the chosen EVM, which are then passed, along with the mean deformation field, to the ASBM. The ASBM in turn uses an algebraic procedure to compute the structure state of the turbulence, including the full Reynolds stress tensor. Such couplings are attractive since they can be easily included in many existing RANS codes, providing a shortcut to the additional information in the ASBM.

In evaluating such couplings, we have noticed differences in terms of both performance and numerical stability of the hybrid models. The motivation of the present study is to compare the couplings of ASBM with a series of popular two-equation EVM. The main test case chosen is the flow over the 'Witch of Agnesi' bump, a steep hill that is often used as a model of complex terrain. The validation of all RANS computations is done against the experiments of Loureiro [9], and also against results from a Large Eddy Simulation (LES), which we have carried out specifically for this purpose.

2 Validation

All computations were performed using an unstructured nodal-based finite-volume code, called CDP. The code uses collocated grids and the Rhie and Chew momentum interpolation technique to evaluate the cell face velocities. The diffusion term is discretized using an Adam-Bashforth scheme, including an upwind differencing. The convective term uses the divergence theorem and the sub-edge concept. For the time-advancement, a Crank-Nickolson scheme is used for diffusion/convective terms. A time-splitting approach is used to solve the mass and momentum equations. Details of the code can be found in [3, 11].

2.1 Channel Flow

First, computations are carried out for fully developed turbulent channel flow at a Reynolds number $Re_\tau = 300$, based on the channel half-height δ and friction velocity u_τ. The profiles are compared to DNS data [4]. A uniform mean pressure gradient is imposed along streamwise direction (x), whereas nonslip boundary conditions are

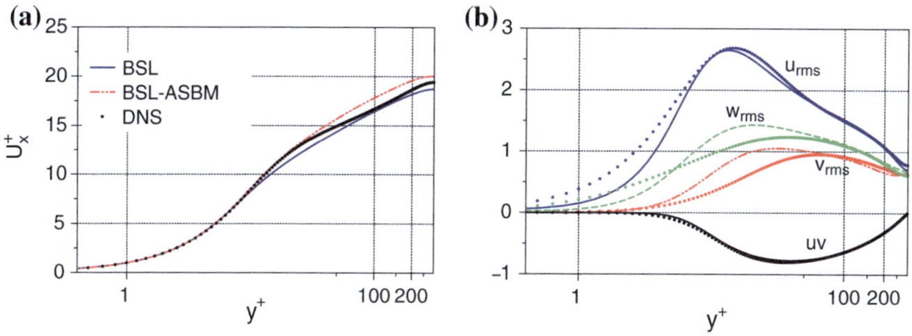

Fig. 1 Model predictions for **a** streamwise mean velocity and **b** rms and shear correlations for BSL and ASBM-BSL closures. Comparison is made to DNS values [4], shown as symbols

applied at the top and bottom walls and periodic conditions along the other directions. The computations involve a $1 \times 200 \times 1$ nonuniform grid. In order to avoid resolution issues in the viscous sublayer, the first cells are placed at a distance of $y^+ \approx 0.25$ form the wall. Results are reported in dimensionless form using the friction velocity u_τ, for example $U_x^+ = U_x/u_\tau$, $u_{rms} = \sqrt{R_{xx}}/u_\tau$, $v_{rms} = \sqrt{R_{yy}}/u_\tau$, $w_{rms} = \sqrt{R_{zz}}/u_\tau$ and $uv = R_{xy}/u_\tau^2$. The normal direction is expressed in the form$y^+ = yu_\tau/\nu$ in order to better highlight the near-wall behavior.

Figure 1a shows that, when compared to the standard BSL model, the ASBM-BSL closure captures more accurately the streamwise mean velocity in the buffer layer. Yet, the coupled closure over-predicts mean velocity values near the channel centerline. The maximum discrepancy between the model predictions and the DNS in the centerline region is estimated to be around 3 %. As shown in Fig. 1b, ASBM-BSL produces fair predictions for the turbulence correlations.

2.2 Flat-Plate Boundary Layer

Next, we consider the case of a zero pressure gradient boundary layer flow, for which many experimental datasets are available. Here, we have chosen to validate the model against the experiments of Loureiro [9], since these are related to the inlet conditions used later for the 'Witch of Agnesi' case. The Reynolds number is $Re_\delta = 4772$, based on free stream velocity u_δ and the boundary thickness δ. The corresponding value of the Reynolds number based on the friction velocity is $Re_\tau = 277$. The mesh involves a $99 \times 59 \times 1$ nonuniform grid, stretched along the transverse direction, whereas uniform spacing is chosen at the streamwise one. In order to ensure adequate resolution, the maximum height of the cells adjacent to the wall was maintained at $y^+ < 0.9$. In contrast to the channel case, here we observe that the correct magnitude of the freestream mean velocity is predicted by both models (Fig. 1a). Figure 1b shows that the ASBM-BSL closure accurately predicts

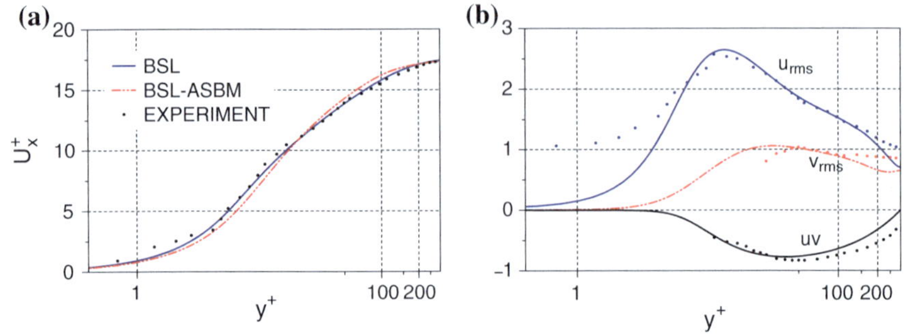

Fig. 2 Model predictions for **a** streamwise mean velocity and **b** rms and shear correlations for BSL and ASBM-BSL closures. Comparison is made to the experiments of Loureiro et al. [9], shown as symbols

both the peak magnitude of the streamwise turbulence intensity and its freestream value, while a fair shear stress profile is obtained. The peak magnitude of the wall-normal turbulence intensity is accurately captured, nevertheless with a displacement of its location closer to the wall (Fig. 2).

3 Witch of Agnesi 2D Bump

3.1 Flow Configuration

Next, we consider the 'Witch of Agnesi' hill that has been studied experimentally by Loureiro et al. [9]. As shown in Fig. 3, the geometry consists of a hill of height H and width $10H$, placed at a distance of $7.5H$ downstream from the inlet and $12.5H$ upstream from the outlet of the computational domain. The hill shape is given by $y_h(x) = H_1[1 + (x/L_H)^2]^{-1} - H_2$,where $H_1 = 75\,\text{mm}$, $H_2 = 15\,\text{mm}$, and $L_H = 150\,\text{mm}$, yielding a hill height of $H = H_1 - H_2 = 60\,\text{mm}$. The origin of the coordinates axes is placed at the middle of the hill. The height and width of the entire computational domain are set to $4H$, and $30H$, respectively.

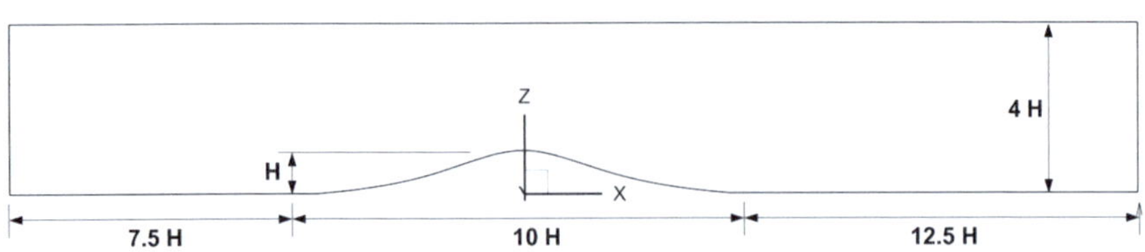

Fig. 3 Geometry for computations of turbulent flow over the 'Witch of Agnesi' hill

3.2 LES Results

A set of LES were motivated by questions raised about the relatively shallow level of the water in the channel, which in turn raised questions about the appropriateness of the free-surface boundary conditions assumed in the RANS simulations. The methodology of inflow generation used in the present work, is a modification of the inflow method of Lund et al. [10]. This method consists of obtaining turbulent data from a plane in a location downstream of the inlet (but not at the outlet of the domain), rescaling the data, and reintroducing them at the inlet. The mesh involves a 220 × 96 × 128 nonuniform grid, sufficiently stretched around the hill region. Profiles of the streamwise component of the mean velocity at different streamwise stations are shown in Fig. 4a–b, for which very good agreement against experimental results is achieved. The agreement is also very good when measuring the size of the recirculation region formed behind the hill and the point where the flow reattaches, as shown in Table 1. Figure 4c–d shows the profiles of the rms values of velocity. This time agreement is also acceptable especially in the near-wall region. Closer to the free surface, higher values of fluctuations were reported experimentally mainly due to the relatively low water depth and the deformation of the free surface, which is not accounted for in the present simulations.

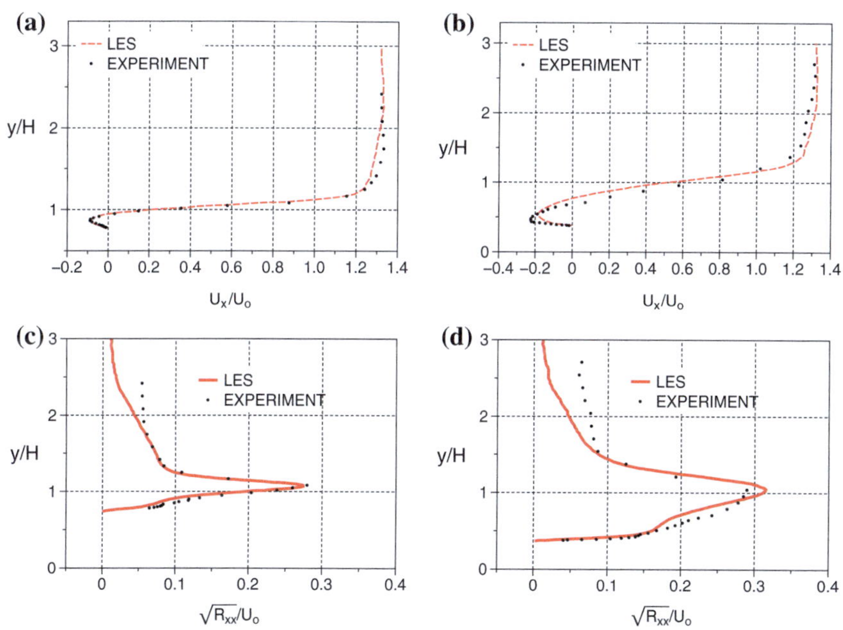

Fig. 4 LES predictions of rms streamwise velocity fluctuations at stations **a** $x/H = 1.25$ and **b** $x/H = 2.50$. Comparison is made to experiments of Loureiro et al. [9]. **c** $x/H = 1.25$. **d** $x/H = 2.50$

RANS Simulations and Results

Based on the good agreement between the LES and experiments, we now proceed to evaluate the model performance in the same flow configuration. The inlet profile is extracted from a flat-plate boundary layer simulation at $Re_\delta = 4772$, where the Reynolds number is based on the freestream velocity U_o and the boundary layer thickness δ. An outflow penalty condition was used at the outlet, a no-slip condition at the bottom wall, a slip condition at the top free surface, and periodic conditions were imposed in the spanwise direction.The mesh involves a $307 \times 119 \times 1$ nonuniform grid, sufficiently stretched around the hill region along both the streamwise and transverse directions. In the figures that follow, all quantities are normalized using the hill height H and the reference inlet freestream velocity U_0. At the inlet, the first cell adjacent to the wall is placed at $y^+ \approx 0.25$.

Stability Issue for the ASBM-v^2-f Closure

When the ASBM closure is coupled to $v^2 - f$ as the carrier model, spurious streaks emerge on leeward side of the hill, as shown in the contour plots of Fig. 5a. For this case, predictions of additional models from the κ-ω family are obtained, both coupled and uncoupled with the ASBM closure. Figure 5b shows the predictions of the hybrid ASBM-BSL model. Clearly, the streaks are absent and smooth converged solutions are obtained. Smooth solutions are obtained for all remaining couplings shown in Table 1, suggesting that the instabilities are caused by the particular coupling in the case of the v^2-f model.

Recirculation Size

Table 1 highlights the predictions related to the recirculation bubble using different models. As shown, the BSL model gives the best results, in both the coupled and the uncoupled case. The essence of this comparison among the models is that the size of the recirculation zone is strongly affected by the choice of the carrier model.

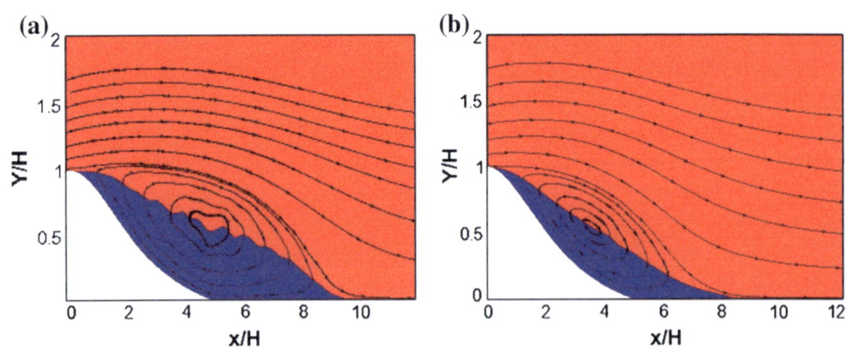

Fig. 5 Contour levels of the streamwise mean velocity for **a** the ASBM—$v^2 - f$ and **b** the ASBM-BSL models on the leeward side of the hill

Table 1 Recirculation zone data

Turbulence closure	$X_s(H)$	$X_r(H)$	$L_s(H)$
$v^2 - f$	0.26	9.18	8.92
k-ω-BSL	0.33	6.67	6.34
k-ω-SST	0.27	8.08	7.81
k-ω-Wilcox	0.27	9.67	9.40
ASBM—$v^2 - f$	0.23	9.70	9.47
ASBM-k-ω-BSL	0.27	7.80	7.53
ASBM-k-ω-SST	0.27	9.1	8.83
ASBM-k-ω-Wilcox	0.27	10.1	9.83
LES	0.50	6.60	6.10
Experiments	0.50	6.67	6.17

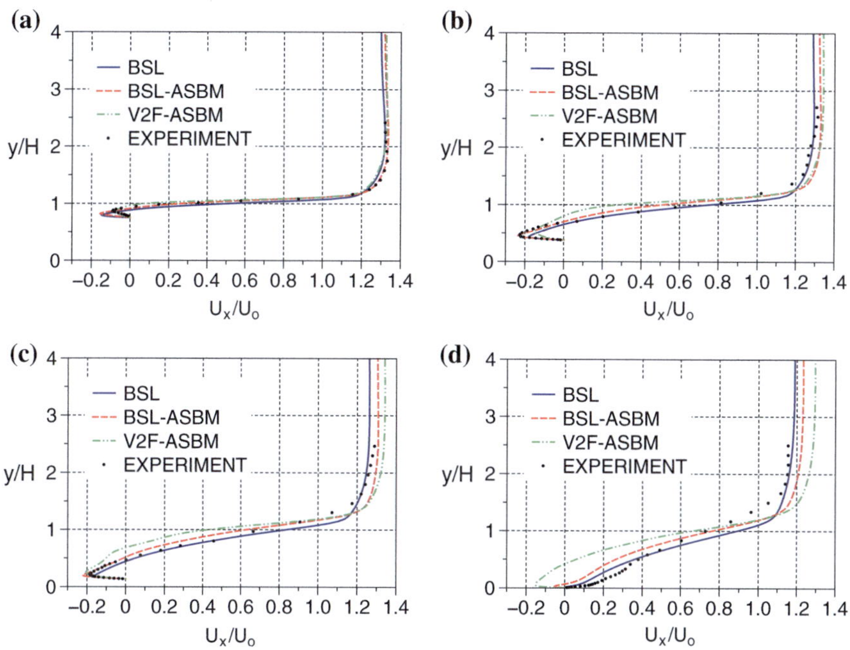

Fig. 6 Model predictions for the streamwise mean velocity U_x at various x-stations for BSL and ASBM-BSL. Comparison is made to experimental values and the predictions of the $v^2 - f$-ASBM model. **a** $x/H = 1.25$. **b** $x/H = 2.5$. **c** $x/H = 3.75$. **d** $x/H = 6.67$

Streamwise Predictions

Figure 6a–d show results for the mean streamwise velocity U_x. In all cases, model predictions are compared to the experimental database. The BSL and ASBM-BSL results are in much better agreement with experiments than ASBM-v^2-f, and this

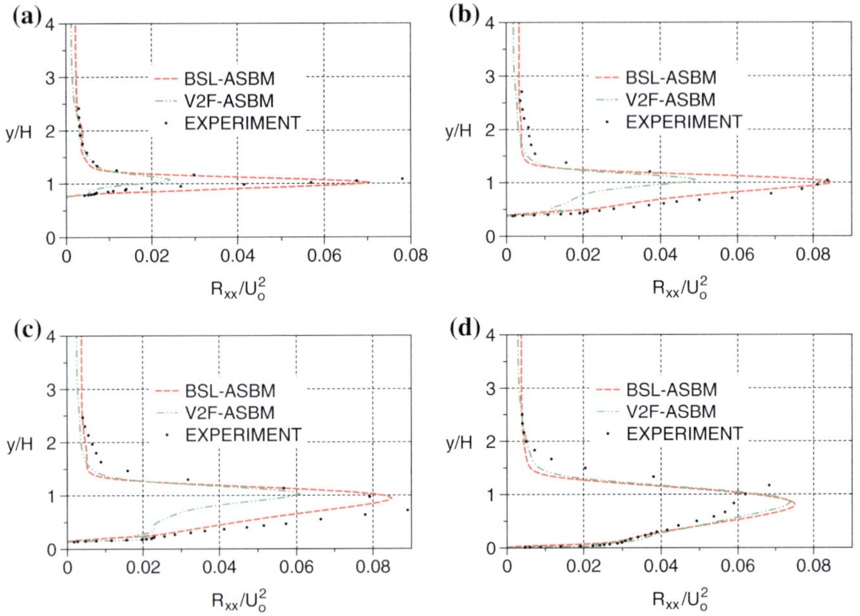

Fig. 7 Same as Fig. 6, but for the streamwise intensity R_{xx}/U_o^2 **a** $x/H = 1.25$. **b** $x/H = 2.5$. **c** $x/H = 3.75$. **d** $x/H = 6.67$

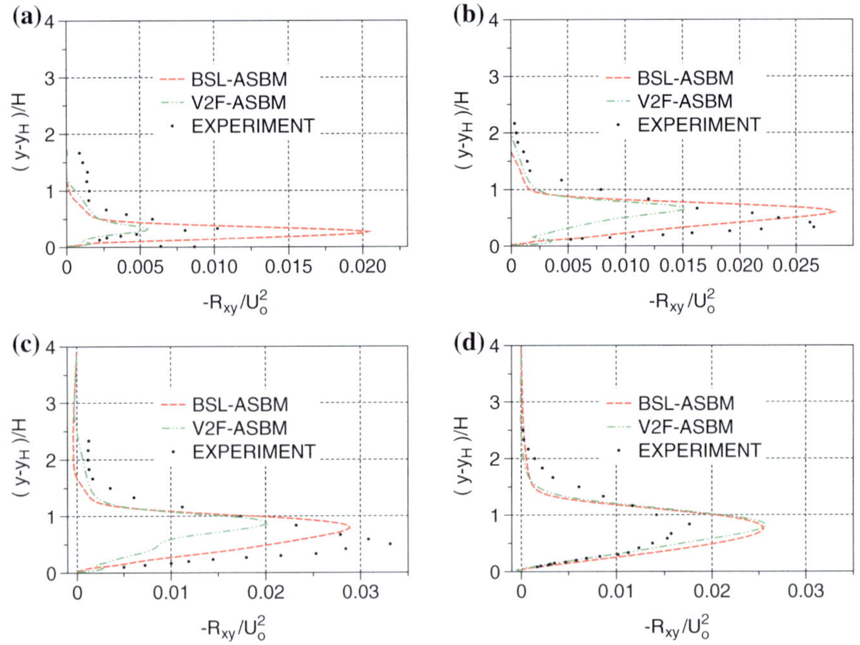

Fig. 8 Same as Fig. 7, but for the turbulent shear stress R_{xy}/U_o^2. **a** $x/H = 1.25$. **b** $x/H = 2.5$. **c** $x/H = 3.75$. **d** $x/H = 6.67$

attributed to the fact that these models are able to capture more accurate the recirculation region.

Figure 7a–d show results for the streamwise intensity R_{xx}/U_0^2. The ASBM-BSL closure exhibits a noticeable improvement relative to ASBM-v^2-f, being able to capture the near-wall peak over the entire range of the recirculation zone, which is consistently under-predicted by ASBM-v^2-f.

Shear Stress Predictions

Figure 8a–d show a comparison of the shear stress predictions of the ASBM-BSL model with the corresponding experimental database and the predictions of ASBM-v^2-f. The ASBM-BSL model again exhibits better agreement with the experiments, especially in the near-wall region where it captures the peak magnitudes. However, at the top of the hill (not shown), we have noticed positive magnitudes for the shear stresses. This deficiency is present for any choice of EVM as a carrier model, implying that this issue is mostly related to the coupling with ASBM.

4 Conclusions

The algebraic structure prediction procedure of ASBM has been successfully coupled with the popular two-equation eddy viscosity models. All hybrid models are able to return fair values for the Reynolds stress tensor. Nevertheless, significant differences are noted in terms both the performance and the stability of the hybrid closures. Based on the results of the current evaluation, the ASBM-BSL hybrid closure seems to be the most promising.

Acknowledgments Support from the US Army International Technology Center and the US Air Force European Office of Aerospace Research and Development (EOARD) under grant W911NF-11-1-0425, and from the Republic of Cyprus through the Research Promotion Foundation Project KOY$\Lambda/\Sigma/0510/01$ is grateful acknowledged.

References

1. B. Aupoix, S. Kassinos, C. Langer, An easy access to the structure based model technology, in *Sixth International Symposium on Turbulence and Shear Flow Phenomena*, Seoul-Korea, 22–24 June 2009
2. J. Benton, Evaluation of v^2-f and ASBM turbulence models for transonic aerofoil RAE2822, in *Proceedings of the WALLTURB International Workshop*, Lille, France, 21–23 April (2009)
3. F. Ham, K. Mattson, G. Iaccarino, Accurate and stable finite volume operators for unstructured flow solvers. Annual Research Briefs, Center for Turbulence Research, NASA Ames/Stanford University, p. 522 (1996) (2006)
4. K. Iwamoto, Y. Suzuki, N. Kasagi, Fully developed 2-D channel flow (2012) http://www.thtlab.t.u-tokyo.ac.jp/DNS. Cited 25 May 2014

5. S.C. Kassinos, W.C. Reynolds, An extended structure-based model based on a stochastic eddy-axis evolution equation. Annual Research Briefs, Center for Turbulence Research, NASA Ames/Stanford University, pp. 133–148 (1995)
6. S.C. Kassinos, C.A. Langer, G. Kalitzin, G. Iaccarino, A simplified structure-based model using standard turbulence scale equations: computation of rotating wall-bounded flows. Int. J. Heat Fluid Flow **27**, 653–660 (2006)
7. S.C. Kassinos, W.C. Reynolds, A structure-based model for the rapid distortion of homogeneous turbulence. Technical Report TF-61, Mechanical Engineering Department, Stanford University (1994)
8. C.A. Langer, W.C. Reynolds, A new algebraic structure-based turbulence model for rotating wall-bounded flows. Technical Report TF-85, Mechanical Engineering Department, Stanford University (1994)
9. J. Loureiro, F. Pinho, A. Silva Freire, Near wall characterization of the flow over a two-dimensional steep smooth hill. Exp. Fluids **42**, 441–457 (2007)
10. T.S. Lund, X. Wu, K.D. Squires, Generation of turbulent inflow data for spatially-developing boundary layer simulations. J. Comput. Phys. **140**, 233–258 (1998)
11. P. Moin, S.V. Apte, Large-eddy simulation of realistic gas turbine combustors. AIAA J. **44**, 698–708 (2006)
12. C. Panagiotou, S. Kassinos, The ASBM-SA turbulence closure: Taking advantage of structure-based modeling in current engineering CFD codes. Int. J. Heat Fluid Fl. **52**, 111–128 (2015)

Computation of High Reynolds Number Equilibrium and Nonequilibrium Turbulent Wall-Bounded Flows Using a Nested LES Approach

Yifeng Tang and Rayhaneh Akhavan

Abstract A new nested LES approach for computing high Reynolds number, wall-bounded turbulent flows is presented. The method is based on nested LES of the full-domain at coarse resolution, coupled with well-resolved LES of a minimal flow unit. The coupling between the two domains is achieved by renormalizing the kinetic energies of components of the mean velocity and the turbulent velocity fluctuations in both domains to that of the minimal flow unit in the near-wall region, and to that of the full-size domain in the outer region, at each time-step. The method can be implemented with a fixed number of grid points, independent of Reynolds number, in any given geometry, and is applicable to both equilibrium and nonequilibrium flows. The proposed method has been applied to LES of equilibrium turbulent channel flow at $1000 \le Re_\tau \le 10{,}000$ and LES of nonequilibrium, shear-driven, three-dimensional turbulent channel flow at $Re_\tau \simeq 2000$. All computations were performed using a spectral patching collocation method, and employed resolutions of $64 \times 64 \times 17/33/17$ in the full-size domain ($L_x \times L_y = 2\pi \times \pi$), and resolutions of $32 \times 64 \times 17/33/17$ and $64 \times 64 \times 17/33/17$ the minimal flow units ($l_x^+ \times l_y^+ \approx 3200 \times 1600$) of equilibrium and non-equilibrium channel flows, respectively. The dynamic Smagorinsky model with spectral cutoff filter was used as the SGS model in all the simulations. The results show that the proposed nested LES approach can predict the friction coefficient to within 5 % of Dean's correlation in equilibrium turbulent channel flow, and the one-point turbulence statistics in good agreement with DNS and experimental data in turbulent channel flow and in shear-driven, three-dimensional turbulent boundary layer.

Y. Tang · R. Akhavan (✉)
Department of Mechanical Engineering, University of Michigan, Ann Arbor, MI 48109-2125, USA
e-mail: raa@umich.edu

Y. Tang
e-mail: yifeng@umich.edu

M. Stanislas et al. (eds.), *Progress in Wall Turbulence 2*,
ERCOFTAC Series 23, DOI 10.1007/978-3-319-20388-1_11

1 Introduction

Large Eddy Simulation (LES) is an approach to computing turbulent flows based on directly computing the large, energy-containing scales of motion and modeling the small scales. Despite numerous advances in LES over the recent years [29], computation of high Reynolds number, complex, nonequilibrium, wall-bounded turbulent flows continues to remain a challenge for LES. The difficulty arises from the stringent resolution requirements of LES in the near-wall region, where the size of energy-containing eddies becomes proportional to their distance from the wall. Resolving these eddies requires $\mathcal{O}(Re_x^{1.8})$ grid points in LES of turbulent boundary layers [4] or $\mathcal{O}(Re_\tau^2)$ grid points in LES of general wall-bounded flows [19]. This is not much different from $\mathcal{O}(Re_\tau^{9/4})$ grid points required for DNS and makes LES of high Reynolds number turbulent flows prohibitively expensive.

A number of wall-modeling approaches have been proposed in recent years to alleviate these stringent resolution requirements. These methods can be broadly classified into two categories: approaches based on formulating off-wall boundary conditions, and approaches based on solving alternative, simplified dynamical equations in the near-wall region. In the first category, a set of approximate boundary conditions are formulated for the first grid point away from the wall. These approximate boundary conditions are either derived from the law of the wall [8, 30], or from integration of simplified momentum equations in the vicinity of the wall [5, 13]. In the second category, the flow in the near-wall region is computed using a set of simplified and less expensive dynamical equations, and matched to the LES solution away from the wall. Commonly used alternative equations include the Reynolds-Averaged Navier-Stokes (RANS) equations [25, 26, 32], and thin boundary layer equations [1, 2]. Approximate off-wall boundary conditions have been shown to provide reasonably accurate predictions in simple, equilibrium wall-bounded flows, but they fail in more complex flows, such as flows with mean three-dimensionality or separation [27]. Methods based on alternative dynamical equations in the near-wall region often face the difficulty of exchanging information between two solutions with disparate spectral content, such as RANS and LES [29]. One well-known manifestation of this difficulty is the "logarithmic law mismatch" observed with most RANS/LES based models in turbulent channel flows [26].

In this study, we present a new LES wall-modeling approach based on nested LES. In this approach, two simultaneous, nested LES are performed at each time-step, one in the full-size domain using a coarse resolution, and the other in a minimal flow unit using a well-resolved grid. Two domains are coupled by renormalizing the kinetic energies of the components of the mean velocity and the turbulent velocity fluctuations in both domains to match those of the minimal flow unit in the near-wall region and those of the full-size domain in the outer region, at each time-step in each wall-normal location. The method can be implemented with a fixed number of grid points, independent of Reynolds number, in any given geometry, and is applicable to both equilibrium and nonequilibrium flows.

The organization of the paper is as follows. In Sect. 2, the nested LES approach is described. In Sect. 3, the results of application of nested LES to equilibrium turbulent channel flow at $Re_\tau \approx 1000$, 2000, 5000, and 10000, and to nonequilibrium, shear-driven, three-dimensional turbulent channel flow at $Re_\tau \approx 2000$ are presented. A summary and conclusion are given in Sect. 4.

2 The Nested LES Approach

It has been recognized, since the pioneering work of Townsend [34, 35], that two classes of structures are involved in the dynamics of wall turbulence: near-wall structures of size proportional to their distance from the wall, and larger, outer-layer structures with size on the order of several boundary layer thicknesses. While earlier studies considered the outer-layer structures to have little influence on the dynamics of wall turbulence [28], more recent studies have shown that these structures have an impact on every aspect of wall turbulence, including the turbulent velocity fluctuations, Reynolds shear stresses, and pressure field in the near-wall region [14, 15, 20, 24], as well as the wall-shear stress [22]. Furthermore, it has been shown that these influences become more pronounced as the Reynolds number increases [23, 31]. More recently, it has been suggested that the influence of the large-scale structures on the near-wall eddies is in the form of an amplitude modulation [24].

From an LES standpoint, the dynamics of the large, outer-layer structures can be computed with a coarse-resolution LES in the full-size domain. However, accurate prediction of these dynamics requires prescription of accurate off-wall boundary conditions, which in turn requires accurate computation of the wall-shear stress and the near-wall turbulence dynamics. Developing such off-wall boundary conditions for general nonequilibrium flows can be extremely difficult. One possible solution may be acquiring this information from a well-resolved LES in a minimal flow unit. However, in the absence of proper large-scale structures, a well-resolved LES in a minimal flow unit cannot predict the correct wall-shear stress [16], even though it can predict the correct normalized turbulence statistics in the near-wall region [11, 17, 18]. Thus, a two-way coupling between the coarse-resolution full-domain LES and the well-resolved LES in the minimal flow unit is required.

The nested LES approach addresses the need for this two-way coupling by performing two coupled, simultaneous LES, one at coarse resolution in the full-size domain, and the other a well-resolved LES in a minimal flow unit. The two-way coupling between the two domains is established by renormalizing the kinetic energies of the components of the mean velocity and turbulent velocity fluctuations in both domains, at each wall-normal location and each time-step, to have a magnitude equal to those in the full-size domain in the outer region, and to those in the minimal flow unit in the near-wall region. In this process, the spectral content and phase information of the rescaled quantities are retained in each domain, and only the magnitudes of the quantities are rescaled.

The application of the proposed nested LES method to homogeneous flows is straightforward. For weakly inhomogeneous flows, one may employ several minimal flow units to provide the near-wall solution in different subdomains, while for strongly inhomogeneous flows, the minimal flow unit may need to be extended along the full length of the direction of inhomogeneity, while still remaining minimal in the nearly homogeneous lateral direction. In either case, the total number of required grid points remains only a weak function of the Reynolds number, thus allowing application of the nested LES approach to high Reynolds number flows.

3 Application to Equilibrium and Nonequilibrium Turbulent Channel Flow

The nested LES approach was applied to LES of equilibrium turbulent channel flow at $Re_\tau \approx 1000$, 2000, 5000, and 10000, and nonequilibrium, shear-driven, three-dimensional turbulent channel flow at $Re_\tau \approx 2000$.

A schematic of the channel and the nested LES geometry is shown in Fig. 1a. In all the simulations, a full-domain channel of size $L_x \times L_y \times L_z = 2\pi h \times \pi h \times 2h$ and a minimal flow unit of size $l_x^+ \times l_y^+ \times l_z^+ \approx 3200 \times 1600 \times 2h^+$ wall units were employed, where h denotes the channel half-height. The Dynamic Smagorinsky Model [12, 21] with spectral cutoff filter was used as the sub-grid scale (SGS) model in all the simulations. For the purpose of the renormalization procedure in the nested LES approach, the near-wall and outer regions were defined as $z/h \leq 0.05$ and $z/h > 0.05$, respectively, in all the simulations.

3.1 Numerical Methods

In both domains, the LES equations were solved using a patching collocation spectral domain-decomposition method [3], employing Fourier series in the streamwise (x) and spanwise (y) directions, and mapped Chebyshev polynomials in the wall-normal

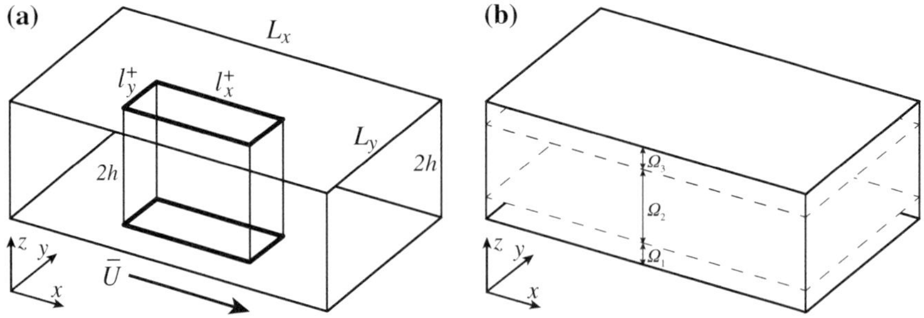

Fig. 1 **a** The full-size domain and the minimal flow unit in the nested LES approach and **b** the domain partition in the patching collocation spectral domain-decomposition method

(z) direction. In the wall-normal direction, the computational domain was partitioned into three nonoverlapping subdomains, Ω_s ($s = 1, 2, 3$), as shown in Fig. 1b. The same grid distribution was used in both the full-size domain and the minimal flow unit in the wall-normal direction, with the interface between the Ω_1/Ω_2 or Ω_3/Ω_2 domains placed at $z^+ \approx 200 - 250$ wall units from the walls in all the simulations. LES computations of equilibrium turbulent channel flow were performed with resolutions of $64 \times 64 \times 17/33/17$ in the full-size domain and $32 \times 64 \times 17/33/17$ in the minimal flow unit at all Reynolds numbers. For the shear-driven turbulent channel flow, resolutions of $64 \times 64 \times 17/33/17$ were used in both the full-size domain and the minimal flow unit.

3.2 Results in Equilibrium Turbulent Channel Flow

Table 1 summarizes the skin-friction coefficients predicted by the nested LES approach compared to Dean's correlation [7]. The predicted skin-friction coefficients were all within 5 % of Dean's correlation. To show the impact of the nested LES approach, the simulations at $Re_\tau \approx 2000$ were repeated in the full-size domain and in the minimal flow unit as uncoupled simulations, at the same resolutions used in the nested LES. The skin-friction coefficients predicted in these two cases had much larger errors than that from the nested LES approach, with the uncoupled coarse-resolution full-domain LES giving errors of 32 % and the uncoupled resolved LES in the minimal flow unit giving errors of 26 % in the skin-friction coefficient compared to Dean's correlation, as shown in Table 1.

Figure 2 shows the comparison of the one-point turbulence statistics predicted by the nested LES approach with available DNS [9, 14], and experimental data [6]. The mean velocity profiles predicted by the nested LES approach are in good agreement with DNS and experimental data, and exhibit the correct behavior in both the logarithmic layer and the wake region. Good agreement is also observed for the rms velocity fluctuations and the Reynolds shear stress. All turbulent stresses reported

Table 1 Predicted skin-friction coefficients using the nested LES approach in comparison with Dean's correlation [7]

$Re_{\tau,nominal}$	Re_b	C_f (present study)	C_f (Dean's)	Error (%)
1,000	40,000	4.904×10^{-3}	5.162×10^{-3}	−5.0
2,000	85,333	4.092×10^{-3}	4.271×10^{-3}	−4.2
5,000	266,667	3.308×10^{-3}	3.212×10^{-3}	+2.9
10,000	533,333	2.688×10^{-3}	2.701×10^{-3}	−0.5
$2{,}000^C$	85,333	2.887×10^{-3}	4.271×10^{-3}	−32.4
$2{,}000^M$	85,333	3.162×10^{-3}	4.271×10^{-3}	−26.0

^{C}uncoupled coarse-resolution LES in the full-size domain
^{M}uncoupled resolved LES in the minimal flow unit

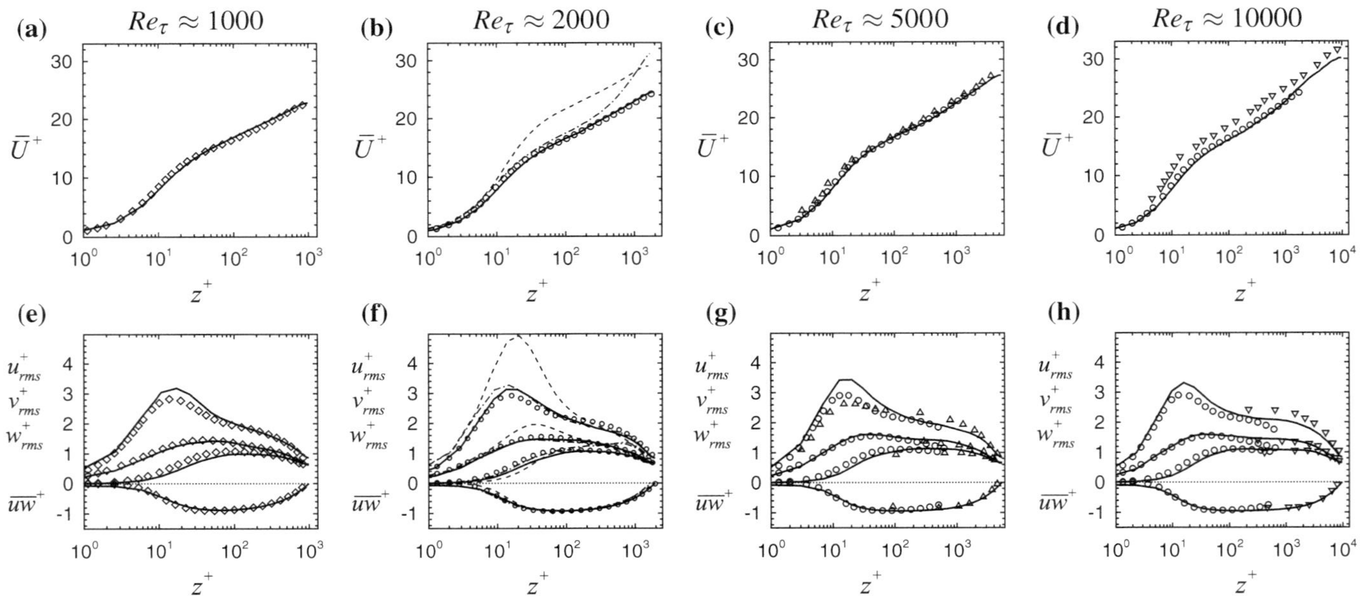

Fig. 2 The mean velocity, turbulent intensities, and Reynolds shear stresses predicted by nested LES at $Re_\tau \approx 1000$, 2000, 5000, and 10,000. —, nested LES; – –, coarse-resolution LES in full-domain at $Re_\tau \approx 2000$; – · –, resolved LES in minimal flow unit at $Re_\tau \approx 2000$; ⋄, DNS at $Re_\tau \approx 950$ [9]; ∘, DNS at $Re_\tau \approx 2000$ [14]; △, experiment at $Re_\tau \approx 5000$ [6]; ▽, experiment at $Re_\tau \approx 8500$ [6]. The DNS data at $Re_\tau \approx 2000$ have been added to plots at $Re_\tau \approx 5000$ and 10,000 to compensate for the lack of availability of detailed experimental data at these higher Reynolds numbers

in this study represent the true Reynolds-averaged stresses and were reconstructed by extending the one-dimensional (1D) energy spectra computed in LES using a theoretical form of the 1D energy spectra for wall flows, integrating the area under the extended spectra to recover the SGS kinetic energy [33], and recovering the individual components of Reynolds stresses using the formulation of Winckelmans et al. [36]. Unlike most existing wall-modeling approaches, in which only the statistics in the outer layer are available, the nested LES approach provides accurate statistics in both the near-wall region and the outer layer. Both the magnitudes and locations of the peaks of turbulent stresses are all well-predicted by nested LES. In contrast, LES performed in the two uncoupled full-size and minimal flow domains at $Re_\tau \approx 2000$ yield one-point statistics in poor agreement with the DNS data.

Analysis of the energy spectra shows that the two-way coupling procedure in the nested LES approach results in correction of the spectral energy distribution in both domains and this is responsible for accurate predictions of the turbulence statistics by the nested LES approach in both the near-wall and outer regions.

3.3 Results in Shear-Driven Three-Dimensional Turbulent Channel Flow

To investigate the performance of the nested LES approach in nonequilibrium flows, the method was applied to a shear-driven, three-dimensional turbulent channel flow subjected to transverse strain produced by impulsive motion of the bottom wall. The LES was set up to reproduce the experimental conditions of Driver and Hebbar [10], shown in Fig. 3, in which an initially two-dimensional turbulent boundary layer (TBL) is sheared by sudden spanwise motion of the wall at speed $V_s = U_\infty$, where U_∞ denotes the free-stream velocity in the original TBL. After the three-dimensional flow travels approximately 40.5 boundary layer thickness downstream, the spanwise shear is removed and the flow reverts back to a two-dimensional flow.

In the present study, the expensive simulation of the long boundary layer was avoided by performing the simulations in a turbulent channel flow at comparable

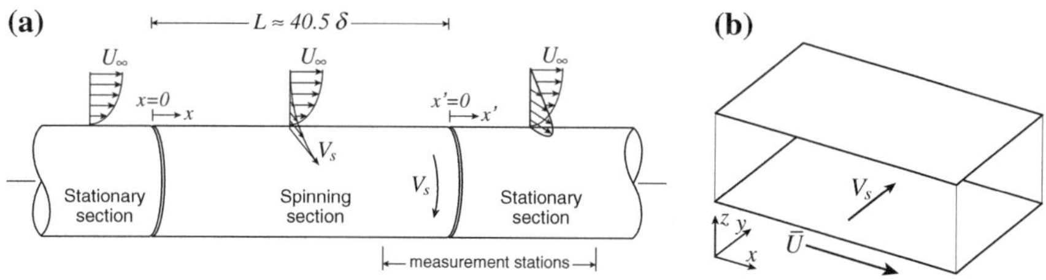

Fig. 3 Configurations of the shear-driven three-dimensional boundary layer in **a** experimental setup with spinning cylinder [10] and **b** present simulation as a shear-driven, three-dimensional channel flow

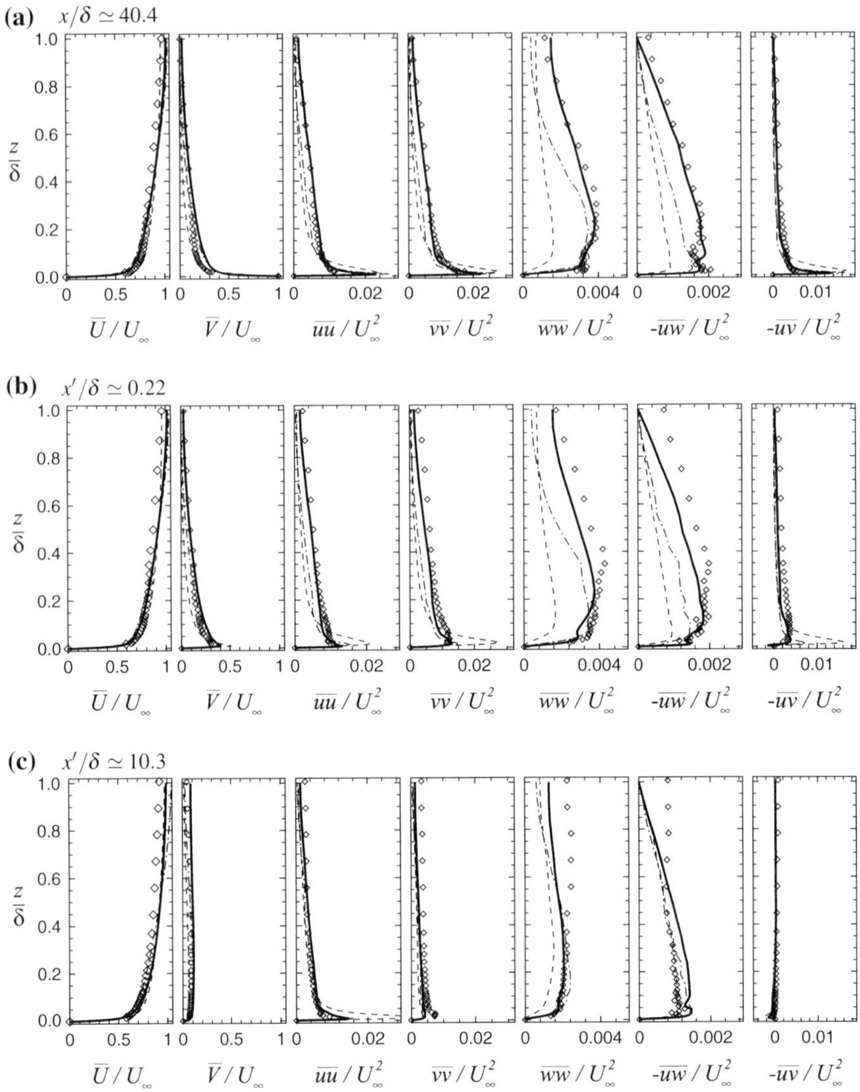

Fig. 4 Evolution of the mean velocities and Reynolds stresses at (**a**) the end of the sheared region, (**b**) the early recovery region, and (**c**) the final recovery region. x/δ denotes the downstream distance measured from the leading edge of the sheared section, and x'/δ denotes the downstream distance measured from the leading edge of the stationary recovery section (see Fig. 3). In plotting the LES results, δ and U_∞ were replaced with h and U_c, respectively. —, nested LES method; – –, coarse-resolution LES in the full-size domain; – · –, resolved LES in the minimal flow unit; ⋄, experimental data of Ref. [10]

$Re_\tau \simeq 2000$ of the boundary layer flow at the end of the sheared region. The lower wall of the channel was impulsively moved in the spanwise direction at a speed of $V_s = U_c$, where U_c denotes the centerline velocity of the base turbulent channel flow. This spanwise motion was later stopped at $t = 63.5$, corresponding to $x/\delta \approx 40.5$ based on Taylor's hypothesis. The spatial development of the flow in the boundary layer was therefore simulated as temporal evolution of the flow in the channel, and downstream distances in the experiments were converted to time in LES of the channel flow. The turbulence statistics in LES were obtained by averaging over the wall-parallel planes and over five independent realizations of the flow.

Figure 4 shows the comparison between the LES predictions and experimental data for the evolution of the wall-parallel mean velocity profiles and Reynolds stresses at the end of the sheared region ($x/\delta \simeq 40.4$), at the beginning of the recovery region ($x'/\delta \simeq 0.22$), and in the final stage of the recovery region ($x'/\delta \simeq 10.3$), where x/δ denotes the downstream distance measured from the leading edge of the sheared section, and x'/δ denotes the downstream distance measured from the leading edge of the stationary recovery section (see Fig. 3). At the end of the sheared region (Fig. 4a), the mean velocities and Reynolds stresses have reached a new equilibrium, in which $\overline{vv}$ and $\overline{uv}$ have become highly pronounced due to the spanwise wall-shear. These features were accurately predicted by nested LES, as shown in Fig. 4a. In the early stages of recovery (Fig. 4b), $\overline{uu}$, $\overline{vv}$, and $\overline{uv}$ quickly drop in the near-wall region. These features were again accurately predicted by nested LES, as shown in Fig. 4b. Toward the end of the recovery region (Fig. 4c), $\overline{V}$ and $\overline{uv}$ have damped out to zero, while the other components of Reynolds stresses relaxed to their equilibrium states in the non-sheared, two-dimensional base flow. Nested LES approach successfully captures all these flow features of the shear-driven boundary layer and provides excellent agreement with the experimental data throughout the sheared, and recovery regions. In contrast, LES performed in the two uncoupled full-size and minimal flow domains predicts Reynolds stresses in poor agreement with the experimental data. These results show that the nested LES approach offers the same advantages in nonequilibrium flows as in the equilibrium flows, and provides a method for computing high Reynolds number nonequilibrium turbulent wall flows at low computational cost.

4 Summary and Conclusions

A nested LES wall-modeling approach for computing high Reynolds number, wall-bounded turbulent flows was presented. This approach keeps the required number of grid point manageable by performing nested LES of the full-domain at coarse resolution, coupled with well-resolved LES of a minimal flow unit in the course of the simulation. At each time-step, the coupling procedure renormalizes the kinetic energies of components of the mean velocity and the turbulent velocity fluctuations in the two domains to that of the minimal flow unit in the near-wall region, and to that of the full-size domain in the outer region. The method is applicable to both equilibrium and nonequilibrium flows and can be implemented with a fixed number of grid points, independent of Reynolds number, in any given geometry. The proposed nested LES approach has been applied to LES of high Reynolds number equilibrium turbulent channel flow and nonequilibrium, shear-driven, three-dimensional turbulent channel flow. The results show excellent agreement with available DNS and experimental data in both equilibrium turbulent channel flow and nonequilibrium, shear-driven, three-dimensional boundary layer. In addition, the present method accurately predicts the statistics in both the near-wall region and the outer layer, and therefore provides a unique advantage in application where detailed near-wall information may be crucial. In conclusion, the proposed nested LES approach provides an accurate and affordable wall-modeling approach for LES of high Reynolds number equilibrium and nonequilibrium wall-bounded turbulent flows.

References

1. E. Balaras, C. Benocci, U. Piomelli, Two-layer approximate boundary conditions for large-eddy simulations. AIAA J. **34**(6), 1111–1119 (1996)
2. W. Cabot, P. Moin, Approximate wall boundary conditions in the large-eddy simulation of high Reynolds number flow. Flow Turbul. Combust. **63**(1), 269–291 (1999)
3. C. Canuto, M.Y. Hussaini, A. Quarteroni, T.A. Zang, *Spectral methods: evolution to complex geometries and applications to fluid dynamics* (Springer, Berlin, 2007)
4. D.R. Chapman, Computational aerodynamics, development and outlook. AIAA J. **17**(12), 1293–1313 (1979)
5. D. Chung, D.I. Pullin, Large-eddy simulation and wall modelling of turbulent channel flow. J. Fluid Mech. **631**, 281–309 (2009)
6. G. Comte-Bellot, Turbulent flow between two parallel walls. Ph.D thesis, University of Grenoble (1963)
7. R.B. Dean, Reynolds number dependence of skin friction and other bulk flow variables in two-dimensional rectangular duct flow. J. Fluids Eng. **100**(2), 215–223 (1978)
8. J.W. Deardorff, A numerical study of three-dimensional turbulent channel flow at large Reynolds numbers. J. Fluid Mech. **41**, 453 (1970)
9. J.C. del Álamo, J. Jimenez, P. Zandonade, R.D. Moser, Scaling of the energy spectra of turbulent channels. J. Fluid Mech. **500**, 135–144 (2004)
10. D.M. Driver, S.K. Hebbar, Three-dimensional turbulent boundary layer flow over a spinning cylinder. NASA Technical Memorandum 102240 (1991)
11. O. Flores, J. Jimenez, Hierarchy of minimal flow units in the logarithmic layer. Phys. Fluids **22**(7), 071704 (2010)
12. M. Germano, U. Piomelli, P. Moin, W.H. Cabot, A dynamic subgrid-scale eddy viscosity model. Phys. Fluids A **3**(7), 1760–1765 (1991)
13. G. Hoffmann, C. Benocci, Approximate wall boundary conditions for large eddy simulations, in *Advances in Turbulence V*, ed. by R. Benzi (Kluwer, Dordrecht, 1995), pp. 222–228
14. S. Hoyas, J. Jimenez, Scaling of the velocity fluctuations in turbulent channels up to Re-tau = 2003. Phys. Fluids **18**(1), 011702 (2006)
15. N. Hutchins, I. Marusic, Large-scale influences in near-wall turbulence. Philos. Trans. R. Soc. A **365**(1852), 647–664 (2007)
16. Y.Y. Hwang, Near-wall turbulent fluctuations in the absence of wide outer motions. J. Fluid Mech. **723**, 264–288 (2013)
17. J. Jimenez, P. Moin, The minimal flow unit in near-wall turbulence. J. Fluid Mech. **225**, 213–240 (1991)

18. J. Jimenez, A. Pinelli, The autonomous cycle of near-wall turbulence. J. Fluid Mech. **389**, 335–359 (1999)
19. J. Jimenez, Computing high-Reynolds-number turbulence: will simulations ever replace experiments? J. Turbul. **4**, N22 (2003)
20. J. Jimenez, S. Hoyas, Turbulent fluctuations above the buffer layer of wall-bounded flows. J. Fluid Mech. **611**, 215–236 (2008)
21. D.K. Lilly, A proposed modification of the Germano-subgrid-scale closure method. Phys. Fluids A **4**(3), 633–635 (1992)
22. I. Marusic, W.D.C. Heuer, Reynolds number invariance of the structure inclination angle in wall turbulence. Phys. Rev. Lett. **99**(11), 114504 (2007)
23. I. Marusic, B.J. McKeon, P.A. Monkewitz, H.M. Nagib, A.J. Smits, K.R. Sreenivasan, Wall-bounded turbulent flows at high Reynolds numbers: recent advances and key issues. Phys. Fluids **22**(6), 065103 (2010)
24. R. Mathis, N. Hutchins, I. Marusic, Large-scale amplitude modulation of the small-scale structures in turbulent boundary layers. J. Fluid Mech. **628**, 311–337 (2009)
25. F.R. Menter, Y. Egorov, A scale-adaptive simulation model using two-equation models. AIAA paper 2005-1095 (2005)
26. N.V. Nikitin, F. Nicoud, B. Wasistho, K.D. Squires, P.R. Spalart, An approach to wall modeling in large-eddy simulations. Phys. Fluids **12**(7), 1629–1632 (2000)
27. U. Piomelli, Wall-layer models for large-eddy simulations. Prog. Aerosp. Sci. **44**(6), 437–446 (2008)
28. S.K. Robinson, Coherent motions in the turbulent boundary-layer. Annu. Rev. Fluid Mech. **23**, 601–639 (1991)
29. P. Sagaut, S. Deck, Large eddy simulation for aerodynamics: status and perspectives. Philos. Trans. R. Soc. A **367**(1899), 2849–2860 (2009)
30. U. Schumann, Subgrid scale model for finite-difference simulations of turbulent flows in plane channels and annuli. J. Comput. Phys. **18**(4), 376–404 (1975)
31. A.J. Smits, B.J. McKeon, I. Marusic, High-Reynolds number wall turbulence. Annu. Rev. Fluid Mech. **43**, 353–375 (2011)
32. P.R. Spalart, W.H. Jou, M. Strelets, S.R. Allmaras, Comments on the feasibility of LES for wings, and on a hybrid RANS/LES approach. In: *Proceedings of 1st AFOSR International Conference on DNS/LES*, Ruston (1997), pp. 137–147
33. Y. Tang, R. Akhavan, Recovery of subgrid-scale turbulence kinetic energy in LES of channel flow, in *Advances in Turbulence XII*, ed. by B. Eckhardt (Springer, Berlin, 2009), p. 949
34. A.A. Townsend, Equilibrium layers and wall turbulence. J. Fluid Mech. **11**(1), 97–120 (1961)
35. A.A. Townsend, *The Structure of Turbulent Shear Flow* (Cambridge University Press, Cambridge, 1976)
36. G.S. Winckelmans, H. Jeanmart, D. Carati, On the comparison of turbulence intensities from large-eddy simulation with those from experiment or direct numerical simulation. Phys. Fluids **14**(5), 1809–1811 (2002)

An Attempt to Describe Reynolds Stresses of Turbulent Boundary Layer Subjected to Pressure Gradient

Artur Dróżdż and Witold Elsner

Abstract The paper is concerned with the issue of scaling of Reynolds stresses and the phenomenon of the outer peak of velocity fluctuations, which appears in adverse pressure gradient conditions. For this purpose, experimental data from favorable and adverse pressure gradient turbulent boundary layers, for Reynolds number varying from $Re_\theta \approx 2300 \div 6200$, have been analyzed. At pressure gradient conditions, the self-similarity cannot be obtained using the scale, which is constant across the boundary layer thickness. In this paper, we also propose a modification of the Alfredsson et al. (Eur J Mech B/Fluids 36, 167–175, 2012, [1]) expression, which is dedicated to ZPG flows. The new formulation, utilizing the shape factor H and pressure gradient parameter Λ, allows an extension of the validity of Alfredsson et al. proposal for pressure gradient flows.

1 Introduction

Recent studies deal with the scaling problem of Reynolds stresses for zero pressure gradient flows and for high Reynolds number. Particular attention is also given to the appearance of a second, so-called outer maximum of the $\overline{uu}$ for sufficiently high Reynolds numbers. The physical basis of the outer peak appearance is still not well understood. Marusic et al. [13] and Alfredsson et al. [1] state that the outer peak appears for high Reynolds number and for high enough Re even overcomes the inner one. Mathis et al. [14], using velocity signal-scale decomposition, demonstrated that the appearance and growth of the outer peak is due to the rise of energy of large-scale motion. Later, Monty et al. [15] showed the similar phenomena for APG turbulent boundary layers proving also that it was mainly a result of large-scale motions. Marusic et al. [13] observed that the intensity of the outer peak grows much more

A. Dróżdż (✉) · W. Elsner
Institute of Thermal Machinery, Czestochowa University of Technology, Armii Krajowej 21, 42-201 Czestochowa, Poland
e-mail: arturdr@imc.pcz.czest.pl

W. Elsner
e-mail: welsner@imc.pcz.czest.pl

M. Stanislas et al. (eds.), *Progress in Wall Turbulence 2*,
ERCOFTAC Series 23, DOI 10.1007/978-3-319-20388-1_12

rapidly than the inner peak and at sufficiently high Reynolds number it may overcome the inner peak. Turbulent boundary layers subjected to adverse pressure gradient (referred hereinafter to as APG) or those subjected to nonzero pressure gradient (favorable—FPG, zero—ZPG and APG) often with surface curvature are considered the most difficult to describe. Performing bursting process analysis Dróżdż and Elsner [4] showed that the reason for the appearance of the outer peak in the APG could also be traced to the continuous growth of the trajectory inclination of vortices. The systematic increase of inclination of the trajectory of the vortex with the pressure gradient may cause, at a certain streamwise location, clustering of hairpin vortex packets in the outer layer, and intensification of large-scale motion. It appears also that rms distribution is dependent on the Reynolds number [12], which makes it more difficult to find the proper scaling.

Recently, Alfredsson et al. [1] proposed a scaling method for streamwise turbulence intensity for ZPG flows (AOS scaling) which takes into account this dependency assuming that the streamwise turbulence intensity u'/U in the outer region appears to collapse on a straight line with a functional dependence on U/U_∞. They showed that linear distribution is invariant with Reynolds number at least for cases analyzed by the author. The parameters of that line for turbulent boundary layers are described by the following equation:

$$\frac{u'}{U} = 0.286 - 0.255\frac{U}{U_\infty}. \tag{1}$$

To obtain the collapse of data in the near wall region Alfredsson et al. [1] proposed the difference function, where U/U_∞ was replaced by U^+ in order to obtain complete u' fluctuation velocity profile.

$$\Delta(U^+) = \frac{u'}{U} - \left(0.286 - 0.255\frac{U}{U_\infty}\right). \tag{2}$$

The resulting new composite profile for the streamwise turbulence intensity is valid for canonical wall-bounded turbulent flows, when it is combined with any composite velocity profile for the mean streamwise velocity component. When this model is extrapolated toward higher Reynolds numbers, it exhibits properly the outer peak in the streamwise turbulence intensity profile.

The aim of the study was devoted to scaling of the streamwise Reynolds stresses. Finding the proper length and velocity scales for Reynolds stresses is very difficult, especially for nonequlibrium flows. The paper presents an attempt to propose a new approach for pressure gradient flows and its verification based on experimental data. The data used in the analysis comes from the experiment performed for Reynolds number varying from $Re_\theta = 2300 \div 6200$ and the pressure gradient conditions representative for practical turbomachinery flows, where sudden change from favorable (FPG) to APG occurs [8].

2 Experimental Set up

The experiment was performed in an open-circuit wind tunnel, where the turbulent boundary layer developed along the flat plate, which was 2807 mm long and 250 mm wide. The test section is located in the rear part of the wind tunnel. The upper wall of the test section was shaped according to the assumed distribution of the pressure gradient corresponding to the conditions encountered on axial compressor blades. The facility was equipped with a computer-controlled 2D traversing system (in streamwise and wall-normal direction). The traverse carriage was driven over a maximum displacement of 180 mm by a servo-motor. The uncertainty on the driver step was 0.001 mm with the smallest step equal to 0.01 mm.

Static pressure measurements were done using 70 pressure holes of 0.5 mm diameter, drilled along the streamwise direction, from 2067 to 2767 mm of the x coordinate. The spacing of pressure taps was equal to 10 mm. Distributions of static pressure and pressure gradient are presented in Fig. 1. The pressure distribution is typical of a turbomachinery case, where after a short region of zero pressure gradient the flow accelerates (from $x_s = 197$ mm) and then (from $x_s = 427$ mm) decelerates. It is seen that pressure gradient values are within the range of $-0.27 \div 0.28$ Pa/mm. To have reference friction velocity along the flow, the fringe skin friction (FSF) technique was also applied [7].

Velocity profiles were measured with a single hot-wire anemometry probe of diameter $d = 3\,\mu$m and length $l = 0.4$ mm (Dantec Dynamics 55P31). The probes were combined with the DISA 55M hot-wire bridge connected to a 14 bit PC card. Acquisition was maintained at frequency 50 kHz with 10 s sampling records. For the assumed sampling frequency the non-dimensional inner scale representation was $f^+ \approx 1$. It is consistent with the assumption of Hutchins et al. [10], stating that for the proper anemometer/probe response cutoff must be in the range of $f^+ > 1/3(t^+ < 3)$. The l/d value does not fulfill the recommendation of Ligrani and Bradshaw [11], however Dróżdż and Elsner [6] showed that the magnitude of the inner peak ($y^+ \approx 15$) increased by 10 % for a miniature probe in comparison with a standard wire probe of $l = 1.25$ mm and $d = 5\,\mu$m.

The closest wall position of the hot-wire probe was determined using the mirrored image. As the flat plate was made of plexiglass it can be treated as a nonconducting wall and wall correction was not used. The positions of 24 measuring traverses are shown in Fig. 1. The distances of traverses from the inlet plane, the corresponding dimensionless distances $Sg = x_s/L$, where L is the length of the test section ($L = 1067$ mm). The favorable pressure gradient covers 8 locations and the adverse pressure gradient 16 locations (dot lines in Fig. 1).

Flow parameters determined at the inlet plane, located in the zero pressure gradient area are the mean velocity in core flow $U_\infty \approx 15$ m/s and turbulence intensity $Tu = 0.4$ %. It may be noticed that tripping boundary layer at the leading edge of the flat plate allowed us to obtain a relatively high value of the characteristic Reynolds number equal $Re_\theta \approx 2500$ at the inlet plane ($Sg = 0$).

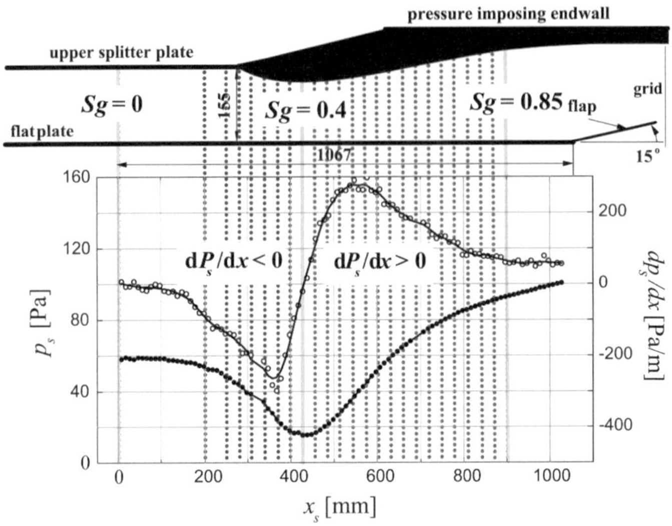

Fig. 1 The shape of the channel *upper* wall with corresponding static pressure and pressure gradient distributions

3 Scaling of Streamwise Reynolds Stresses

Before analyzing the Reynolds stress scaling, the major parameters of the boundary layer are presented. Figure 2 shows the downstream evolution of friction velocity u_τ and shape factor H. Distributions are typical for a turbulent boundary layer with nonzero pressure gradient conditions, the values of u_τ and H show that the turbulent boundary layer has not yet separated.

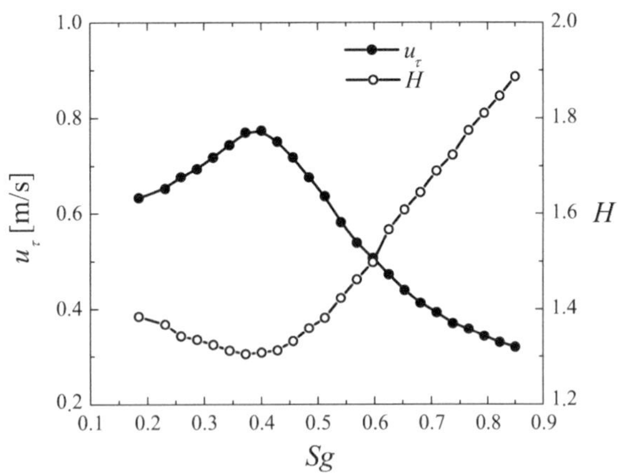

Fig. 2 Distribution of the friction velocity u_τ and shape factor H

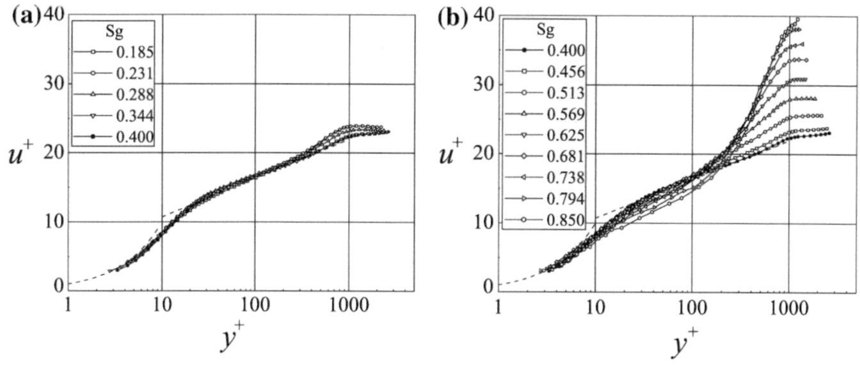

Fig. 3 Developement of the mean streamwise velocity U in the FPG (**a**) and APG (**b**) and fluctuating velocity u' in the FPG (**c**) and APG (**d**)

Fig. 4 Streamwise turbulence intensity profiles scaled on u_τ

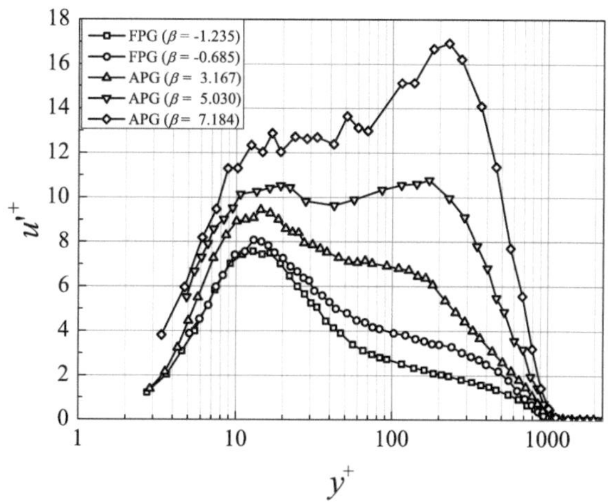

The distributions of mean velocity profiles for the FPG and APG regions in semi-logarithmic coordinates are shown in Fig. 3a–b. In both figures the same profile $Sg = 0.400$ (bold black line) is shown as a reference, for which $\mathrm{d}p_S/\mathrm{d}x \approx 0$. It is worth noting that in the APG between the inner and outer layers the build-up of $\mathrm{d}U/\mathrm{d}y$ gradient is observed, and is accompanied by a larger drop of velocity in the inner layer in comparison with the outer one.

Figure 4 shows the streamwise turbulence intensity scaled on u_τ for selected cross-sections characterized by different values of Clauser pressure gradient parameter β. As can be seen for $\beta > 5$ the outer peak ($y^+ \approx 200$) overcomes the inner one ($y^+ \approx 15$) which is in agreement with the data of Nagano et al. [16] and Monty et al. [15]. It can also be concluded that the inner peak is no longer present for the strong APG. Summing up, it seems that an analyzed turbulent boundary layer reacts differently under FPG and APG conditions. Analysis of the mean flow field shows that the APG causes a strong reaction of velocity fluctuation generally by damping and enhancing of the inner and outer peaks respectively, which coincides well with changes of $\mathrm{d}U/\mathrm{d}y$ gradient.

Dróżdż and Elsner [5], among others, confirmed the role of large-scale motion, in that case, by calculating the energy spectra E scaled on u_τ, as a function of the y^+ for ZPG and APG conditions for the analyzed boundary layer. Despite the relatively small-scale separation of the inner and outer peak, the latter one clearly appears at $y^+ = 120$. The outer peak is formed for the large scales ($\lambda_x \approx 3\delta$), which indicates similar phenomenon to that observed recently by Harun et al. [9] but for higher Reynolds number. It can be concluded that in the presence of APG, the second peak of turbulent velocity fluctuations appears due to the energy increase of large-scale vortices present in the outer region, which indicates the more pronounced contribution of the outer region to the downstream development of the turbulent boundary layer.

As has been already stated, the scaling of Reynolds stresses has been attempted by many authors and for the present analysis the modified scaling for streamwise Reynolds stress $\overline{uu}$, based on the AOS approach is proposed. As per the analysis performed in [8], original AOS scaling could not be treated as universal, especially for pressure gradient flows. To improve its universality, we consider applying the shape factor $H = \delta^*/\theta$. The scaling by the shape factor seems to be beneficial for boundary layers with a pressure gradient because H depends weakly on Reynolds number and strongly upon the pressure gradient. Furthermore, for the APG case velocity decreases at the given distance for the wall, while the shape factor increases.

The profiles of $(\overline{uu}/(U^2H))^{1/2}$ for the present experiment are presented in Fig. 5. It is seen that the data converge, although this convergence takes place for three lines of different slope. It is suggested [8] that these differences are due to the sequence of ZPG, FPG and APG, present in the experiment. The boundaries among these states are defined by locations of distinct minimum or maximum pressure gradient (see Fig. 1).

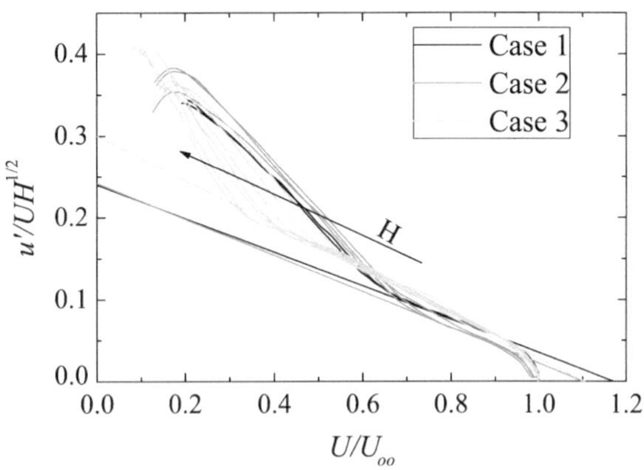

Fig. 5 Square root of streamwise Reynolds stress $(\overline{uu}/U^2H)^{1/2}$ scaled with modified AOS scaling

Following Alfredsson et al. [1] argument, it was decided to propose the modified version of the difference function (Eq. 2) for streamwise Reynolds stresses $\overline{uu}$ written in the following form:

$$\Delta_H(U^+H) = \frac{\overline{uu}}{U^2H} - \left(A + B\frac{U}{U_\infty}\right)^2 \tag{3}$$

where A and B where derived for three regions (for comparison see Eq. 2). As result from Table 1 the values of A and B depend upon the sequence of upstream pressure gradient conditions.

Application of the new difference function for our experimental data is shown in Fig. 6, where similar behavior to the one obtained by Alfredsson et al. [1] may be observed. All profiles correspond to the three flow states included in Table 1. The shape of the complete difference function varies for flows with sudden changes of pressure gradient. Case 3 is divided into three groups of different shapes. What is interesting is that in each group the constant pressure gradient parameter Λ, introduced by Castillo and George [2] defined as:

$$\Lambda = \frac{\delta}{\rho U_\infty^2 d\delta/dx}\frac{dp_\infty}{dx}, \tag{4}$$

is preserved.

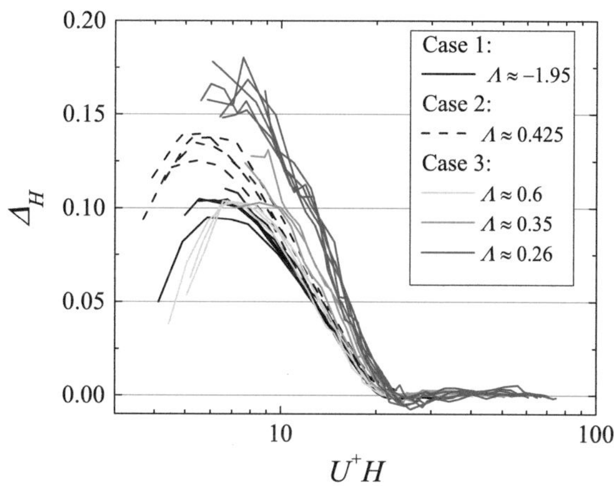

Fig. 6 Complete difference function $\Delta_H(U^+H)$ for FPG and APG regions

Table 1 Parameters of the difference function

Case	Conditions of PG	B	A
1	FPG following ZPG, $d^2P/dx^2 < 0$	−0.205	0.24
2	APG following FPG, $d^2P/dx^2 > 0$	−0.22	0.24
3	APG following FPG, $d^2P/dx^2 < 0$	−0.27	0.30

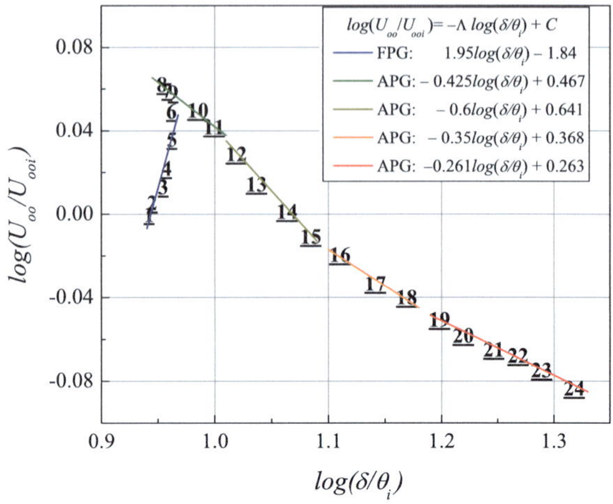

Fig. 7 Pressure gradient parameter Λ: distribution of $\log(U_\infty/U_{\infty i})$ versus $\log(\delta/\theta_i)$ b)

Finally, one can see that the profiles are grouped in five bundles, where each corresponds to a different local equilibrium state, i.e., local equilibrium, defined by constant pressure gradient parameter Λ (see Fig. 7). In each local equilibrium, the profiles collapse well across boundary layer thickness.

It may be assumed that the collapse occurs because in these regions the self-similar profiles of velocity deficit were obtained when they were scaled by Zagarola-Smits scaling [3]. While looking at the lines in Fig. 6, it can be seen that the maxima of difference function decreases with the increasing of pressure gradient parameter Λ. Taking into account this behavior the further modification of the relation (Eq. 2) is proposed:

$$\Delta_H(U^+H)\Lambda^{n/2} = \frac{\overline{uu}}{U^2H} - \left(A + B\frac{U}{U_\infty}\right)^2 \tag{5}$$

where n is the sign of Λ.

As can be noticed (Fig. 8), a very good convergence of all profiles has been achieved. The differences are visible only very close to the wall, in the viscous sub-layer, which may be due to the thermal effect of the wall on the hot-wire probe.

To be consistent, the following formula describing the streamwise Reynolds stresses of analyzed turbulent boundary layer in pressure gradient conditions could be proposed:

$$\overline{uu} = \left(\Delta_H(U^+H)\Lambda^{n1/2} + \left(A + B\frac{U}{U_\infty}\right)^2\right)U^2H \tag{6}$$

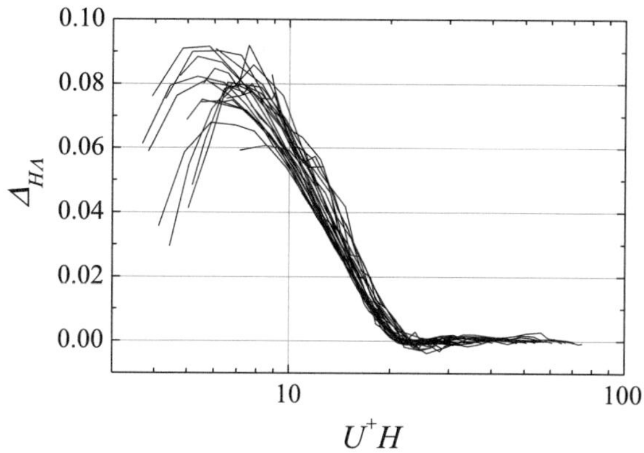

Fig. 8 Self-similar complete difference function $\Delta_H(U^+H)\Lambda^{n/2}$ for FPG and APG regions

4 Conclusions

The substantial change of fluctuation distributions, which may be attributed to a complexity of the analyzed case, is the reason for the lack of the self-similarity of Reynolds stress profiles. At given conditions the self-similarity cannot be obtained using the scale, which is constant across the boundary layer thickness. The analyzed flow is characterized by a strong APG region which is preceded by a strong FPG region. It results in few local equilibrium regions defined by constant pressure gradient parameter Λ. The new proposal of streamwise Reynolds stresses scaling completed with difference function, which is based on [1] concept, was introduced. It extends the applicability of the AOS scaling to pressure gradient turbulent boundary layers by means of an additional scaling factor, which is the product of U^2 and shape factor H. This expression takes into account the change of the mean velocity profile and corrects the streamwise Reynolds stress in the outer region, which is especially important for APG conditions. Pressure gradient parameter Λ further corrects the complete difference profiles especially close the wall. Finally, the profiles collapse across turbulent boundary layer thickness both in favorable and adverse pressure gradients.

Acknowledgments The investigation was supported by National Science Centre under Grant no. DEC-2012/07/B/ST8/03791.

References

1. P.H. Alfredsson, R. Örlü, A. Segalini, A new formulation for the streamwise turbulence intensity distribution in wall-bounded turbulent flows. Eur. J. Mech. B/Fluids **36**, 167–175 (2012). doi:10.1016/j.euromechflu.2012.03.015. http://linkinghub.elsevier.com/retrieve/pii/S0997754612000490
2. L. Castillo, W.K. George, Similarity analysis for turbulent boundary layer with pressure gradient: outer flow. AIAA J. **39**(1), 41–47 (2001)

3. S. Drobniak, W. Elsner, A. Dróżdż, M. Materny, Experimental analysis of turbulent boundary layer with adverse pressure gradient corresponding to turbomachinery conditions, in *Progress in Wall Turbulence: Understanding and Modeling*, ed. by M. Stanislas, J. Jimenez, I. Marusic. ERCOFTAC Series, vol. 14. (Springer, Dordrecht, 2011), pp. 143–150. doi:10.1007/978-90-481-9603-6. http://www.springerlink.com/index/10.1007/978-90-481-9603-6
4. A. Dróżdż, W. Elsner, Detection of coherent structures in a turbulent boundary layer with zero, favourable and adverse pressure gradients. J. Phys. Conf. Ser. **318**(6), 062007 (2011). doi:10.1088/1742-6596/318/6/062007
5. A. Dróżdż, W. Elsner, Amplitude modulated near-wall cycle in a turbulent boundary layer under an adverse pressure gradient. Arch. Mech. **65**(6), 511–525 (2013)
6. A. Dróżdż, W. Elsner, Comparison of single and X-wire measurements of streamwise velocity fluctuations in turbulent boundary layer. J. Theory Appl. Mech. **52**(2), 499–505 (2014)
7. A. Dróżdż, W. Elsner, S. Drobniak, Application of oil-fringe interferometry for measurements of wall shear stress. Turbomachinery **133**, 1–8 (2008)
8. A. Dróżdż, W. Elsner, S. Drobniak: Scaling of streamwise Reynolds stress for turbulent boundary layers with pressure gradient. Eur. J. Mech. B/Fluids (2014) (In press)
9. Z. Harun, J.P. Monty, R. Mathis, I. Marusic, Pressure gradient effects on the large-scale structure of turbulent boundary layers. J. Fluid Mech. **715**, 477–498 (2013)
10. N. Hutchins, T.B. Nickels, I. Marusic, M.S. Chong, Hot-wire spatial resolution issues in wall-bounded turbulence. J. Fluid Mech. **635**, 103 (2009). doi:10.1017/S0022112009007721. http://www.journals.cambridge.org/abstract_S0022112009007721
11. P. Ligrani, P. Bradshaw, Spatial resolution and measurement of turbulence in the viscous sublayer using subminiature hot-wire probes. Exp. Fluids **5**, 407–417 (1987)
12. I. Marusic, G.J. Kunkel, Streamwise turbulence intensity formulation for flat-plate boundary layers. Phys. Fluids **15**(8), 2461–2465 (2003). doi:10.1063/1.1589014
13. I. Marusic, R. Mathis, N. Hutchins, Reynolds number dependence of the amplitude modulated near-wall cycle, in *Progress in Wall Turbulence: Understanding and Modeling. ERCOFTAC Series*, vol. 14, ed. by M. Stanislas, J. Jimenez, I. Marusic (Springer, 2011), pp. 105–112
14. R. Mathis, N. Hutchins, I. Marusic, Evidence of large-scale amplitude modulation on the near-wall turbulence, in *Australasian Fluid Mechanics Conference*, December (2007), pp. 1442–1448
15. J. Monty, Z. Harun, I. Marusic, A parametric study of adverse pressure gradient turbulent boundary layers. Int. J. Heat Fluid Flow **32**(3) (2011). doi:10.1016/j.ijheatfluidflow.2011.03.004
16. Y. Nagano, T. Tsuji, T. Houra, Structure of turbulent boundary layer subjected to adverse pressure gradient. Int. J. Heat Fluid Flow **19**(5), 563–572 (1998). doi:10.1016/S0142-727X(98)10013-9

The Temporal Coherence of Prograde and Retrograde Spanwise Vortices in Zero-Pressure Gradient Turbulent Boundary Layers

Callum Atkinson, Vassili Kitsios and Soria

Abstract Spatial and temporal statistics associated with spanwise aligned vortical structures are extracted from high repetition rate particle image velocimetry (HR-PIV) experimental measurements of a zero-pressure gradient turbulent boundary layer. Measurements were performed in the LTRAC water tunnel with a momentum thickness-based Reynolds number of $\mathrm{Re}_\theta = 2{,}250$. Streamwise wall-normal planes of the field were recorded at rate of $\Delta t = 0.008\delta/U_\infty$, spanning a streamwise domain of 3.2δ. This enables a single structure to be sampled approximately 400 times for a duration of $3.2\delta/U_\infty$ as it convects downstream. A model Oseen vortex is fit to each local peak in swirling strength, in order to detect and classify the radius, centroid velocity, circulation, and centroid location of each spanwise vortex. Attempts to track the evolution of these vortices show that on average these Oseen vortices only appear to remain temporally coherent for a time of $0.02\delta/U_\infty$.

1 Introduction

Vortical structures and their stretching and interaction form a fundamental part of all turbulent flows and play a significant role in the transfer of energy between different scales. The lifetime and rate at which these structures propagate is therefore highly relevant to the study of the dynamics of wall-bounded turbulence. Analysis of the mean and scale-dependent convection velocities of streamwise and wall-normal velocity fluctuations, in both channel [1] and boundary layer flows [2], have shown that both fluctuations convect at different velocities with respect to the local mean velocity and with respect to each other. These convection velocities are not only

C. Atkinson (✉) · V. Kitsios · Soria
Laboratory for Turbulence Research in Aerospace and Combustion,
Department of Mechanical and Aerospace Engineering, Monash University,
Melbourne, Vic, Australia
e-mail: callum.atkinson@monash.edu

Soria
Department of Aeronautical Engineering, King Abdulaziz University,
Jeddah, Kingdom of Saudi Arabia

M. Stanislas et al. (eds.), *Progress in Wall Turbulence 2*,
ERCOFTAC Series 23, DOI 10.1007/978-3-319-20388-1_13

functions of streamwise and spanwise scale but may also vary for streaks, sweeps, ejections, and individual vortices.

Spatially and temporally coherent streamwise and spanwise vortical structures play an important role in the conceptual models of the organization of wall-bounded turbulence. These models are generally divided into those that assume that streamwise vortices form the basis for self-sustaining turbulence [3, 4] and those that propose that the dominant feature is the hairpin vortex or a packet of hairpin vortices [5, 6]. Understanding the potential role of these models therefore depends on an understanding of the nature and dynamics of these vortices.

Considerable work has been done to statistically classify the distribution, size, and strength of individual vortices. This work has typically involved the use of 2D particle image velocimetry (PIV) or stereo-PIV (SPIV) measurements in either, streamwise wall-normal (x–y), spanwise wall-normal (y–z), or dual planes at various angles to the flow [7–9]. Similar studies have been performed for turbulent channel flows using data from Direct Numerical Simulations (DNS) [10, 11]. Results have shown that the PDFs of the radii and peak vorticity of these vortices show only weak dependence on Reynolds number and wall-normal position when scaled by the local Kolmogorov length η and time-scales τ.

Less is understood about the dynamics and evolution of individual vortices. Using time-series DNS of a turbulent channel flow ($Re_\tau \approx 800$), Kag et al. [12] were able to track clusters of fine-scale vortices from their generation at the base of the log layer to their growth in scale as they convected downstream; however, no statistics are given about the change in radius of individual vortices, their convection velocity or their time-scale. Using a similar time-series channel flow, DNS Lozano-Durán and Jimenéz [13] identified and tracked individual vortex clusters (or volumes) of positive discriminant. Described as "sponges of strings" these clusters consist of interconnected tubes with a radius of 7η, corresponding to the mean vortex radius reported by [14] for both spanwise and streamwise 2D vortices. The growth and decay of these clusters is investigated with and without taking into account the merging and splitting of individual clusters. Clusters appeared to be advected by the local mean velocity with only a slight bias toward moving away from the wall.

Recently LeHew et al. [15] used HR-PIV measurements of wall-parallel planes (x–z) to investigate the convection and evolution of predominately wall-normal vortical structures at $\mathrm{Re}_\tau = 410$. Vortices were detected by thresholding regions of wall-normal swirling strength λ_i, then classified by area, swirling strength weighted centroid, and average velocity. The average trajectory velocities showed a greater tendency toward velocities lower than the local mean $U(y)$, when compared with the average velocity in the vortex core. The majority of these structures were found to exist for less than one eddy turnover time δ/U_∞, where δ is the boundary layer thickness and U_∞ the freestream velocity. A small number of structures persisted for more than $5\delta/U_\infty$. LeHew et al. [15] state that this time of $5\delta/U_\infty$ is similar to the mean time spent in stable focus stretching, as predicted by the periodic nonlocal mean topological flow evolution cycle [16]. However, that comparison fails to account for the additional $5\delta/U_\infty$ time spent in unstable focus contraction, during which time the swirling strength should remain nonnegligible. Using boundary layer

DNS data Atkinson et al. [17] showed that this periodic mean topological evolution is only associated with relatively weak flow gradients and large scale bulk motions. For the stronger gradients associated with intense vortex structures, Atkinson et al. [17] showed mean focal time-scales between 2 to $4\delta/U_\infty$, much closer to the longer time-scales observed by LeHew et al. [15].

In the present paper, the analysis methodology of Herpin et al. [14] is applied to time-resolved measurements of a turbulent boundary layer, performed in a large water tunnel facility using HR-PIV. Statistics of individual prograde and retrograde spanwise vortices are calculated from instantaneous fields and compared with existing databases. The centroid velocities are calculated based on the velocity relative to the frame of reference of each vortex. Vortices are tracked in time at a resolution of $\Delta t = 0.008\delta/U_\infty$. The spatial coherence and lifetime of the detected vortices are then explored.

2 Experimental Database

Analysis of the evolution of individual spanwise vortices is performed using velocity fields obtained from HR-PIV measurements that were performed in a streamwise, wall-normal plane (x–y) at a station approximately 4 m downstream of the contraction in the Laboratory for Turbulence Research in Aerospace and Combustion's (LTRAC) large horizontal water tunnel at Monash University. Flow was tripped at the outlet of the contraction, after which a turbulent boundary layer develops along the tunnel floor. The properties of the boundary layer at the measurement domain are given in Table 1. Further details of the experiment can be found in Atkinson et al. [2]. Velocity fields were computed with interrogation window size of 32×32 pixels (or $23^+ \times 23^+$) with a light sheet thickness corresponding to 23^+ wall units and with a 50 % window overlap resulting in a grid spacing of $\Delta_{x,y} = 11.6^+$.

Table 1 Parameters of the HR-PIV boundary layer measurements

Boundary layer thickness	$\delta^+ = 840$
Reynolds number	$Re_\theta = 2{,}250$
Friction velocity	$u_\tau = 0.018\ \mathrm{ms}^{-1}$
Kinematic viscosity	$\nu = 0.84 \times 10^{-6}\ \mathrm{m^2/s}$
Field-of-view (L_x, L_y)	3.2δ, 1.7δ
	2680^+, 1424^+
Spatial resolution (W_x,W_y,W_z)	0.03δ, 0.03δ, 0.03δ
	23^+, 23^+, 23^+
Vector spacing	$\Delta_{x,y} = 11.6^+$
Temporal resolution	$\Delta_t = 0.008\delta/U_\infty$
	$\Delta_t = 0.3^+$

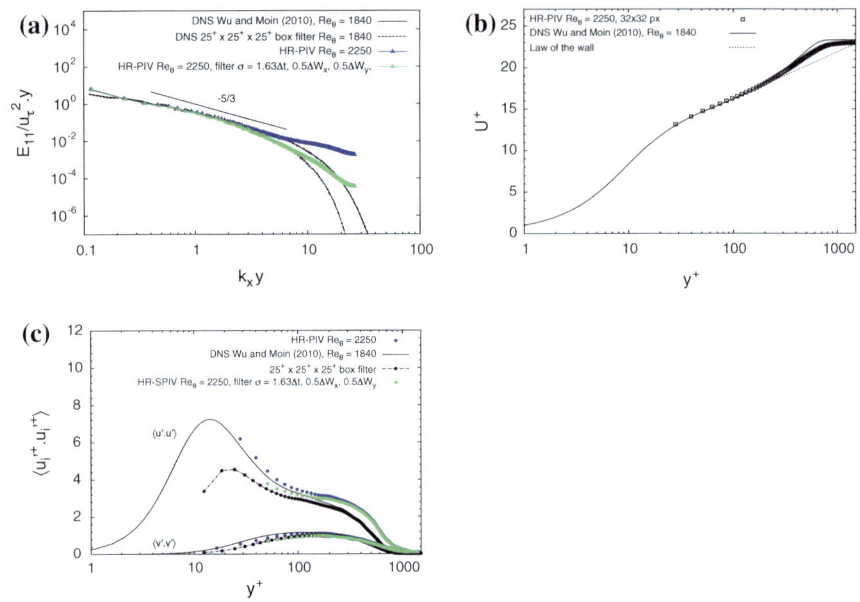

Fig. 1 (**a**) Longitudinal velocity power spectra at $y^+ = 100$ for unfiltered and filtered HR-PIV data, compared with full resolution boundary layer DNS [19] and DNS filtered at the spatial resolution of the HR-PIV; (**b**) Mean velocity profile; (**c**) Reynolds stresses

In order to reduce the influence of measurement noise, the PIV velocity data was Gaussian filtered in both space and time with temporal and spatial cut-offs selected to match the frequencies beyond which the velocity power spectra indicated that the measurement noise was greater than the energy of the turbulence [2]. As detailed in Atkinson et al. [18], it is important to compare the spectra and statistics obtained from the measurement with the quantities obtained from lower noise measurements or DNS data at the same spatial resolution. The spatial averaging inherent in PIV measurements can otherwise mask the presence of measurement noise and produce one-point statistics that appear to be in better agreement with higher spatial resolution data than they should. Figure 1 shows the longitudinal velocity power spectra estimated from the PIV data before and after filtering, compared with DNS data sampled at the full resolution and at the same spatial resolution as the HR-PIV. Filtered HR-PIV profiles are shown to be in good agreement with DNS at the same resolution.

3 Vortex Detection and Classification

For planar data, where only the 2D velocity gradient tensor (VGT) is available $\frac{\partial u_i}{\partial x_j}$, $i = 1, 2$, the majority of methods used to detect vortices are equivalent. In this case, the imaginary part of the complex eigenvalues of the VGT, termed the swirling strength λ_i, is used to indicate regions of swirling flow. To account for the higher swirling strength near the wall and to enable a thresholding operation to be performed to identify locally significant vortices, the swirling strength is first normalized by the standard deviation of the swirling strength $\sigma_{\lambda i}$ for each wall-normal height, such that

$$\hat{\lambda_i}(x, y) = \frac{\lambda_i(x, y)}{\sigma_{\lambda i}(y)}. \tag{1}$$

A 3×3 point sliding average was applied to the normalized swirling strength field to reduce the influence of measurement noise, before the detection of local extrema with a minimum normalized value greater than 1.5 following [8, 14]. An Oseen vortex was then fitted to each extrema using a Levenberg–Marquardt minimization over a domain corresponding to the radius of the vortex, with the vortex model defined in polar coordinates (r, θ) as:

$$\mathbf{u}(r, \theta) = \mathbf{u}_c + \frac{\Gamma}{2\pi r}\left(1 - e^{-\left(\frac{r}{r_0}\right)^2}\right). \tag{2}$$

where r_0 is the vortex radius, Γ is the circulation and $\mathbf{u}_c$ is the centroid velocity, or the velocity relative to the vortex frame of reference. The fit was performed in two passes with an assumed initial vortex radius of three points or $\approx 35^+$ wall units, with the estimated radius from the first pass used to determine the fitting domain of the second. To ensure that the region corresponds to a spanwise vortex, only regions where the correlation between the model and the instantaneous vector field is greater than 75 % are accepted as detected vortices.

In practice, approximate 80–90 % of the local swirling strength peaks show a correlation with the Oseen model above 75 % with approximately 70 vortices detected in each field. Figure 2 shows the mean velocity field conditioned on a local peak in swirling strength at $(x^+, y^+) = (0{,}175)$ and positive or negative vorticity. The

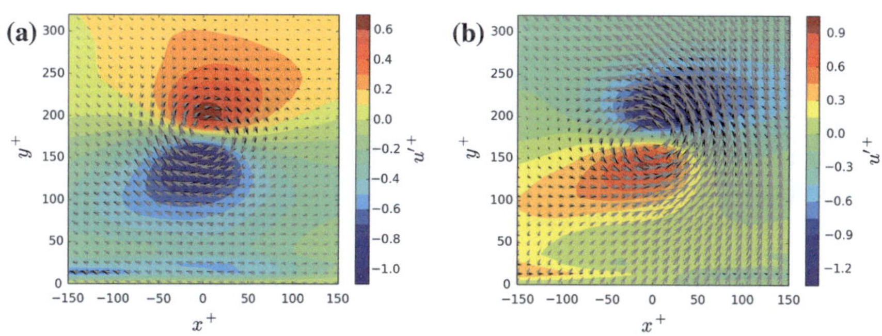

Fig. 2 Fluctuating velocity field conditioned on a local peak in swirling strength at $(x^+, y^+) = (0{,}175)$ for: (**a**) $\Gamma < 0$ (prograde vortex); and (**b**) $\Gamma > 0$ (retrograde vortex). *Black arrows* represent the conditional mean field; *gray* arrows represents the Oseen vortex field

conditional fields indicate the presence of what appear to be a spanwise vortex at their center. Fitting the Oseen model to these conditional fields returns spanwise vortices with radii of $r_o \approx 25.2^+$ and $r_o \approx 39.9^+$ for the negative vorticity (prograde vortex) and the positive vorticity (retrograde vortex), respectively. The term prograde vortex is used to describe a vortex with a negative spanwise vorticity $\omega_z < 0$ and circulation ($\Gamma < 0$), which has a rotation in the same direction as the mean shear. In both cases, the correlation between the model and the conditional field is approximately 99 %, over the radius of the vortex. Naturally, the flow field further away from the vortex starts to differ from that of the Oseen model; however, the large percentage of successful fits and the strong agreement between the conditional field and the vortex model, suggest that the Oseen vortex is a good representation of the flow field in the vicinity of local peaks in the swirling strength.

4 Vortex Statistics and Instantaneous Distribution

PDFs for the wall-normal position, radius and circulation of detected vortices, as well as the mean radius of prograde and retrograde vortices detected at each wall-normal height are shown in Fig. 3. Statistics are accumulated over 25,220 time-steps or approximately 200 eddy turnover times. Figure 3a shows a lower number of retrograde vortices near the wall with a peak at around $y^+ = 300$, with a local minimum at $y^+ = 600$. In contrast, the number of prograde vortices is almost an order of magnitude higher near the wall where the influence of the mean shear is strongest. The number of prograde vortices decreases with wall-normal position, with an equal likelihood of finding a prograde or a retrograde vortex for $y^+ \geq \delta^+$, where $\delta^+ \approx 840$. Figure 3c shows a similar trend for the distribution of circulation magnitude, with retrograde vortices tending to have a weaker circulation, distributed over a much narrower range.

The difference between the radii of the two vortex types is less significant. The PDFs of both prograde and retrograde vortices (Fig. 3b) peak at $r^+ \approx 25$ before decreasing logarithmically. In Fig. 3d, both vortices show an increase in mean radius with wall-normal position up to a maximum of $r^+ \approx 50$ at $y^+ = \delta^+$. Prograde vortices are shown to be larger on average for $y^+ < \delta^+$. For $y^+ \geq \delta^+$ both the prograde and retrograde vortices are of the same size. These results are in good agreement with those of Herpin et al. [14], with the exception of a slightly larger mean radii in the present results. This is likely due to the lower spatial resolution of the present measurement and it is inability to capture vortices with a radius less than 11^+.

JPDFs of the velocity at the core of each vortex are shown in Fig. 4 as a function of wall-normal position. For comparison, the local mean velocity is illustrated by the overlaid black line. The core velocity represents the convection velocity relative to the frame of reference of the vortex (or Oseen model). Core velocities are influenced by the presence of local velocity fluctuations and also by the swirl generated by surrounding vortices. Streamwise core velocities tend to be close to the local mean

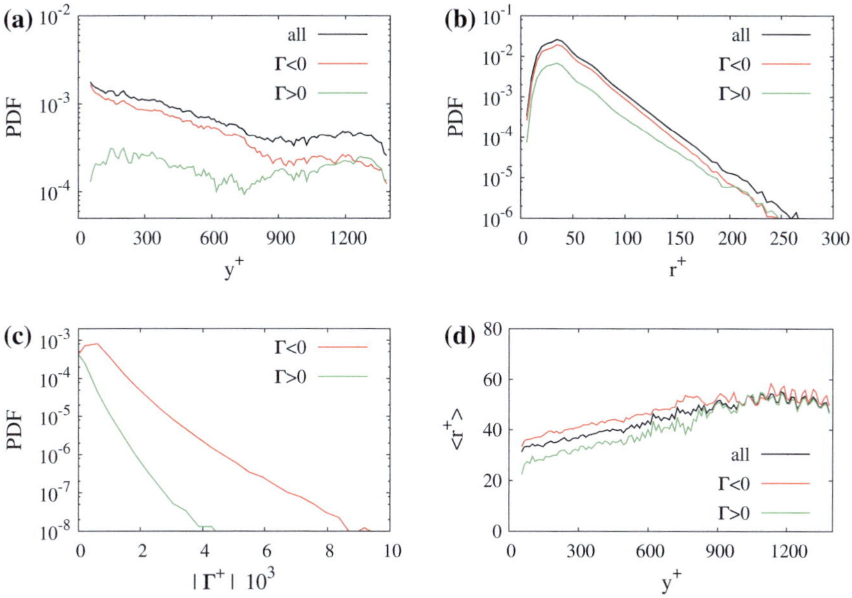

Fig. 3 PDFs of detected prograde ($\Gamma < 0$) and retrograde ($\Gamma > 0$) spanwise Oseen vortices: (**a**) wall-normal vortex position; (**b**) vortex radius; (**c**) circulation; and (**d**) mean radius

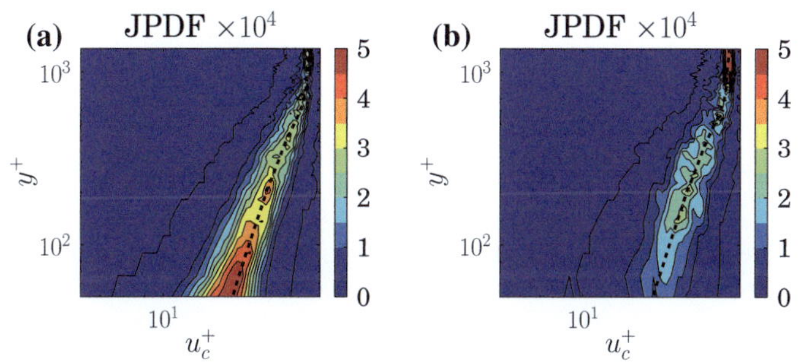

Fig. 4 JPDFs of instantaneous streamwise vortex centroid velocity with wall-normal position y^+ for both: (**a**) prograde ($\Gamma < 0$); and (**b**) retrograde ($\Gamma > 0$) spanwise Oseen vortices. *Dashed black line* represents the local mean velocity $U(y)$

velocity; however, the JPDFs show a greater skew toward lower velocities for the prograde vortices. This results in a slightly lower average centroid velocity for the prograde vortices at $y^+ < 200$. Negligible differences were observed for the wall-normal velocities.

The instantaneous distribution of these detected vortices is shown in Fig. 5. Consistent with the hairpin packet model, the prograde vortices show a tendency to align along the top surface of regions of low streamwise velocity. In contrast, the retrograde vortices are most frequently found on the lower surface of these low speed

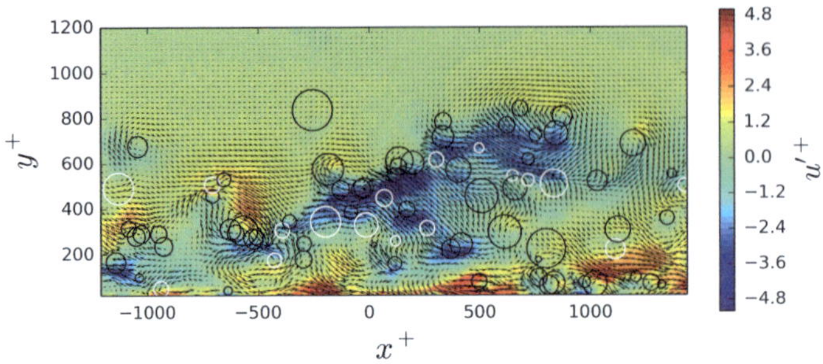

Fig. 5 Iso-contours of streamwise velocity fluctuations u' with overlaid *circles* representing the radii and location of the detected prograde (*black*) and retrograde (*white*) vortices

regions. Figure 5 also highlights the tight clustering of many of the detected vortices. During detection, any two vortices whose centroids were less than one grid spacing ($\Delta_x = 11.6^+$) away were merged, with their centroids and radii averaged and their circulation combined. This merging was performed to prevent the fitting to multiple swirling strength peaks from returning multiple copies of the same vortex. No further restriction was place on the overlapping of these vortices. However, the presence of overlapping vortices does reduce the correlation between the model and the instantaneous velocity field. This will subsequently limit the detection to regions where overlapping vortices still show a strong correlation with the Oseen vortex model over their radius.

5 Temporal Tracking and Coherence of Spanwise Vortices

To investigate the evolution of these spanwise vortices and determine their convection velocity, it is first necessary to track their position through each time-step. In the present work, vortices were detected in each time-step based on peaks in the instantaneous swirling strength field (see Sect. 3), independent of any other time-step information. For each vortex at time t_o, their approximate position at time $t_o + \Delta_t$ is calculated based on the local mean velocity. The two vortices that are closest to the expected location of the original vortex are then identified as potential candidate trajectories. These potential vortex locations are then validated, assuming the displacement of the vortex centroid from the previous vortex location is within 0 to $2U(y)\Delta_t$ in the x-direction and within $\pm 2U(y)\Delta_t$ for the y-direction, where $U(y)$ is local mean streamwise velocity. If both nearest vortices are deemed to be valid then the original vortex is assumed to correspond to the one with the circulation closest to that at the previous time-step.

Table 2 shows the mean, standard deviation, and maximum percentage of time that a vortex is successfully located, relative to the number of time-steps where it should remain in the experimental field-of-view (3.2δ), assuming no significant out-of-plane motion. Statistics are also given for the number of consecutive time-steps over which

Table 2 Statistics of tracked vortex success rate and vortex lifetime

Threshold		Time detected/Total (%)			Consecutive time detected (%)			Vortex duration (δ/U_∞)		
$\hat{\lambda}_i$	R_{model}	mean	σ	max	mean	σ	max	mean	σ	max
1.5	0.75	7.96	15.06	93.0	8.30	10.10	83.33	0.020	0.040	0.8*
1.5	0.5	7.84	15.63	95.7	8.42	10.55	95.65	0.022	0.044	0.8*
1.0	0.75	6.95	12.61	93.0	6.33	8.74	82.95	0.019	0.038	0.8*
1.0	0.5	7.32	13.25	93.0	6.46	9.08	82.95	0.019	0.042	0.8*
0.5	0.75	6.21	12.43	93.0	5.89	8.80	82.95	0.017	0.036	0.8*
0.5	0.5	6.59	12.96	93.0	5.94	8.95	82.95	0.019	0.040	0.8*
0.0	0.75	6.23	12.39	93.0	5.60	8.44	82.95	0.017	0.035	0.8*
0.0	0.5	6.47	12.49	93.0	5.55	8.08	82.95	0.016	0.037	0.8*
1.5	0.0†	11.10	20.92	98.86	8.17	15.41	99.43	0.063	0.146	0.8*

*Only 100 time-steps were processed for this comparison

†Local peaks in swirling strength are tracked without fitting vortices

a vortex is tracked, along with the vortex duration or predicted lifetime. The vortex duration is longer than the consecutive tracking length as it allows for a single vortex to be missed for a maximum of two consecutive time-steps. The results show that for local peaks is swirling strength above a threshold of $\hat{\lambda}_i > 1.5$ and a correlation between the Oseen vortex model and the instantaneous field of $R_{model} > 0.75$, a successful track is only recorded for 8 % of the time when averaged over all the vortices with the mean vortex trace only lasting for $0.02\delta/U_\infty$. The longest vortex track lasts for the duration of the 100 time-steps considered; however, only 0.4 % of the vortex traces last for more than 25 steps or $0.2\delta/U_\infty$. This suggests the lifetime of spanwise vortices is considerably shorter than the $5\delta/U_\infty$ observed in the conditional mean evolution of the flow topology [17].

To test if the minimal temporal coherence predicted by these results is the consequence of either the threshold or the required model correlation, tracking was performed using a range of parameters. As shown in Table 2, reducing the swirling strength threshold and relaxing the model correlation did not increase the time over which the detected vortices were tracked. Examination of consecutive fields of detected vortices shows a tendency for some detected vortices to show rapid variation in size, which subsequently results in large jumps in the detected vortex centroid and occasional rearward movement. This is often the result of vortices moving toward or away from others, resulting in a superposition of the swirl from both vortices which can drastically influence the field around each. Tracking of local peaks in swirling strength without fitting the vortex model results in a higher percentage of successful traces and mean lifetimes approximately three times longer than that of the model. At this stage, it is not clear whether most spanwise vortices only remain predominantly spanwise for less than one eddy turnover time and only resemble streamwise aligned Oseen vortices for a very short time, or whether the results are limited by our inability to extract an individual vortex from 2D slices of 3D fields that consist of the superposition of many in plane and out-of-plane vortices.

6 Conclusions

Streamwise wall-normal velocity fields from HR-PIV measurements of a zero-pressure gradient turbulent boundary layer flow are analyzed to detect and track spanwise vortices at a temporal resolution of $0.008\delta/U_\infty$. To classify the centroid position, velocity, radius, and circulation of prograde and retrograde spanwise vortices, an Oseen vortex model is fitted to each local peak in swirling strength. Results show good agreement with previously published vortex statistics; however, temporal tracking shows that the mean lifetime of the spanwise Oseen vortices is only on the order of $0.02\delta/U_\infty$. Comparison of the convection times for swirling strength peaks show that the short lifetimes are at least partially due to restricting the detection to an assumed vortex shape, which appears to be only weakly coherent in time.

Acknowledgments This work was supported by funding from the Australian Research Council.

References

1. J.C. del Álamo, J. Jiménez, J. Fluid Mech. **640**, 5 (2009)
2. C. Atkinson, N. Buchmann, O. Amili, J. Soria, in *Proceedings of the 8th International Symposium Turbulence and Shear Flow Phenomena*, Poitiers, France (2013)
3. F. Waleffe, J. Kim, Self-sustaining Mechanisms of Wall Turbulence (Computational Mechanics Publications, Southampton, 1997), Chap. How streamwise rolls and streaks self-sustain in a shear flow, pp. 309–332
4. W. Schoppa, F. Hussain, J. Fluid Mech. **453**(1), 57 (2002)
5. J. Zhou, R. Adrian, S. Balachandar, T. Kendall, J. Fluid Mech. **387**, 353 (1999)
6. R. Adrian, Phys. Fluids **19**(4), 041301 (2007)
7. J. Carlier, M. Stanislas, J. Fluid Mech. **535**, 143 (2005)
8. Y. Wu, K. Christensen, J. Fluid Mech. **568**, 55 (2006)
9. B. Ganapathisubramani, E. Longmire, I. Marusic, Phys. Fluids **18**, 05510501 (2006)
10. S. Das, M. Tanahashi, K. Shoji, T. Miyauchi, Theor. Comput. Fluid Dyn. **20**, 55 (2006)
11. J. DelAlamo, J. Jiménez, P. Zandonade, R. Moser, J. Fluid Mech. **561**, 329 (2006)
12. S.J. Kang, M. Tanahashi, T. Miyauchi, J. Turbul. (8) (2007)
13. A. Lozano-Durán, J. Jiménez, in *Journal of Physics: Conference Series*, vol. 318 (IOP Publishing, 2011), p. 062016
14. S. Herpin, M. Stanislas, J. Soria, J. Turbul. **11** (2010)
15. J. LeHew, M. Guala, B. McKeon, Exp. Fluids **54**(4), 1 (2013)
16. G. Elsinga, I. Marusic, Phys. Fluids **22**(1), 015102 (2010)
17. C. Atkinson, S. Chumakov, I. Bermejo-Moreno, J. Soria, Phys. Fluids **24**(10), 105104 (2012)
18. C. Atkinson, N.A. Buchmann, O. Amili, J. Soria, Exp. Fluids **55**(1), 1 (2014)
19. X. Wu, P. Moin, Phys. Fluids **22** (2010)

Boundary Layer Vorticity and the Rise of "Hairpins"

Peter S. Bernard

Abstract The downstream evolution of the vorticity field in the vicinity of hairpin-shaped regions of rotational motion appearing in the transitioning boundary layer is examined. It is shown that the dynamics of hairpins is inseparable from that of the nonrotational vorticity out of which they develop in a self-reinforcing process of ejection and reorientation. Widening the concept of structure to include the complete localized vorticity that produces hairpins, allows for a more complete and self-contained explanation of the boundary layer physics.

1 Introduction

By their nature, boundary layers in high-speed flow contain prodigious amounts of vorticity produced at the solid bounding surface by the action of viscosity. A variety of measurement techniques applied to either physical experiments or numerical computations of boundary layers suggest the presence of "coherent" objects within the vorticity field that make essential contributions to the dynamical behavior of the flow. The possibility of arriving at a precise understanding of the boundary layer structure depends on what is meant by "coherency," a concept that is intrinsically difficult to define. In recent times, this has come to mean distinctive "regions of rotational motion" and the coherent objects discovered by this criterion are generally hairpin-shaped [1], meaning a flow volume with one or two streamwise oriented "legs" attached to a spanwise "arch."

Since much of the vorticity in the boundary layer does not lie in regions where rotational flow is occurring, the assertion that rotational regions are structures begs the question as to what role the remaining, nonrotational, vorticity has both in the dynamics of the boundary layer in general, and the rotational regions in particular. The importance of this vorticity has been recognized previously, for example in [5] where isosurfaces of spanwise vorticity reveal structural details that conform to the

P.S. Bernard (✉)
Department of Mechanical Engineering, University of Maryland,
College Park, MD 20742, USA
e-mail: bernard@umd.edu

M. Stanislas et al. (eds.), *Progress in Wall Turbulence 2*,
ERCOFTAC Series 23, DOI 10.1007/978-3-319-20388-1_14

rotational field in the form of hairpins. In addition, several recent studies [2–4] using a vortex filament scheme visualized vortex structure in the boundary layer without the *a priori* decision that the structures must occupy regions of rotational motion. This showed that hairpin-like regions of rotation have a natural association with uplifting furrow-like folds in the surface vorticity layer that develop downstream into mushroom-like shapes prior to descending into chaotic forms. It is evident from results such as these that there is much to be gained by examining the vorticity environment surrounding individual hairpins with the goal of exposing the role played by all of the vorticity that acts coherently to create the hairpins and that is essential to the physics of the boundary layer. This is the focus of this study.

2 Vortex Filament Scheme

A vortex filament simulation of a spatially growing boundary provides the numerical data for this study. Past work [3, 4] has described the numerical algorithm in detail. It suffices for the present to mention that this is a hybrid approach in the sense that vorticity determined from a finite volume solution to the viscous flow equations on a thin wall mesh is converted to vortex filaments that represent the flow outside the near-wall region. The mesh calculation is both more efficient and accurate at resolving the largely 2D regions of intense vorticity diffusing out of the wall surface than can be achieved using filaments. Within the filament field, hairpin removal provides spatial and temporal intermittent dissipation at inertial range scales as well as limiting growth in the number of vortices. The velocity is calculated using the Biot–Savart law that takes into account the contributions from the vorticity in the mesh and filaments. A velocity potential on the surface triangularization is used to enforce the nonpenetration boundary condition.

The simulation considered here occupies a larger spatial region and higher Reynolds number than in the previous work [4]. The boundary layer is computed on the top and bottom surfaces of a flat plate with rounded edges of dimensionless length 4 in the streamwise direction, 2.5 in the spanwise direction, and 0.05 thick. The region from the front of the plate until $x = 1$ is kept as an inviscid surface so that the boundary layer starts at $x = 1$. Thin regions on the sides and rear of the plate are also taken to be inviscid. The advantage of setting up the flow this way is that it successfully stabilizes the boundary layer against the large-scale spanwise and wall-normal flow asymmetries that may appear in the unfettered viscous motion past a finite plate.

The Reynolds number at the end of the boundary layer is $R_e = 225{,}000$. Just after the start of the viscous computation at $x = 1$, the boundary layer is smooth and is in excellent agreement with the Blasius boundary layer as shown in Fig. 1 comparing the computed velocity field with the Blasius result. Filaments produced in this region are exactly aligned in the spanwise direction. Transition is induced by the response of the many filaments and vortex elements in the mesh to slight perturbations that are inherent in the discretization of the Biot–Savart law. Though the perturbations are

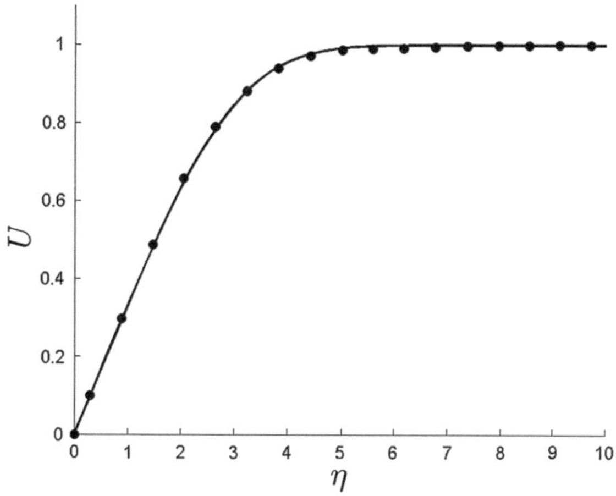

Fig. 1 Velocity in the Blasius region at $x = 1.3$. –, similarity solution; •, computed. The similarity variable $\eta = y\sqrt{R_e/(x - 0.95)}$ includes a slight adjustment to the virtual origin

very small initially, and originate entirely within the near-wall mesh, once provoked they grow quickly due to the mutual interactions between vortices. The end result is transition to turbulence, though at Reynolds numbers they are somewhat more typical of a heavily forced flow. For the present simulation, this means that the fully turbulent state is achieved at $R_e \approx 100{,}000$. With finer discretization this trend can be reversed, though at significant increase in the cost of the simulation and without qualitative change in the observed vortex structure.

To obtain information about the vorticity field for this study, the velocity was computed on a fine mesh covering the flow domain and then substituted into second-order accurate finite difference formulas to get the derivatives needed to compute vorticity. The same dataset was used to compute λ_2 whose isosurfaces mark the presence of rotational regions in the flow field including the hairpins.

3 Vorticity and Structure

The filament calculation produces a Klebanoff-type transition bridging the gap between the Blasius boundary layer and a fully turbulent flow. This transition mode is marked by the presence of low-speed streaks and rotational structures as revealed by λ_2 in the form of hairpins. Figure 2 gives an overhead view of the computed $\lambda_2 = -30$ isosurfaces from $x = 1.4$ until the onset of the fully developed turbulent field. Apart from some noise at the upstream end caused by the locally low amplitude of λ_2 as well as the coarse discretization used in its computation, the visualization of λ_2 reveals Λ-like vortices preceding the formation of rotational regions in the shape of hairpins spaced approximately $\Delta z^+ = 350$ that are not unlike those seen in more traditional grid-based simulations.

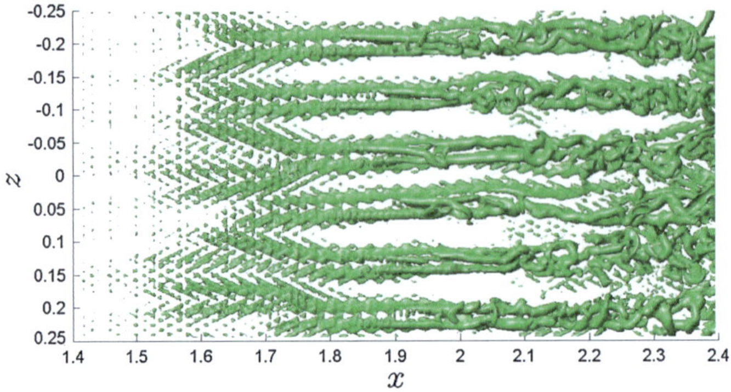

Fig. 2 Overview of $\lambda_2 = -30$ isosurfaces

To make sense of the vorticity field associated with the appearance of hairpins, it is useful to consider the streamwise behavior of the maximum amplitudes of the vorticity components at fixed distances above the wall. Such data is shown in Fig. 3 at a point inside ($y^+ = 24.3$) and a point outside ($y^+ = 51.3$) the near-wall viscous region. In these plots, the maxima are restricted to the local spanwise region of a particular Klebanoff streak, in this case the one located between $0.15 \leq z \leq 0.25$. Other streaks give qualitatively the same result with some small shifting in the streamwise direction. For each figure, the appropriate Blasius spanwise vorticity ω_3 is plotted as a dashed line that shows how the boundary layer would behave in the same location if it had remained laminar. In all cases, the turbulent solution departs smoothly from the Blasius values as the flow transitions.

Vertical lines in Fig. 3 indicate three relatively distinct zones in the evolution of the vorticity as it impacts the hairpins. The first region, between $x = 1.4$ and 1.7, is characterized by a steady increase in the streamwise (ω_1) and wall-normal (ω_2) vorticity components near the wall. This is significant if for no other reason than the fact that such nonspanwise vorticity is absent in the Blasius boundary layer. Downstream of the first zone there commences a number of significant changes to the vorticity amplitudes including a dramatic drop in the magnitude of ω_3 that effectively ends any remaining connection it has to the laminar form. The region $1.7 \leq x \leq 1.9$ is singled out as being distinctive because of the structural transformation accompanying the vorticity behavior in Fig. 3. The end result of the developments in the second zone is to enable the appearance of the fully formed hairpin-like rotational regions located downstream of $x > 1.9$ that characterize transition until the breakdown to fully turbulent flow.

To give more context to the ensuing discussion, the trends in the vorticity magnitude both closer and further from the wall than is considered in Fig. 3 are displayed in Fig. 4. Figure 4a shows that after transition there is a very large increase in the spanwise vorticity at $y^+ = 8.1$ to values much higher than in the equivalent laminar boundary layer. This state persists indefinitely downstream and underlies all the activity that produces turbulent structure. In particular, this reservoir of very high vor-

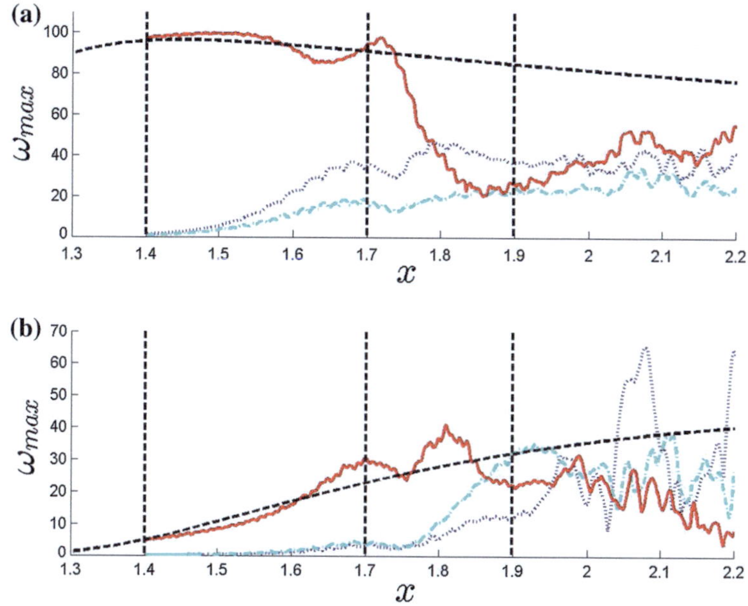

Fig. 3 Streamwise dependence of the maximum vorticity amplitude $0.15 \le z \le 0.25$. **a** at $y^+ = 24.3$; **b** $y^+ = 51.3$. ω_1, ($\cdots$, *blue*); ω_2, ($- \cdot -$, *cyan*); ω_3, (—, *red*). *Dashed line* is the equivalent ω_3 for the Blasius boundary layer

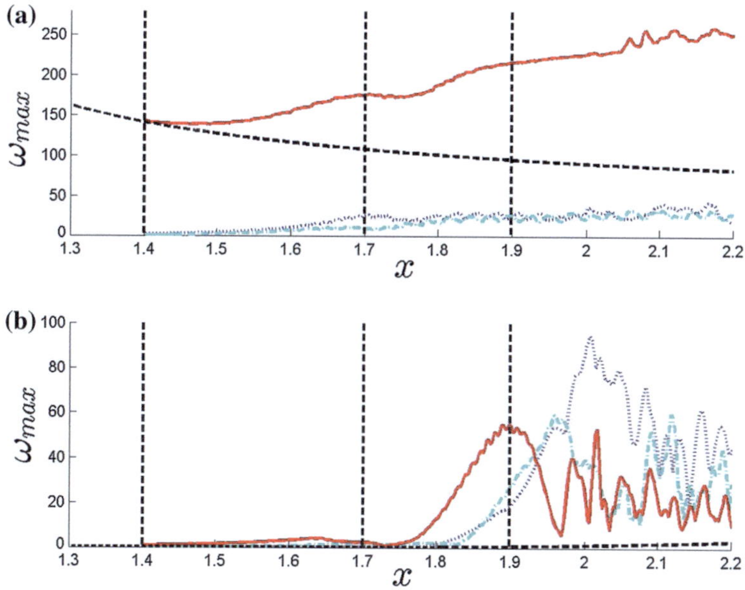

Fig. 4 Streamwise dependence of the maximum vorticity amplitude $0.15 \le z \le 0.25$. **a** at $y^+ \approx 8.1$; **b** $y^+ = 89.1$. Definitions of curves are the same as in Fig. 3

ticity is undoubtedly the source of all "new" vorticity that enters into the creation of downstream structures after transition. The vorticity trends in Fig. 4b at $y^+ = 89.1$—a region well beyond the direct reach of the viscous boundary layer—shows the last stages in the creation of structure during transition.

Zone 1. The onset of the streamwise structure in Fig. 2 is in the same location as the steady growth in streamwise vorticity near the wall in Fig. 3a. Upstream of $x = 1.4$ the magnitude of ω_1 is a very slight fraction of the ambient spanwise vorticity, but after $x = 1.4$ it grows significantly becoming more than one third of the magnitude of ω_3 by $x = 1.7$. There is also a somewhat smaller, but non-negligible, rise in wall-normal vorticity accompanying that of ω_1. The location where these vorticity components grow is very close to the boundary, entirely within the viscous region of large spanwise vorticity out of which ω_1 and ω_2 develop by reorientation. The physical boundary no doubt has an influence on suppressing ω_2 in this region.

An idea of the structural form taken by ω_1 as it appears in the flow is given in Fig. 5a showing its isosurfaces at four locations in zone 1. The streamwise vorticity has formed into oppositely signed concentrated pairs consistent with the λ_2 isosurfaces in Fig. 2, and is strengthening with downstream distance. Overlying the ω_1 isosurfaces, though not shown, is the smaller wall-normal vorticity.

It can be anticipated that there is some loss of ω_3 as ω_1 and ω_2 develop from its reorientation. On the other hand, the counterrotating velocity created by ω_1 causes significant convection of ω_3 away from the wall. This effect is apparent in Fig. 5b giving a visualization of the ω_3 contours at locations in zone 1. Spanwise vorticity decreases near the wall accompanied by its simultaneous rise away from the wall. As the spanwise vorticity is propelled upward, its place is filled by low-speed fluid forming a streak.

The trends in zone 1 of Fig. 3 are consistent with the outward ejection process. For example, in the region just beyond $x = 1.55$ the loss of spanwise vorticity shows up at $y^+ = 24.3$, while at $y^+ = 51.3$ there is a gain. Note, as well, that there is little

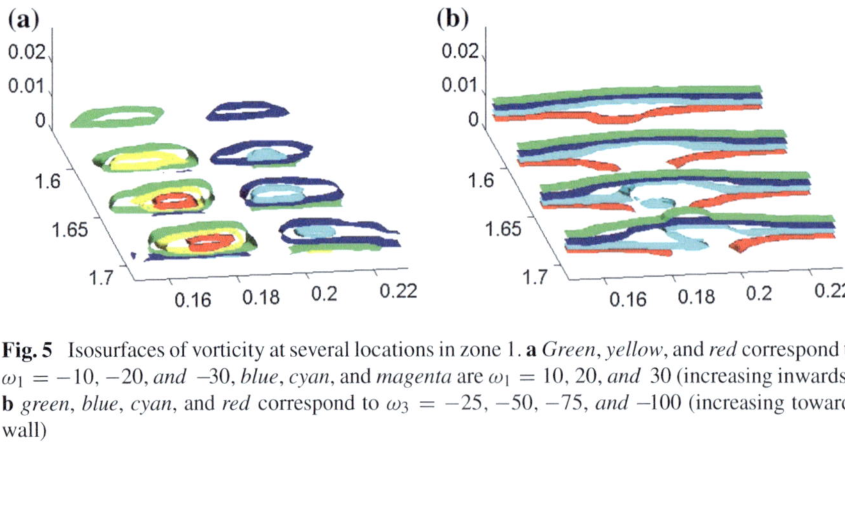

Fig. 5 Isosurfaces of vorticity at several locations in zone 1. **a** *Green*, *yellow*, and *red* correspond to $\omega_1 = -10, -20,$ *and* -30, *blue*, *cyan*, and *magenta* are $\omega_1 = 10, 20,$ *and* 30 (increasing inwards); **b** *green*, *blue*, *cyan*, and *red* correspond to $\omega_3 = -25, -50, -75,$ *and* -100 (increasing towards wall)

nonspanwise vorticity at $y^+ = 51.3$ suggesting that the reorientation process that produces such vorticity has yet to rise up to this level above the wall.

Zone 2. Now consider the vorticity trends in Fig. 3 for $1.7 \le x \le 1.9$ that is taken to be the second development zone of the structures. A prominent feature is a precipitous drop—to one-fifth of its maximum value—of the spanwise vorticity at $y^+ = 24.3$. A significant decrease in ω_3 also occurs at $y^+ = 51.3$, though delayed until $x = 1.8$. Near the wall the streamwise vorticity increases and levels off, while there is a sudden and substantial growth in the wall-normal vorticity at $y^+ = 51.3$ until it exceeds the local amplitude of ω_3. The picture that emerges from these trends is that near the wall the reorientation process by which streamwise vorticity develops from ω_3 has run its course. At the same time, the prodigious amount of ejected spanwise vorticity, now outside the near-wall domination of viscous diffusion, rapidly reorients to establish a significant presence of wall-normal vorticity that reaches a good distance from the wall. The latter process is also visible in Fig. 4b at points even further downstream where a sudden and significant growth in spanwise vorticity is followed immediately by the production of wall-normal and streamwise vorticity. With its rapid conversion to ω_1 and ω_2, the local dominance of spanwise vorticity ends at these distances from the wall.

Some of the trends in zone 2 are illustrated in the isosurfaces of ω_2 and ω_3 shown in Fig. 6. ω_2 grows to prominence as an oppositely signed pair with a concentration at its upper end coinciding with the λ_2 isosurfaces at the same location. Though not shown, there is a growing presence of ω_1 in this region as well. Figure 6b shows that the continued ejection of spanwise vorticity results in its collection in the region

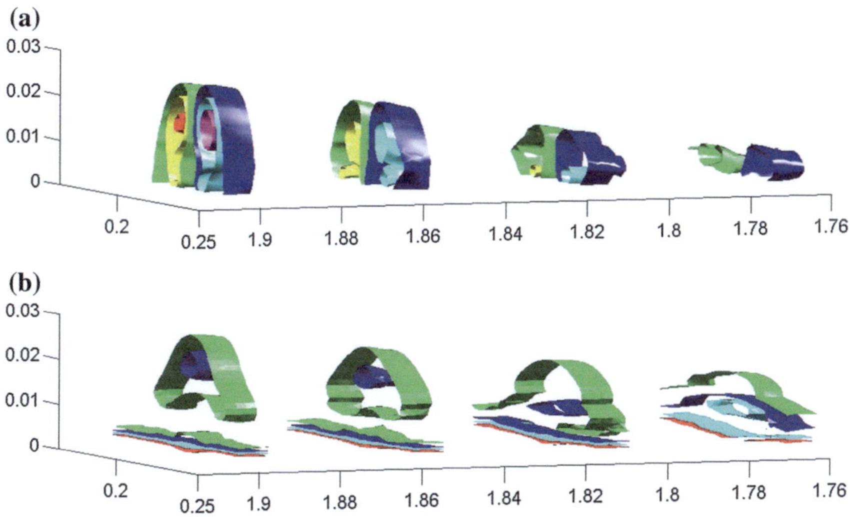

Fig. 6 Isosurfaces of vorticity at several locations in zone 2. **a** *green*, *yellow*, and *red* correspond to $\omega_2 = -10, -20,$ and -30, *blue*, *cyan*, and *magenta* are $\omega_2 = 10, 20,$ and 30 (toward the center); **b** *green*, *blue*, *cyan*, and *red* correspond to $\omega_3 = -20, -45, -70,$ and -95 (toward the wall)

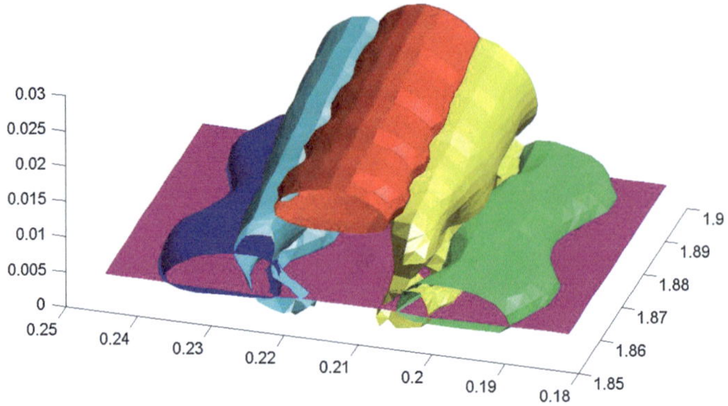

Fig. 7 Isosurfaces of vorticity at the end of zone 2: $\omega_1 = \pm 30$, *blue* and *green*; $\omega_2 = \pm 20$, *yellow* and *cyan*; and ω_3 = -40, *red* and *magenta*

between the counterrotating motions. Near the wall, ω_3 continues to fall in magnitude even though it strengthens in the fluid sublayer closest to the boundary as seen in Fig. 4a.

A summary of the state of affairs at the end of zone 2 is given in Fig. 7 that shows, in one plot, isosurfaces of the three vorticity components. Prominent features include the strong streamwise vorticity near the wall that forms a counterrotating pair, the wall-normal vorticity that has developed outside of the near-wall flow, and finally, the significant spanwise vorticity that collects within the region between the emerging "legs" of the hairpin.

Zone 3. Coincident with $x \approx 1.9$ in Fig. 2 rotational regions with the character of hairpin "legs" emerge out of the λ_2 isosurfaces. Further downstream, arch-like structures can be found that cross between the streamwise structures. With increasing x, the isosurfaces reveal a breakdown of the rotational regions into a wide range of more complicated rotational forms. A notable aspect of the vorticity maxima in this region, seen in Figs. 3b and 4b, is the development of relatively large sustained peaks in the streamwise vorticity that now dominate the other components. Evidently, this is the vorticity that is responsible for the presence of hairpin "legs," and it appears as the final phase of the conversion of the spanwise vorticity that has been ejecting outwards from the wall.

Previous work that concentrated on examining the vortex filament field in transition [4] showed a direct connection between hairpin "legs" and the lobes of mushroom-shaped structures in the vortex filaments that emerged out of the surface vorticity layer. Such patterns are present among the developing hairpin regions in the current simulation as well, and an example of this is shown in Fig. 8b. Accompanying this image, in Fig. 8a, is a view of the isosurfaces of the three vorticity components at the same location. The streamwise vorticity occupies the "legs" that are coincident with the lobes of the filament structure, while the wall-normal

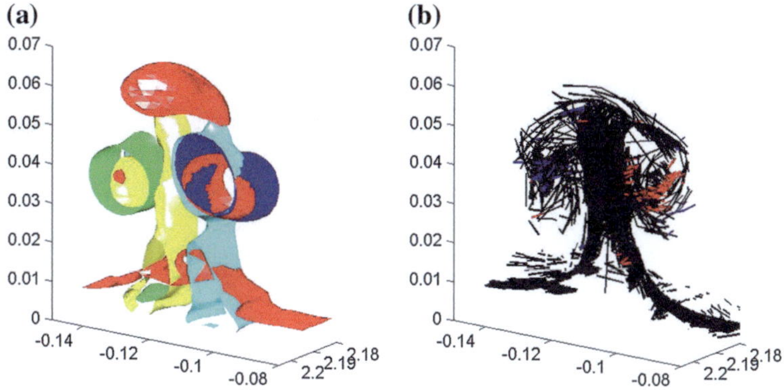

Fig. 8 Comparison of vorticity isosurfaces in a hairpin with the local vortex filament field. **a** Isosurfaces: $\omega_1 = \pm 30$, *blue* and *green*; $\omega_2 = \pm 15$, *yellow* and *cyan*; $\omega_3 = -15$, *red*; **b** vortex filaments intersecting $2.18 \leq x \leq 2.2$, *red* and *blue* filaments are close to the streamwise direction

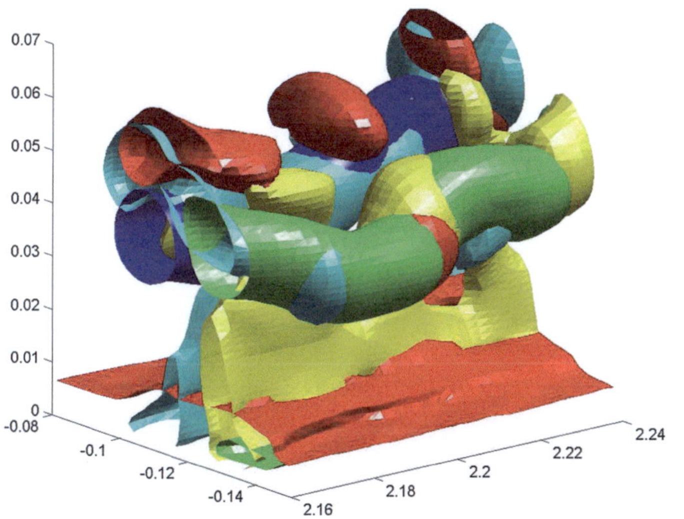

Fig. 9 Isosurfaces of vorticity associated with a hairpin as they develop in zone 3: isosurfaces have the same interpretation as in Fig. 8

vorticity extends upwards through the boundary layer encompassing the stem of the mushroom. Finally, some streamwise vorticity persists at the top of the structure.

An extended streamwise view of the vorticity isosurfaces for this structure is given in Fig. 9 that may be taken as a summary statement of what is entailed in a complete view of the vorticity field associated with hairpins. Spanwise and wall-normal vorticity accompany the prominent counterrotating streamwise vorticity as a legacy of the mechanisms by which the hairpin-like region developed. It should be emphasized that because the mushroom-like form is unstable, it often falls to one side or the other producing single-legged hairpins [4]. For this common occurrence, elements of the vorticity arrangement in Fig. 9 are present even if they do not fully resemble the somewhat idealized case depicted here.

Locations above the hairpin "legs" where the spanwise vorticity is concentrated are where the λ_2 signal is likely to indicate the presence of an arch vortex. This agrees with the observation in previous work [2, 4] that the filament field forms concentrated spanwise structure at the locations of arches.

4 Summary and Conclusions

Some essential aspects of the development of the vorticity field in a transitional boundary layer have been examined as it relates to the presence of hairpin-like structures. It is seen that by considering the complete local vorticity field surrounding hairpins, and not focusing exclusively on the rotational motion by which they are defined, the dynamics of the hairpins as well as those of the boundary layer itself can be more clearly understood.

It is found that counterrotating regions of streamwise vorticity develop in the viscous sublayer via a self-reinforcing process in which they are strengthened by the spanwise vorticity that they cause to eject. Thus they fuel their own development. As the spanwise vorticity extends beyond the viscous sublayer, it shears to create wall-normal vorticity that concentrates away from the boundary. Out of this vorticity, streamwise vorticity appears and accumulates to form the hairpins. The ejection process near the wall proceeds until the available spanwise vorticity is depleted. Remnants of the ejected spanwise vorticity, perhaps aided by a roll-up process, persists in the outward flowing regions to produce isosurfaces of rotation that give the impression of arch-type vortices connecting the "legs" of the structures.

A main conclusion is that while it is tempting to view the streamwise isosurfaces of λ_2 as forming hairpin "structure," without also taking into account the active role of the surrounding nonrotational vorticity, the analysis of hairpins misses essential clues to the physics of the boundary layer. For example, the initial appearance of streamwise rotation near the wall is only as a perturbation upon the dominant and "invisible" spanwise vorticity field. Downstream, largely missed by λ_2 isosurfaces, is ejecting spanwise vorticity that provides the "fuel" out of which the hairpins appear. Shearing of nonrotational wall-normal and spanwise vorticity explains the emergence of hairpin-shaped rotational regions that constitute just one aspect of a much more intricate vortical structure than is visible as a "hairpin."

After the ejection process depletes spanwise vorticity, it regenerates from the very high vorticity that persists at all times at the wall surface. It is likely that low-speed streaks in the fully turbulent region pinpoint locations where some variants of the ejection mechanism described here act to produce additional vortical structures that work their way to the outer flow. Vortical structures and their remnants accumulate away from the wall and fill out the growing turbulent boundary layer.

Acknowledgments This research was supported in part by the National Science Foundation through XSEDE resources provided by the Pittsburgh Supercomputing Center.

References

1. R.J. Adrian, Hairpin vortex organization in wall turbulence. Phys. Fluids **19**, 041301 (2007)
2. P.S. Bernard, P. Collins, M. Potts, Vortex filament simulation of the turbulent boundary layer. AIAA J. **48**, 1757–1771 (2010)
3. P.S. Bernard, The hairpin vortex illusion. J. Phys: Conf. Ser. **318**, 060024 (2011)
4. P.S. Bernard, Vortex dynamics in transitional and turbulent boundary layers. AIAA J. **51**, 1828–1842 (2013)
5. U. Rist, Visualization and tracking of vortices and shear layers in the late stages of boundary-layer laminar-turbulent transition. AIAA Paper 2012-0084 (2012)

On the Extension of Polymer Molecules in Turbulent Viscoelastic Flows: Statistical and Tensor Investigation

Anselmo Soeiro Pereira, Ramon Silva Martins, Gilmar Mompean, Laurent Thais and Roney Leon Thompson

Abstract In the present work, direct numerical simulations of turbulent channel flow of a viscoelastic FENE-P fluid, at zero-shear friction Reynolds number equal to 180, are used to analyze the polymer extension mechanism. As a primary focus, the relative polymer stretch and the probability distribution function of the alignment between the conformation tensor and other relevant entities are investigated. In near-wall regions, polymers present a strong tendency to orient along the streamwise direction of the flow. Furthermore, the polymer extension seems to be strongly correlated to the alignment between both conformation tensor and the velocity fluctuations product tensor, τ' (defined as $u'_i u'_j$). Joint probability density functions show that large positive polymer work fluctuations, E'_x, are closely related to the positive growth rate of the product of streamwise velocity fluctuations, $\partial_t u'^2_x$. In contrast, small negative fluctuations of polymer work are observed in the regions of negative rate of u'^2_x. However, in both cases, polymers are predominantly oriented along the principal direction of τ', which indicates the relevance of this tensor for the polymer-turbulence interaction mechanism.

A.S. Pereira (✉) · R.S. Martins · G. Mompean · L. Thais
Laboratoire de Mécanique de Lille (LML), CNRS, UMR 8107,
École Polytechnique Universitaire de Lille, Université Lille Nord de France,
F59655 Villeneuve D'ascq, France
e-mail: anselmo.pereira@polytech-lille.fr

R.S. Martins
e-mail: ramon.martins@polytech-lille.fr

G. Mompean
e-mail: gilmar.mompean@polytech-lille.fr

L. Thais
e-mail: laurent.thais@polytech-lille.fr

R.L. Thompson
Laboratório de Mecânica Teórica Aplicada (LMTA),
Department of Mechanical Engineering, Universidade Federal Fluminense,
Rua Passo da Pátria 156, Niterói, RJ 24210-240, Brazil
e-mail: rthompson@id.uff.br

M. Stanislas et al. (eds.), *Progress in Wall Turbulence 2*,
ERCOFTAC Series 23, DOI 10.1007/978-3-319-20388-1_15

1 Introduction

The addition of a small amount of high molecular weight polymers can lead to a pressure drop decrease in turbulent flows. Since this first observation [1–3] numerous experimental studies have been conducted in attempts to make practical use of polymer-induced drag reduction including long-distance transportation of liquids, oil well operations, firefighting, transport of suspensions and slurries, and biomedical applications [4]. In a remarkable and pioneering work, Virk et al. [5] performed careful analyses with an experimental turbulent pipe flow apparatus and showed that whether the friction drag for pipe flows is plotted in Prandtl-Kármán coordinates, polymer-induced drag reduction (DR) departs from Prandtl-Kármán law (onset of DR) to its bound, so-called maximum drag reduction (MDR) or Virk asymptote, as a result of Reynolds number, polymer concentration or polymer molecular weight increases. Over the years, researchers have successfully analyzed relevant aspects of this phenomenon [6]. However, up to now, there has been no definitive consensus concerning the interactions between the turbulent energy and polymer deformations.

Phenomenological polymer drag reduction explanations gravitate around two major theories. According to the viscous theory independently proposed by Lumley [7] and Seyer and Metzner [8], and supported by Ryskin [9], polymer stretching in a turbulent flow produces an increase in the effective viscosity in the region outside of the viscous sublayer and in the buffer layer which suppress turbulent fluctuations, increasing the buffer layer thickness and reducing the wall friction. The elastic theory postulated by Tabor and de Gennes [10] assumes that the elastic energy stored by the polymer becomes comparable to the kinetic energy in the buffer layer. Since the corresponding viscoelastic length scale is larger than the Kolmogorov scale, the usual energy cascade is inhibited, which thickens the buffer layer and reduces the drag.

In an attempt to quantify the viscous scenario, L'vov et al. [11] used conservation principles to show that an additional effective viscosity growing linearly with the distance from the wall in the buffer layer has similar effects to those observed by the addition of flexible polymers in turbulent flows. This theoretical prediction was later confirmed by De Angelis et al. [12], who performed DNS of Newtonian turbulent flows with an added viscosity profile obtaining results previously observed in viscoelastic FENE-P simulations. Additionally, the authors showed that, using this simple linear viscosity model, they were capable to predict the maximum drag reduction asymptote, a point discussed in detail by Benzi et al. [13]. It is also important to note that the elastic theory has been actively explored. Min et al. [14] conducted DNS of turbulent drag reducing channel flows in which the dilute polymer solution is simulated using the Oldroyd-B model. Their results showed good agreement with previous theoretical and experimental predictions of the onset of DR at specific friction Weissenberg numbers, which is interpreted based on elastic theory. Min et al. [14], as well as Dallas et al. [15], describe an elastic scenario in which the elastic energy stored in the near-wall region due to the uncoiling of polymer chains is transported to and, in some portion, released in the buffer and log layers. This storage of energy around near-wall vortices was confirmed by Dubief et al. [16], who performed DNS of turbulent polymer solutions in a channel using the FENE-P model, although, in contrast with Min et al. [14] and Dallas et al. [15], they proposed an autonomous regeneration cycle of polymer wall turbulence in which coherent release of energy occurs in the very near-wall region, just above the viscous layer.

Despite the discrepancies between the two most prominent theories, what seems to be in accordance with both scenarios is the relevance of the coil–stretch polymer process, which further imposes a transient behavior on drag reduction as well as subsequent polymer degradation as a consequence of high polymer elongation. In the present work, we investigate the process of polymer coil–stretch with the aid of direct numerical simulations of turbulent channel flow of a viscoelastic FENE-P fluid, at zero-shear friction Reynolds number equal to 180. Tensor and statistical analyses are developed. The relative polymer stretch and the alignment between the conformation tensor and other relevant entities are studied. Additionally, joint probability density functions are used in order to correlate the polymer–turbulence exchanges of energy and polymer orientations.

2 Methodology

A turbulent channel flow of an incompressible dilute polymer solution is considered. The channel streamwise direction is x_1=x, the wall-normal direction is $x_2 = y$, and the spanwise direction is $x_3 = z$. The instantaneous velocity field in the respective directions is $(u_1, u_2, u_3) = (u, v, w)$. The governing equations are scaled with the channel half-gap h, the bulk velocity U_b, and the fluid density ρ. The scaled momentum equations get the form

$$\frac{\partial u_i}{\partial t} + u_j \frac{\partial u_i}{\partial x_j} = \frac{\partial p}{\partial x_i} + \frac{\beta_0}{Re_b} \frac{\partial^2 u_i}{\partial x_j^2} + \frac{(1-\beta_0)}{Re_b} \frac{\partial \Xi_{ij}}{\partial x_j} + e_i \delta_{i1}. \tag{1}$$

In Eq. 1, β_0 is the ratio of the Newtonian solvent viscosity (ν_N) to the total zero-shear viscosity ($\nu_0 = \nu_N + \nu_{p0}$), and the bulk Reynolds number is $Re_b = \rho U_b h/\nu_0$.The extra-stress tensor is denoted by Ξ_{ij} and the quantity $e_i \delta_{i1}$ represents the non-periodic pressure gradient driving the flow in the streamwise direction. The formalism of Eq. 1 includes the assumption of a uniform polymer concentration which is governed by the viscosity ratio β_0 where $\beta_0 = 1$ yields the limiting behavior of the Newtonian case.

The extra-stress tensor (Ξ_{ij}) in Eq. 1 holds the polymer contribution to the solution tension. Such contribution is accounted for with a single spring-dumbbell model. We consider here the kinetic theory Finitely Extensible Non-linear Elastic in the Peterlin approximation (FENE-P) model. The FENE-P model is mostly preferred due to its physically realistic finite extensibility of the polymer molecules and to its relatively simple second-order closure. This model considers the phase-averaged conformation tensor $c_{ij} = \langle q_i q_j \rangle$, where q_i are the components of the end-to-end vector of each individual polymer molecule. The extra-stress tensor is then $\Xi_{ij} = \{f\,(tr\,(\mathbf{C}))\,c_{ij} - \delta_{ij}\}/We_b$ in which $We_b = \lambda U_b/h$ is the bulk Weissenberg number (λ being the relaxation time scale), δ_{ij} is the Kronecker delta operator and $f\,(tr\,(\mathbf{C}))$ is given by the Peterlin approximation $f\,(tr\,(\mathbf{C})) = \left(L^2 - 3\right) / \left(L^2 - tr\,(\mathbf{C})\right)$, where L is the maximum polymer molecule extensibility and $tr(.)$ represents the trace operator. The governing equation for the conformation tensor is

$$\frac{\partial C_{ij}}{\partial t} + u_k \frac{\partial C_{ij}}{\partial x_k} - \frac{\partial u_i}{\partial x_k} C_{kj} - \frac{\partial u_j}{\partial x_k} C_{ki} + \frac{f\,(tr\,(\mathbf{C}))\,c_{ij} - \delta_{ij}}{We_b} = \left(\frac{1}{Pr_c Re_b}\right) \frac{\partial^2 C_{ij}}{\partial x_k^2}. \tag{2}$$

in which $Pr_c = \nu_0/\rho\kappa_c$ is a stress Prandtl number defined as the ratio of the total kinematic zero-shear rate viscosity (ν_0/ρ) to an artificial stress diffusivity κ_c. This explicit elliptic diffusion term included in 2 is necessary to remove non-physical high wave-number instabilities typically induced by the chaotic nature of viscoelastic turbulent flows. This dissipative term was first introduced in this context by Sureshkumar and Beris [17], and the methodology subsequently validated under a variety of flow and material parameter values [18]. The numerical scheme for DNS used here was carefully detailed by Thais et al. [19].

3 Results and Discussion

A viscoelastic fluid can have significantly different mean and turbulent statistical behavior than a Newtonian fluid. For a given turbulence level as parameterized by the zero-shear friction Reynolds number, Re_τ (being $Re_\tau = \rho u_\tau h/\nu_0$, where u_τ is the friction velocity), this effect can vary with the friction Weissenberg number, We_τ and maximum polymer extension length, L. In this work, one Newtonian flow and four viscoelastic FENE-P flows are examined keeping $Re_\tau = 180$ fixed and using two different values of We_τ (being $We_\tau = \lambda \rho {u_\tau}^2/\nu_0$), and L ($We_\tau = 50$; $We_\tau = 115$; $L = 30$; $L = 100$). Our main results are separated into two parts. In Sect. 3.1, we analyze the evolution of polymer stretch along the wall distance, y^+, of which the effects on near-wall vortices and dependence on L and We_τ are investigated as well. Tensor and statistical investigations, and polymer–turbulence energy exchange analysis are conducted in Sect. 3.2 in an attempt to verify the alignment between the conformation tensor and other relevant entities and its effects on turbulent energy transfers.

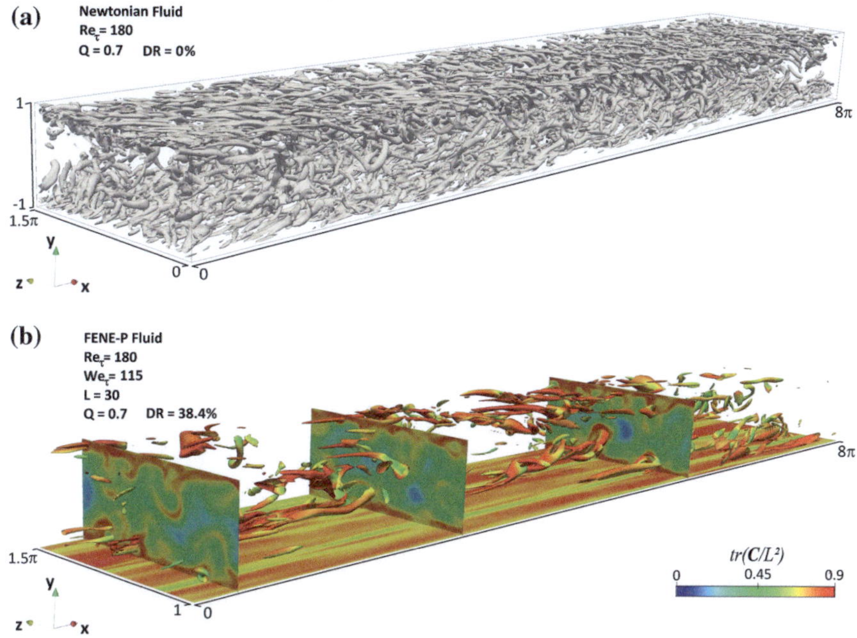

Fig. 1 The three-dimensional structures represent isosurfaces of vortical regions defined as the positive second invariant of velocity gradient tensor, $\nabla\mathbf{u}$. The colors in figures indicate the polymer stretch, $tr\left(\frac{\mathbf{C}}{L^2}\right)$

3.1 Polymer Stretch

The three-dimensional structures showed in Fig. 1 represent isosurfaces of vortical regions defined as the positive second invariant of velocity gradient tensor, $\nabla\mathbf{u}$, in Newtonian (a) and viscoelastic (b) flows. For incompressible flows, the second invariant of $\nabla\mathbf{u}$, the so-called *Q-criterion* [20], is simplified $Q = 1/2\left(||\mathbf{W}||^2 - ||\mathbf{D}||^2\right)$, which indicates spatial regions where the Euclidean norm of the rate of rotation tensor, $||\mathbf{W}||$, dominates that of the rate of strain, $||\mathbf{D}||$. Both $\mathbf{D}$ and $\mathbf{W}$ are defined as $\frac{1}{2}\left(\nabla\mathbf{u} + \nabla\mathbf{u}^T\right)$ and $\frac{1}{2}\left(\nabla\mathbf{u} - \nabla\mathbf{u}^T\right)$, respectively.

Comparing Fig. 1a, b it is observed that the number of vortices with a value of Q criterion equal to 0.7 decreases with increasing of elasticity (We_τ and L). In viscoelastic flows the vortical structures are significantly weaker than in the Newtonian flow, which is considered a fundamental evidence of the polymer–turbulence interactions and the consequently drag reduction [6, 21, 22]. As the elasticity increases, some vortices characteristic change. Their thicknesses and streamwise lengths increase, while their strengths weaken. Furthermore, vortices become more parallels to the wall. It has been experimentally and numerically shown that, in drag reducing flows, the streamwise component of the Reynolds normal stresses increase relative to the Newtonian case, while the other components of the Reynolds stress tensors decrease

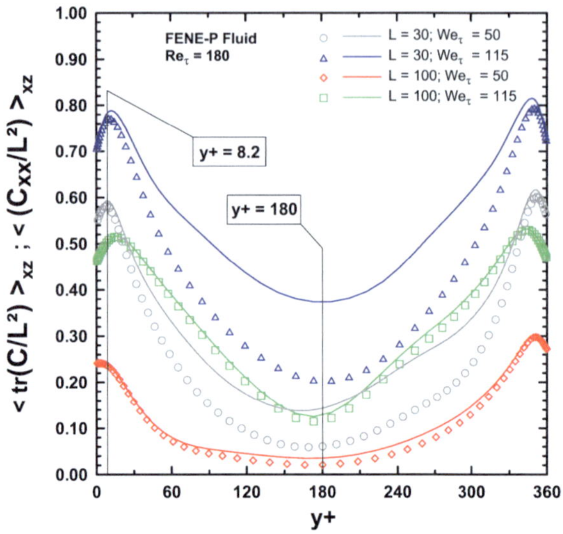

Fig. 2 Evolution of xz-plane average polymer stretch, $\langle tr\left(\frac{\mathbf{C}}{L^2}\right)\rangle_{xz}$, along the wall distance, y^+, normalized with the local friction velocity, u_τ

[23, 24]. These variations seem to be closely connected with the coil–stretch polymer transition and the following vortex structural changes [25]. The colors in Fig. 2b indicate the relative polymer stretch, $tr\left(\frac{\mathbf{C}}{L^2}\right)$. The yz-planes show that polymers are more stretched close to the wall (yellow and red regions). In contrast, polymer extensions are less pronounced in the middle of channel (blue regions). The isosurface colors and those of the intersections between vortical structures and yz-planes show that polymers present a more significant extension around the near-wall vortices.

The polymer stretch can be seen more clearly in Fig. 2, where the evolution of xz-plane average normalized trace of the instantaneous conformation tensor, $\langle tr\left(\frac{\mathbf{C}}{L^2}\right)\rangle_{xz}$, along the wall distance, y^+, is displayed (solid lines) together with the normalized streamwise normal component of the conformation tensor, $\langle\left(\frac{C_{xx}}{L^2}\right)\rangle_{xz}$ (open symbols). The polymer extension percentage, $\langle tr\left(\frac{\mathbf{C}}{L^2}\right)\rangle_{xz}$, is relatively high at the wall, achieving a peak in the near-wall region, of which the exact location varies with L and We_τ. This peak is commonly associated with the streamwise vortices [15, 16, 25]. After this point, $\langle tr\left(\frac{\mathbf{C}}{L^2}\right)\rangle_{xz}$ starts to decrease until reaches a minimum value at the channel center.

In comparing both gray and red solid lines in Fig. 2, it is clearly observed that $\langle tr\left(\frac{\mathbf{C}}{L^2}\right)\rangle_{xz}$ decreases with increasing L, keeping fixed Re_τ and We_τ, which suggests that large polymer molecules could be less susceptible to chain scission degradation [26, 27]. A further comparison of red and green solid lines reveals that the relative polymer extension becomes greater as the friction Weissenberg number increases, since higher values of polymer time scale are influenced from a wider spectrum of flow time scales [15]. Figure 2 also shows that the dominant contribution in conformation tensor trace comes from C_{xx}, i.e., $\langle tr\left(\frac{\mathbf{C}}{L^2}\right)\rangle_{xz} \approx \langle\left(\frac{C_{xx}}{L^2}\right)\rangle_{xz}$ (especially near the wall). This distribution suggests a significant stretching of the polymeric chain in the streamwise direction. The analysis of the conformation tensor trace reveals two locations of interest: $y^+ = 8.2$, the approximate position where $\langle tr\left(\frac{\mathbf{C}}{L^2}\right)\rangle_{xz}$ is maximum; $y^+ = 180$, where the conformation tensor trace reaches its minimum value.

3.2 Tensor and Statistical Analysis of Polymer Orientation and Polymer–Turbulence Exchanges of Energy

Figure 3 shows the evolution of xz-plane average cosine of the angle Φ between the principal direction of relevant entities, e_1 (the eigenvector related to the largest eigenvalue) and the streamwise direction, e_x, along y^+. These relevant entities correspond to the conformation tensor, **C**, the velocity fluctuations product tensor, τ', the rate of strain tensor, **D**, and the vorticity vector, ω, which is defined as the non-null components of **W**.

For all viscoelastic fluids investigated here, the conformation tensor (open gray balls) exhibits an important alignment along the streamwise direction, which is more pronounced at the wall, achieving a minimum value in the middle of channel (a). This minimum value grows with increasing of both We_τ and L. Consequently, in the most viscoelastic case (b), $\langle\cos\left(e_1^{\mathbf{C}}, e_x\right)\rangle_{xz} \approx 1$ for any wall distance.

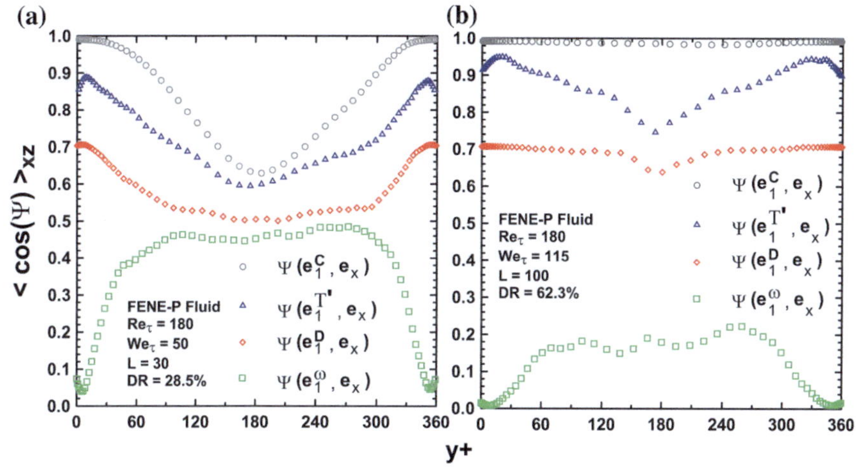

Fig. 3 Xz-plane average cosines of the angles between the principal direction of a given tensor and the unit vector e_x along the normalized wall distance, y^+

The rate of strain tensor (red diamonds) presents a similar behavior of that described for **C**. The average angle between **D** and e_1 is maximum and equal to 45° within the viscous sublayer ($0 < y^+ < 5$) since the main component of the strain rate in this flow region comes from the wall-normal shear $\frac{\partial u}{\partial y}$. As the wall distance increases, $\frac{\partial u}{\partial y}$ becomes less important and $\langle \cos\left(e_1^{\mathbf{D}}, e_x\right)\rangle_{xz}$ decreases smoothly from the *buffer layer* ($5 < y^+ < 30$) to the *outer layer* ($y^+ > 50$). However, as We_τ and L rise, this 45 degrees angle is maintained, even at y^+ far from the wall (see Fig. 3b), a fact linked whit the extension of the *buffer layer* region into the channel caused by the polymers [24].

The orientation of τ' along the streamwise direction is relatively accentuated at the wall, achieving a peak in the near-wall region, of which the exact location vary with L and We_τ as well as $\langle tr\left(\frac{\mathbf{C}}{L^2}\right)\rangle_{xz}$. After this point, $\langle \cos\left(e_1^{\tau'}, e_x\right)\rangle_{xz}$ stars to decrease until reaching a minimum value at the channel center. Comparing Fig. 1a, b, it is clearly observed that $\langle \cos\left(e_1^{\tau'}, e_x\right)\rangle_{xz}$ increases when the elasticity is incremented.

Looking at the relative orientation between the vorticity vector and the streamwise direction (green squares), it can be concluded that the peak magnitude of $\langle \cos\left(e_1^{\omega}, e_x\right)\rangle_{xz}$ is smaller than that for the other curves in Fig. 1, presenting a zero wall value and growing with increasing wall distance. This peak magnitude falls with increasing elasticity since ω becomes more aligned along the spanwise direction with increasing We_τ and L (not shown here).

It is worth noting that the preferential orientation of both **C** and τ' along the x direction reveals a strong coaxiality between these two tensors. This alignment seems to play an important role in both coil–stretch process and polymer–turbulence exchanges of energy, which is clarify in Fig. 4. For this figure, let us consider the Reynods stress equation,

$$\underbrace{\frac{1}{2}\frac{\partial u_x'^2}{\partial t}}_{T_x'} = \underbrace{-u'\frac{\partial\left(u_x' u_j'\right)}{\partial x_j}}_{A_x'} \underbrace{-u_x'\frac{\partial p'}{\partial x}}_{P_x'} + \underbrace{\frac{\beta_0}{Re_b}u_x'\frac{\partial u_x'^2}{\partial^2 x_j}}_{V_x'} + \underbrace{\frac{(1-\beta_0)}{Re_b}u_x'\frac{\partial \Xi_{xj}'}{\partial x_j}}_{E_x'}, \tag{3}$$

where the instantaneous amount of energy which is stored ($E_x' < 0$) or released ($E_x' > 0$) by polymers from the streamwise velocity fluctuation, u_x', is represented by E_x. The complementary work terms denote de advection A_x', the pressure redistribution P_x', and the viscous stress V_x', all of them in the streamwise direction. The sum $A_x' + P_x' + V_x'$ is referred here as to Newtonian work, N_x' [16].

Figure 4 shows three different joint probability density functions (JPF) considering the xz-plane located close to the wall, at $y^+ = 8.2$ (approximately where the relative polymer extension is maximum), for the less viscoelastic case analyzed here ($We_\tau = 50$, $L = 30$). The black solid line refers to the JPF of E_x' versus T_x', whereas the dashed line indicates the JPF linking E_x' and N_x'. Lastly, the dot-dash line represents the JPF which considers the polymer work fluctuation and the cosine of the angle

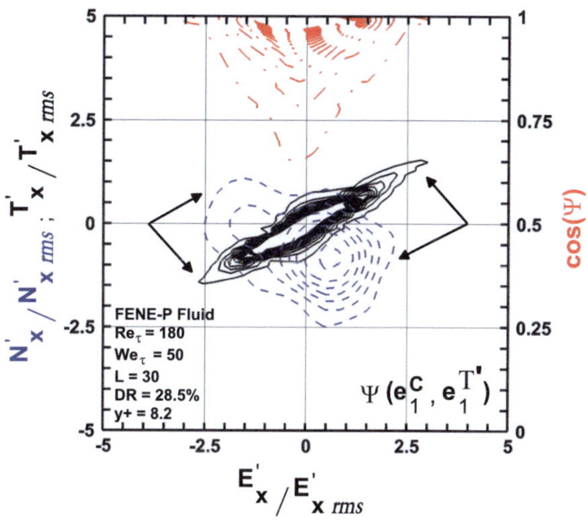

Fig. 4 Joint probability density functions of polymer work versus other relevant quantities at $y^+ = 8.2$. For each work term, fluctuations are normalized by the respective standard global deviation

between the principal directions **C** and τ'. The work terms were normalized by their respective spatial root mean square (*rms*, which considers the whole channel).

First, with regard to Fig. 4, it is important to observe that polymers are allowed to coil within this region due to the reduction of turbulent kinetic energy (T'_x) by viscous dissipation (V'_x), as exposed by Dubief et al. [16]. In comparing Newtonian work to polymer work, it is apparent that large positive polymer work fluctuations occur in regions where the Newtonian turbulent work is negative. In contrast, E'_x is closely related to the positive growth rate of the product of streamwise velocity fluctuations, $\partial_t u'^2_x$, indicating an important injection of energy into the flow. Quite surprisingly, in these both opposite scenarios, polymers are predominantly oriented along the principal direction of τ' ($\cos\left(e_1^{\mathbf{C}}, e_1^{\tau'}\right) \approx 1$), which reveals the relevance of this tensor for the polymer–turbulence interactions. These observations are equally valid for the other viscoelastic cases (not shown here).

4 Final Remarks

We investigated the process of polymer coil–stretch with the aid of direct numerical simulations of turbulent channel flows. One Newtonian flow and four viscoelastic FENE-P flows were examined keeping $Re_\tau = 180$ fixed and using two different values of We_τ and L ($We_\tau = 50$; $We_\tau = 115$; $L = 30$; $L = 100$). Polymers present a strong tendency to orient along the streamwise direction of the flow in near-wall regions (where their extension are accentuated), as well as the velocity fluctuations product tensor, τ'. Joint probability density functions show that large positive polymer work fluctuations, E'_x, are closely related to the positive growth rate of the product of streamwise velocity fluctuations, $\partial_t u'^2_x$. In contrast, small negative fluctuations of polymer work are observed in regions of negative rate of u'^2_x. However, in both cases, polymers are predominantly oriented along the principal direction of τ'.

Acknowledgments The authors are grateful to Dr. Enrico Calzavarini and Dr. Stefano Berti from the Laboratoire de Mécanique de Lille of Université Lille Nord de France for their useful comments and suggestions. This work was granted access to the HPC resources of IDRIS under the allocation 2014-i20142b2277 made by GENCI. The authors would also like to express their acknowledgment and gratitude to the Brazilian Scholarship Program *Science Without Borders*, managed by CNPq (National Council for Scientific and Technological Development), for the partial financial support for this research.

References

1. F. Forrest, G.A. Grierson, Pap. Trade J. **92**, 39 (1931)
2. B.A. Toms, *Proceedings of the International Congress of Rheology*, Section II (Holland, North-Holland, Amsterdam, 1948), pp. 135–141
3. K.J. Mysels, US Patent 2 **492**, 173, 27 December 1949
4. R.H.J. Sellin, J.W. Hoyt, J. Poliert, O. Scrivener, J. Hydraul. Res. **20**, 235 (1982)
5. P.S. Virk, H.S. Mickley, K.A. Smith, J. Fluid Mech. **22**, 22 (1967)
6. C.M. White, M.G. Mungal, Annu. Rev. Fluid Mech. **40**, 235 (2008)
7. J.L. Lumley, Annu. Rev. Fluid Mech. **11**, 367 (1969)
8. F.A. Seyer, A.B. Metzner, AIChE J. **492**, 426 (1949)
9. G. Ryskin, Phys. Rev. Lett. **59**, 2059 (1987)
10. M. Tabor, P.G. de Gennes, Europhys. Lett. **7**, 519 (1986)
11. V.S. L'vov, A. Pomyalov, I. Procaccia, V. Tiberkevich, Phys. Rev. Lett. **92**, 244503 (2004)
12. E. De Angelis, C. Casciola, V.S. L'vov, A. Pomyalov, I. Procaccia, V. Tiberkevich, Phys. Rev. E **70**, 055301 (2004)
13. R. Benzi, E.D. Angelis, V.S. L'vov, I. Procaccia, Phys. Rev. Lett. **95**, 194502 (2005)
14. T. Min, J.Y. Yoo, H. Choi, D.D. Joseph, J. Fluid Mech. **486**, 213 (2003)
15. V. Dallas, J.C. Vassilicos, G.F. Hewitt, Phys. Rev. E **82**, 066303 (2010)
16. Y. Dubief, C.M. White, V.E. Terrapon, E.S.G. Shaqfeh, P. Moin, S.K. Lele, J. Fluid Mech. **514**, 271 (2004)
17. R. Sureshkumar, A.N. Beris, J. Non-Newton. Fluid Mech. **60**, 53 (1995)
18. K.D. Housiadas, A.N. Beris, Phys. Fluids **15**(8), 2369 (2003)
19. L. Thais, A. Tejada-Martinez, T.B. Gatski, G. Mompean, Comput. Fluids **43**, 134 (2011)
20. J.C.R. Hunt, A.A. Wray, P. Moin, in *Proceedings of Summer Program. Center for Turbulence Research*, Report CTR-S88 (1988) p. 193
21. K. Kim, C.F. Li, R. Sureshkumar, L. Balachandar, R.J. Adrian, J. Fluid Mech. **584**, 281 (2007)
22. K. Kim, R.J. Adrian, L. Balachandar, R. Sureshkumar, Phys. Fluids **100**, 134504 (2008)
23. M.D. Warholic, H. Massah, T.J. Hanratty, Exp. Fluids **27**, 461 (1999)
24. L. Thais, T.B. Gatski, G. Mompean, J. Turbul. **13**, 1 (2012)
25. C.D. Dimitropoulos, Y. Dubief, E.S.G. Shaqfeh, P. Moin, S.K. Lele, Phys. Fluids **17**, 1 (2005)
26. A.S. Pereira, E.J. Soares, J. Non-Newton. Fluid Mech. **179**, 9 (2012)
27. A.S. Pereira, R.M. Andrade, E.J. Soares, J. Non-Newton. Fluid Mech. **202**, 72 (2013)

Velocity of Line Plumes on the Hot Plate in Turbulent Natural Convection

Vipin Koothur and Baburaj A. Puthenveettil

Abstract We measure the velocity field in a horizontal field near the hot plate in turbulent convection using stereo PIV for $10^6 < Ra_w < 10^9$ and $5.2 < Pr < 4$. We then extract the line plumes from this velocity field using a divergence criterion using the PIV technique on the obtained plume structures, which gives us the velocity field of the plume motion. The statistical analysis of this velocity field of the plume motion shows the coexistence of two different kinds of motion of the plumes, lateral merging and motion along the plumes.

1 Introduction

Natural convection over a heated horizontal plate, where a fluid layer of height H is heated from the bottom is extensively studied as a simple system of buoyancy generated turbulence in laboratories. Such flows are characterized by the Rayleigh number and Prandtl numbers, given by

$$Ra = g\beta \Delta T_w H^3/\nu\alpha, \quad Pr = \nu/\alpha \tag{1}$$

and a geometric parameter $AR = L/H$. The nondimensional flux Nusselt number $Nu = q/(\alpha \Delta T_w/H)$, depends on the above parameters. Here, g is the acceleration due to gravity, β is the coefficient of thermal expansion, ΔT_w is the temperature difference between the hot plate and the bulk fluid, ν is the kinematic viscosity, α is the thermal diffusivity, L is horizontal dimension of the fluid layer, and q is the kinematic heat flux. The flow near the hot plate is characterized by rising sheets of fluids called line plumes which are separated by the surrounding bulk fluids which

V. Koothur (✉) · B.A. Puthenveettil
Department of Applied Mechanics, Indian Institute of Technology Madras, Chennai, India
e-mail: vipink159@gmail.com

B.A. Puthenveettil
e-mail: apbraj@iitm.ac.in

M. Stanislas et al. (eds.), *Progress in Wall Turbulence 2*,
ERCOFTAC Series 23, DOI 10.1007/978-3-319-20388-1_16

entrain into them. Dynamics of these line plumes on the wall is important since they play a major role in the transport of heat from the hot plate to the bulk.

Spangenberg and Rowland [10] were the first to study these structures experimentally at the free surface in evaporative convection, they found the transient nature of forming, random motion, and plunging down of line plumes. Sparrow, Husar, and Goldstein [5] also studied the patterns of line plumes on a horizontal plate with no sidewalls. Theerthan and Arakeri [11] visualized these structures using liquid crystal sheet and found that these line plumes move about randomly and the adjacent plumes merge with each other most of the time. Zocchi et al. [13] used encapsulated liquid crystal particles to visualize and measure the horizontal velocities of these plumes. Puthenveettil and Arakeri [8] captured the planform of near wall structures in concentration driven convection and found that the PDF of the spacing between these plumes showed a common log-normal form at different *Ra* and *Pr*. The mean plume spacing at any *Ra* and *Pr* was shown to be proportional to the near wall length scale in turbulent convection

$$Z_w = (\nu\alpha/g\beta\Delta T_w)^{1/3} = H/Ra_w^{1/3} \quad (2)$$

(Puthenveettil and Arakeri, Theertan and Arakeri [8, 11]). Puthenveettil et al. [9] gave the relation for the total length of the plume in any given area *A* at any Ra_w and *Pr* as

$$L_p = A/C_1 Pr^{n_1} Z_w \quad (3)$$

where $C_1 = 47.5$ $n_1 = 0.1$.

Gunasegarane and Puthenveettil [4] have studied the dynamics of the near plate and showed plume merging motion along plumes and initiation of plumes to be the dominant dynamics. There were other minor motions too, with all these motions being dependent on Ra_w and *Pr*. They showed that the merging velocities V_m at all the Ra_w had a common log-normal distribution and were an order lower than shear velocity V_{sh}. The Reynolds number Re_H based on this merging velocity was found to scale as $Ra_w^{1/3}$. In the present work we study the coexistence of these two dominant dynamics near the hot plate. We obtain the velocity field of the motion of these plume structures using PIV. We then study the probability distribution of these motions and their dependence on the flow parameters and compare our results with that given by [4].

2 Experimental Setup

The measurements were carried out in an open tank of size 300 × 300 × 250 mm with four glass sidewalls [7]. The bottom hot plate was 10 mm thick and was made of copper. The heat was supplied from resistance heating of a Ni–Cr wire placed below an aluminum plate. The Al plate was separated from the Cu plate by a glass plate. The temperature difference measured across the glass plate gives the heat flux supplied to the copper plate. The temperature difference ΔT_w between the bottom plate and bulk fluid was measured by T-type thermocouples with precision limit of 0.02°.

2.1 Stereo PIV System

The velocity field in a horizontal plane close to the bottom plate, for Ra_w ranging from 10^6–10^9 and Pr ranging from 5.2–4 were obtained using Stereo PIV technique [1] with forward scattering as shown in Fig. 1. The horizontal laser sheet (532 nm Nd:YAG laser at 100 mJ/pulse) was passed above the hot plate parallel to it in such a way that it was inside the velocity boundary layer thickness $\delta_v \approx \sqrt{\nu L/u_c}$ as given by Puthenveettil et al. [8] for a natural convection boundary layer. Since $\delta_{vb} \gg \delta_v$ for all Ra_w and Pr, the laser sheet was also within the Blasius boundary layer thickness given by Ahlers et al. [2] as $\delta_{vb} = 0.482H/Re_L^{1/2}$. The seeding particles used were polyamide spheres with mean diameter of ~55 μm and density of 1.023 g/cm^3. The

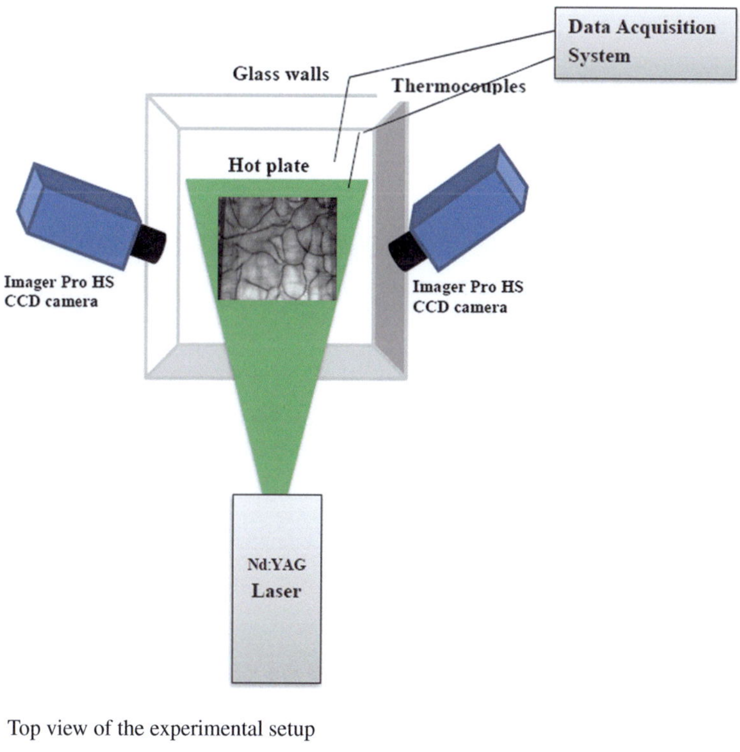

Fig. 1 Top view of the experimental setup

particle images were captured by two Imager ProHS CCD cameras (1280 × 1024) px which were kept at 25° to a plane perpendicular to the laser sheet. Scheimpflug adapters were used to focus the camera lenses on the laser plane. The magnification and resolution were changed by changing the focal length of the lens attached to the cameras keeping the distance between the object and camera to be fixed. The depth of field was set to be 2 mm larger than the laser sheet thickness with the lens aperture set at $f = 3.5$. The stereo PIV was calibrated using a two-level calibration plate provided by LaVision. In a plane parallel to the hot plate, the highest velocity would be the centerline out-of-plane rise velocity of the plume, we set the separation between the laser pulses so that the displacement due to this velocity is at least half the laser sheet thickness. The imaging area was chosen in the center of the plate. The Stereo PIV images were processed by 'DaVis 7.2' software using a multipass cross-correlation method with a fixed window size of 32 × 32 px with 50 % overlap to get the velocity vector maps.

3 Identification of Line Plumes

The velocity field of the near plate regions in turbulent convection is complex as shown in Fig. 2. It has rising line plumes accompanied by flow in the boundary layer on either sides of it at any instant. Along with this, the effect of spatially varying external shear due to the large-scale flow is also seen in the velocity field, which becomes predominant at higher Rayleigh numbers. Due to this complex nature, not all plumes can be identified from a qualitative observation of the velocity field. In order to extract the line plumes from the velocity field, we tried to use some of the criteria used in near-wall shear turbulence to extract coherent structures, such as the swirling strength method [12] and λ_2 method [6]. However, these methods were found to fail since the total length of the structures obtained from these did not match the values obtained from (3).

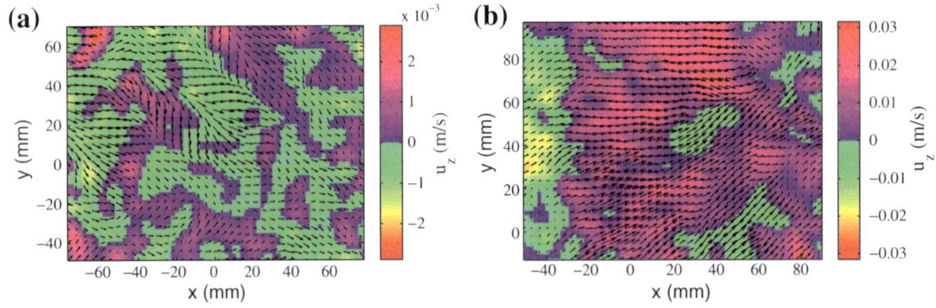

Fig. 2 The 2D vector field in a plane overlaid over the vertical component of velocity for **a** $Ra_w = 3.89 \times 10^6$ and $Pr = 5.2$, **b** $Ra_w = 1.37 \times 10^9$ and $Pr = 4$

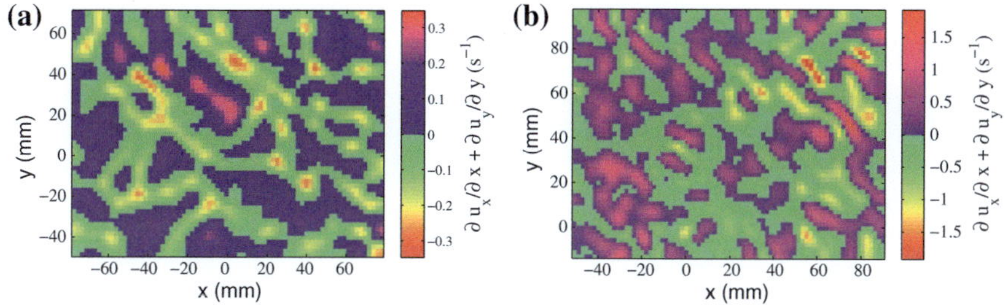

Fig. 3 The 2D divergence field of the velocity fields in Fig. 2; **a** $Ra_w = 3.89 \times 10^6$ and $Pr = 5.2$, **b** $Ra_w = 1.37 \times 10^9$ and $Pr = 4$

We hence propose a new technique to extract line plumes from the velocity field in horizontal plane close to the plate. Figure 3 shows the 2D divergence field obtained from the velocity field in Fig. 2. The divergence field separates into regions with negative and positive divergences. Since the divergence field shows the distribution of $du/dx + dv/dy = -dw/dz$, where z is the direction perpendicular to plane of measurement and w is the velocity component in that direction, the negative divergence regions are regions where fluid has a positive vertical spatial acceleration. We expect such regions to be plumes since it is known that vertical velocity increases with height inside the plumes [3]. As could be seen from Fig. 3a, b, these regions with negative divergence also look like lines. The total length of the negative divergence regions measured from the divergence field in Fig. 3a was 1134 mm and was close to the plume length obtained from the expression (3) which was 1066 mm. Comparing Figs. 2 and 3 it is clear that the downflow regions in between the plumes that have positive divergence show a velocity field that is usually directed toward the plume lines that bound these regions. Hence, these regions show the velocity field within the boundary layers that become unstable to give rise to the line plumes.

3.1 Velocity of Line Plumes

The horizontal velocity field of the line plumes in the plane of laser, as against the complete velocity field in the plane is obtained by subjecting the divergence field to a PIV analysis. We apply a threshold on the images of the divergence field by setting the positive divergence values to zero. The plume structures are turned to white pixels and we find the velocity field of these structures using PIVlab algorithm in MATLAB. An adaptive multipass cross-correlation method is used to obtain the vector field. The largest window gives the mean velocity of the largest structure in that window. The window size is reduced in the next steps with their position determined based on the displacement vector obtained in the previous step. This is continued until velocity vectors of small groups of structures are obtained with higher spatial resolution. The time difference between each frames were chosen as twice the time difference of the

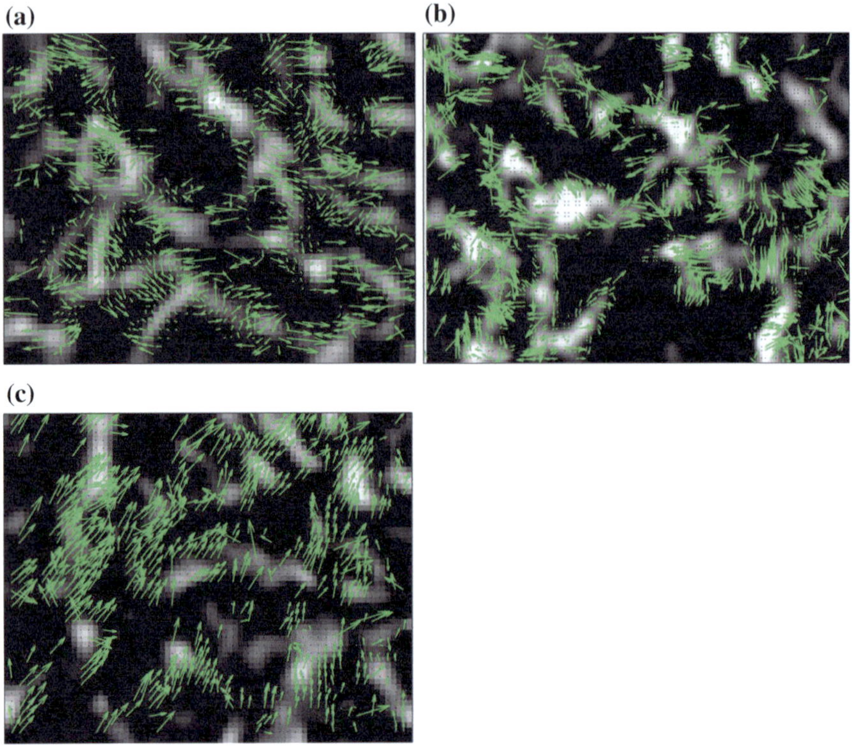

Fig. 4 The 2D vector field of the plume motion overlaid over the plume structures which are obtained as the negative regions of the divergence field at **a** $Ra_w = 3.84 \times 10^6$ and $Pr = 5.2$, **b** $Ra_w = 7.29 \times 10^7$ and $Pr = 4.75$, and **c** $Ra_w = 1.43 \times 10^9$ and $Pr = 4$. The image sizes are of the same as in Fig. 2

original particle images from which the divergence field were obtained. In the post-processing step, we remove spurious vectors by excluding the range of velocities that are questionable from the scatter plot of the velocity field. Figure 4 shows the velocity field obtained due to the motion of plume structures near the wall.

4 Results

As seen in Fig. 5, the velocity field for the lower Ra_w has large number of smaller vectors as compared to the higher Ra_w cases. The smaller vectors are more along the edges of the plumes as compared to the larger vectors. We analyze the horizontal velocity field of the structures by taking the histogram of the magnitudes of velocity $U = \sqrt{u^2 + v^2}$, where u and v are the velocity components in the horizontal plane, at each Ra_w. Figure 5 shows the respective histograms for the velocity fields in Fig. 4.

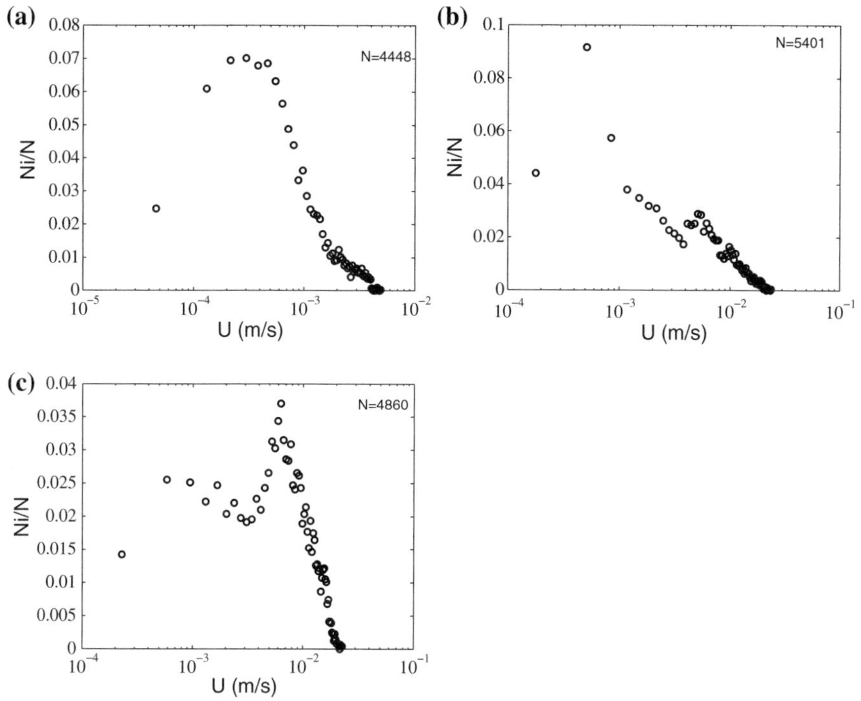

Fig. 5 Histogram of velocities obtained from the 2D vector field of the plume motions at **a** $Ra_w = 3.84 \times 10^6$ and $Pr = 5.2$, **b** $Ra_w = 7.29 \times 10^7$ and $Pr = 4.75$, and **c** $Ra_w = 1.43 \times 10^9$ and $Pr = 4$

It is clear from the histograms that velocity field of plumes at the lower Ra_w flow has a normal distribution with a single peak. As the Ra_w increases, the histogram of the velocity distribution becomes bimodal due to the rise of a second peak and the relative strength of this second peak with respect to the first peak increases with Ra_w. Such an evolution of the histogram with Ra_w shows that the plumes predominantly have a single type of motion at the lower Ra_w but with increase in Ra_w, a second type of motion becomes more dominant. The single peak at the lower Ra_w could be because there is hardly any large-scale flow present at this Ra_w. The rise of the second peak at higher Ra_w could be due to the increasing effect of the large-scale flow on the plumes.

We now separate the two modes from the bimodal histogram shown in Fig. 5 and study the nature of PDF associated with these modes. The velocities under each of these modes are normalized with their median $\bar{U}$ and the PDF is plotted for the nondimensional parameter χ where

$$\chi = (log(U/\bar{U}) - \overline{(logU/\bar{U})})/\sigma log(U/\bar{U}) \tag{4}$$

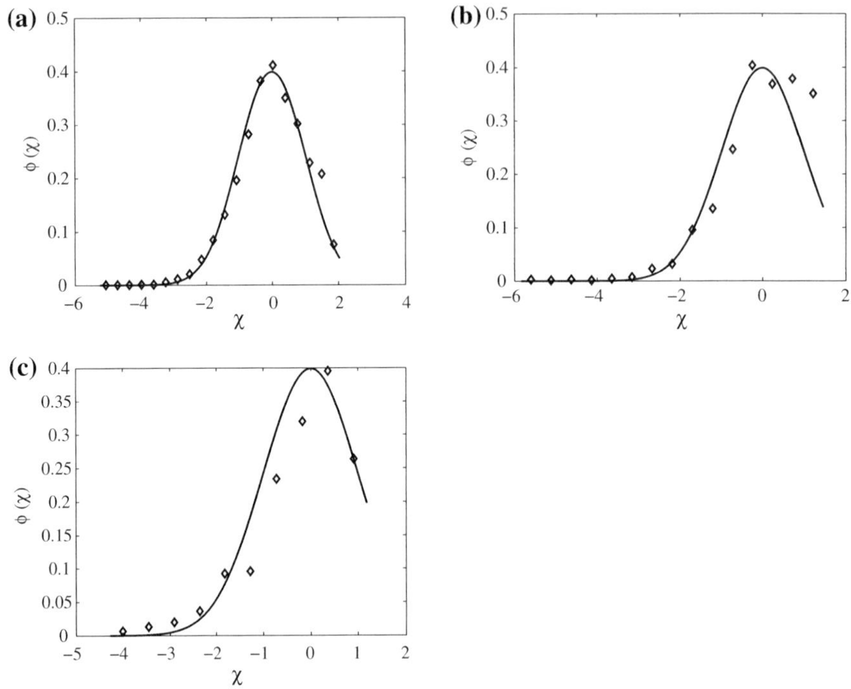

Fig. 6 The probability density function ϕ of χ, the logarithm of the dimensionless velocities from the first peak of the histograms in Fig. 5, in their standardised forms at **a** $Ra_w = 3.84 \times 10^6$ and $Pr = 5.2$, **b** $Ra_w = 7.29 \times 10^7$ and $Pr = 4.75$, and **c** $Ra_w = 1.43 \times 10^9$ and $Pr = 4$. The *solid lines* indicate the standard Gaussian curve

At all Ra_w, a normal distribution curve fits the PDFs obtained from both the modes in Fig. 5, as shown in Figs. 6 and 7. Thus the velocities of the dominant dynamics in a plane near the plate in turbulent convection have a log-normal distribution at all Ra_w.

The mean peak values obtained from these curves were used to find the Reynolds number Re_H and Re_{sh} in terms of the merging velocity V_m and shear velocity V_{sh} respectively. Gunasegarane and Puthenveettil [4] had given the relation for these Reynolds numbers in terms of Ra_w and Pr as

$$Re_H = V_m H/\nu = 0.55 Ra_w^{1/3} Pr^{-3/4} \tag{5}$$

and

$$Re_{sh} = V_{sh} H/\nu = 0.55 Ra_w^{4/9} Pr^{-2/3} \tag{6}$$

We used the velocity corresponding to the first peak as V_m and that corresponding to the second peak as V_{sh} to calculate Re_H and Re_{sh} from our measurements. The

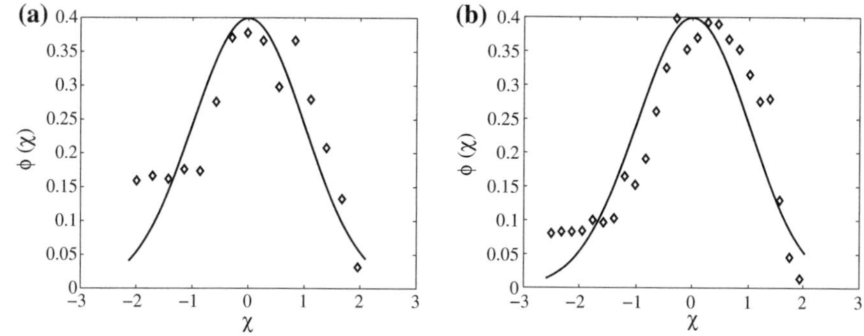

Fig. 7 The probability density function ϕ of χ, the logarithm of the dimensionless velocities from the second peak of the histograms in Fig. 5, in their standardised forms at **a** $Ra_w = 7.29 \times 10^7$ and $Pr = 4.75$, **b** $Ra_w = 1.43 \times 10^9$ and $Pr = 4$. The *solid lines* indicate the standard Gaussian curve

Fig. 8 Variations of : o, Re_H and +, Re_{sh} with Ra_w. −. − .− $0.55Ra_w^{4/9}Pr^{-2/3}$ and − − −− $0.55Ra_w^{4/9}Pr^{-2/3}$

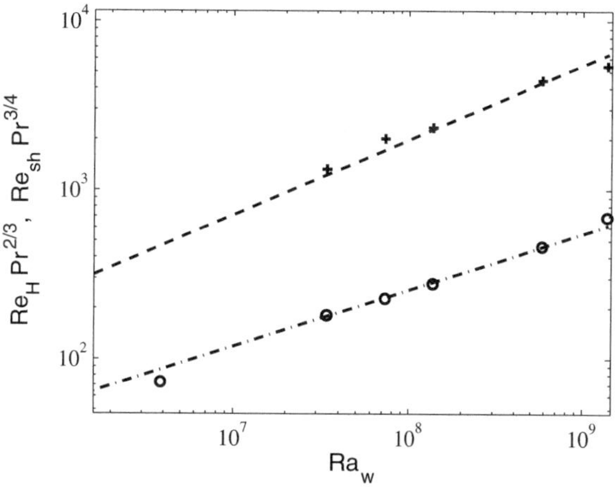

variation of these Re_H and Re_{sh} with Ra_w are shown in Fig. 8 along with the curves corresponding to (5) and (6). The excellent match obtained between the experimental Re_H and Re_{sh} and the relations (5) and (6) in Fig. 8 shows that the first mode in Fig. 5 is due to the merging motion of the plumes, while the second mode is due to the longitudinal motion of the plumes.

5 Conclusion

Stereo PIV technique was used to obtain the velocity field in a horizontal plane very close to the hot plate in turbulent natural convection. We showed that 2D divergence criterion extracts the line plumes from these velocity fields satisfactorily. A spatial cross-correlation of the divergence field gave the velocity field of the motion of plumes. The histograms of the velocity of line plumes have bimodal distributions with each of modes being distributed log-normally. The dependence of Reynolds number obtained from these modes with Ra_w agree well with the theoritical relations of in [4].

References

1. R.J. Adrian, J. Westerweel, *Particle Image Velocimetry* (Cambridge University Press, New York, 2011)
2. G. Ahlers, S. Grossman, D. Lohse et al., Heat transfer and large scale dynamics in turbulent Rayleigh-Benard convection. Rev. Mod. Phys. **81**, 503 (2009)
3. B. Gebhart, L. Pera, A.W. Schorr, Steady laminar natural convection plume above a horizontal line heat source. Int. J. Heat Mass Transf. **13**, 161–171 (1970)
4. G.S. Gunasegarane, B.A. Puthenveettil, Dynamics of line plumes on horizontal surfaces in turbulent convection. J. Fluid Mech. **749**, 37–78 (2014)
5. R.B. Husar, E.M. Sparrow, Patterns of free convection flow adjacent to horizontal heated surfaces. Int. J Heat Mass Transf. **11**, 1208–1211 (1968)
6. J. Jeong, F. Hussain, On the identification of a vortex. J. Fluid Mech. **285**, 69–94 (1995)
7. V. Koothur, B.A. Puthenveettil, Horizontal velocity field near the hot plate in turbulent natural convection, in *Proceedings of the 14th ACFM*, Hanoi-Halong, Vietnam, ISBN: 978-604-913-146-2:90 (2013)
8. B.A. Puthenveettil, J.H. Arakeri, Plume structure in high Rayleigh number convection. J. Fluid Mech. **542**, 217–249 (2005)
9. B.A. Puthenveettil, G.S. Gunasegarane, K. Yogesh, D. Schmeling, J. Bosbach, J.H. Arakeri, Length of near wall plumes in turbulent convection. J. Fluid Mech. **685**, 335–364 (2011)
10. W.G. Spangenberg, W.G. Rowland, Convection circulation in water induced by evaporative cooling. Phys. Fluids **4**(6), 743–750 (1961)
11. S.A. Theerthan, J.H. Arakeri, A model for near wall dynamics in turbulent Rayleigh-Benard convection. J. Fluid Mech. **373**, 221–254 (1998)
12. J. Zhou, R.J. Adrian, S. Balachandar, T.M. Kendall, Mechanisms for generating coherent packets of hairpin vortices in channel flow. J. Fluid Mech. **387**, 353–396 (1999)
13. G. Zocchi, E. Moses, A. Libchaber, Coherent structures in turbulent convection, an experimental study. Phys. A **166**, 387–407 (1990)

LES of a Converging–Diverging Channel Performed with the Immersed Boundary Method and a High-Order Compact Discretization

Mariusz Ksiezyk and Artur Tyliszczak

Abstract The paper presents the results of numerical simulations of an incompressible flow in a converging–diverging channel performed with Large Eddy Simulation (LES) combined with the immersed boundary (IB) method. The computations are carried out using a high-order code with the spatial discretization based on the compact difference method for half-staggered meshes. IB method is implemented in the so-called direct forcing approach with a second-order interpolation near the boundaries. Two relatively new subgrid models are used in the simulations, i.e. the model proposed by Vreman, Phys Fluids 16:3670–3681, 2004, [1] and the model proposed by Nicoud et al., Phys Fluids 23:193–202, 2011, [2]. It is demonstrated that both of them perform well and there is no evident advantage for either of them. The mean and r.m.s velocity profiles agree with exemplary DNS data.

1 Introduction

Undoubtedly, from the point of view of a solution accuracy none of the discretization methods may compete with the spectral and pseudo-spectral methods which are regarded as the most accurate [3]. The weak point of these approaches is that they can only be applied in rather simple computational domains and with nodes distribution and boundary conditions enforced by the type of the method. The high-order compact difference methods [4] seem to give more possibilities regarding non-uniformity of the computational meshes, selection of the boundary conditions or shapes of computational domains. They are successfully applied on non-uniform meshes and in irregular domains [5–7]. However, such applications require domain division, normalisation, coordinate transformations, etc., which are not trivial tasks. Possibly the easiest solution allowing to use the compact methods in complicated domains is to combine them with the so-called Immersed Boundary (IB) method. Application of

M. Ksiezyk (✉) · A. Tyliszczak
Czestochowa University of Technology, Czestochowa, Poland
e-mail: ksiezykm@imc.pcz.czest.pl

A. Tyliszczak
e-mail: atyl@imc.pcz.czest.pl

M. Stanislas et al. (eds.), *Progress in Wall Turbulence 2*,
ERCOFTAC Series 23, DOI 10.1007/978-3-319-20388-1_17

this approach seems to be relatively easy and very efficient [8]. The Navier–Stokes equations are solved on Cartesian regular grids with arbitrary boundaries or arbitrary objects embedded directly on the grid points. The influence of such objects on the flow field is enforced by body force terms added to the Navier–Stokes equations [8, 9]. The present work focuses on application of the high-order compact method with IB approach for LES of incompressible flows. The computations are performed using two well-known and relatively new subgrid models proposed by Vreman [1] and Nicoud et al. [2] and the obtained solutions agree very well with DNS data.

2 Mathematical Model and Numerical Algorithm

Flow of an incompressible fluid is governed by the continuity equation and Navier–Stokes equations which in the framework of LES combined with the IB method are given as:

$$\frac{\partial \bar{u}_i}{\partial x_i} = 0 \tag{1}$$

$$\frac{\partial \bar{u}_i}{\partial t} + \bar{u}_j \frac{\partial \bar{u}_i}{\partial x_j} = -\frac{1}{\rho}\frac{\partial \bar{p}}{\partial x_i} + \frac{\partial}{\partial x_j}\left((\nu + \nu_t)\left(\frac{\partial \bar{u}_i}{\partial x_j} + \frac{\partial \bar{u}_i}{\partial x_j}\right)\right) + f_{IB} \tag{2}$$

where the bar symbol denotes spatial filtering [10], u_i are velocity components, ρ is constant density, p—pressure, ν, ν_t—kinematic and eddy viscosity and f_{IB} denotes source terms which will be used to force zero values of velocity at the domain boundaries or inside the bodies embedded on the computational nodes.

2.1 Subgrid Modelling

In this paper we compare the results obtained using the models proposed by Vreman [1] and Nicoud et al. [2]. These models belong to the family of the so-called eddy viscosity models and hence their implementation relies on calculation of the ν_t which is then added to the kinematic viscosity as in Eq. (2). In the case of the model proposed by Vreman [1] ν_t is computed as:

$$\nu_t = C\sqrt{\frac{B_\beta}{\alpha_{ij}\alpha_{ij}}} \tag{3}$$

$$B_\beta = \beta_{11}\beta_{22} - \beta_{12}^2 + \beta_{11}\beta_{33} - \beta_{13}^2 + \beta_{22}\beta_{33} - \beta_{23}^2 \tag{4}$$

$$\alpha_{ij} = \frac{\partial \overline{u}_j}{\partial x_i} \quad , \quad \beta_{kl} = \Delta^2 \alpha_{mk}\alpha_{ml} \tag{5}$$

where the constant in (3) is taken as $C = 2.5 \times 10^{-2}$ [1]. The LES filter width is computed as $\Delta = (\Delta x \Delta y \Delta z)^{1/3}$ where Δx, Δy, Δz are the mesh spacings.

The model proposed by Nicoud et al. [2] is commonly known as σ-model. In the case we compute ν_t as follows:

$$\nu_t = (C_\sigma \Delta)^2 D_\sigma \tag{6}$$

$$D_\sigma = \frac{\sigma_3(\sigma_1 - \sigma_2)(\sigma_2 - \sigma_3)}{\sigma_1^2} \tag{7}$$

where the model constant is $C_\sigma = 1.35$ [2] and $\sigma_1 \geq \sigma_2 \geq \sigma_3 \geq 0$ are the singular values of the matrix

$$G = g^t g \quad \text{with} \quad g_{ij} = \frac{\partial \overline{u}_i}{\partial x_j} \tag{8}$$

Above models share features desirable in modelling of turbulent flows, i.e. they yield zero eddy viscosity close to a solid wall, in laminar flows or in pure shear regions.

2.2 Description of the Flow Solver

The set of Eqs. (1)–(2) is solved using the numerical code (SAILOR) which is an academic high-order flow solver based on the low Mach number approximation. The solution algorithm in the SAILOR code is based on the projection method in which the pressure is computed from the Poisson equation. The time advancement of Eq. (2) is performed with a predictor–corrector method with the help of the second-order Adams–Bashforth and Adams–Moulton methods. The spatial discretization is based on sixth-order compact differencing developed for half-staggered meshes (see Fig. 1) in the Cartesian coordinate system. In the present paper, the SAILOR code is used together with the IB method which is briefly presented in the next subsection.

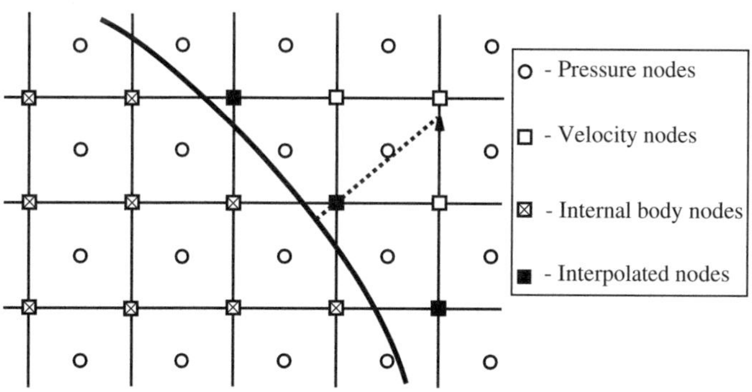

Fig. 1 Linear velocity interpolation method

2.3 Immersed Boundary (IB) Method

In general, there are two options of the IB method called feedback forcing method [11] and direct forcing method [8]. They differ in the evaluation of the forcing term. In this work the latter approach is implemented which seems to be simpler and more efficient. Interested reader is referred to [12] where all details and variants of IB methods may be found. Here, we limit the description to the definition of the forcing term used in the predictor step together with the second-order Adams–Bashforth method. In this case the term f_{IB} is defined as:

$$f_{IB} = \frac{u_{WALL} - \bar{u}_i^n}{\Delta t} - \left[\frac{3}{2}Res(\bar{u}^n) - \frac{1}{2}Res(\bar{u}^{n-1})\right] + \frac{1}{\rho}\frac{\partial \bar{p}^n}{\partial x_i} \tag{9}$$

where $Res(\bar{u}^n)$, $Res(\bar{u}^{n-1})$ represent the convection and diffusion terms in (2) discretized on the time levels n and $n-1$. The symbol u_{WALL} stands for the velocity at the wall which is a part of the computational domain as shown in Fig. 1 by black bold line. The velocity on that boundary is known and this allows to estimate the values of velocity in its closest vicinity, i.e. in the computational nodes shown in Fig. 1 by black squares. In the present approach the velocity in these nodes is obtained from a second-order linear interpolation based on the velocity values from the second node line from the boundary (shown by high arrow in Fig. 1) and the desired boundary values. Inside the immersed body, i.e. in the nodes with crosses, the velocity is set equal to zero.

3 Results

The accuracy of the SAILOR-IB code has been validated by computations of laminar flows in a lid-driven skewed cavity and over a backward facing step [13]. The obtained results were in very good agreement with the literature data obtained using the classical body fitted meshes. Most likely, in the present implementation the errors due to the approximate treatment of the walls are compensated by the high-order approximation far from the boundaries.

In this paper we deal with the turbulent flow in the converging–diverging channel in which the solution accuracy in near wall regions is of crucial importance. The computational domain for this test case is shown in Fig. 2. The dotted lines indicate the locations in which we will compare our results with DNS data obtained from Laboratoire de Mécanique de Lille (LML). The length of the domain (dimensionalised by a half domain height) is $L = 4\pi$, its height is equal to $H = 2$ and width is $W = 2\pi$. The computations are performed for $Re_\tau = u_t h/\nu = 395$ where u_τ is the friction velocity and $h = H/2$. At the inlet the fully developed turbulent flow is prescribed using the solution obtained from the simulations of periodic channel flow for the same Re_τ.

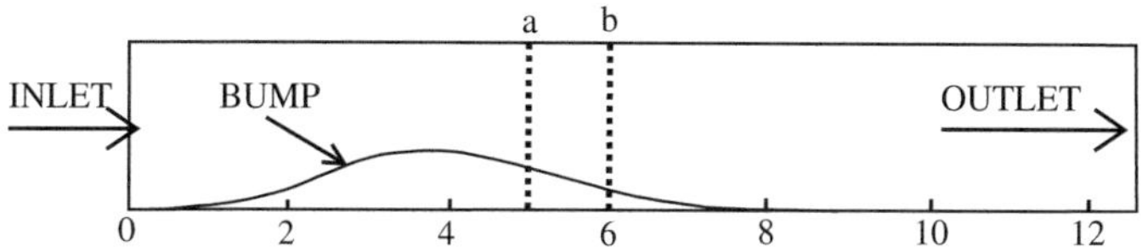

Fig. 2 Computational domain. The *vertical dotted lines* indicate the locations at which the solutions are compared with DNS data

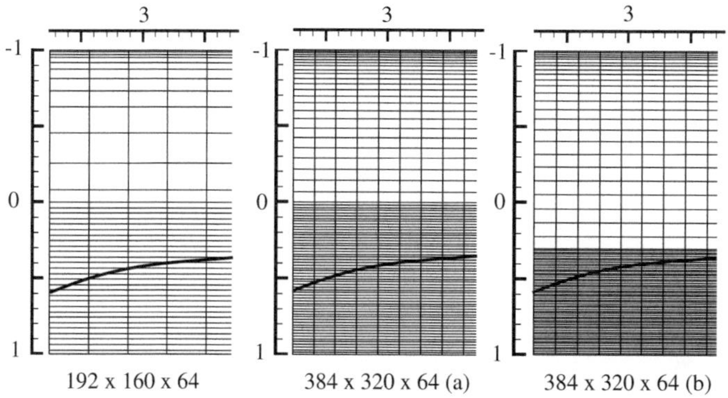

Fig. 3 Three types of the meshes used in the computations, every fourth grid lines are shown

The computational meshes used in the simulations consist of $192 \times 160 \times 64$ nodes and $384 \times 320 \times 64$ nodes. The denser mesh is additionally used with two different grid refinements. This is shown in Fig. 3 where additionally in a zoomed region of the bump the bold line shows location of the wall of the bump. The dense meshes will be denoted as (a) and (b). In all the cases, near the upper wall the grid nodes are distributed such that the first node is at $y^+ = 0.95$. On the bump side, the meshes are uniform and y^+ on the top of the bump is equal to 3.2 for the coarse mesh and 1.76 and 1.23 for the dense meshes (a) and (b), respectively.

3.1 Mesh Sensitivity Study

In this subsection, we analyse the influence of the mesh density based on the results obtained using the subgrid model proposed by Vreman. The reference DNS data were obtained on the mesh with $1536 \times 257 \times 384$ nodes which is approximately 77 and 20 times more nodes than in our coarse and dense meshes, respectively. Sample results obtained using the coarse mesh are presented in Fig. 4 showing an isosurface of the Q-parameter. It can be seen that the large turbulent scales vanish on the left-hand side of the bump and they reappear again on the falling side of the bump. This behaviour coincides with the regions where the flow accelerates and decelerates. These regions can be easily found in Fig. 5 showing instantaneous and time-averaged

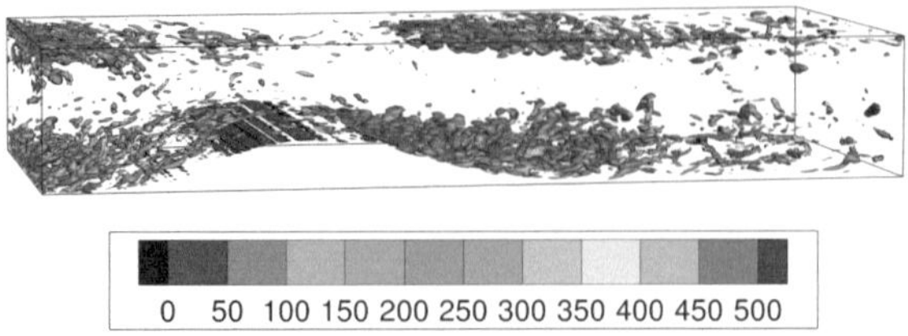

Fig. 4 Isosurface of Q-parameter ($Q = 2000(u_\tau/h)^2$) coloured by vorticity magnitude

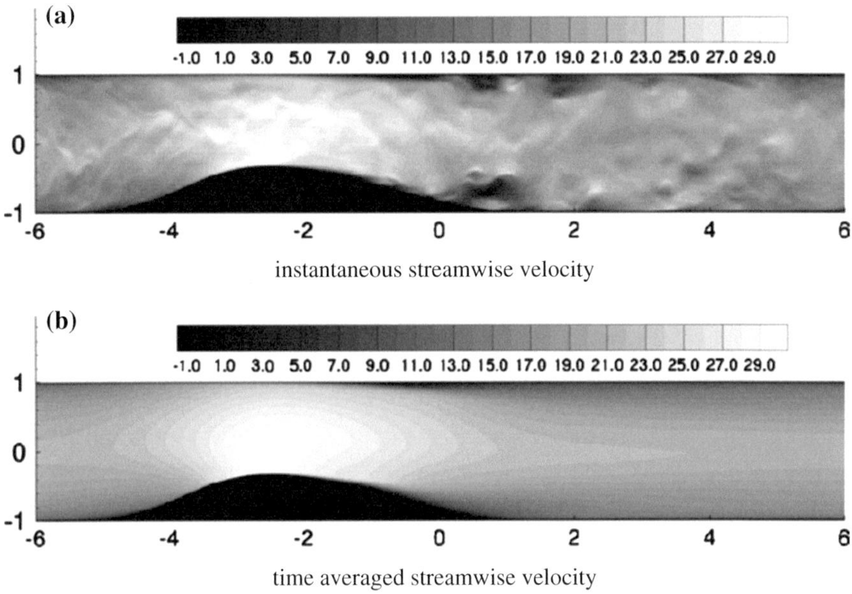

Fig. 5 Contours of the streamwise velocity normalised by u_τ: **a** mean values and **b** time-averaged values

contours of the streamwise velocity. We remind that in the IB method the velocity near the boundary of the bump is computed from the interpolation whereas in the nodes located inside the bump the velocity is explicitly set to zero every time step. Careful analysis of instantaneous velocity field near the wall of the bump shows small unphysical wrinkles. They are limited to the first two layers of the nodes, however they are not seen at all in the time-averaged solutions.

Detailed comparison and verification of the results are performed based on the velocity profiles at different locations in the channel. In all the cases the present solutions are very close to DNS data. In general, it cannot be said that the denser meshes provide significantly better results than using the coarse mesh. A sample comparison is presented in Fig. 6 showing the profiles of the mean and fluctuating components of the streamwise and the wall normal velocity along the line ‘a’ from Fig. 2. At this particular location it seems that the best solution is obtained using the dense mesh (a), though the results on the coarse mesh are also correct.

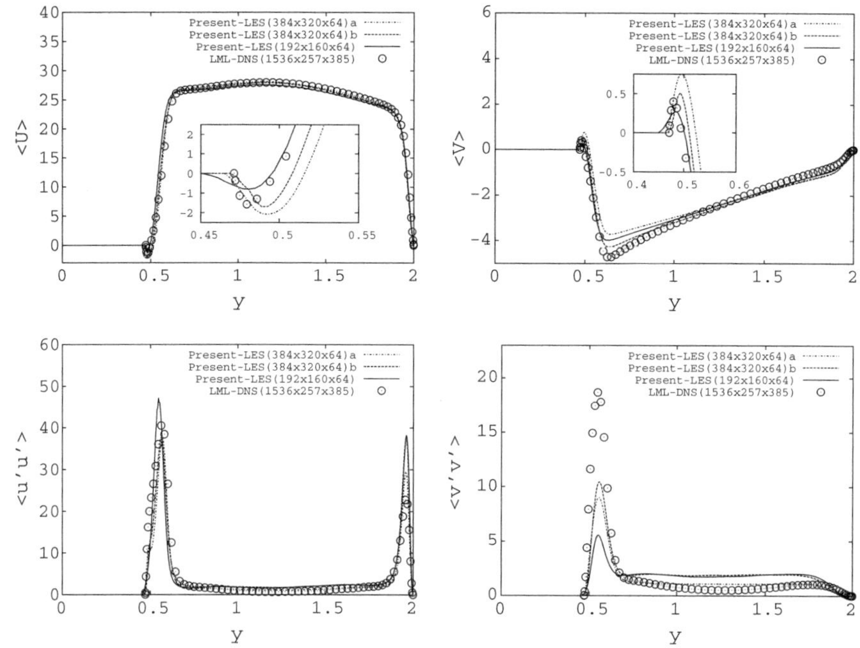

Fig. 6 Profiles of the streamwise and wall normal velocity components along line 'a'

3.2 Comparison of the SGS Models

The comparison of the subgrid models is performed using the mesh $384 \times 320 \times 64$(a). Figure 7 shows the contours of the total turbulent kinetic energy and the subgrid kinetic energy defined as $k_{sgs} = \frac{\nu_t^2}{(C_\nu \Delta)^2}$ with $C_\nu = 0.1$ [14]. These results were obtained using the σ-model but we note that the solution obtained using the Vreman model is very similar and practically indistinguishable by visual inspection. The maxima of kinetic energy are located in the regions of separation existing near the wall of the bump and close the upper wall. The ratio of k_{sgs} to the total kinetic energy is maximally 0.15. Hence, according to Pope's criterion [15] the mesh used in these simulations ensure the proper resolution. Detailed comparison of the solutions is presented in Figs. 8 and 9. It can be seen that both subgrid models provide accurate and similar solutions and it cannot be said which one performs better. In the centre of the channel both the mean and fluctuating velocity profiles match the DNS results almost exactly. Only closer to the walls some discrepancies are observed. This can be caused by the IB method as well as by the errors due to the subgrid modelling. Nevertheless, it can be seen that the location of the velocity extrema in separation

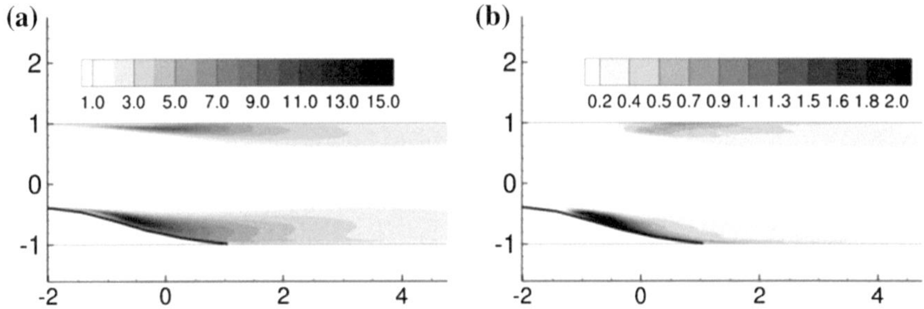

Fig. 7 Contours of the time-averaged subgrid and total kinetic energy. **a** Total turbulent kinetic energy. **b** Subgrid turbulent kinetic energy

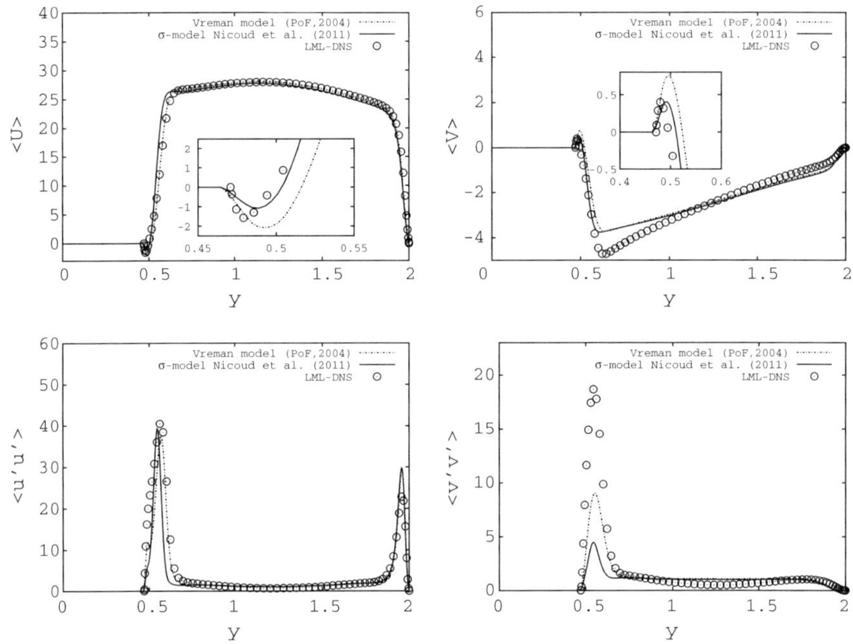

Fig. 8 Profiles of the streamwise and wall normal velocity components along the line 'a' obtained using two subgrid models

zones (the regions where the mean streamwise velocity near the walls is negative) is predicted relatively well by both models. The velocity fluctuations are also computed with good accuracy. Both the shapes of their profiles and their maximum values are close to DNS data.

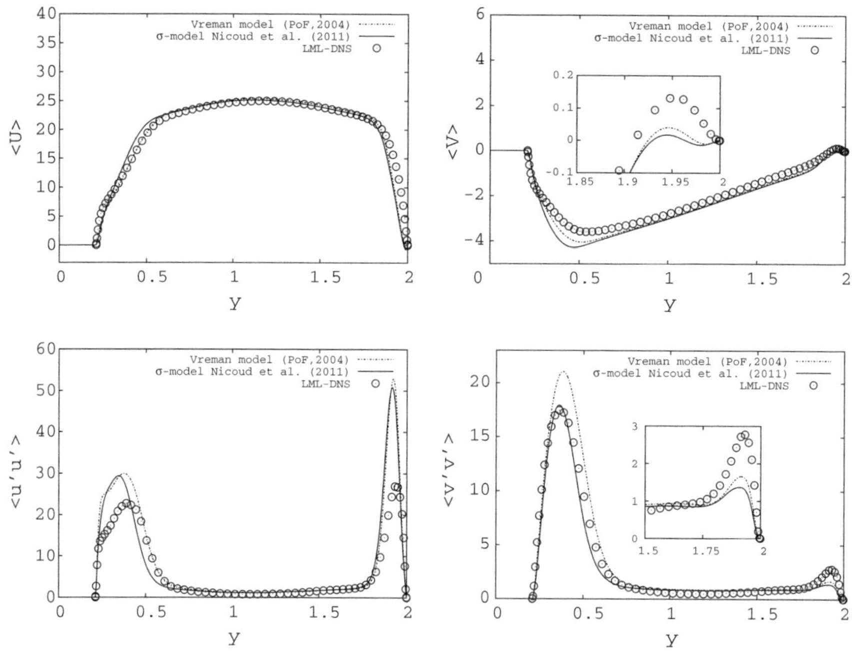

Fig. 9 Profiles of the streamwise and wall normal velocity components along the line 'b' obtained using two subgrid models

4 Conclusions

This paper shows results of flow modelling in a converging–diverging channel using LES combined with the Immersed Boundary (IB) method. With respect to exemplary literature DNS computations the application of LES allows significant reduction of the computational costs by using relatively coarse computational meshes. The use of the IB method allows to use the high-order code in the non-Cartesian computational domain without significant modifications of the solution algorithm. The LES computations were performed with the help of two subgrid models, proposed by Vreman [1] and Nicoud et al. [2] and in both the cases the obtained results were accurate. It could be seen that the better or worse agreement with DNS data was dependent on which quantity was compared and at which location in the channel. Hence, we conclude that both models are suitable for LES computations using IB method.

Acknowledgments Authors thank to Prof. W. Elsner for fruitful discussion. These research were supported by Polish National Science Centre under grant no. DEC-2012/07/B/ST8/03791. The computations were performed using PL-Grid Infrastructure (Poland).

References

1. A.W. Vreman, An eddy-viscosity subgrid-scale model for turbulent shear flow: algebraic theory and applications. Phys. Fluids **16**, 3670–3681 (2004)
2. F. Nicoud, H. Baya Toda, O. Cabrit, S. Bose, J. Lee, Using singular values to build a subgrid-scale model for large eddy simulations. Phys. Fluids **23**, 193–202 (2011)
3. C. Canuto, M.Y. Hussaini, A. Quarteroni, T.A. Zang, *Spectral Methods in Fluid Dynamics* (Springer, Berlin, 1988)
4. S.K. Lele, Compact finite schemes with Spektral-like resolution. J. Comput. Phys. **103**, 16–42 (1992)
5. T.K. Sengupta, A. Dipankar, A. Kameswara Rao, A new compact scheme for parallel computing using domain decomposition. J. Comput. Phys. **220**, 654–677 (2007)
6. S.K. Pandit, J.C. Kalita, D.C. Dalal, A transient higher order compact scheme for incompressible viscous flows on geometries beyond rectangular. J. Comput. Phys. **225**, 1100–1124 (2007)
7. L. Kuban, J.-P. Laval, W. Elsner, A. Tyliszczak, M. Marquillie, LES modelling of converging-diverging turbulent channel flow. J. Turbul. **13**, 1–19 (2012)
8. J. Mohd-Yosuf, Combined immersed-boundary/B-spline methods for simulations of flow in complex geometries. Ann. Res. B. Cent. Turbul. Res. (1997)
9. C.S. Peskin, Flow patterns around heart valves: a numerical method. J. Comput. Phys. **10**, 252–271 (1972)
10. B.J. Geurts, *Elements of Direct and Large-Eddy Simulation* (R.T. Edwards, Philadelphia, 2003)
11. R. Verzicco, G. Iaccarino, Immersed boundary technique for turbulent flow simulations. Appl. Mech. Rev. **56**, 331–347 (2003)
12. R. Mittal, G. Iaccarino, Immersed boundary methods. Annu. Rev. Fluid Mech. **37**, 239–261 (2005)
13. M. Ksiezyk, A. Tyliszczak, Immersed boundary method combined with a high order compact scheme on half-staggered meshes. J. Phys. Conf. Ser. **530**, 012066 (2014)
14. S.B. Pope, *Turbulent Flows* (Cambridge University Press, Cambridge, 2000)
15. S.B. Pope, Ten questions concerning the large-eddy simulation of turbulent flows. New J. Phys. **6**, 35 (2004)

On Minimum Aspect Ratio for Experimental Duct Flow Facilities

Ricardo Vinuesa, Eduard Bartrons, Daniel Chiu, Jean-Daniel Rüedi, Philipp Schlatter, Aleksandr Obabko and Hassan M. Nagib

Abstract To the surprise of some of our colleagues, we recently recommended aspect ratios of at least 24 (instead of accepted values over last few decades ranging from 5 to 12) to minimize effects of sidewalls in turbulent duct flow experiments, in order to approximate the two-dimensional channel flow. Here we compile available results from hydraulics and civil engineering literature, where this was already documented in the 1980s. This is of great importance due to the large amount of computational studies (mainly Direct Numerical Simulations) for spanwise-periodic turbulent channel flows, and the extreme complexity of constructing a fully developed duct flow facility with aspect ratio of 24 for high Reynolds number with adequate probe resolution. Results from this nontraditional literature for the turbulence community are compared to our recent database of DNS of turbulent duct flows with aspect ratios ranging from 1 to 18 and $Re_{\tau,c} \simeq 180$ and 330, leading to very good agreement between their experimental and our computational results.

1 Introduction and Literature Survey

Direct Numerical Simulation (DNS) studies of canonical turbulent channel flows have provided the turbulence community with insight on the understanding of wall-bounded turbulence, and the wide range of mechanisms it involves. A good review of the contributions from DNS to the community by using the simplified geometry of pressure-driven flow between two infinite flat plates was provided by Jiménez [1]. However, recent experimental (Vinuesa et al. [2]) and computational (Vinuesa et al.

R. Vinuesa · E. Bartrons · D. Chiu · H.M. Nagib (✉)
MMAE Department, Illinois Institute of Technology, Chicago, IL 60616, USA
e-mail: nagib@iit.edu

J.-D. Rüedi
II Facoltà di Ingegneria, Università di Bologna, 47100 Forlí, Italy

R. Vinuesa · P. Schlatter
Linné FLOW Centre, KTH Mechanics, Osquars Backe 18, Stockholm, Sweden

A. Obabko
MCS, Argonne National Laboratory, Argonne, IL 60439, USA

M. Stanislas et al. (eds.), *Progress in Wall Turbulence 2*,
ERCOFTAC Series 23, DOI 10.1007/978-3-319-20388-1_18

[3]) studies carried out by our group demonstrate that the flow in an experimental facility of a particular aspect ratio AR (where AR is defined as the total width of the duct W divided by its total height H) cannot be reproduced by DNSs of spanwise-periodic channel flows. The role of the sidewalls, which includes three-dimensional effects through secondary vortices, and sidewall boundary layers, leads to a more complicated configuration, which differs from the simplified model considered in the DNS approach. Therefore, we reserve the term "channel" for spanwise-periodic DNSs, and denote the experimental flow by the term "duct".

The wall-bounded turbulence community witnessed how the reported minimum aspect ratio required for an experimental duct flow facility to approach "nominaly two-dimensional conditions" has increased from a value of 5 in the 60s (Bradshaw and Hellens [4]), to 7 one decade later (Dean [5]), and to approximately 12 in two studies from 2003 (Zanoun et al. [6]) and 2005 (Monty [7]). Our recent experiments [2], in a Reynolds number range representative of the one spanned by a significant number of currently available DNS studies of turbulent channel flow, show that an aspect ratio of at least 24 is required for the conditions at the centerplane of the duct to become independent of the aspect ratio. We carried out Oil Film Interferometry (OFI) and static pressure measurements in an adjustable-geometry duct flow facility. The conditions at the centerplane of the duct were characterized through the local skin friction from the OFI and the centerline velocity at four different streamwise locations, and through the wall shear based on the streamwise global pressure gradient. The inner-scaled centerline velocity $U_c^+ = U_c/u_\tau$ measured in that study is shown here in Fig. 1 as a function of the friction Reynolds number Re_τ (defined in terms of

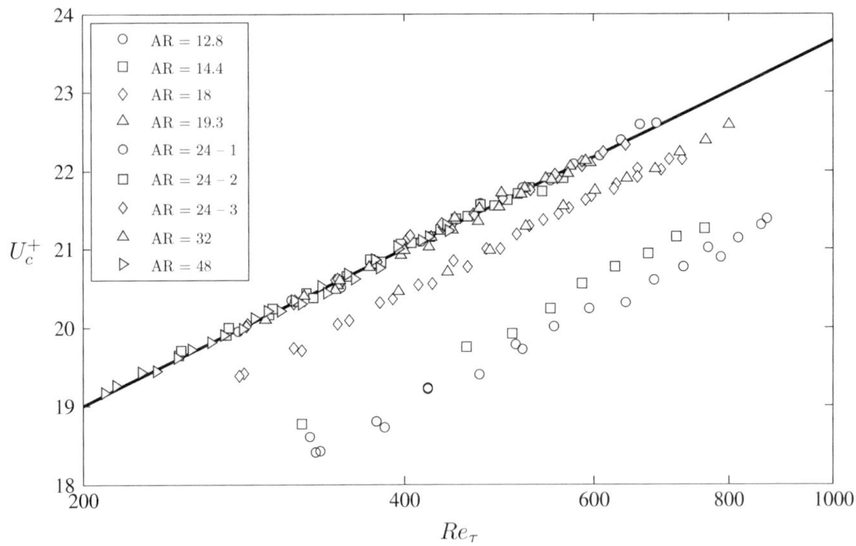

Fig. 1 Fully developed Oil Film Interferometry measurements at the centerplane for duct aspect ratios from 12.8 to 48. This figure corresponds to Fig. 10 in Vinuesa et al. [2]

the friction velocity u_τ and the duct half-height h). The skin friction obtained from the pressure gradient overestimated the local shear measurements obtained from the OFI, and did not reproduce the same AR dependence observed with OFI. Also note that the flow features measured in duct flows with aspect ratios larger than 24 do not exactly match the ones exhibited by spanwise-periodic computations.

Interestingly, the impact of the sidewalls in duct flows and the three-dimensional mechanisms introduced by them have received more attention in other research areas, such as hydraulic engineering (Knight and Patel [8], Rhodes and Knight [9]), civil engineering (Yang [10]) and even applied mathematics (Kotsovinos [11]). The dependency of mean flow features such as wall shear stress with aspect ratio well beyond the previously reported values ranging from 5 to 12 is evident from these studies, with some of them from the 1980s. Here we will highlight some of their most relevant findings, and connect them with the recent experimental and computational results obtained by us. In their 1985 paper, Knight and Patel [8] discuss the relation between aspect ratio and boundary shear from a practical point of view, since their aim was to obtain empirical relations for engineering purposes. In order to do that, they carried out experimental measurements of wall shear distributions along the four walls of a 9.25-m long duct with variable aspect ratio, which was modified from 1 to 10. They used Preston tubes for the measurements, which are commonly used to obtain such boundary shear distributions. Although Preston tubes are not capable of accurately capturing the spanwise distribution of wall shear stress, due to spatial averaging effects on their pressure readings (especially in the tubes used for this study, with diameters ranging from 3.15 to 5 mm), they provide a good qualitative representation of this distribution and values averaged over the whole perimeter of the duct are generally accurate. Their experimental database was extended to higher aspect ratios ranging from 10 to 50 by Rhodes and Knight [9] in 1994, using the same measurement technique (this time using smaller Preston tubes with diameters of 2.05 mm) in a water flow duct which was 13.1 m long. Their wall shear measurements were recently compiled by Yang [10], who developed empirical relations describing the average wall shear stress over the duct sidewalls $\overline{\tau}_w$ and over the bed $\overline{\tau}_b$ (which in civil engineering notation refers to the upper and lower walls of the duct). His results are reproduced here in Fig. 2, where b/h corresponds to the duct aspect ratio AR $= W/H$, ρ is the fluid density, g is the acceleration of gravity, h is the duct half-height, and S is the energy slope (this is another term more often used in civil engineering applications than in wall-bounded turbulence). Also note that the term $\rho g h S$ corresponds to the wall shear stress $\overline{\tau}_{w,\mathrm{PG}}$ defined as:

$$\overline{\tau}_{w,\mathrm{PG}} = -h\frac{\mathrm{d}P}{\mathrm{d}x}, \tag{1}$$

where $\mathrm{d}P/\mathrm{d}x$ is the streamwise pressure gradient. As discussed by Monty [7] and Vinuesa et al. [2], this expression is obtained by assuming an infinitely wide channel, without any sidewalls, and with a homogeneous spanwise distribution of wall shear throughout its span (which are also the hypotheses considered in spanwise-periodic simulations of turbulent channel flows). The result (1) follows from a control vol-

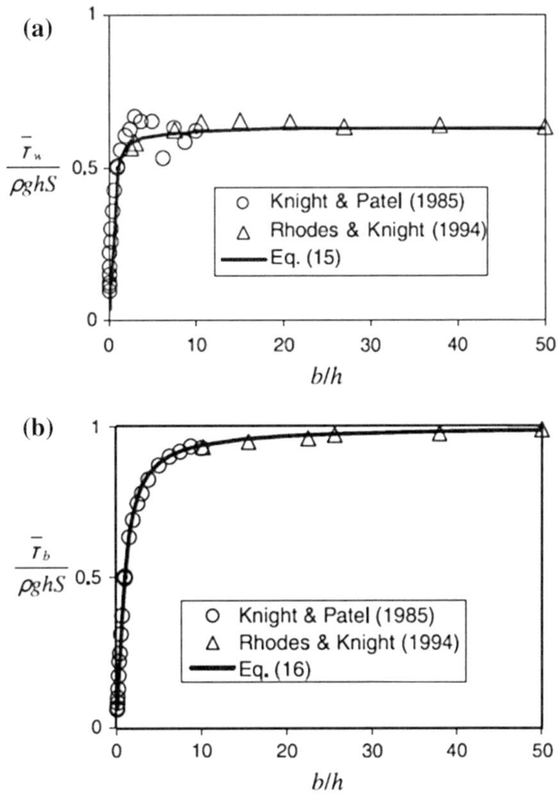

Fig. 2 Normalized average wall shear stress along the duct sidewalls $\overline{\tau}_w$ and upper and lower walls $\overline{\tau}_b$ (also denoted by duct bed). This figure corresponds to Fig. 3 in Yang [10]

ume analysis carried out at the centerplane of the channel. The fact that the ratio $\overline{\tau}_b/(\rho g h S) \equiv \overline{\tau}_b/\overline{\tau}_{w,\mathrm{PG}}$ is lower than 1 even for aspect ratios as large as 20 or 25 indicates that the flow is in fact nonhomogeneous in the spanwise direction well beyond classically reported values, and supports our recent findings. Also note that, although asymptotically $\overline{\tau}_b/\overline{\tau}_{w,\mathrm{PG}} \rightarrow 1$, our experiments show that the conditions at the *centerplane* of the duct do not exactly match the ones exhibited by a spanwise-periodic channel.

Another parameter analyzed by Rhodes and Knight [9] as a function of aspect ratio is the proportion of total shear force acting on the sidewalls $\%\mathrm{SF}_w$. Note that the absolute shear force on one sidewall is $\mathrm{SF}_w = \overline{\tau}_w H$, the shear force acting on the upper or lower wall is $\mathrm{SF}_b = \overline{\tau}_b W$, and the total shear force acting on the duct boundary would be $\mathrm{SF}_T = 2\mathrm{SF}_w + 2\mathrm{SF}_b$. As a consequence, $\%\mathrm{SF}_w = 200\mathrm{SF}_w/\mathrm{SF}_T$. Figure 3 shows that the evolution of $\%\mathrm{SF}_w$ as a function of aspect ratio obtained by Rhodes and Knight [9] is consistent with the results shown in Fig. 2, since only at aspect ratios larger than around 30 the sidewall shear force represents less than 1 % of the total. However, another result is remarkable in this figure, since the shear force measurements are compared with the geometric line. This line essentially predicts

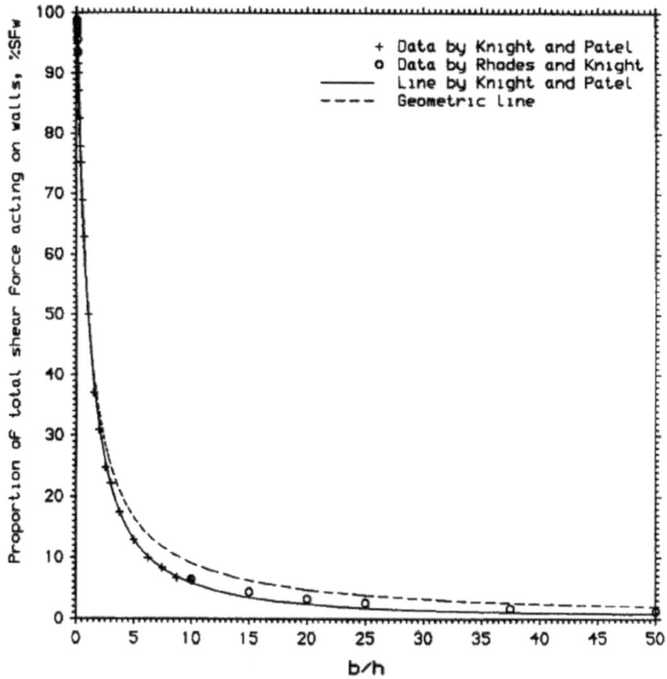

Fig. 3 Proportion of total shear force acting on the sidewalls as a function of the duct aspect ratio. This figure corresponds to Fig. 4 in Rhodes and Knight [9]

the expected $\%\mathrm{SF}_w$ from the geometric configuration of the duct, i.e., by taking into account the relation between the length of both sidewalls and the total perimeter. Although at very low aspect ratios these trends seem to collapse, aspect ratios larger than 3 show differences between both trends, with the experimental $\%\mathrm{SF}_w$ curve lying below the expected trend from the geometry up until very large aspect ratios beyond 30. This is again a consequence of the inhomogeneity throughout the span of the duct, and since the $\tau_b(z)$ distribution is not a constant, the geometrical relations do not apply exactly.

The inhomogeneities in wall shear distribution along the perimeter of the duct, which impact integrated quantities as manifested in Figs. 2 and 3, are further explored by Knight and Patel [8] by performing Preston tube measurements of shear stress at a number of locations on the upper, lower, and sidewalls. These results are shown in Fig. 4 in the forms of normalized wall and bed shear stress distributions ($\tau_w/\overline{\tau}_w$ and $\tau_b/\overline{\tau}_b$) for a number of aspect ratios ranging from the square duct case up to $\mathrm{AR} = 10$. It is interesting to observe how in their measurements they obtain several local maxima and minima in the shear distributions along the walls, so the increase in friction from the sidewall to the centerplane is not monotonous and therefore there must be additional physical mechanisms leading to this particular trend. Knight and Patel argue that the streamwise secondary vortices produced by the spanwise inho-

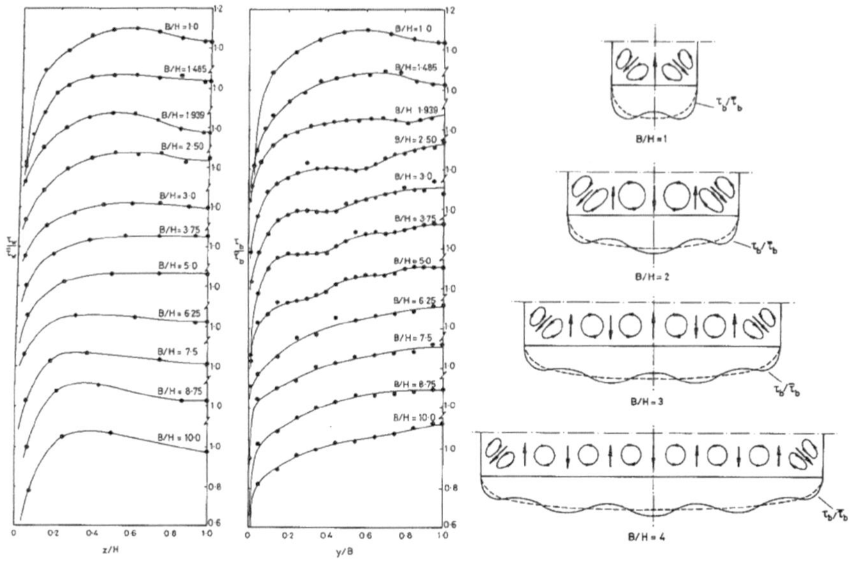

Fig. 4 Normalized distributions of sidewall shear $\tau_w/\overline{\tau}_w$, bed shear $\tau_b/\overline{\tau}_b$ and proposed interactions between secondary flow and shear stress distributions; Figs. 2–4 in Knight and Patel [8]

mogeneity of the Reynolds stress tensor are responsible for this, and even propose a model to predict the number of secondary flow cells that appear in a duct as a function of aspect ratio, as well as their interaction with the bed shear stress distribution as shown here in Fig. 4. According to Knight and Patel [8], the number of secondary vortices n exhibited by a turbulent duct flow of aspect ratio $\mathrm{AR} = W/H$ is given by:

$$n = \left(\frac{2W}{H} + 1\right) \pm 1, \tag{2}$$

where the lower value given by (2) accounts for the vortices located on the duct bed, and the larger value also accounts for the ones on the sidewall at the duct corners.

2 Comparison with DNS of Turbulent Duct Flows

After reviewing some key results showing that, in fact, aspect ratio values much larger than previously recommended in the turbulence literature are required to obtain AR-independent conditions in turbulent duct flows, here we use the computational database generated by Vinuesa et al. [3] to reproduce some of the trends observed in the experiments by Knight and Patel [8] and Rhodes and Knight [9]. In this database, turbulent duct flows are computed by means of Direct Numerical Simulation (DNS), using the spectral element code Nek5000 [12]. All the ducts are computed assuming

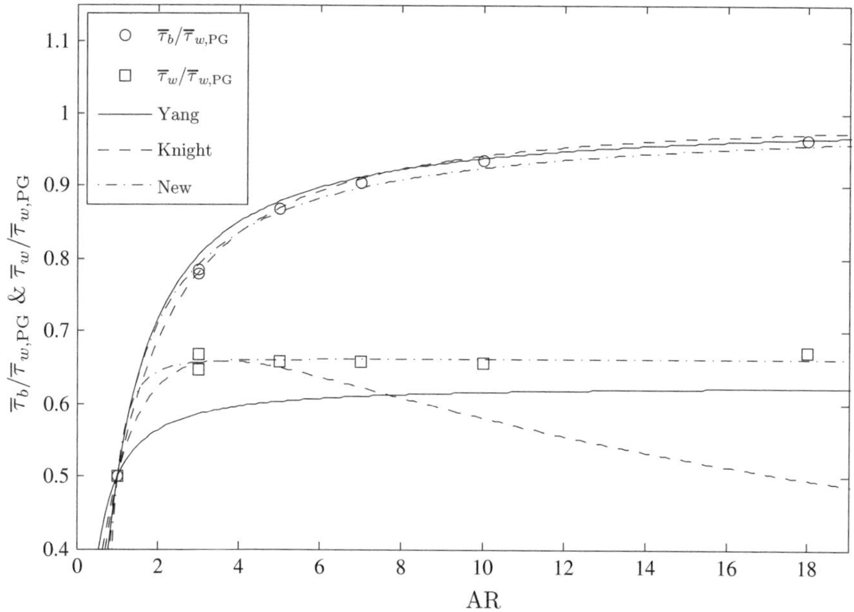

Fig. 5 Average bed and wall shear stress from the DNS database normalized with the shear stress computed using (1). Empirical relations proposed by Yang [10] and Knight and Patel [8], together with new forms proposed in the present study, are shown for comparison

periodicity in the streamwise direction, and number of aspect ratios ranging from 1 to 18 are considered at two Reynolds numbers based on centerplane friction velocity: $Re_{\tau,c} \simeq 180$ and 330.

Figure 5 shows average bed and wall shear stresses normalized by the stress calculated from the streamwise pressure gradient as defined in (1) from our computations. Although the aspect ratio range under investigation is narrower than the one shown in Fig. 2, it is interesting to note that the trend observed in $\overline{\tau}_b/\overline{\tau}_{w,\mathrm{PG}}$ agrees very well with the empirical correlations proposed by both Yang [10] and Knight and Patel [8], therefore supporting the observations made in relation to Fig. 2. Note also that in the case of the sidewall shear distribution, qualitative agreement is obtained with the correlation proposed by Yang [10] (which is based on the whole aspect ratio range up to 50), and the different behavior obtained from the equation by Knight and Patel [8] is attributed to the fact that their correlation is based on a narrower range of aspect ratios only up to 10. Quantitative differences between our computational results and the relation by Yang [10] may arise from the fact that it is more complicated to measure accurately sidewall shear distributions using Preston tubes. Based on these results, we propose the following empirical relations describing both trends:

$$\frac{\overline{\tau}_b}{\overline{\tau}_{w,\mathrm{PG}}} = -0.5125\ \mathrm{AR}^{-0.7808} + 1, \tag{3}$$

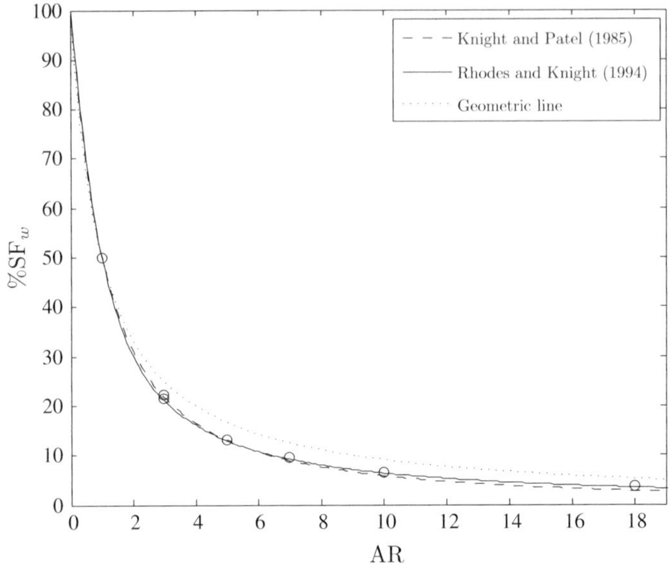

Fig. 6 Percentage of shear force acting on the sidewalls obtained from the DNS database compared with empirical relations by Knight et al. [8, 9] and the geometric line

$$\frac{\overline{\tau}_w}{\overline{\tau}_{w,\mathrm{PG}}} = -0.1623\ \mathrm{AR}^{-3.166} + 0.6621, \qquad (4)$$

where also here $\overline{\tau}_b/\overline{\tau}_{w,\mathrm{PG}} \rightarrow 1$ for infinite aspect ratios. Following this analysis, in Fig. 6 we show the percentage of shear force acting on the sidewalls in all our computations compared with the empirical formulas by Knight and Patel [8] and Rhodes and Knight [9], and also with the geometric line as in Fig. 3. In this case both empirical trends represent our data very well, and the same deviation with respect to the geometrical trend is observed. The empirical relation reported by Rhodes and Knight [9], which is based on the full dataset with aspect ratios ranging from 1 to 50 is given by:

$$\%\mathrm{SF}_w = \frac{100}{1 + \left(\frac{1+1.345/\mathrm{AR}}{1+1.345\ \mathrm{AR}}\right)^{-1.057}}. \qquad (5)$$

Finally, we compare some of the results shown in Fig. 3 with our numerical simulations in Figs. 7 and 8. Spanwise distributions of bed wall shear normalized by the integrated value, as well as wall-normal variations of sidewall shear stress show similar trends, and differences around the local maxima and minima can be attributed to two main factors. First, the effect of the spanwise averaging experienced by Preston tube measurements, especially the ones of higher diameter. Second, differences in the Reynolds number between the two sets of results: the range of our DNSs

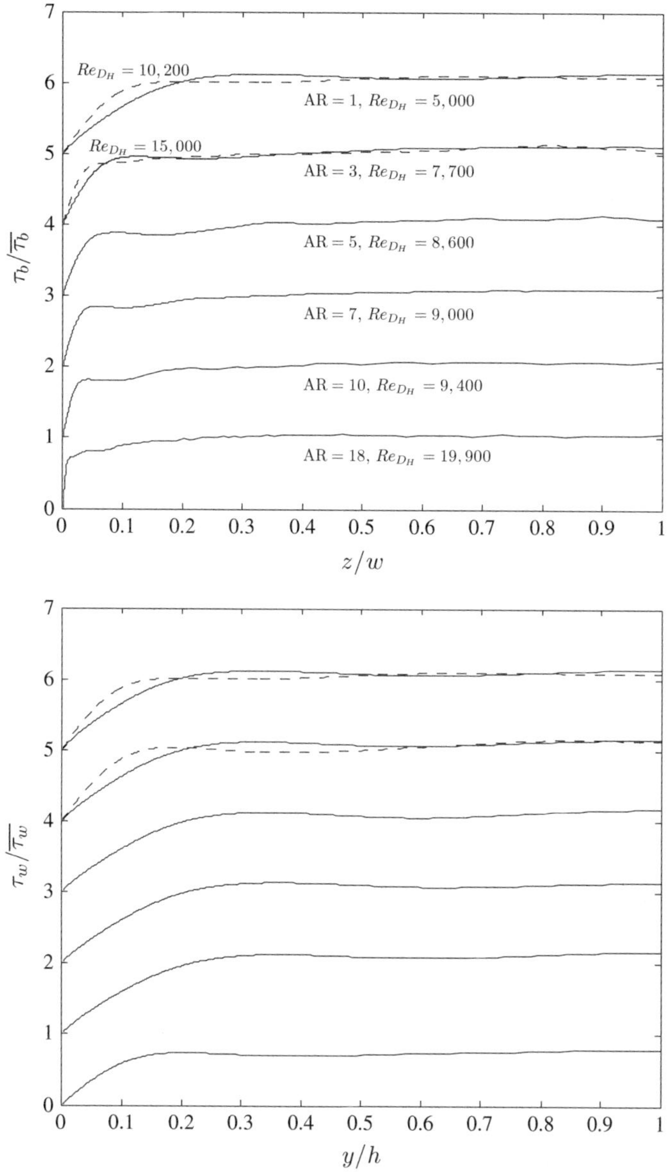

Fig. 7 Spanwise distribution of bed shear stress normalized by the average value (*top*) and wall-normal variation of wall shear divided by integrated value along sidewall (*bottom*). Aspect ratio and Reynolds number based on bulk velocity U_b and hydraulic diameter D_H are listed. For the cases with more than one Reynolds number, the Re_{D_H} values closer to the left correspond to *dashed lines*. Data obtained from DNS results, and profiles shifted up vertically one unit for clarity

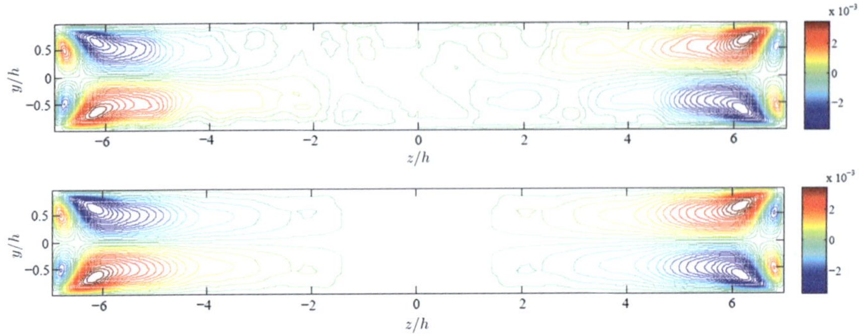

Fig. 8 Streamlines corresponding to the aspect ratio 7 duct computed with $Re_{\tau,c} \simeq 180$ after averaging times of 2,200 (*top*) and 9,000 (*bottom*) convective time units. Note that the two symmetries in the flow were also applied on the *bottom* figure

in Re_{D_H} (based on bulk velocity U_b and hydraulic diameter D_H) is from 5,000 to 20,000, whereas in the experiment the minimum Reynolds number is 50,000 and the maximum is 10^5. It is also important to note that in the experimental dataset lower Reynolds numbers correspond to higher aspect ratios (due to physical limitations of the facility the U_b is reduced as the cross-sectional area increases), whereas in our simulations the Reynolds number increases with aspect ratio (since U_b is held constant as the hydraulic diameter increases). In any case, the experimental database shows interesting trends, especially in terms of interactions with secondary motions. The models proposed by Knight and Patel [8] in Fig. 4 and in Eq. (2) for the number of secondary cells are checked in Fig. 8 with our aspect ratio 7 duct computed with $Re_{\tau,c} \simeq 180$. In this figure we show the streamlines obtained in the inplane section of the duct after averaging in the homogeneous streamwise direction, and in time. Two averaging periods were considered: first a shorter one of 2,000 convective time units (note that nondimensionalization is done in terms of bulk velocity U_b and duct half-height h), then a much longer one of 9,000 (where also the two symmetries in the flow were applied). A number of secondary cells are noticeable in the case with shorter averaging period, and although the number of individual cells is lower than the values predicted by Eq. (2), it is interesting to observe that this equation is based on the hypothesis that all these vortices are of size $h \times h$. This is roughly the size of some of vortices located close to the duct centerplane. If one considers Fig. 8 (bottom), it can be inferred that after long averaging times only the vortical structures at the corners remain in the main flow, with the ones on the horizontal walls being significantly stretched in the spanwise direction. But what is remarkable is that all the other secondary vortices exist instantaneously in the flow, and either move or live for a certain period before eventually dissipating through viscosity. It is also important to note that the effect of the secondary vortices in the mean flow is significant, leading to the modification of the mean velocity profile by transporting momentum from the duct core toward its bisectors.

3 Conclusion

The need of aspect ratios of at least 25–30 in experimental duct flow facilities in order to minimize aspect ratio effects in the flow is a well documented fact since the 80 s in the hydraulics and civil engineering literature, as can be observed in studies like the one by Knight and Patel [8]. The wall-bounded turbulence community has not been aware of this until very recently, in the experimental study by Vinuesa et al. [2] where a minimum aspect ratio of 24 is recommended, well beyond traditionally reported values from 5 to 12. In this study we compile some of the classic and more recent results in areas of hydraulics and civil engineering, and compare them to our recent DNSs of turbulent duct flows with aspect ratios ranging from 1 to 18 [3].

References

1. J. Jiménez, Cascades in wall-bounded turbulence. Annu. Rev. Fluid Mech. **44**, 27–45 (2012)
2. R. Vinuesa, E. Bartrons, D. Chiu, K.M. Dressler, J.-D. Rüedi, Y. Suzuki, H.M. Nagib, New insight into flow development and two dimensionality of turbulent channel flows. Exp. Fluids (2014) (In press)
3. R. Vinuesa, A. Noorani, A. Lozano-Durán, G.K. El Khoury, P. Schlatter, P.F. Fischer, H.M. Nagib, Aspect ratio effects in turbulent duct flows studied through direct numerical simulation. J. Turbul. (2014) (In press)
4. P. Bradshaw, G.E. Hellens, The N.P.L. 59 in $\times$ 9 in boundary layer tunnel. NPL Aero Report 1119 (1964)
5. R.B. Dean, Reynolds number dependence of skin friction and other bulk flow variables in two dimensional rectangular duct flow. Phys. Fluids **100**, 215–223 (1978)
6. E.-S. Zanoun, F. Durst, H.M. Nagib, Evaluating the law of the wall in two-dimensional fully developed turbulent channel flows. Phys. Fluids **15**, 3079–3089 (2003)
7. J.P. Monty, Developments in smooth wall turbulent duct flows. Ph.D. thesis, University of Melbourne (2005)
8. D.W. Knight, H.S. Patel, Boundary shear in smooth rectangular ducts. J. Hydraul. Eng. **111**, 29–47 (1985)
9. D.G. Rhodes, D.W. Knight, Distribution of shear force on boundary of smooth rectangular duct. J. Hydraul. Eng. **120**, 787–807 (1994)
10. S. Yang, Mechanisms for initiating secondary currents in channel flows. J. Civ. Eng. **36**, 1506–1516 (2009)
11. N.E. Kotsovinos, Secondary currents in straight wide channels. Appl. Math. Model. **12**, 22–24 (1988)
12. P.F. Fischer, J.W. Lottes, S.G. Kerkemeier, NEK5000: open source spectral element CFD solver (2008). http://nek5000.mcs.anl.gov

Riblets Induced Drag Reduction on a Spatially Developing Turbulent Boundary Layer

Amaury Bannier, Eric Garnier and Pierre Sagaut

Abstract Large eddy simulations have been conducted to gain further insight into the drag-reducing mechanisms of riblets in zero-pressure gradient turbulent boundary layer. The retained groove geometry achieves 9.8 % drag reduction on the controlled zone developing from $Re_\theta = 670$ to 975. It is shown that the turbulent contribution to the drag—as defined by Fukagata et al. Phys. Fluids, 14(11):L73, 2002 [7]—is the most affected. In the light of the obtained results, energy and enstrophy budgets will finally conduct to isolate a key mechanism involved in the riblets drag reduction.

1 Introduction

Making grooves on a surface increases the wetted area and, counter-intuitively, may reduce the drag for turbulent flows [22]. For the last three decades, those so-called "riblets" have remained under active consideration. They have been studied through experiments [2, 22], numerical simulations [4, 9], and also flight tests [5]. In their oil channel facility, Bechert et al. [2] achieve 8.2 % friction reduction with a technologically feasible design. However, a lack of understanding persists regarding the underlying mechanisms.

Based on viscous analyses, Luchini et al. [14] observed that, above both clean and ribbed surfaces, the Stokes flows converge asymptotically to the same far-field solution, providing the appropriate relative position of the two walls. For a given riblet geometry, one can thus compute the "virtual origin" h at which a flat plate should be located to produce asymptotically the same Stokes flow. The length $h_\parallel$

A. Bannier (✉) · E. Garnier
ONERA - The French Aerospace Lab, Applied Aerodynamics Department, Meudon, France
e-mail: amaury.bannier@onera.fr

E. Garnier
e-mail: eric.garnier@onera.fr

P. Sagaut
CNRS, Centrale Marseille, Aix-Marseille Université, M2P2 UMR 7340 Marseille, France

M. Stanislas et al. (eds.), *Progress in Wall Turbulence 2*, ERCOFTAC Series 23, DOI 10.1007/978-3-319-20388-1_19 can be defined for a longitudinal flow over the grooves (see Fig. 3 for illustration) and, similarly, $h_\perp$ for the transverse Stokes flow above the same riblets. Luchini's "protrusion height" is defined as the difference between the two obtained virtual origins: $\Delta h = h_\perp - h_\parallel$. They describe this quantity as a measure of "how much the corrugated wall impedes the cross-flow more than it does the longitudinal flow."

García-Mayoral and Jiménez [8] identified drag reduction regimes that incorporated previously proposed theories [4, 9, 10, 14]. For low Reynolds number or small riblets, effects are well characterized by Δh. When size or Reynolds increases, breakdown mechanisms, based on Kelvin–Helmotz instabilities, cancel out the beneficial viscous effect. In their analysis, García-Mayoral and Jiménez provide a scaling law for designing riblets within the appropriate regime.

Although the drag reduction is well characterized by a Stokes flow analogy in the first regime, the mechanism itself remains turbulent per se: riblets always increase drag in laminar flows [3].

To further understand the drag reduction engendered by riblets, the present work will focus at identifying its cause. Fukagata, Iwamoto, and Kasagi (FIK) [7] presented a decomposition of the skin friction. In their formulation, they split the contribution to the drag into different terms according to their physical origin. The drag-producing mechanisms can be isolated and their intensities spatially observed along the boundary layer.

After having described the retained riblets' geometry and computational parameters in the next section, the FIK decomposition will constitute the starting point of the present analysis Sect. 3. Each of the "bulk" and the turbulent contributions to the drag will be analyzed in Sects. 4 and 5. The observation of turbulent energy Sect. 6 and enstrophy Sect. 7 budgets will point out mechanisms responsible for the drag reduction.

2 Flow Configuration and Riblets Geometry

Throughout this paper, the longitudinal, normal, and spanwise directions will be, respectively, denoted as x, y, and z. U_0, ρ, and ν stand for the reference fluid velocity, density, and kinematic viscosity. The superscript + denotes quantity scaled with the friction velocity $u_\tau = \sqrt{\tau_w/\rho}$ and with ν, where τ_w is the wall shear stress. The overbar is used for time-averaged variables and Reynolds decomposition yields $f = \overline{f} + f'$.

Under the constraint of realizable grooves, the most efficient riblets so far are the trapezoidal ones studied by Bechert et al. [2]. They have been shown to achieve up to 8.2 % drag reduction. The rib angle is $\alpha = 30°$ and the height-over-width ratio is fixed to $h_{rib}/s_{rib} = 0.5$. This riblets design leads to a "protrusion height" of $\Delta h \approx 0.10 s_{rib} = 0.20 h_{rib}$, a value which can be compared to the optimal limit $\Delta h_{max} \approx 0.13 s_{rib}$ computed analytically by Luchni et al. [14]. Further increasing our Δh requires us to sharpen the riblets even more and would lead to a technologically unfeasible design. The retained geometry is depicted in Fig. 1.

Fig. 1 Riblets geometry and mesh representation

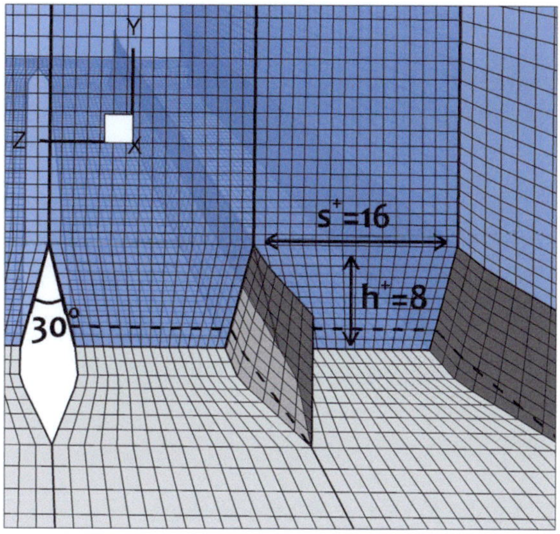

Regarding the sizing of the groove, García-Mayoral and Jiménez [8] provide a simplified stability model aimed at estimating the optimal riblet dimension for a given geometry. With our design, h_{rib} has to be close to approximately 8 wall-units.

The bottom surface of the computational domain is not entirely covered by riblets. Smooth surfaces are placed up and downstream of the controlled zone. The question of the relative vertical position of the riblets with respect to the one of the flat plates has to be addressed. We set the ordinate origin $y = 0$ at the flat wall location and define Δy_0 the (likely negative) vertical position of the riblet valley (*cf* Fig. 3 right-hand side axis). Among other possibilities, ribets can be set such as either the valleys or the tips are located at the flat plate level, corresponding, respectively, to $\Delta y_0 = 0$ and $\Delta y_0 = -h_{rib}$. Alternatively, the use of the previously defined virtual origin $h_{\parallel}$ is a consistent choice with respect to viscous effects. Here, the decision is made to cancel, or at least restrict, the net pressure force acting at the emergence and at the disappearance of the grooves. The riblets' relative position is set to $\Delta y_0 = -\frac{h^2}{s}\tan(\frac{\alpha}{2}) \approx -0.13\, h_{rib}$ so as to preserve the same cross-sectional area.

The simulation is conducted in a zero-pressure gradient turbulent boundary layer (ZPGTBL). The Reynolds number based on the momentum thickness develops from $Re_\theta = 670$, at the beginning of the grooved area, to $Re_\theta = 925$. The corresponding friction Reynolds number ranges from $Re_\tau = 260$ to $Re_\tau = 350$. The Synthetic eddy method (SEM) [16] is used to generate the turbulent inflow. Flow properties just upstream the work area are successfully compared to reference DNS [19, 20].

The ONERA in-house flow solver FUNk which has been used, for which details and validations can be found in [11, 15], resolves the Direct Navier–Stokes (DNS) and the Large Eddy Simulations (LES) equations for compressible flows on multi-block structured grids. The simulation Mach number is set to 0.2. In this condition, density and viscosity fluctuations are negligible and the flow will be considered incompressible. The explicit selective mixed-scale model [13] is used to compute the subgrid viscosity for LES simulations.

The tiny size of the riblets strongly constrains the cells' dimensions. In order to accurately reproduce their shape, Δy^+ and Δz^+ are maintained around 1 close to the wall. Δx^+ is set to 25. Such a resolution goes beyond state-of-the-art DNS baseline refinements in the normal and spanwise directions. The longitudinal resolution justifies the qualification of LES and the use of a subgrid-scale model. The size of the computational domain is $38\delta_0 \times 10\delta_0 \times 3.3\delta_0$, where δ_0 denotes the inflow boundary layer thickness. Two-point correlations have been computed to ensure that the domain is wide enough. The computational domain counts a total of $413 \times 143 \times 673$ cells.

3 Analysis of the Wall Friction

The time- and spatial-averaged wall friction coefficient in Fig. 2 is defined by $c_f = \frac{1}{L_z}\int_z \left(\overline{\tau}_w/\frac{1}{2}\rho U_0^2\right) \mathrm{d}z$. Over the ribbed plate, the spanwise averaging has to be changed into $\frac{1}{L_z}\int_{\tilde{L}_z} \cdot \, \mathrm{d}l$, where $\tilde{L}_z$ denotes the wetted span at the riblets' surface and L_z the corresponding span of the flat plate.

Upstream the ribbed zone, the brief transitory burst of c_f is due to the sudden geometric outbreak of riblets out the planes. The performance loss caused by the riblets' appearance is approximately overcome by the afterward persistent benefit. Taking into account these up and downstream effects, the global drag reduction adds up to 9.8 % of the baseline drag on the controlled domain. The pressure drag, due to

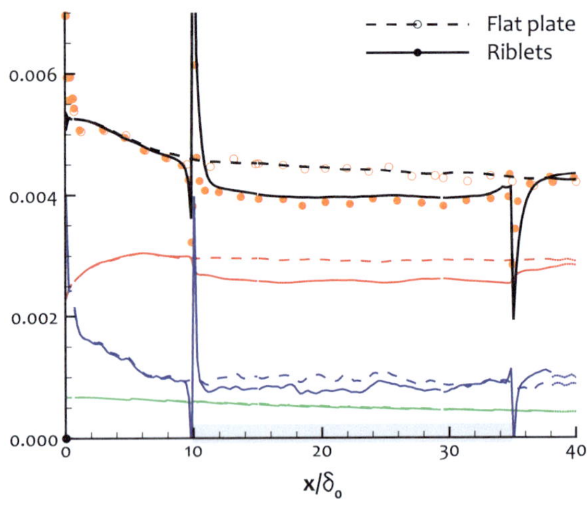

Fig. 2 Skin friction coefficient and FIK decomposition: c_f from numerical flux (*black*); $c_{f,1}$ (*green*); $c_{f,2}$ (*red*); $c_{f,3}$ (*blue*); $c_{f,1}+c_{f,2}+c_{f,3}$ (*symbols*)

the three-dimensional shape of the riblet leading and trailing edges, is found to be negligible, justifying the previously discussed choice of riblets-versus-plate relative vertical position.

The FIK decomposition is applied here. Using the notation of Deck et al. [6], the friction coefficient is split as $c_f = c_{f,1} + c_{f,2} + c_{f,3}$. Because of the geometric complexity of our flow, the original FIK formulation cannot be applied as is. Following Peet and Sagaut [17] who derived the equivalent of FIK's channel flow decomposition for a three-dimensional complex wall shape, a similar formula can be obtained from FIK's boundary layer formulation:

$$c_{f,1}(x) = \frac{4}{\alpha Re_\delta} \iint_{\Sigma(x)} \frac{\overline{u}_x}{U_0} \frac{\mathrm{d}\Sigma}{L_z \delta} \tag{1}$$

$$c_{f,2}(x) = \frac{4}{\alpha} \iint_{\Sigma(x)} \left(1 - \frac{y}{\delta}\right) \frac{-\overline{u'v'}}{U_0^2} \frac{\mathrm{d}\Sigma}{L_z \delta} \tag{2}$$

$$c_{f,3}(x) = -\frac{4}{\alpha} \iint_{\Sigma(x)} \frac{1}{2} \left(1 - \frac{y}{\delta}\right)^2 \frac{\delta}{\rho U_0^2} \overline{\left(\frac{\partial \rho u_x}{\partial t} + I_x + \frac{\partial p}{\partial x}\right)} \frac{\mathrm{d}\Sigma}{L_z \delta} \tag{3}$$

$$\alpha(x) = \int_{\tilde{L}_z(x)} \left(1 - \frac{y}{\delta}\right)^2 \sqrt{1 - n_x^2} \frac{\mu\, \partial \overline{u}_x / \partial n}{F_w} \frac{\mathrm{d}l}{L_z} \tag{4}$$

$$I_x(x) = \rho \overline{u}_i \frac{\partial \overline{u}_x}{\partial x_i} - \mu \frac{\partial^2 \overline{u}_x}{\partial x^2} + \rho \frac{\partial \overline{u'_x u'_x}}{\partial x} \tag{5}$$

where $\Sigma(x)$ stands for the transverse plane at constant x between the wall and the boundary layer height and n_x is the x-component of the wall-normal local unit vector. The coefficient α, which takes into account the wall stress distribution along the normal direction, is close to unity as soon as h^+_{rib}/Re_τ is negligible. For the present simulation, $1 - \alpha$ is of the order of 3 %.

In Fig. 2, the "bulk" term $c_{f,1}$ is largely unaffected by the riblets. This term is related to the mean velocity $\overline{u}_x$ which is examined in the following section. The second term, $c_{f,2}$, accounting for the turbulent contribution to the drag, is the most affected by the riblets presence. However, ratios $c_{f,2}/c_f$ for both the flat and the ribbed cases perfectly collapse, supporting a scaling in wall-units of the turbulence mechanisms. Turbulence quantities will be investigated Sect. 5. Finally, term $c_{f,3}$ is associated with the streamwise growth of the boundary layer. In the presence of riblets, both the term itself and its relative contribution in c_f are reduced. Although the evolution of this last term is not well understood yet, it has to satisfy $(c_{f,1}+c_{f,3})/c_f = 1 - (c_{f,2}/c_f)$ to fulfill previous requirements on $c_{f,1}$ and $c_{f,2}$.

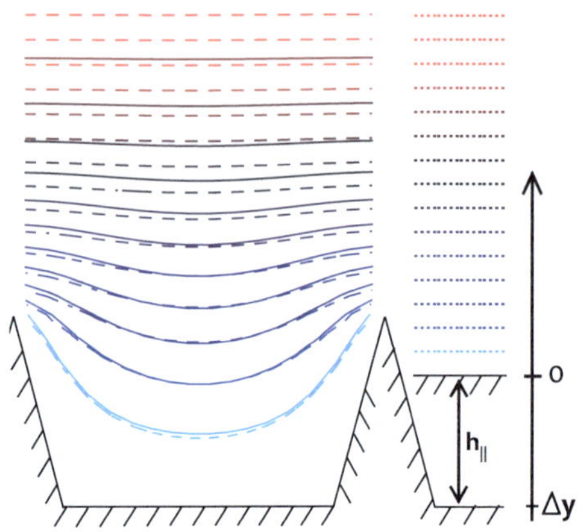

Fig. 3 Transverse view of mean streamwise velocity iso-contours over riblets: LES turbulent flow at $Re_\theta = 750$ (*solid*); Stokes flow (*dashed*). *Right-sided dotted* iso-contours denote the Stokes flow over a flat plate located $h_\parallel$ above riblets valleys

4 Mean Velocity Field

The “bulk” term $c_{f,1}$ is directly linked to the mean streamwise velocity and is proportional to the mean flow rate within the boundary layer.

Before depicting the mean velocity profile, the question of the vertical ordinate origin must be clarified. The vertical origin $y = 0$ in figures still corresponds to the smooth wall location. However, for the ribbed wall, at what location Δy should the riblets’ valley be located? The choice of Δy_0 has already been made, Sect. 2, to answer this question regarding the geometrical setup. However, to ease the comparison with the smooth wall flow, another ordinate origin $\Delta y_{(u)}$ may be more relevant to plot the mean flow profile $\overline{u}_x$. Among the different options listed in Sect. 2, the choice based on the longitudinal Stokes flow, leading to $\Delta y_{(u)} = -h_\parallel \approx -0.63\, h_{rib}$, appears to be the most pertinent. Indeed, since riblets are quasi-immersed in a viscous sublayer, the mean flow field is expected to be very similar to the Stokes flow solution. Figure 3 illustrates this point. From the riblets floor up to some few wall-units above the tips, the turbulent and the Stokes viscous profile closely fit.

Since the mean velocity depends on z nearby the riblets, spanwise averaging of the velocity is performed to allow an intelligible comparison. Two different scalings are used for the nondimensionalization in Fig. 4. The first one is based upon the actual wall friction velocity $u_\tau = \sqrt{c_f/\rho}$. This scaling is used to compare terms in usual wall-units and highlight variables whose magnitude is proportional to the wall shear stress. Alternatively, to assess the absolute evolution with respect to the uncontrolled—smooth wall—case, another scaling is used, based on the friction velocity $u_\tau^{nominal}$ of the baseline clean plate.

Above the flat surface, the velocity profile is fitted by the log-law with coefficients $\kappa = 0.39$ and $C^+ = 5.1$. Riblets cause a thickening of the viscous layer and a significant upward shift of the log-law [1]: $\Delta u^+ = \sqrt{2/c_f^{rib}} - \sqrt{2/c_f^{clean}} \approx 1.14$.

The nominal friction velocity scaling highlights the minor impact of the riblets on $\overline{u}_x$ outside the viscous sublayer: the two velocity profiles neatly collapse above

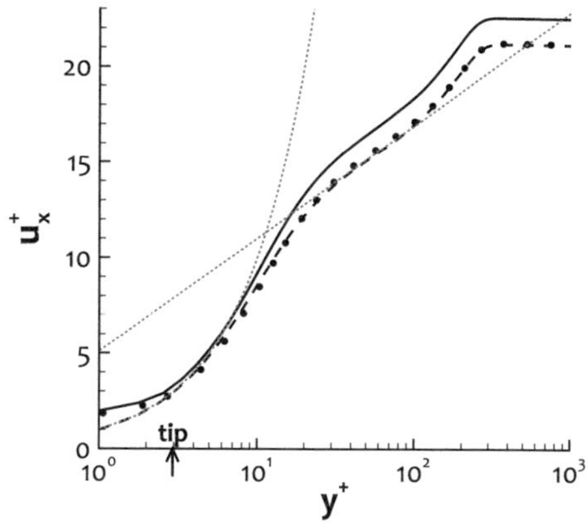

Fig. 4 Mean spanwise-averaged streamwise velocity profile at $Re_\theta = 750$ over flat plate (*dash*) and riblets in actual (*solid*) and nominal (*symbols*) wall-units. Linear ($u^+ = y^+$) and logarithmic ($u^+ = \log(y^+)/0.39 + 5.1$) profiles (*dotted*). The *arrow* indicates the riblets tip location

$y^+ \approx 10$. This very spatially limited modification of the mean velocity explains why riblets do not significantly impact the "bulk" coefficient $c_{f,1}$. Nevertheless, since the global c_f is reduced in the presence of riblets, the relative contribution of $c_{f,1}$ is increased.

Conversely to $c_{f,1}$, the relative turbulent contribution to the drag $c_{f,2}/c_f$ remains unchanged by the riblets, while the absolute value for $c_{f,2}$ decreases. The impact of riblets on the turbulence properties is investigated in the following sections.

5 Turbulent Statistics

The root-mean-square turbulent velocity fluctuations and the normal Reynolds stress $\overline{u'v'}$ are displayed in Fig. 5. On the whole, all intensities are reduced in absolute terms (*i.e.,* scaled using nominal wall friction velocity), whereas, modifications are much limited in (actual) wall-unit scaling. More specifically, the u_{rms} profile is only affected by the riblets below $y^+ = 17$. This results in a slight lowering and upward shift of its maximum peak. The profiles for v_{rms} and $\overline{u'v'}$ above the ribbed and smooth walls collapse almost perfectly, consistently with their close role with the wall friction generation. Finally, w_{rms} is somewhat increased above $y^+ = 10$.

Note that only the spanwise component w_{rms} is damped in the riblets' vicinity, in agreement with Luchini's vision of riblets as a device selectively impeding the transverse over the longitudinal flow [14].

Because $c_f = 2(U_0^+)^{-2}$, the ratio $c_{f,2}/c_f$ explicitly reveals a weighted integral of $\overline{u'v'}^+$, scaled in wall-units. Owing to the negligible impact of riblets on the wall-unit-scaled profile for the Reynolds normal shear stress, the turbulent contribution $c_{f,2}$ and the drag coefficient c_f are de facto proportional.

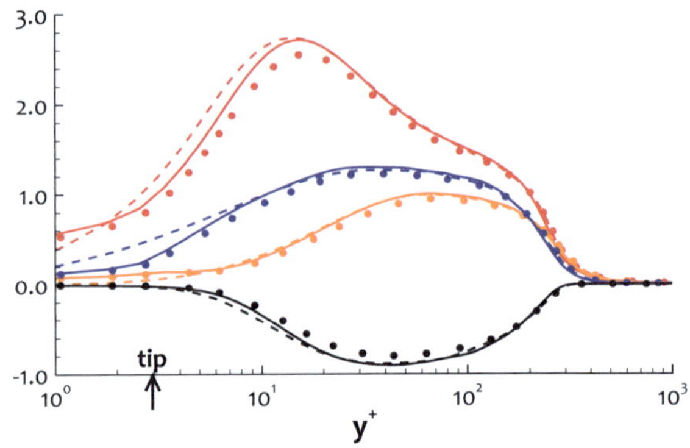

Fig. 5 Rms turbulent intensities u^+_{rms} (*red*), v^+_{rms} (*orange*), w^+_{rms} (*blue*), and Reynolds normal stress $\overline{u'v'}^+$ (black), at $Re_\theta = 750$, over flat plate (*dash*) and riblets in actual (*solid*) and nominal (*symbols*) wall-units. The arrow indicates the riblets tip location

6 Budget of Turbulent Kinetic Energy

The impact of riblets on turbulence is now investigated. It has been shown in Sect. 5 that, overall, fluctuation intensities scale in wall-units.

The same scaling will be adapted for energy budget, and beyond that, most term intensities are essentially unchanged by riblets, assuming that the appropriate vertical origin ordinate is chosen. The previously used Stokes virtual origin $\Delta y_{(u)} = -h_\parallel$ does not lead, here, to a meaningful collapse of the curves. However, a striking superimposition is observed if one shifts the riblets further approximately one wall-unit downward. This coordinate shift can be related to the velocity shift of the log layer ($\Delta u^+ \approx 1$) computed in Sect. 4. The ordinate origin $\Delta y^+_{(K)} = \Delta y^+_{(u)} - \Delta u^+ \approx -0.75\, h^+_{rib}$ allows us to make the log-law region of both flows start at the same location. This positioning will be used for the following energy budgets of this article.

The transport equation for the turbulent kinetic energy $k = \frac{1}{2}\overline{u'_i u'_i}$ reads:

$$\frac{\partial k}{\partial t} = \underbrace{-\overline{u}_i \frac{\partial k}{\partial x_i}}_{1} \underbrace{- 2\nu \overline{s_{ij} \frac{\partial u'_i}{\partial x_j}}}_{2} \underbrace{- \overline{u'_i u'_j} \frac{\partial \overline{u}_j}{\partial x_i}}_{3} - \frac{\partial}{\partial x_i}\left(\underbrace{\frac{1}{\rho}\overline{p' u'_i}}_{4} \underbrace{- 2\nu \overline{s_{ij} u'_j}}_{5} + \underbrace{\overline{u'_i u'_j u'_j}}_{6} \right) \quad (6)$$

where Einstein implicit summation is used. The wall-normal distributions of those terms are shown in Fig. 6.

Terms 1 and 4, standing for the transport work done by the total dynamic pressure, have little impact on the budget. The production term 3 and the turbulent diffusion term 6 exhibit profiles almost identical between flat and ribbed plates. The main difference between flows is an enhancement of the dissipation (term 2) in the $y^+ \in [6, 12]$ layer. To balance the budget, a tiny downward shift of the production level (term 3) in the viscous sublayer ($y^+ < 6$) provides the additional energy. The viscous diffusion term 5 ensures the appropriate energy transport between those two zones.

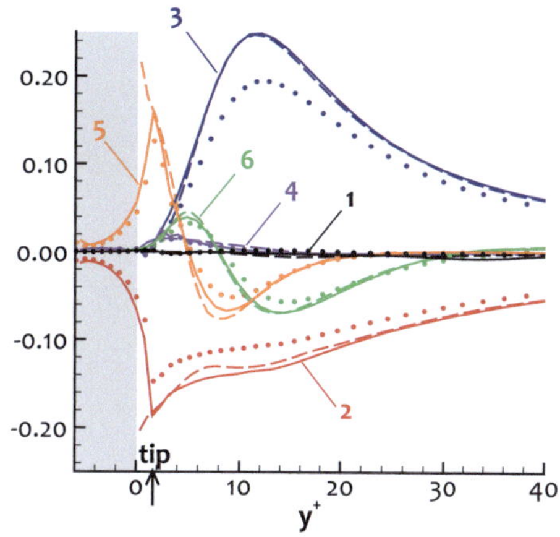

Fig. 6 Profiles for the turbulent kinetic energy budget terms, at $Re_\theta = 750$, over flat plate (*dash*) and riblets in actual (*solid*) and nominal (*symbols*) wall-units. The arrow indicates the riblets tip location

Compared to the smooth wall case in wall-units, more energy is pumped from the mean flow to be dissipated into heat. In absolute scaling however, dissipation levels are greatly reduced. Thus, the key role in the riblets' drag-reducing mechanisms seems to be played by the dissipation term 2. It can be rewritten into:

$$2\nu \overline{s_{ij} \frac{\partial}{\partial x_j} u_i'} = \nu \, \overline{\omega_i' \omega_i'} + 2\nu \frac{\partial^2}{\partial x_i \partial x_j} \overline{u_i' u_j'}, \tag{7}$$

with ω_i the flow vorticity components. The first term, called turbulent enstrophy, is observed to be the predominant one [21]. The turbulent dissipation is therefore directly related to the turbulent enstrophy. Its budget is now broken up.

7 Budget of Turbulent Enstrophy

The detailed equation for the enstrophy budget of an incompressible flow can be found in [21], Eq. (3.3.38). Labels in Fig. 7 correspond to the following terms: Term 1 accounts for the convection by the mean flow. Terms 2 through 5 are production terms. Term 2 represents the gradient production and is the enstrophy analog for the kinetic energy budget term 3 in Eq. (6). Terms 3 and 4 denote the production (or removal) caused by the stretching of vorticity fluctuations by the turbulent and the mean strain rate. Term 5 is a mixed production term also appearing in the budget for $\frac{1}{2}\overline{\omega}_i\overline{\omega}_i$. Terms 6 and 7 represent the turbulent and the viscous diffusion. Finally, term 8 expresses the enstrophy dissipation. Terms 7 and 8, both accounting for viscous effects, have been gathered for clarity's sake in Fig. 7.

The riblets' main effect is an enhancement of most of the terms in the enstrophy budget. More specifically, a striking alteration occurs with production term 2, whose expression is : $\overline{u_j' \omega_i'} (\partial \overline{\omega_i} / \partial x_j)$. It accounts for the transfer between the mean and

Fig. 7 Profiles for the turbulent enstrophy budget terms. For line legend, refer to Fig. 6 above

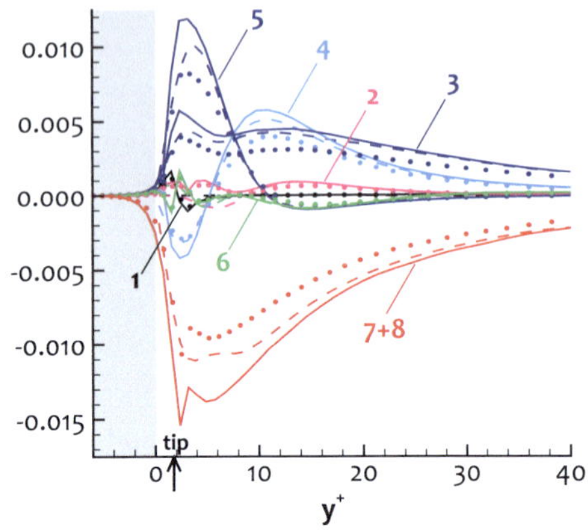

the turbulent squared vorticity. Term 2 acts as a sink term below $y^+ = 9$ for the smooth wall, but becomes a production term for the ribbed wall. This change is due to the creation, by the riblets, of mean vorticity in the normal direction $\overline{\omega}_y$ in order to satisfy the nonslip condition on the grooves' quasi-vertical sides. By producing mean normal vorticity, riblets induce extra production in the enstrophy budget—production which is counter-balanced by viscous effect—and enhance the enstrophy level. Since enstrophy translates into turbulent kinetic energy dissipation, the turbulent activity drops because of the increased dissipative nature of the flow. This drop explains the absolute reduction of all turbulence intensities, Fig. 5. A new equilibrium establishes with weakened turbulence levels. All levels of the turbulent budget (6) are dumped in absolute terms, including the dissipation which has been initially enhanced. As long as the riblets maintain an important level of enstrophy production, the flow budget remains balanced with small turbulence levels.

The reduced level of turbulence intensities, and especially for the normal Reynolds shear stress, involves a drop of the momentum diffusion toward the wall and a fall of the wall friction. Similar drag mechanisms are observed as well for other drag-reducing devices, such as transverse wall oscillation [18] and suboptimal control theory [12].

8 Outlook

Skin friction has been decomposed and the turbulent contribution was shown to be the most affected by the riblets. The modification of turbulence properties was investigated through a dissection of the budgets of the turbulent kinetic energy first, and of the main part of its dissipation—the enstrophy—thereafter. The mean normal vorticy created on the riblets' lateral sides turns out to be the cause of an extra dissipation of the turbulence intensities, which eventually reduce the drag.

Further investigations of the energetic exchanges involved in the turbulence above riblets, and, more especially, within the transient area downstream of the forefront of the ribbed wall, may bring further details and help clarify the drag reduction mechanisms.

Acknowledgments This work was granted access to the HPC resources from IDRIS under the allocation 2014-100392 made by the GENCI. The thesis of Amaury Bannier is partly funded by Ecole Polytechnique.

References

1. B. Aupoix, G. Pailhas, R. Houdeville, Towards a general strategy to model riblet effects. AIAA J. **50**(3), 708–716 (2012)
2. D.W. Bechert, M. Bruse, W. Hage, J.G.T. Van Der Hoeven, G. Hoppe, Experiments on drag-reducing surfaces and their optimization with an adjustable geometry. J. Fluid Mech. **338**, 59–87 (1997)
3. H. Choi, P. Moin, J. Kim, On the effect of riblets in fully developed laminar channel flows. Phys. Fluids A **3**(8), 1892–1896 (1991)
4. H. Choi, P. Moin, J. Kim, Direct numerical simulation of turbulent flow over riblets. J. Fluid Mech. **255**, 503–539 (1993)
5. E. Coustols, A. Savill, *Turbulent Skin-Friction Drag Reduction by Active and Passive Means: Parts 1 and 2. Special Course on Skin-Friction Drag Reduction* (Brussels, Belgium, 1992)
6. S. Deck, N. Renard, R. Laraufie, P.E. Weiss, Large-scale contribution to mean wall shear stress in high-Reynolds-number flat-plate boundary layers up to $Re_\delta = 13650$. J. Fluid Mech. **743**, 202–248 (2014)
7. K. Fukagata, K. Iwamoto, N. Kasagi, Contribution of Reynolds stress distribution to the skin friction in wall-bounded flows. Phys. Fluids **14**(11), L73 (2002)
8. R. García-Mayoral, J. Jiménez, Drag reduction by riblets. Philos. Trans. R. Soc. A **369**(1940), 1412–1427 (2011)
9. D.B. Goldstein, T.C. Tuan, Secondary flow induced by riblets. J. Fluid Mech. **363**, 115–151 (1998)
10. J. Jiménez, Turbulent flows over rough walls. Annu. Rev. Fluid Mech. **36**, 173–196 (2004)
11. L. Larchevêque, P. Sagaut, T.H. Lê, P. Comte, Large-eddy simulation of a compressible flow in a three-dimensional open cavity at high Reynolds number. J. Fluid Mech. **516**, 265–301 (2004)
12. C. Lee, J. Kim, H. Choi, Suboptimal control of turbulent channel flow for drag reduction. J. Fluid Mech. **358**, 245–258 (1998)
13. E. Lenormand, P. Sagaut, L. Ta Phuoc, P. Comte, Subgrid-scale models for large-eddy simulation of compressible wall bounded flows. AIAA J. **38**(8), 1340–1350 (2000)
14. P. Luchini, F. Manzo, A. Pozzi, Resistance of a grooved surface to parallel flow and cross-flow. J. Fluid Mech. **228**, 87–109 (1991)
15. I. Mary, P. Sagaut, Large eddy simulation of flow around an airfoil near stall. AIAA J. **40**(6), 1139 (2002)
16. M. Pamiès, P.E. Weiss, E. Garnier, S. Deck, P. Sagaut, Generation of synthetic turbulent inflow data for large-eddy simulation of spatially-evolving wall-bounded flows. Phys. Fluids **21**(4), 045103 (2009)
17. Y. Peet, P. Sagaut, Theoretical prediction of turbulent skin friction on geometrically complex surfaces. Phys. Fluids **21**(10), 105105 (2009)
18. P. Ricco, C. Ottonelli, Y. Hasegawa, M. Quadrio, Changes in turbulent dissipation in a channel flow with oscillating walls. J. Fluid Mech. **700**, 77–104 (2012)

19. P. Schlatter, R. Örlü, Assessment of direct numerical simulation data of turbulent boundary layers. J. Fluid Mech. **659**, 116–126 (2010)
20. P.R. Spalart, Direct simulation of a turbulent boundary layer up to $Re_\theta = 1410$. J. Fluid Mech. **187**, 61–98 (1988)
21. H. Tennekes, J.L. Lumley, *A First Course in Turbulence* (MIT Press, Cambridge, 1972)
22. M.J. Walsh, L.M. Weinstein, Drag and heat transfer on surfaces with small longitudinal fins. AIAA Paper 78-1161 (1978)

Characterization of Pipe-Flow Turbulence and Mass Transfer in a Curved Swirling Flow Behind an Orifice

N. Fujisawa, R. Watanabe, T. Yamagata and N. Kanatani

Abstract This paper deals with the extraction of turbulent structure correlated with the wall mass transfer in a curved swirling pipe flow behind an orifice. The cross-sectional velocity field behind the orifice is measured by the Stereo Particle Image Velocimetry (SPIV) and the results are analyzed by the proper orthogonal decomposition (POD). The instantaneous velocity field shows the asymmetric vortex structure in the cross section due to the combined effect of the swirling flow and the secondary flow generated at the upstream elbow. The POD analysis indicates that the highly turbulent flow is generated on the upper left-hand side of the pipe in the lower POD modes suggesting the occurrence of high wall-thinning rate due to the mass transfer enhancement, while that of the higher modes do not show such asymmetry. This result suggests that the lower POD modes of the velocity field contribute to the non-axisymmetric pipe-wall thinning behind an orifice in a curved swirling flow.

1 Introduction

Pipe-wall thinning due to the flow accelerated corrosion (FAC) is one of the important issues in the safety management of the steel pipeline in the nuclear/fossil power plant. The mechanism of FAC is the corrosion of the wall materials of the pipeline, which is accelerated by the turbulence in the flow field, while it is also influenced by the chemical aspect of the fluid, such as temperature, pH, oxygen concentration, and so on [4]. Such complex phenomenon of FAC often occurs in the pipe wall behind the orifice, T-junction, and in the elbow, where the turbulence is highly generated by the flow separation and the related flow phenomenon [10].

N. Fujisawa (✉) · T. Yamagata
Visualization Research Center, Niigata University, Niigata, Japan
e-mail: fujisawa@eng.niigata-u.ac.jp

T. Yamagata
e-mail: yamagata@eng.niigata-u.ac.jp

R. Watanabe · N. Kanatani
Graduate School of Science and Technology, Niigata University, Niigata, Japan

M. Stanislas et al. (eds.), *Progress in Wall Turbulence 2*,
ERCOFTAC Series 23, DOI 10.1007/978-3-319-20388-1_20

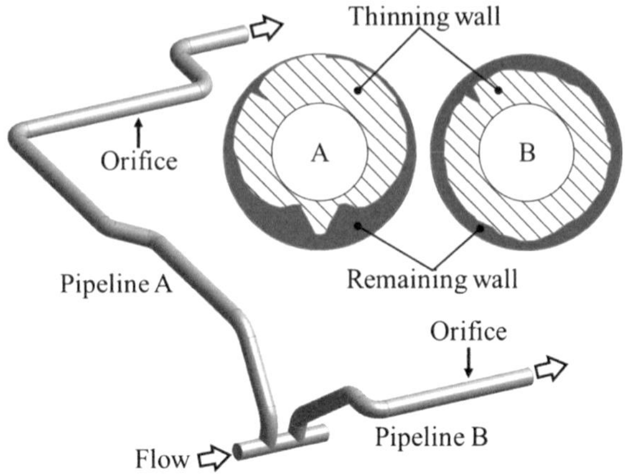

Fig. 1 Illustration of pipeline geometry

Non-axisymmetric pipe-wall thinning due to FAC is a topic of interest, since the pipeline break accident in the Mihama nuclear power plant in 2004 [14]. The pipeline layout consists of an elbow, a straight pipe, and an orifice, as shown in Fig. 1. The pipeline break happened at one diameter downstream of the orifice. It should be mentioned that the swirling flow was observed in the scaled model experiment [14]. The resulting wall thinning rate was not uniform in the circumferential direction, which accelerated the pipeline accident. Since then, several studies have been carried out to elucidate the mechanism of the non-axisymmetric pipe-wall thinning due to FAC. Ohkubo et al. [15] and Fujisawa et al. [6] indicate the influence of orifice bias on the non-axisymmetric pipe-wall thinning and Takano et al. [20] suggest the influence of the elbow, but the main cause of the non-axisymmetry has not been fully understood yet due to the complexity of the flow field and the related mechanism of FAC in the actual pipelines. To understand the flow mechanism of non-axisymmetric pipe-wall thinning, the velocity field behind the orifice has been measured by the stereo PIV, which allows the three components of instantaneous velocity field in the two-dimensional cross section of interest [7]. The results indicate that the mean velocity field and the turbulent energy distribution become non-axisymmetric, suggesting the correlation of the flow field and the mass transfer. In the past, the flow parameters, such as the vertical velocity [10], turbulent energy [19, 22, 23], and wall shear stress [21] are considered as major parameters contributing to the wall mass transfer [17, 20], but the structure of turbulence has not been identified yet in literature.

Proper orthogonal decomposition (POD) is one of the statistical methods for analyzing the low-dimensional representation of the multidimensional flow field of interest. The snapshot POD is very useful for recognizing the coherent structure of turbulent flow [2]. There are several successful examples of the snapshot POD analysis applied to the turbulent flow in the literature, such as the jet in counter flow [3], channel flow [13], backward-facing step flow [12], complex unsteady flow [1, 9], and so on. By introducing the POD analysis, the most energetic structure of the flow has been extracted from the planar PIV data by decomposing the fluctuating properties of the turbulent velocities into the linear sum of orthogonal eigenfunctions of temporal and spatial correlations.

The purpose of this paper is to introduce the snapshot POD analysis into the velocity field measured by stereo PIV to localize the high mass transfer rate in the pipe wall behind an orifice under the influence of swirling flow combined with the elbow.

2 Experimental Method

2.1 Experimental Setup

The experiment on the velocity field behind an orifice in a curved swirling flow has been carried out in a closed-circuit water tunnel [6]. The water tunnel used in this experiment consists of a pump, settling chamber, honeycomb, flow-developing section, and test section of the pipeline, which have been described in [6]. The measurements of velocity field and mass transfer rate are carried out by using the stereo PIV and benzoic acid dissolution methods, respectively. Figure 2 shows the details of the test section, which consists of a swirl generator, elbow, straight pipe, orifice, and downstream pipe. It should be mentioned that the length of the straight pipe between the elbow and the orifice is set to $10d$ to meet with the Mihama case [14], where d is a pipe diameter. The diameter of the pipe is $d = 56$ mm and the radius to diameter ratio of the elbow is 1.4. The test section is made of an acrylic resin material for flow visualization. The swirling flow was generated by a swirl generator, which is made of 6 plane vanes having an angle of 45° to the flow axis [6]. The swirl generator was placed $13d$ downstream of the honeycomb and $3d$ upstream of the elbow. It should be mentioned that the swirl generator produces a swirl flow

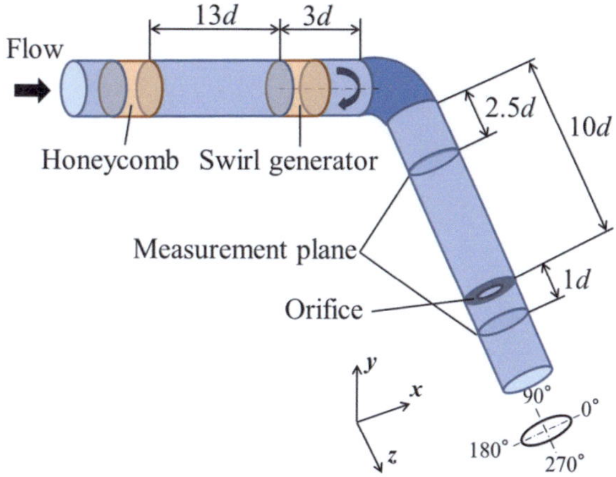

Fig. 2 Details of experimental test section

having a swirl intensity S defined by the circumferential momentum to axial one $S = 0.35$ at 3 diameters upstream of the orifice [5], while it is estimated as $S = 0.26$ in the Mihama case from the velocity measurement in the scaled model experiment [14]. In the downstream of the elbow, the straight pipe section of $10d$ and the circular orifice having a diameter ratio 0.6 are placed before entering into the downstream pipe. Note that the diameter ratio of the orifice agrees with that of the Mihama case [14]. The working fluid of water is kept at 298 K during the experiment. Therefore, the Reynolds number of the flow was set to $Re(=Ud/\nu) = 3 \times 10^4$, while that of the Mihama case was $Re = 5.8 \times 10^6$ [14], where U is the bulk velocity and ν is the kinematic viscosity of water.

2.2 Measurement of Velocity Field

The velocity field behind the orifice in the pipeline under the influence of curved swirling flow was measured by the stereo PIV system, which can measure the three-dimensional velocity field in the cross section of interest using the two oblique observations through the water jackets on both sides of the pipe. Figure 3 shows an illustration of the experimental test section and the stereo PIV system for the three-dimensional velocity measurement. The stereo PIV system consists of two CCD cameras with the frame straddling function (1018 × 1008 pixels with 8 bits in gray level), double pulsed Nd:YAG lasers (maximum output power 50 mJ/pulse), and a pulse controller. The light sheet illumination for flow visualization was provided from the top of the test section normal to the pipe axis. The thickness of the light sheet was about 2 mm. The flow visualization was carried out using nylon tracer particles 40 μm in diameter having a specific gravity of 1.02, which was supplied to the flow in the tank located upstream of the pump. The two CCD cameras are located on both sides of the pipe in a Scheimpflug configuration.

Fig. 3 Stereo PIV system

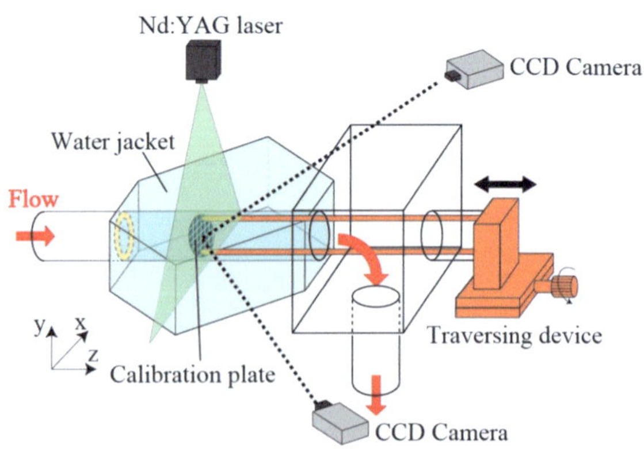

The stereo PIV analysis of the two sequential images requires the calibration in a three-dimensional thin volume of interest in the light sheet plane. The calibration plate was traversed axially using the mechanical traversing gear driven manually and allowed the out-of-plane displacement of 0.5 mm in interval with an accuracy of 5 μm. Then, a total of 5 images are captured in an axial distance of 2 mm. These calibration images were used for the three-dimensional reconstruction for stereo PIV [8, 18]. The interrogation between the sequential two images was carried out using the direct cross-correlation algorithm with the sub-pixel interpolation technique. Then, the particle displacements from each camera were combined to produce the three-dimensional velocity vectors by solving the system equations for stereo cameras with a least square method. It should be mentioned that the interrogation window size was set to 31 × 31 pixels with 50 % overlap in the analysis [11]. The details of the stereo PIV analysis can be found in Raffel et al. [16].

2.3 Proper Orthogonal Decomposition Analysis

The snapshot POD analysis is introduced into the statistical analysis of 1000 instantaneous velocity fields in the curved swirling flow, which is measured by stereo PIV at one diameter downstream of the orifice. The basic idea of snapshot POD is that it yields a set of orthogonal eigenfunctions that are optimal in energy representing temporal and spatial correlations of instantaneous velocities. It is assumed that the instantaneous velocity vector $U(x, t_k)$ is acquired at time t_k, where $k = 1, 2, -, M$. The POD analysis allows the evaluation of the mean velocity field (0th POD mode) and the fluctuating velocity field (1st, 2nd, —). In the POD analysis, three velocity components u, v, w are arranged in the velocity matrix U and the two-point correlation matrix C_{jk} is expressed by the velocity matrix U as follows:

$$C_{jk} = \frac{1}{M} \int U(x, t_j) U(x, t_k) dx \quad (j, k = 1, 2, -, M) \tag{1}$$

The POD mode Φ_k is obtained by solving the following eigenvalue problem of the two-point correlation matrix C_{jk},

$$C_{jk} a = \lambda_k \tag{2}$$

where a is the eigenvector and λ_k is the eigenvalue of C_{jk}. The eigenvectors and eigenvalues can be obtained by solving these equations numerically.

The POD mode Φ_k can be expressed by the linear combination of the eigenvector a and the instantaneous velocity vector U, as follows:

$$\Phi_k(x) = \sum_{k=1}^{N} a_n^k U(x, t_k) \tag{3}$$

The fluctuating energy E_k of the corresponding POD mode is expressed by the eigenvalue λ_k divided by the total fluctuating energy E_t, which is written as follows:

$$E_k = \frac{\lambda_k}{E_t} \tag{4}$$

where

$$E_t = \sum_{k=1}^{N} \lambda_k = \int U(x, t_k)^2 dx \tag{5}$$

The reconstructed velocity field can be expressed by using the eigenfunction and eigenvector as follows:

$$\widetilde{U}(x, t_k) = \sum_{n=1}^{N_{POD}} a_k^n \Phi_k(x) \tag{6}$$

where N_{POD} is the number of POD mode. The details of the snapshot POD analysis have been described in the literature [1–3, 12, 13]. It should be mentioned that sufficiency of the number of PIV data for the analysis was confirmed by the analysis with a small number of PIV data of 500.

3 Results and Discussions

Figure 4 shows the first three POD modes of the flow behind the orifice in a curved flow without swirl. The 0th mode (a) corresponds to the mean flow, while the 1st mode (b) and 2nd mode (c) show the fluctuating velocity fields. The 0th mode of the analysis agrees with that of the mean velocity field, which suggests the validity of the present analysis. On the other hand, the periodic pattern appears in circumferential direction in the fluctuating velocity modes. The energy levels of the 1st and 2nd POD modes are 3.8 and 3.6 %, respectively. The 1st POD mode shows the periodic pattern on the left-hand side and the 2nd POD mode shows the similar pattern on the lower side. These results indicate that the total energy of the periodic pattern is distributing almost uniformly in energy in the cross section of the pipe, suggesting the presence of axisymmetric flow structure in the pipe flow behind the curved flow.

Figure 5 indicates the corresponding three POD modes of the flow behind a curved swirling flow. The 0th POD mode shows the high axial velocity on the left-hand side of the pipe center and the high circumferential velocity on the left-hand pipe wall, suggesting the occurrence of non-axisymmetric flow behind the orifice. This flow pattern suggests the presence of a secondary flow in the pipe. On the other hand, 1st and 2nd POD modes show the presence of positive and negative sign in the cross-sectional distribution of fluctuating velocities, while the magnitude of the POD mode decreases with an increase in the mode number. The energy levels of the 1st and 2nd POD modes are 13 and 7.4 %, respectively. It should be mentioned that the presence

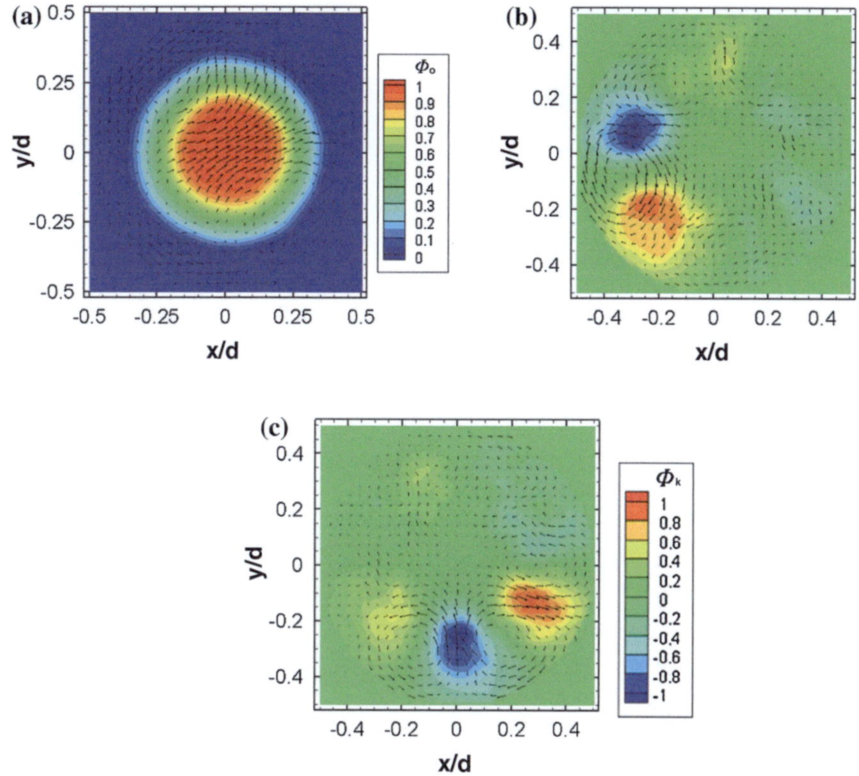

Fig. 4 Cross-sectional distributions of POD analysis without swirl. **a** Mean velocity. **b** 1st mode. **c** 2nd mode

of positive and negative sign in the neighborhood of the pipe center indicates the formation of vortex structure in the pipe. These results show the complexity of the mean and fluctuating velocity field downstream of the orifice under the influence of curved swirling flow.

Figure 6 shows the cross-sectional turbulent energy contours downstream of the orifice, which are reconstructed from the velocity field from the first 19 POD modes (a) and that from the rest of the POD modes (b). Note that the first 19 modes occupy 50 % of the total energy and the rest of the POD modes the other 50 %. The results indicate that the reconstructed turbulent energy by the first 19 modes shows non-axisymmetric distribution along the pipe wall and the magnitude of turbulent energy is high on the upper side and is low on the right-hand side, while the reconstructed turbulent energy contour in the rest of POD modes shows almost uniform along the wall. This result suggests that the non-axisymmetric feature of the turbulent energy in the pipe comes from the lower POD modes, which contains the high turbulent energy.

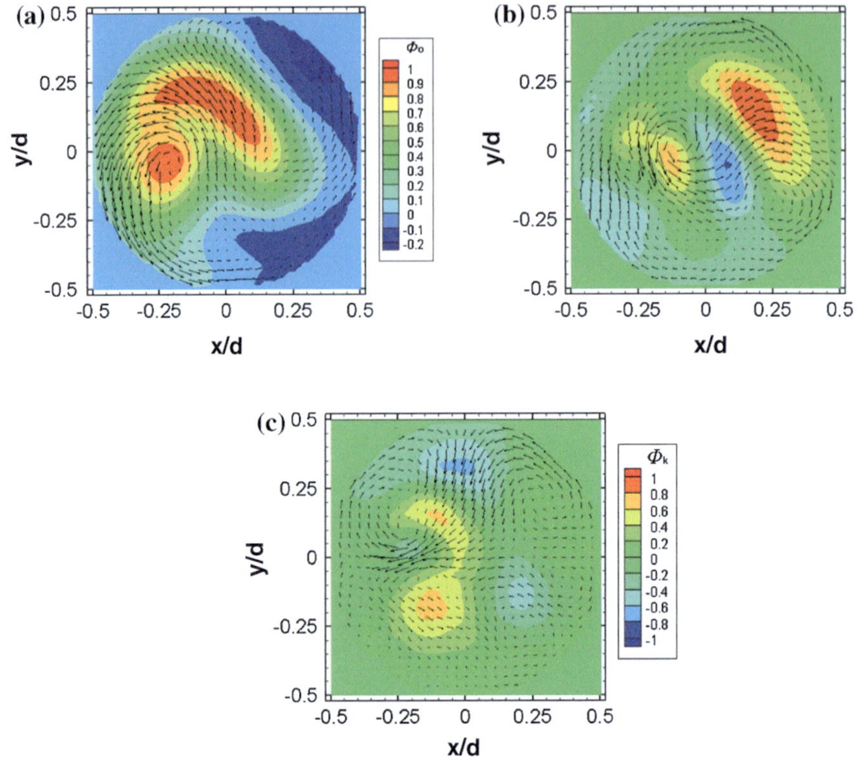

Fig. 5 Cross-sectional distributions of POD analysis with swirl. **a** Mean velocity. **b** 1st mode. **c** 2nd mode

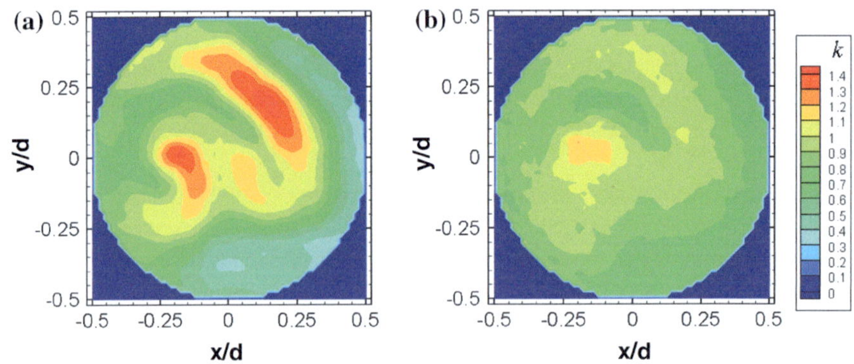

Fig. 6 Reconstruction of turbulent energy with swirl. **a** Lower POD modes. **b** Higher POD modes

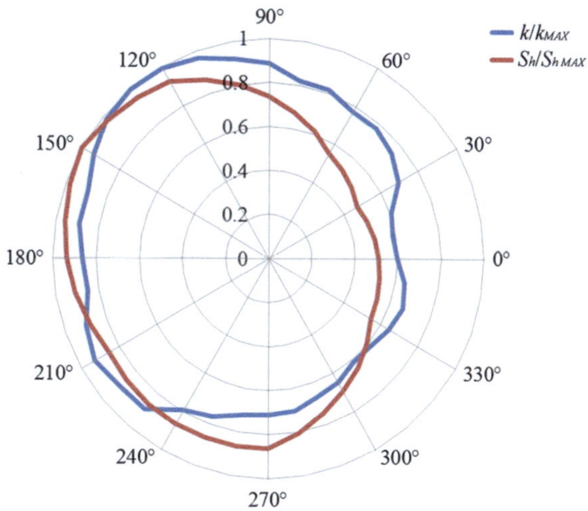

Fig. 7 Distributions of wall mass transfer rate and turbulent energy

Figure 7 compares the distributions of turbulent energy k reconstructed from the velocity in lower POD modes and the mass transfer rate S_h along the pipe wall in the curved swirling flow, which is measured by the benzoic acid dissolution method [20]. They are normalized by maximum value. The result indicates that the high mass transfer rate is found on the top left side of the pipe and both results are consistent with each other. This result suggests that the turbulent energy distribution in the lower POD modes is highly correlated with that of the mass transfer rate.

4 Conclusions

The cross-sectional velocity field behind an orifice under the influence of curved swirling flow was measured by Stereo Particle Image Velocimetry, and the results were analyzed by the proper orthogonal decomposition to extract the turbulent structure correlated with the wall mass transfer. The instantaneous velocity field showed the vortical structure due to the swirling flow combined with the secondary flow caused by the upstream elbow, which is generated by non-axisymmetric flow field behind the orifice. The POD analysis showed the variation of turbulent structure in the lower POD modes due to the influence of the curved swirling flow. The turbulent energy reconstructed from the lower POD modes indicates the similar distribution to that of the mass transfer rate. This result suggests the correlation of turbulent energy in the lower POD modes with the mass transfer rate.

References

1. P. Aranyi, G. Janiga, K. Zahringer, D. Thevenin, Analysis of different POD methods for PIV; measurements in complex unsteady flows. Int. J. Heat Fluid Flow **43**, 204–211 (2013)
2. G. Berkooz, P. Holmes, J.L. Lumley, The proper orthogonal decomposition in the analysis of turbulent flows. Annu. Rev. Fluid Mech. **25**, 539–575 (1993)
3. S. Bernero, H.E. Fielder, Application of particle image velocimetry and proper orthogonal decomposition on the study of a jet in counterflow. Exp. Fluids **29**, S274–S281 (2000)
4. R.B. Dooley, Flow-accelerated corrosion in fossil and combined cycle/HRSG plants. Power Plant Chem. **10**, 68–89 (2008)
5. M. Escudier, Vortex breakdown, observations and explanations. Prog. Aerosp. Sci. **25**, 189–229 (1988)
6. N. Fujisawa, T. Yamagata, S. Kanno, A. Ito, T. Takano, The mechanism of asymmetric pipe-wall thinning behind an orifice by combined effect of swirling flow and orifice bias. Nucl. Eng. Des. **252**, 19–26 (2012)
7. N. Fujisawa, T. Yamagata, T. Takano, N. Kanatani, K. Iwata, A. Ishizuka, Experimental and numerical study on pipe-wall-thinning by swirling flow through complex pipeline geometry, in *Proceedings of 12th Asian Symposium on Visualization*, Tainan, Taiwan, ASV12-K3 (2013)
8. S. Funatani, N. Fujisawa, Simultaneous measurement of temperature and three velocity components in planar cross section by liquid-crystal thermometry combined with stereoscopic particle image velocimetry. Meas. Sci. Technol. **13**, 1197–1205 (2002)
9. L. Graftieaux, M. Michard, N. Grosjean, Combining PIV, POD and vortex identification algorithms for the study of unsteady turbulent swirling flows. Meas. Sci. Technol. **12**, 1422–1429 (2001)
10. K.M. Hwang, T.E. Jin, K.H. Kim, Identification of relationship between local velocity components and local wall thinning inside carbon steel piping. J. Nucl. Sci. Technol. **46**, 469–478 (2009)
11. M. Kiuchi, N. Fujisawa, S. Tomimatsu, Performance of PIV system for combusting flow and its application to spray combustor model. J. Vis. **8**, 269–276 (2005)
12. J. Kostas, J. Soria, M.S. Chong, A comparison between snapshot POD analysis of PIV velocity and vorticity data. Exp. Fluids **38**, 146–160 (2005)
13. Z.-C. Liu, R.J. Adrian, T.J. Hanratty, Large-scale modes of turbulent channel flow: transport and structure. J. Fluid Mech. **448**, 53–80 (2001)
14. NISA, Secondary piping rupture accident at Mihama power station, unit 3, of the Kansai Electric Power Company, Inc. (2005)
15. M. Ohkubo, S. Kanno, T. Yamagata, T. Takano, N. Fujisawa, Occurrence of asymmetrical flow pattern behind an orifice in a circular pipe. J. Vis. **14**, 15–17 (2011)
16. M. Raffel, C. Willert, J. Kompenhans, *Particle Image Velocimetry*, vol. 174 (Springer, Heidelberg, 1998)
17. F. Shan, A. Fujishiro, T. Tsuneyoshi, Y. Tsuji, Effects of flow field on the wall mass transfer rate behind a circular orifice in a round pipe. Int. J. Heat Mass Transf. **73**, 542–550 (2014)
18. S.M. Soloff, R.J. Adrian, Z.C. Liu, Distortion compensation for generalized stereoscopic particle image velocimetry. Meas. Sci. Technol. **8**, 1441–1454 (1997)
19. T. Sydberger, U. Lotz, Relation between mass transfer and corrosion in a turbulent pipe flow. J. Electrochem. Soc. **129**, 276–283 (1982)
20. T. Takano, T. Yamagata, Y. Sato, N. Fujisawa, Non-axisymmetric mass transfer phenomenon behind an orifice in a curved swirling flow. J. Flow Control Meas. Vis. **1**, 1–5 (2013)
21. Y. Utanohara, Y. Nagaya, A. Nakamura, M. Murase, Influence of local flow field on flow accelerated corrosion downstream from an orifice. J. Power Energy Syst. **6**, 18–33 (2012)

22. T. Yamagata, A. Ito, Y. Sato, N. Fujisawa, Experimental and numerical studies on mass transfer characteristics behind an orifice in a circular pipe for application to pipe-wall thinning. Exp. Therm. Fluid Sci. **52**, 239–247 (2014)
23. K. Yoneda, R. Morita, M. Satake, I. Inada, Quantitative evaluation of effective factors on flow accelerated corrosion (part 2), Modelling of mass transfer coefficient with hydraulic features at wall. CRIEPI Research Report, No. L07015 (2008) pp. 1–33

Turbulent Structure of a Concentric Annular Flow

Sina Ghaemi, Majid Bizhani and Ergun Kuru

Abstract Turbulent flow in the annular gap between two concentric tubes of 38 and 95 mm diameter at Reynolds number of 79'000 is experimentally investigated. Measurements are conducted using planar particle image velocimetry (PIV) with spatial resolution of 23 μm/pix and interrogation windows of $0.74 \times 0.74\,\text{mm}^2$. The experiments are aimed at scrutinizing the location of the extremums of the asymmetric profiles of velocity and turbulent statistics along with the relevant turbulent structures. The location of maximum average streamwise velocity $<U>_{max}$ and zero Reynolds shear stress $<uv>$ are observed to be apart. Local minimum of $<u^2>$ and $<v^2>$ is also observed to coincide with $<uv> = 0$ and different from $<U>_{max}$. The experiments also demonstrate that the ejection events originating from the inner and outer walls play a dominant role in transport of turbulence toward the midsection of the annulus.

Keywords Turbulent annular flow · Reynolds stress · Particle image velocimetry

1 Introduction

Turbulent flow in the annular gap between two concentric tubes has been frequently investigated in the literature due to several engineering applications. For example, the removal of cutting material from the annular passage between the drill shaft and the well casing is dominated by the transport mechanisms of turbulent flow (Nouri et al. [15]). The convective heat transfer in a double-pipe heat exchanger is another widespread application of turbulent annular flow (Crookston et al. [7]). The asymmetric velocity profile of turbulent annular flow and its' deviations from

S. Ghaemi (✉)
Department of Mechanical Engineering, University of Alberta,
Edmonton T6G 2G8, Canada
e-mail: ghaemi@ualberta.ca

M. Bizhani · E. Kuru
School of Mining and Petroleum Engineering, University of Alberta,
Edmonton T6G 2W2, Canada

M. Stanislas et al. (eds.), *Progress in Wall Turbulence 2*,
ERCOFTAC Series 23, DOI 10.1007/978-3-319-20388-1_21

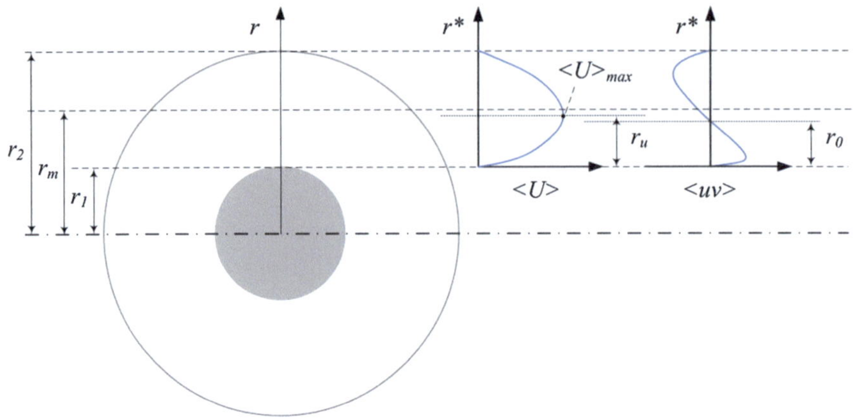

Fig. 1 Turbulent flow in a concentric annulus and schematics profiles of average velocity ($<U>$) and Reynolds shear stress ($<uv>$)

turbulent pipe flow are also of interest to fundamental turbulence research. Flow in an annulus presents a transitional configuration from channel flow to pipe flow. As the ratio of the radius of the inner pipe (r_1) to the radius of the outer pipe (r_2) approaches one, the annular flow would resemble a channel flow. On the other hand, when r_1/r_2 approaches zero, the flow tends toward pipe flow. In spite of several past investigations, there are still major deficiencies in our understanding of turbulent annular flow.

The asymmetry of the average flow field relative to the midpoint ($r_m = 1/2 \times (r_1 + r_2)$) of the annulus has been associated with unique turbulence characteristics. It has been confirmed since the early investigation of Brighton and Jones [2] that the locations of maximum average velocity $<U>_{\max}$ and zero shear stress ($<uv> = 0$) are both at $r < r_m$ as shown in Fig. 1. However, whether these two locations overlap in a turbulent annular flow is still under debate. This is in contrary to fully developed pipe flow and symmetric channel flow in which r_u (location of $<U>_{\max}$) and r_0 (location of $<uv> = 0$) overlap. Asymmetric channel flows with a rough wall on one side have also shown similar asymmetric behavior and possible separate r_u and r_0 locations (Maubach and Rehme [13]).

The noncoincidence of r_u and r_0 appears contradictory with respect to our presumption of negative $<uv>$ in a positive d$<U>$/dy shear flow. As explained by Kundu et al. [10] in an one-dimensional shear flow with d$<U>$/d$y>$ 0, a fluid element at location y_0 would carry on average a negative streamwise fluctuation ($u < 0$) upon traveling to $y_0 + dy$ through a positive wall normal fluctuation ($v > 0$). In the opposite direction, if the fluid element at location y_0 travels to $y_0 - dy$, the negative wall normal fluctuation ($v < 0$) would carry on average a positive streamwise fluctuation ($u > 0$). In both cases, $<uv>$ is negative while dU_m/dy is positive. However, if r_u and r_0 do not coincide in a turbulent annular flow, in the small region between r_u and r_0, there would be a contradictory region of positive $<uv>$ and positive d<U>/dy. The possible shortcoming of this conceptual model would preclude both the eddy viscosity model and the Prandtl's mixing length theory. The eddy viscosity (ϵ) is defined in analogy with molecular transport in gases (Tennekes and Lumley 1972) according to

$$-\rho\langle uv\rangle=\epsilon\frac{\partial\langle u\rangle}{\partial y}. \tag{1}$$

In this equation, positive <uv> and positive dU_m/dy would result in a negative which lacks any physical meaning. In addition, the Prandtl's mixing length (l) cannot be estimated in this region following

$$-\rho\langle uv\rangle=\rho l^2\frac{\partial U_m}{\partial y}\left|\frac{\partial U_m}{\partial y}\right|. \tag{2}$$

Although the success of these models in accurate modeling of turbulent flows has always been limited to only a number of simple flows, the fundamental concept used to develop these models is expected to be valid as it is frequently discussed in our classical fluid mechanics literature.

The early investigation of Brighton and Jones [2] obtained r_u from measurement of radial velocity gradient using two adjacent Pitot tubes spaced 5.08 mm apart and compared it with r_0 measured using a hot-wire probe. They reported that the deviation of the two points is within the measurement uncertainty. Lawn and Elliott [11] also used a combination of double Pitot tube (tubes 6.3 mm apart) and an X-wire probe and observed a distance of about 1–2 mm between r_m and r_0 in annular gaps of 87.2–131.6 mm at *Re* number range of 3×10^4–3×10^5. The deviation is observed to increase as the asymmetry of the profile increases with reduction of r_1/r_2. No significant variation was observed with Reynolds number at a fixed r_1/r_2. Rehme [18] carried out an experimental investigation to identify the location of r_u and r_0 using double Pitot tube and hot-wire anemometry demonstrating their noncoincidence at $Re=2\times10^4$–2×10^5 and $r_1/r_2=0.02$–0.1. Nouri et al. [15] observed that the deviation of r_u and r_0 is within the measurement precision of the Laser Doppler velocimetry conducted at Reynolds numbers of 8'900 and 26'600. The recent experiment of Rodriguez–Corredor et al. [6] shows noncoincidence of the two points using PIV measurement in Re of 17'700 up to 66'900. The consensus of the experimental investigations is toward noncoincidence of r_u and r_0. However, uncertainty sources such as the spatial resolution of the measurement technique, intrusion of the spacers holding the inner tube, and vibration of the inner tube have been debated.

Numerical simulations have also shown contradictory results on the relative location of r_u and r_0. Chung et al. [4] carried out a direct numerical simulation (DNS) at the same condition as the experiment of Nouri et al. [15] and observed noncoincidence of r_u and r_0. However, Boersma and Breugem [1] recently observed coincidence of the two positions based on a DNS of fully developed turbulent flow in an annuli flow with r_1/r_2 of 0.1. Boersma and Breugem [1] associated the noncoincidence observed by Chun et al. (2002) to poor radial resolution.

The location of r_u and r_0 is related to the turbulent mechanisms at the midsection of the annulus. To the authors' knowledge, the only physical explanation for the noncoincidence is provided by Rehme [19] who proposed turbulence diffusion as the cause of the noncoincidence. He conjectures that diffusion of turbulence plays a stronger role in an asymmetric turbulent channel flows relative to tubes and parallel plates. It is interesting to also mention that all the investigations which reported noncoincidence of the two points, r_0 is smaller than r_u, suggest a systematic mechanism. Brighton and Jones [2] investigated the profiles of normal Reynolds stress and observed that the location of minimum $<u^2>$, $<v^2>$, and $<w^2>$ is within the measurement uncertainty of r_u. Chung et al. [4] studied the turbulent structure of annuli flow and budget of Reynolds stress using DNS. However, there is still no clear understanding on the possibility and the cause of noncoincidence of r_m and r_0 in a turbulent annular flow.

The present investigation is carried out to investigate the structure of the turbulent flow at the midsection of an annulus. The experiments are conducted in a high Reynolds facility using particle image velocimetry (PIV) measurement to scrutinize the relative location of $<U>_m$ and $<uv> = 0$ along with the structure of turbulent fluctuations.

2 Experimental Setup

The experiments are conducted in a flow loop with approximately 20 m of pipe flow driven by a centrifugal pump equipped with a variable frequency drive (VFD). The pump inlet is fed from a 600-liters mixing tank and the outlet is connected to a 2-inch pipe. The annular test section is 9 m long and is located on the return route of the loop after a short vertical section as shown in Fig. 2. The annular section has two concentric glass tubes and is constructed from 6 sections each 1.5 m long. The smaller tube has an outer diameter of $d_1 = 38$ mm and the larger tube has an inner diameter of $d_2 = 95$ mm giving a ratio of $d_1/d_2 = 0.4$. The corresponding hydraulic diameter is $d_h = 57$ mm. The six annular sections are connected by stainless steel joints ensuring a smooth connection. The wall thickness of the inner tube is selected to provide a buoyant inner tube in water and minimize deflection and vibration

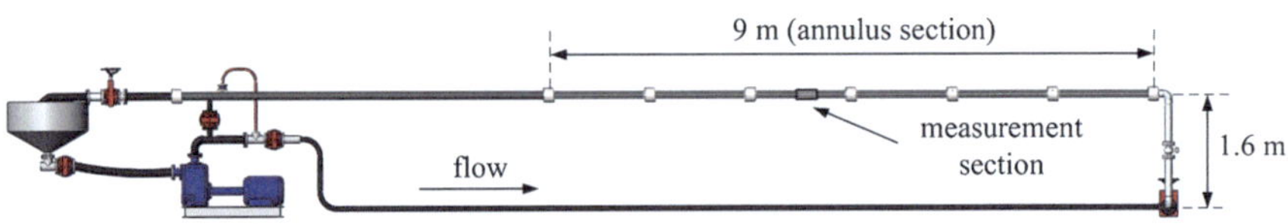

Fig. 2 Schematic of the flow loop and the annulus section formed by two concentric tubes

Fig. 3 The PIV setup consisting of a camera imaging perpendicular to the laser sheet. The test section of the flow loop is contained with a glycerol box minimizing distortion effects caused by the round glass pipe

effects. The inner tube is held in place at the joints using three threaded rods spaced 120° apart. Measurements are conducted at an average velocity of $U_{avg} = 1.11\,\text{m/s}$ corresponding to Reynolds number ($\text{Re} = U_{avg}\, d_h / \nu$) of 79'000 at a flow rate of 323 L/min. The reference point of the radial coordinate $r*$ is defined at the outer surface of the inner tube (see Fig. 1) and is normalized based on the annulus gap according to

$$r^* = \frac{r - r_1}{r_2 - r_1}. \tag{3}$$

The PIV measurements are conducted at 5.5 m downstream of the start of the annulus section and one meter downstream of the closest joint holding the threaded rods. As it is shown in Fig. 3, the test section is submerged in a rectangular box made of cast acrylic and filled with glycerol to reduce image distortion due to the curved pipe surface. Streamwise and radial velocities are denoted by U and V while their fluctuations by u and v, respectively.

The PIV setup consists of a double cavity Nd-YAG laser (SoloPIV III, New Wave Research) and a CCD camera (Imager Intense, LaVision). The laser produces 50 mJ/pulse and is equipped with a combination of spherical and cylindrical lenses forming a 1 mm thick laser sheet. The camera views the laser sheet at 90° angle through a 105 objective lens (Sigma) at $f/8$ aperture setting. The magnification of the imaging system is $M = 0.277$ resulting in a spatial resolution of 23 μm/pix. The CCD sensor has a pixel pitch of 6.45 μm and is cropped to 1376 × 608 pixels corresponding to a field-of-view of 32.092 × 14.180 mm. A set of 12,000 double-frame recordings with pulse separation of 270 μs is acquired at an approximate acquisition frequency of 10 Hz (cropped sensor). The recorded images are processed in Davis 8.0 (LaVision) with final interrogation areas of 32 × 32 pixels (0.736 × 0.736 mm) with 75 % overlap yielding a vector spacing of 184 μm.

3 Asymmetric Boundary Layer and the Extremums

The average streamwise velocity of turbulent annular flow shows an asymmetric profile with $<U>_{max}$ closer to the inner wall as shown in Fig. 4a. The asymmetry is caused by the smaller surface area of the inner wall producing smaller net friction force. As a result, the flow is faster in the vicinity of the inner wall shifting $<U>_{max}$ toward the inner wall. The measurements also show a small radial velocity toward the outer wall and therefore a slight deviation from fully developed channel flow ($\mathrm{d}<V>/\mathrm{d}y \neq 0$).

The asymmetry is also present in the profiles of Reynolds stresses shown in Fig. 4b. As expected the local maxima of $<u^2>$ and $<v^2>$ are in the vicinity of the walls. The local maxima of the $<v^2>$ is at $r* = 0.09$ and 0.83 while the exact location of local maxima of $<u^2>$ is not visible due to lack of spatial resolution in the vicinity of the walls. Both normal stresses have a higher intensity peak at the inner wall side ($r* < 0.46$) while their local minimum is also closer to the inner wall. This trend is also accompanied by larger spatial gradient of normal stresses at the inner wall side of the annular gap. The location of $<uv> = 0$ is at the inner wall side of the midsection. The Reynolds shear stress does not follow a linear distribution in contrary to fully developed channel flow with slightly larger intensities at the inner wall side of the annulus.

The locations of $<U>_{max}$, $<uv> = 0$, $<u^2>_{min}$, and $<v^2>_{min}$ at the midsection of the annular gap are scrutinized in the magnified views of Fig. 5. As it is observed, the maximum streamwise velocity $<U>_{max}$ is at $r* = 0.456$ while the $<uv> = 0$ is at $r* = 0.438$. Figure 5 also shows that the noncoincidence is not only limited to $<uv> = 0$ and $<U>_{max}$. The location of $<u^2>_{min}$ and $<v^2>_{min}$ also overlaps with $<uv> = 0$ and it does not coincide with $<U>_{max}$.

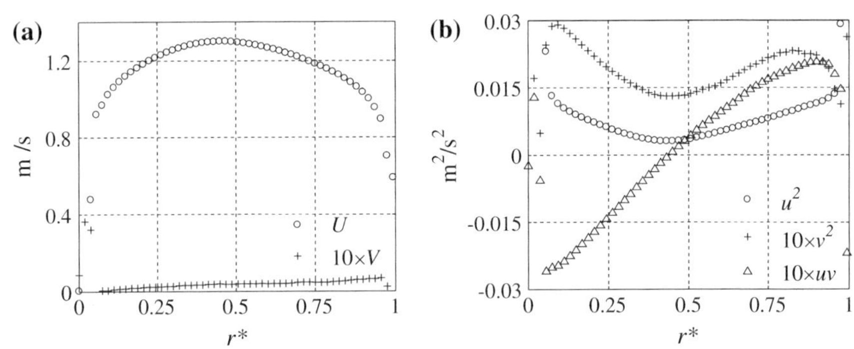

Fig. 4 **a** Average streamwise and radial velocities and **b** Reynolds stresses across the annular gap at $Re = 79'000$. One every three data points is shown for clarity

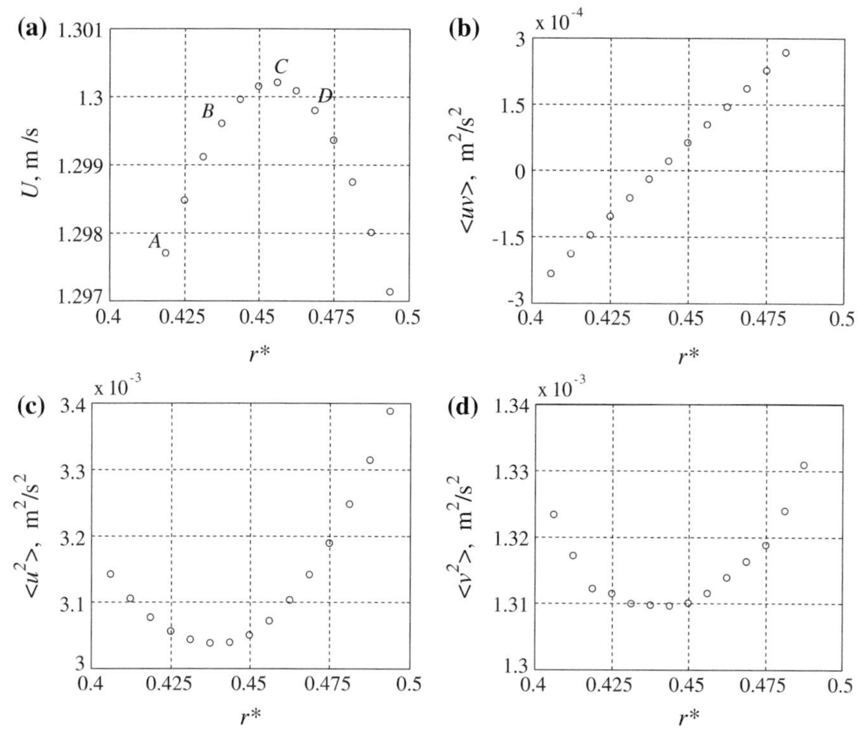

Fig. 5 Profiles of **a** $\langle U \rangle$, **b** $\langle uv \rangle$, **c** $\langle u^2 \rangle$, and **d** $\langle v^2 \rangle$ at the center region of the annular gap

4 Transport of Turbulence

The transport direction of $\langle u^2 \rangle$ is characterized by investigating the triple products of $\langle u^3 \rangle$ and $\langle vu^2 \rangle$ in Fig. 6. Figure 6a shows that $\langle u^3 \rangle$ is negative across the annuli while $\langle vu^2 \rangle$ changes from positive to negative with increase of $r*$. This observation is valid except very close to the walls where there is a lack of spatial resolution. The magnified view of Fig. 6b shows that the local minimum of $\langle u^3 \rangle$ in the centerline region of the annulus is located at $r* = 0.425$ and the sign change of $\langle vu^2 \rangle$ occurs at $r* = 0.431$. As a result, the dominant turbulent fluctuation on average is ($u < 0$ and $v > 0$) at $r* < 0.431$ and ($u < 0$ and $v < 0$) at $r* > 0.431$. These fluctuations are both ejection events originating from the inner wall for $r* < 0.431$ and from the outer wall for $r* > 0.431$. Figure 7 demonstrates that dominant transport of $\langle v^2 \rangle$ is by negative u across the channel. The radial fluctuation v changes from positive to negative at $r* = 0.438$. The trend also shows that $\langle v^2 \rangle$ is mainly transported by ejection events of the inner and outer walls.

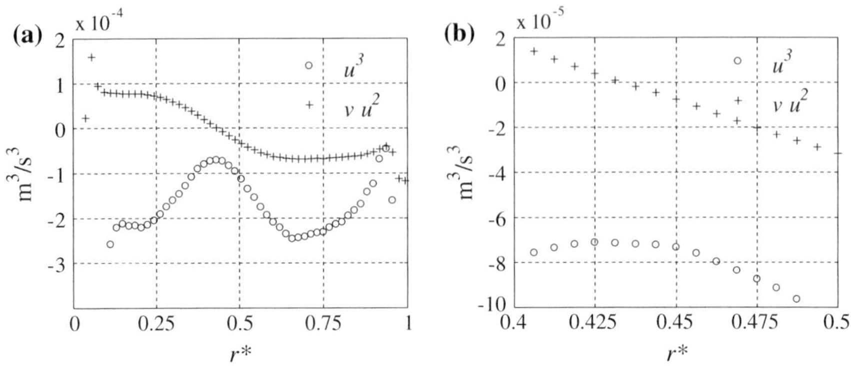

Fig. 6 Turbulent transport by $< u^3 >$ and $< vu^2 >$ **a** across the annulus and **b** magnified view of the centerline region

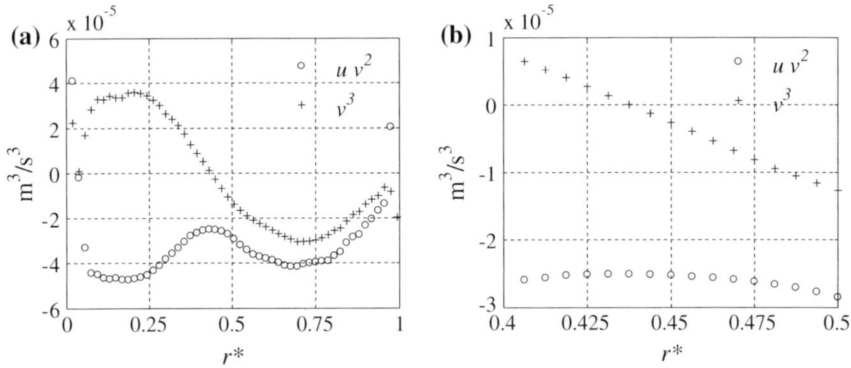

Fig. 7 Turbulent transport by $< v^3 >$ and $< uv^2 >$ **a** across the annulus and **b** magnified view at the centerline

5 Error Analysis

In order to further assure the accuracy of PIV measurements in determining the location of $< U_{max} >$, $< uv > = 0$, $< u^2 >_{min}$, and $< v^2 >_{min}$, the cumulative averages of points A ($r* = 0.4187$), B ($r* = 0.4375$), C ($r* = 0.4562$), and D ($r* = 0.4688$) as indicated in Fig. 5a are plotted versus the number of samples in Fig. 8. The plots demonstrate both low and high frequency variation of averages with the number of samples. The low frequency variations relative to the absolute values within the sample range of Fig. 8 are about 0.3 % for $< U >$, 7 % for $< uv >$ and less than 2 % for $< u^2 >$ and $< v^2 >$. The error is mainly associated with both statistical convergence and unsteadiness of the pump during the PIV acquisition (approximately 20 min). The fluctuations of the average flow cause slight deviation from stationary flow assumption. The high frequency fluctuations of the average values are smaller than

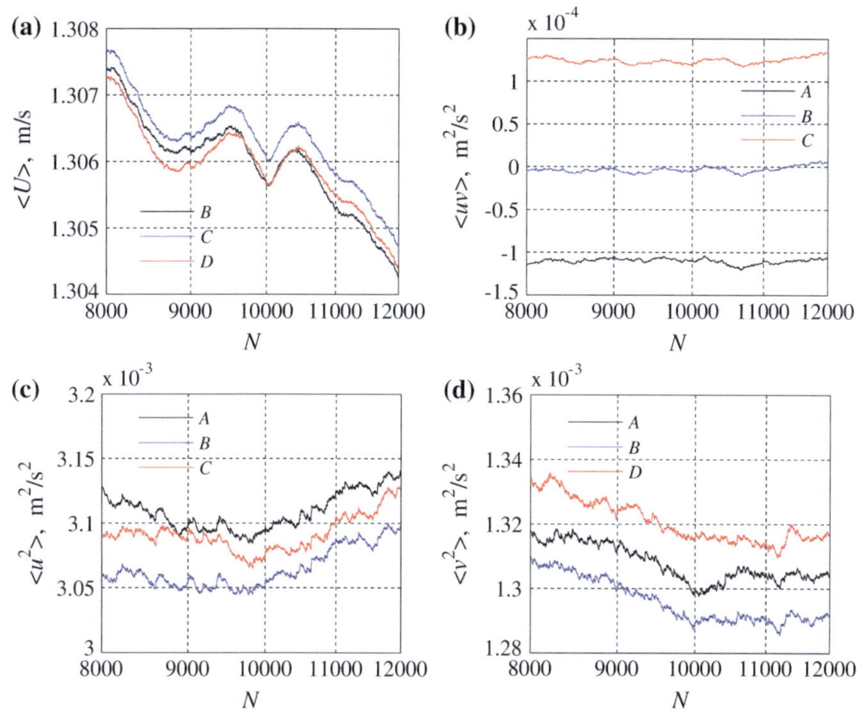

Fig. 8 The convergence of **a** $<U>$, **b** $<uv>$, **c** $<u^2>$, and **d** $<v^2>$ versus number of samples for radial positions A, B, C, D located at $r* = 0.4187$, 0.4375, 0.4562, and 0.4688, respectively

the low frequency fluctuations and are due to random error of the PIV measurement. Although the averages vary with the number of samples, it is observed that the relative magnitude of the values stay in the same order for the considered neighboring locations of A, B, C, and D. For example, in Fig. 8a, point C remains as the maximum velocity while point B as $<uv> = 0$ for the range of samples.

6 Conclusion

The measurements using PIV technique in the annulus at $Re = 79\text{'}000$ shows the location of maximum average streamwise velocity $<U>_{max}$ and zero Reynolds shear stress $<uv>$ do not overlap. In addition, the local minima of $<u^2>$ and $<v^2>$ are also observed to coincide with $<uv> = 0$ and different from $<U>_{max}$ location. Error analysis shows that the difference of the locations is larger than the uncertainty of the measurement system. The experiments also demonstrate that the ejection events originating from the inner and outer walls play the dominant role in transport of turbulence toward the midsection of the annulus.

References

1. B.J. Boersma, W.P. Breugem, Numerical simulation of turbulent flow in concentric annuli. Flow, Turbul. Combust. **86**(1), 113–127 (2011)
2. J.A. Brighton, J.B. Jones, Fully developed turbulent flow in annuli. J. Basic Eng. **86**(4), 835–842 (1964)
3. S.Y. Chung, H.J. Sung, Direct numerical simulation of turbulent concentric annular pipe flow: part 2: heat transfer. Int. J. Heat Fluid Flow **24**(3), 399–411 (2003)
4. S.Y. Chung, G.H. Rhee, H.J. Sung, Direct numerical simulation of turbulent concentric annular pipe flow: part 1: flow field. Int. J. Heat Fluid Flow **23**(4), 426–440 (2002)
5. S.W. Churchill, C. Chan, Turbulent flow in channels in terms of turbulent shear and normal stresses. AIChE J. **41**(12), 2513–2521 (1995)
6. F.E.R. Corredor, M. Bizhani, M. Ashrafuzzaman, E. Kuru, An experimental investigation of turbulent flow in concentric annulus using particle image velocimetry technique. J. Fluid Eng. **136**(051203), 1–11 (2014)
7. R.B. Crookston, R.R. Rothfus, R.I. Kermode, Turbulent heat transfer in annuli with small cores. Int. J. Heat Mass Transf. **11**(3), 415–426 (1968)
8. S.Y. Jung, H.J. Sung, Characterization of the three-dimensional turbulent boundary layer in a concentric annulus with a rotating inner cylinder. Phys. Fluids (1994-present), **18**(11), 115102 (2006)
9. M. Kaneda, B. Yu, H. Ozoe, S.W. Churchill, The characteristics of turbulent flow and convection in concentric circular annuli. Part I: flow. Int. J. Heat Mass Transf. **46**(26), 5045–5057 (2003)
10. P. Kundu, I. Cohen, D. Dowling, *Fluid Mechanics*, 5th edn. (Elsevier Science, Amsterdam, 2012), p. 557
11. C.J. Lawn, C.J. Elliott, Fully developed turbulent flow through concentric annuli. J. Mech. Eng. Sci. **14**(3), 195–204 (1972)
12. N.S. Liu, X.Y. Lu, Large eddy simulation of turbulent concentric annular channel flows. Int. J. Numer. Methods Fluids **45**(12), 1317–1338 (2004)
13. K. Maubach, K. Rehme, Negative eddy diffusivities for asymmetric turbulent velocity profiles? Int. J. Heat Mass Transf. **15**(3), 425–432 (1972)
14. N. Nikitin, H. Wang, S. Chernyshenko, Turbulent flow and heat transfer in eccentric annulus. J. Fluid Mech. **638**, 95–116 (2009)
15. J.M. Nouri, H. Umur, J.H. Whitelaw, Flow of Newtonian and non-Newtonian fluids in concentric and eccentric annuli. J. Fluid Mech. **253**, 617–641 (1993)
16. F. Ogino, T. Sakano, T. Mizushina, Momentum and heat transfers from fully developed turbulent flow in an eccentric annulus to inner and outer tube walls. Wärme-und Stoffübertragung **21**(2–3), 87–93 (1987)
17. A. Quarmby, An experimental study of turbulent flow through concentric annuli. Int. J. Mech. Sci. **9**(4), 205–221 (1967)
18. K. Rehme, Turbulent flow in smooth concentric annuli with small radius ratios. J. Fluid Mech. **64**(02), 263–288 (1974)
19. K. Rehme, Turbulence measurements in smooth concentric annuli with small radius ratios. J. Fluid Mech. **72**(01), 189–206 (1975)

Reconstruction of Wall Shear-Stress Fluctuations in a Shallow Tidal River

Romain Mathis, Ivan Marusic, Olivier Cabrit, Nicole L. Jones and Gregory N. Ivey

Abstract In this paper, we investigate the applicability of the predictive wall shear-stress model, recently developed by Mathis et al. J. Fluid Mech. 715, 163–180, 2013, [17], to environmental flows where near-wall information is typically inaccessible. This wall-model, which embeds the scale interaction mechanisms of superposition and modulation, is able to reconstruct the instantaneous wall (bed) shear-stress fluctuations in turbulent boundary layers. The database considered here comes from field measurements using acoustic Doppler velocimeters carried out in a shallow tidal channel (Suisun Slough in North San Francisco Bay). The model is first applied to a selected subset of data sharing common properties with the canonical turbulent boundary layer. Statistics and energy content of these predictions are found to be consistent with laboratory predictions and DNS results. The model is then used on the whole dataset, whose some of them having properties far from the canonical case. Even for these situations, the model is able to preserve the overall Reynolds trend. This study shows the great capability of the model for environmental applications, which is the only one able to predict both the correct energetic content and probability density function.

R. Mathis (✉)
Laboratoire de Mécanique de Lille, CNRS UMR 8107, Université Lille Nord de France, 59655 Villeneuve d'Ascq, France
e-mail: romain.mathis@univ-lille1.fr

I. Marusic · O. Cabrit
Department of Mechanical Engineering, University of Melbourne, Melbourne, VIC 3010, Australia
e-mail: imarusic@unimelb.edu.au

O. Cabrit
e-mail: o.cabrit@unimelb.edu.au

N.L. Jones · G.N. Ivey
School Civil, Environmental and Mining Engineering and UWA Oceans Institute, University of Western Australia, Crawley, WA, Australia
e-mail: nicole.jones@uwa.edu.au

G.N. Ivey
e-mail: greg.ivey@uwa.edu.au

M. Stanislas et al. (eds.), *Progress in Wall Turbulence 2*,
ERCOFTAC Series 23, DOI 10.1007/978-3-319-20388-1_22

1 Introduction

In wall-bounded turbulent flows the wall shear stress τ_w constitutes a key parameter for accurate prediction of the flow behaviour. Over the years, many studies have been devoted toward understanding and modelling the Reynolds number dependency of the mean, time-averaged, value $\bar{\tau}_w$, which is used in boundary layer inner scaling via the friction velocity $U_\tau = \sqrt{\bar{\tau}_w/\rho}$, where ρ is the fluid density [18, 21]. However, little is known about the fluctuating component, τ'_w, which can be responsible for extreme and destructive events, such as wind gusts in atmospheric flows or scouring and mechanical damage on an aircraft [19]. In environmental flows, the wall shear stress is of great ecological importance where it is linked to erosion, bed formation, sediment and nutrient transportation, etc. [6, 20]. Unfortunately, the wall shear stress is largely inaccessible in field measurements, which prompts the need for predictive models to reconstruct the missing information. Here, the fluctuating component is defined as $\tau'_w(\mathbf{x}, t) = \tau_w(\mathbf{x}, t) - \bar{\tau}_w(\mathbf{x})$, where $\tau_w(\mathbf{x}, t)$ and $\bar{\tau}_w(\mathbf{x})$ are the total and mean values of the wall shear stress, respectively, and $\mathbf{x} = (x, y)$ denotes the position vector in the plane of the wall. The coordinates x, y and z refer to the streamwise, spanwise and wall-normal directions. The respective fluctuating velocity components are denoted by u, v and w. Overbars indicate time-averaged values, and the superscript "+" is used to denote viscous scaling of length $z^+ = zU_\tau/\nu$ and velocities $u^+ = u/U_\tau$, where ν is the kinematic viscosity of the fluid.

Recently, Mathis et al. [17] (hereafter denoted M13) proposed a novel conceptual approach to build up a predictive model able to reconstruct the fluctuating wall shear stress based on a single point measurement taken in the log-layer away from the wall. The model is based on many years of empirical observations, both experimental and numerical, that have clearly established that strong interactions exist between the near-wall region and motions in the outer region. Namely, the Reynolds number effects are closely related to the increasingly energetic content of the large-scale structures associated with the log-layer [1, 7, 11], through superposition and modulation effects [2, 8, 14]. The wall shear-stress model was originally derived from the streamwise velocity model developed by Marusic et al. [12] and Mathis et al. [15], where an algebraic relationship between the streamwise velocity component and the wall shear stress is known, and is of the form:

$$\tau'^{+}_{wp}(t^+) = \tau'^{*}_{w}(t^+)\left[1 + \alpha\, u'^{+}_{OL}(t^+)\right] + \alpha\, u'^{+}_{OL}(t^+), \tag{1}$$

where τ'^{+}_{wp} is the predicted time series normalised by wall variables, $\tau'^{+}_{wp} = \tau'_{wp}/(\rho U_\tau^2)$ and $t^+ = tU_\tau^2/\nu$. The time series τ'^{*}_{w}, which is normalised in wall units, represents the statistically "universal" wall shear-stress signal that would exist in the absence of any inner–outer interactions. The parameters τ'^{*}_{w} and α are determined from a once-off calibration experiment at an arbitrarily chosen Reynolds number, and are hypothesised to be Reynolds number independent. The only user input required for the model is a characteristic signal of the large scales from the log-region, u'^{+}_{OL}, taken nominally at the geometric centre of the log-layer, $z^+_O = \sqrt{15Re_\tau}$ [14, 15], where $Re_\tau = U_\tau \delta/\nu$ is the friction Reynolds number and δ the boundary layer thickness. The M13 model consists of two parts. The first part in Eq. 1 models the amplitude modulation of the small scales, here $\tau_w^{\prime *}$, by the large-scale log-region motions, $u_{OL}^{\prime +}$. The second term, $\alpha u_{OL}^{\prime +}$, models the superposition of the large-scale motions felt at the wall. The underlying idea is that the near-wall small-scale motions are universal (i.e. they do not change with Reynolds number), and therefore are only influenced by large-scale log-region events (the intensity of the influence increasing with increasing Reynolds number). Therefore, the Reynolds number effects are confined to the large-scale log-region input signal, $u_{OL}^{\prime +}$.

In this paper, we test and apply the M13 model on a tidal channel flow in order to assess its applicability to environmental flows, where it is known that external parameters such as roughness or stratification may also play an important role. The objective of this work is to demonstrate the potential application and benefits in environmental conditions. It should be emphasised that M13 model was originally developed and calibrated for the flat-plate smooth-wall turbulent boundary layer, and therefore its applicability in environmental flows is then not straightforward. Results, including statistical properties and spectral content, are analysed and compared with flat-plate smooth-wall turbulent boundary layer data from the literature, as well as with previous predictions using laboratory measurements.

2 Experimental Dataset

The dataset used here comes from the field experiments of [10] carried out in Suisun Slough in North San Francisco Bay from 30 August to 15 September 2005. The measurement location was relatively shallow, with the water depth at the measurement site ranging from 2.5 to 4.0 m with the semidiurnal tide. Measurements of the three velocity components and pressure were made using Nortek acoustic Doppler velocimeters (ADV) at four different heights, 0.15 m, 0.30 m, 0.45 m and 1.45 m, above the seabed (ASB). The dataset consists of 330 records, or bursts, sampling for 20 minutes every half-hour at 16 Hz. Throughout the remainder of the paper we simply refer to these segments of data as *bursts*. A full description of the experiment and measurement procedure is available in [10].

3 Data Analysis and Selection

The tidal channel environment is subjected to natural variability, in the mean flow direction that in a long-term trend have to be taken into account for the data analysis. We determine for each burst the mean flow angle θ to return the true streamwise and spanwise velocity fluctuations. Then, a filtering process is applied to separate the long-term trend related to the tidal channel environment (non-turbulence related) from the original signal. For each velocity component and on each burst, the velocity

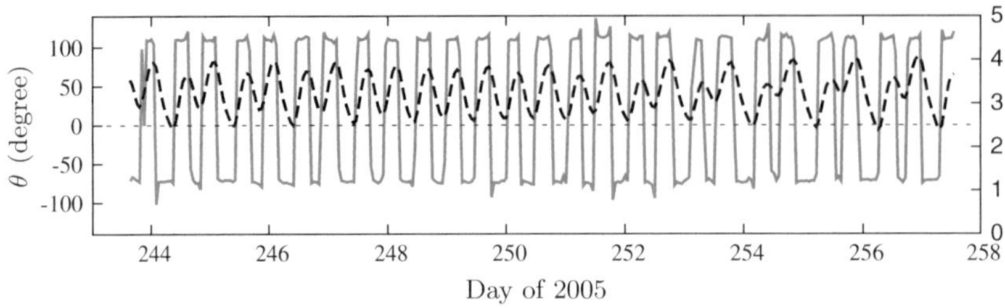

Fig. 1 Mean flow angle θ (*grey solid line*) and depth of water (*dashed line*) (Figure adapted from Mathis et al. [16])

signals from the four bed-normal locations are first averaged together. The resulting averaged signal is then low-pass filtered at a cut-off wavelength of 20δ to extract the long-term trend, which is then removed from the original signal. The cut-off wavelength has been chosen based on our knowledge that the largest scale motions developing in turbulent boundary layers are typically about $10\delta - 15\delta$ in length [5]. A complete description of the procedure to determine the mean flow angle and the filtering process are available in [16].

As shown in Fig. 1, the site's hydrodynamics are dominated by tidal flow, producing a cyclic change in the mean flow direction and water depth. The depth of water is calculated as the median value of the pressure sensor data from the four bed-normal locations, and it is used as the boundary layer thickness δ. The tide produces a wide range of flow conditions with a corresponding high variability in Reynolds numbers, as reported in Sect. 3.

The objective of our study is to use the M13 model of Mathis et al. [17] to predict bed shear-stress information in the tidal channel, and to show its potential applicability to environmental flows where accessing such information remains highly challenging. The model was originally developed for a smooth-wall zero-pressure-gradient turbulent boundary layer and it is noted that the model's parameters might not be fully adequate for the present flow conditions (due to roughness effects, for example). However, if there was variability of these "universal parameters" between different types of wall-bounded flows this should only affect the accuracy of the model, and not the overall Reynolds number trends [15, 17].

Our first approach in the present study is to select the subset of burst sharing common properties with the canonical turbulent boundary layer. To do so, data are first normalised in wall unit using the boundary layer thickness δ given in Fig. 1, and the mean friction velocity U_τ, which is determined using the Reynolds stress peak, i.e. $U_\tau = \max(\sqrt{-\overline{uw}(z)})$. It is acknowledged that this estimation is somewhat inaccurate [4], particularly due to the limited range and number of points in the wall-normal direction, but it is justified in the present case by the fact that the first ADV measurement is very close to the Reynolds shear-stress peak (see Mathis et al. [16] for further discussion). The variation of friction Reynolds number Re_τ during the whole experimental campaign is presented in Fig. 2a. The average Re_τ is around 30,000

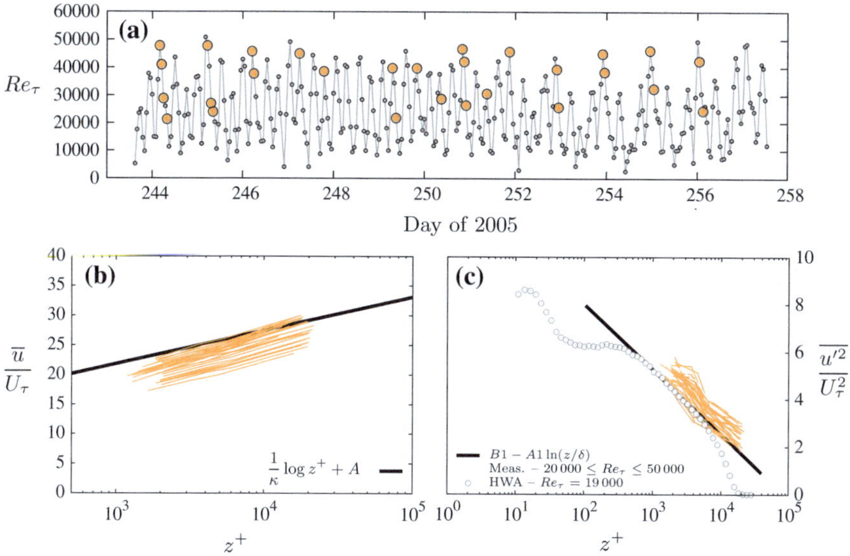

Fig. 2 **a** Reynolds number Re_τ of each burst (*red dots* denotes the selected cases); **b** Mean velocity profiles, the *black solid line* shows the log-law of the wall using $\kappa = 0.41$ and $A = 5.0$; **c** Streamwise turbulence intensity profiles of the selected cases compared to laboratory hot-wire anemometer (HWA) measurements of [9] ($Re_\tau = 19{,}000$), the *black solid line* corresponds to the logarithmic profile with $A_1 = 1.19$ and $B_1 = 1.71$ (Figure adapted from Mathis et al. [16])

but there is a wide range of Reynolds numbers produced by the tidal flow conditions, which provides the opportunity to investigate the Reynolds number dependency of the bed shear-stress fluctuations. To extract the "canonical" segments of data, we then apply the following selection criteria:

1. the vertical profile of the mean streamwise velocity, $\overline{u}$, monotonically increases away from the bed (Fig. 2b);
2. the vertical profile of the turbulent shear stress, $-\overline{u'w'}$, monotonically decreases away from the bed;
3. the vertical profile of the streamwise turbulence intensity, $\overline{u'^2}$, monotonically decreases away from the bed (Fig. 2c);
4. the variability of the points in the mean velocity profile is within 5 % of the mean fitted logarithmic slope (to avoid seesaw trend).

The first three criteria reflect the intrinsic nature of the turbulent boundary layer and the fact that the first wall-normal point is located beyond the Reynolds shear stress and outer peaks (see Sect. 4 for the discussion related to the outer-peak). The last criterion ensures that the data do not deviate excessively from the classical log-law $\overline{u^+} = \frac{1}{\kappa}\ln z^+ + A$. Applying all four selection criteria yields 28 usable cases amongst the 330 original bursts, covering a range of Reynolds numbers from $Re_\tau \simeq 20{,}000$ to 50,000. The streamwise mean and turbulence intensity profiles of the selected cases are shown in Fig. 2b, c, in which the monotonic behaviour of criteria 1 en 3 are clearly visible. The mean velocity profiles are consistently below the smooth-wall log-law, as expected for flows with bed roughness, but they do not collapse on one another as might be expected since they were acquired at the same location. This could indicate a change in the bed roughness, perhaps caused by modifications in the behaviour of the benthic community, which includes clams, and/or changing bedforms in the mud. The overall trend of the turbulence intensity profile (Fig. 2c) agree remarkably well with logarithmic behaviour $\overline{u'^2}^+ = B_1 - A_1 \ln(z/\delta)$ [13] and the laboratory hot-wire measurement of Hutchins et al. [9].

In summary, Fig. 2 shows clear evidence that the retained bursts behave similarly to a canonical turbulent boundary layer, with clear roughness effects being the only significant difference compared to smooth-wall boundary layer flow.

4 Large-Scale Input Signal

To apply the M13 model, given in Eq. 1, the requirement for input is a large-scale velocity signal from the logarithmic centre of the log-layer, $z_O^+ = \sqrt{15Re_\tau}$ [14]. This location was originally chosen as it corresponds to most energetic signature point of the large scales, i.e. to the outer-spectral-peak location, as well as to the peak of the large-scale streamwise turbulence intensity profile $\overline{u_L'^2}^+$. In environmental flows, this optimal wall-normal location to capture the necessary large-scale information is not always easily accessible, particularly as U_τ and δ are also subject to variation. The present dataset is collected at a fixed bed-normal location, where the lowest measurement point ($z = 0.15$ m) is located $2.2 - 2.5$ times above the optimal bed-normal location $z_O^+ = \sqrt{15Re_\tau}$. Cabrit et al. [3] demonstrated that variations in the predicted wall shear stress might occur if the location of the input large-scale information deviates excessively from the middle of the log-layer (any variation from the optimal wall-normal location induces a reduction in the energy of the large-scale signal). We have shown, however, that the large-scale turbulence intensity variation in the range $z_O^+ \leq z^+ \leq 2.5z_O^+$ is less than 5 %, which should not influence significantly the predicted bed shear stress (the error will be within the measurement uncertainty). Hence, we use here the first off-bed measurement point ($z_1 = 0.15$ m, i.e. $z_1^+ \sim 2.2 - 2.5 \times z_O^+$) to determine the input large-scale velocity, $u_{OL}'^+$. The velocity signal at z_1^+ is first low-pass filtered at the non-dimensional frequency $f^+ = 2.65 \times 10^{-3}$ to retain only the large-scale component. Then, the filtered signal is shifted forward in the streamwise direction to account for the large-scale structure angle (see Mathis et al. [17] for full details of the procedure).

Figure 3 shows the streamwise turbulence intensity and energy spectra of the raw and filtered (large-scale) signal for the selected canonical cases, along with laboratory and atmospheric measurements. Overall, both the unfiltered and filtered signals show an energy deficit (about 35 %) compared to smooth wall results, but they do follow the same Reynolds number trend. This deficit is unlikely related to

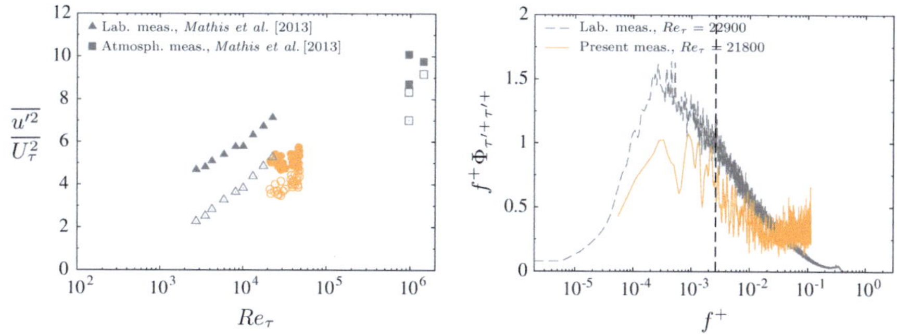

Fig. 3 (*Left*) Reynolds number dependency of the streamwise turbulence intensity of the log-region unfiltered signal $u_O'^+$ (*filled symbols*), and the filtered large-scale component $u_{OL}'^+$ (*opened symbols*). (*Right*) Pre-multiplied energy spectra of the input streamwise velocity signals for the laboratory [17] and for a single burst of the present measurements (first bed-normal point); The *vertical dashed line* shows the location of the cut-off frequency $f^+ = 2.65 \times 10^{-3}$ (Figure adapted from Mathis et al. [16])

the miss-matched optimal location (which should account only for about 5 %). The discrepancy between the laboratory and field measurements is more likely due to the changing environment conditions due to the dynamics of the tidal channel rather than measurement uncertainty, as differences are seen across all frequencies.

Overall, despite some differences in the large-scale input information between laboratory and field measurements, the first off-bed measurements point in the tidal channel located within the log-region allows us to apply the bed shear-stress model.

5 Bed Shear-Stress Prediction

In this section, we present bed shear-stress predictions $\tau_{wp}'^+$, using Eq. 1 and input large-scale information $u_{OL}'^+$ constituted from the first off-bed location z_1. Two cases are considered. First, the M13 model is applied to the selected canonical cases in order to assess if the model applied in environmental flows comply with characteristics of a canonical turbulent boundary layer. Second, the M13 model is applied to the whole database in order to assess if, even in cases far from canonical, the relevance of the bed shear-stress predictions are more generally applicable.

The left-hand side of Fig. 4 shows statistics of the reconstructed bed shear-stress signal. The fluctuations magnitude $\overline{\tau_{wp}'^{+2}}$ agree well with available DNS data and predictions made using laboratory and atmospheric measurements. Overall, a slight underestimation is observed, which is not surprising, considering the aforementioned lower trend of the energy content of the input large-scale streamwise velocity (Fig. 3). Similar observation is made on the spectral energy content (not shown here) where the overall spectral range is well captured. Comparisons of higher order statistics

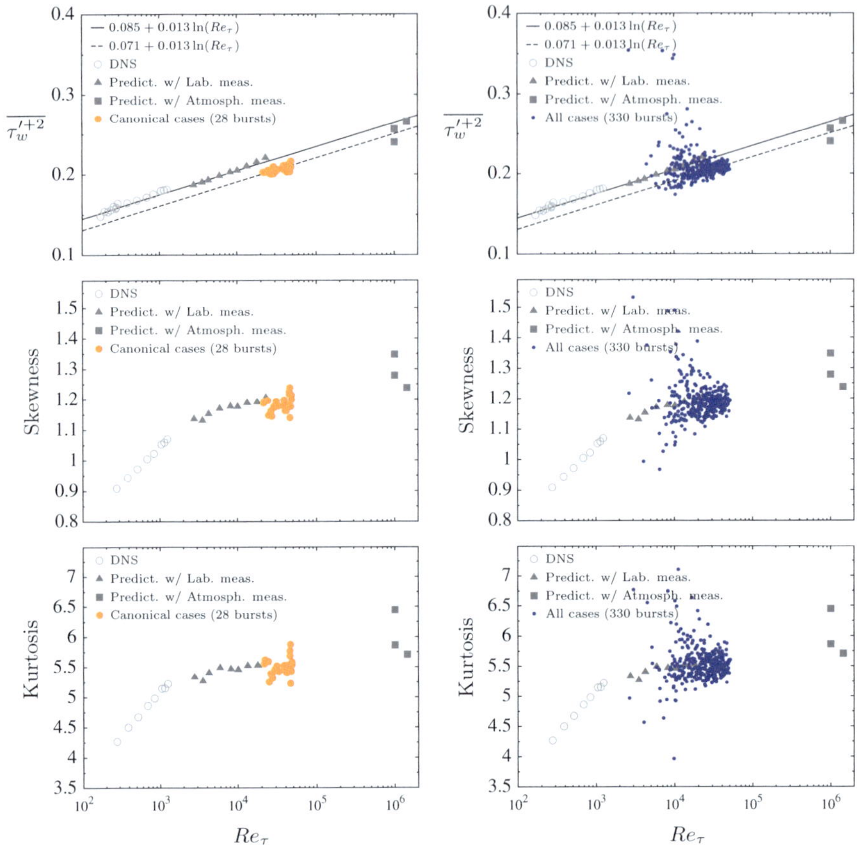

Fig. 4 Fluctuation magnitude $\overline{\tau_{wp}'^{+2}}$, Skewness and Kurtosis of the predicted bed shear-stress signal versus the Reynolds number Re_τ, compared to available data for DNS of zero-pressure-gradient turbulent boundary layer and former predictions using laboratory measurements. (*Left*) For the selected canonical cases. (*Right*) For the whole dataset (330 bursts. Note that the *solid line* corresponds to the trend obtained from previous work [17], and the *dashed line* is a fit of the canonical cases' predictions (Figure adapted from Mathis et al. [16])

(skewness and kurtosis factors in Fig. 4, left) show an increasing Reynolds number trend consistent with previous laboratory predictions. However, the rate of increase with *Re* of the M13 model is generally lower that DNS results, which remains an open question and the subject of ongoing study. Despite some incertitude in the dataset, either due to the measurement uncertainty (e.g. inadequate convergence of the large-scale, ...) and/or the tidal channel environment (e.g. roughness effect, ...), our analysis shows the potential applicability of the M13 model to environmental flows.

To generalise the ability of the M13 model to environmental flows, independently of canonical properties, predictions for the whole dataset are given in the right-hand

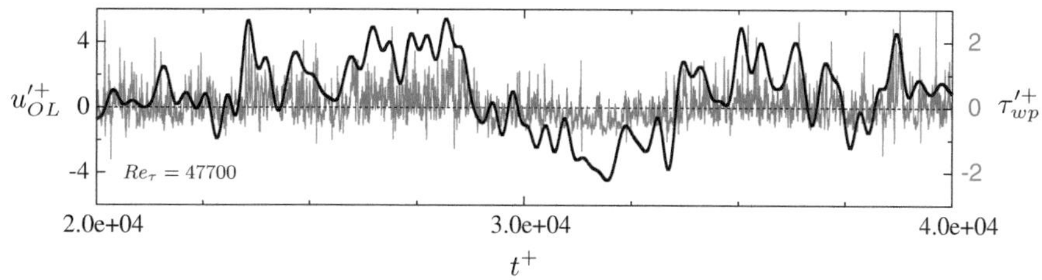

Fig. 5 Example of predicted instantaneous bed shear-stress signal τ'^+_{wp} (*grey thin line*), with the input large-scale streamwise velocity u'^+_{OL} (*black thick line*) (Figure adapted from Mathis et al. [16])

side of Fig. 4. Despite a noticeable scatter, the overall trend of the statistics remains similar. By analysis the residual of the predicted energy intensity, compared to the overall trend (dashed line in the figure) it is found that 60 % of the overall predictions lay between ±10 % of the expected trend. This result confirms the good behaviour of the model, even when the input information does not fully comply with a canonical turbulent boundary layer.

Finally, an example of a reconstructed instantaneous wall shear-stress signal, along with the input outer large-scale component, is depicted in Fig. 5. It is interesting to note long and intense periods of positive and negative excursions, characteristic of very long large-scale motions developing within the log-region [7]. The highly skewed and amplitude modulated character of τ'^+_{wp} is also obvious.

6 Conclusions

In this paper, the wall shear-stress model developed by Mathis et al. [17] is tested and applied to environmental flows, here, a shallow tidal channel. It is shown that a significant portion of the data behaves like a canonical turbulent boundary layer, at least in the logarithmic layer. Predictions based on the selected canonical cases show a good agreement with results obtained from laboratory experiments and DNS. The model is well behaved even for the other "non-canonical" cases, in the sense that the overall Reynolds number trend remains consistent with laboratory and DNS. The relative discrepancy observed on the canonical predictions is attributed to external conditions such as roughness effects, rather than measurement uncertainty. It is however noted that for the less than optimal sampling to converge properly the large-scale input information necessary for the model remains an important question when considering environmental applications. Finally, significant improvements to the model could be achieved by considering the external environment parameters, such as roughness effects due to changing bed form or vegetation, stratification, unsteadiness and pressure gradient, to name only a few. Nevertheless, the present results show that the lack of near-wall or near-bed information in field measurements, yet crucial for understanding ecosystem dynamics, can be overcome by this model.

Acknowledgments The Australian Research Council and the University of Melbourne McKenzie Fellowship program are gratefully acknowledged for their support. The field work was supported by the CALFED Bay Delta Authority Restoration Program (ERP02P22) and The Foundation for Young Australians Centenary Scholarship Award. Thanks to F. Parchaso and B. Richards from the U.S. Geological Survey for assistance with the field experiment.

References

1. J.C. del Álamo, J. Jiménez, Spectra of the very large anisotropic scales in turbulent channels. Phys. Fluids **15**(6), L41–L44 (2003). doi:10.1063/1.1570830
2. P.R. Bandyopadhyay, A.K.M.F. Hussain, The coupling between scales in shear flows. Phys. Fluids **27**(9), 2221–2228 (1984)
3. O. Cabrit, R. Mathis, I. Marusic, Towards a statistically accurate wall-model for large-eddy simulation, in ed. by P. Brandner, B. Pearce *18th Australasian Fluid Mechanics Conference* (Australasian Fluid Mechanics Society, Launceston, Australia, 2012)
4. D.B. DeGraaff, J.K. Eaton, Reynolds number scaling of the flat-plate turbulent boundary layer. J. Fluid Mech. **422**, 319–346 (2000)
5. D. Dennis, T. Nickels, Experimental measurement of large-scale three-dimensional structures in a turbulent boundary layer. Part 2. Long structures. J. Fluid Mech. **673**, 218–244 (2011)
6. W.D. Grant, O.S. Madsen, The continental-shelf bottom boundary layer. Annu. Rev. Fluid Mech. **18**, 265–305 (1986)
7. N. Hutchins, I. Marusic, Evidence of very long meandering features in the logarithmic region of turbulent boundary layers. J. Fluid Mech. **579**, 1–28 (2007). doi:10.1017/S0022112006003946
8. N. Hutchins, I. Marusic, Large-scale influences in near-wall turbulence. Philos. Trans. R. Soc. Lond. A **365**, 647–664 (2007). doi:10.1098/rsta.2006.1942
9. N. Hutchins, T. Nickels, I. Marusic, M.S. Chong, Spatial resolution issues in hot-wire anemometry. J. Fluid Mech. **635**, 103–136 (2009). doi:10.1017/S0022112009007721
10. N.L. Jones, J.K. Thompson, K.R. Arrigo, S.T. Monismith, Hydrodynamic control of phytoplankton loss to the benthos in an estuarine environment. Limnol. Oceanogr. **54**(3), 952–969 (2009)
11. K.C. Kim, R.J. Adrian, Very large-scale motion in the outer layer. Phys. Fluids **11**, 417–422 (1999)
12. I. Marusic, R. Mathis, N. Hutchins, Predictive model for wall-bounded turbulent flow. Science **329**(5988), 193–196 (2010). doi:10.1126/science.1188765
13. I. Marusic, J.P. Monty, M. Hultmark, A.J. Smits, On the logarithmic region in wall turbulence. J. Fluid Mech. **716**, R3 (2013)
14. R. Mathis, N. Hutchins, I. Marusic, Large-scale amplitude modulation of the small-scale structures in turbulent boundary layers. J. Fluid Mech. **628**, 311–337 (2009). doi:10.1017/S0022112009006946
15. R. Mathis, N. Hutchins, I. Marusic, A predictive inner-outer model for streamwise turbulence statistics in wall-bounded flows. J. Fluid Mech. **681**, 537–566 (2011). doi:10.1017/jfm.2011.216
16. R. Mathis, I. Marusic, O. Cabrit, N.L. Jones, G.N. Ivey, Modeling bed shear-stress fluctuations in a shallow tidal channel. J. Geophys. Res. Oceans **119** (2014).doi:10.1002/2013JC009718
17. R. Mathis, I. Marusic, S.I. Chernyshenko, N. Hutchins, Estimating wall-shear-stress fluctuations given an outer region input. J. Fluid Mech. **715**, 163–180 (2013). doi:10.1017/jfm.2012.508
18. P.A. Monkewitz, K.A. Chauhan, H.M. Nagib, Self-consistent high-Reynolds-number asymptotics for zero-pressure-gradient turbulent boundary layers. Phys. Fluids **19**, 115101 (2007). doi:10.1063/1.2780196

19. R. Örlü, P. Schlatter, On the fluctuating wall-shear stress in zero-pressure-gradient turbulent boundary layers. Phys. Fluids **23**(1–4), 021704 (2011)
20. P. Rowiński, J. Aberle, A. Mazurczyk, Shear velocity estimation in hydraulic research. Acta Geophys. Pol. **53**(4), 567–583 (2005)
21. H. Schlichting, K. Gersten, *Boundary Layer Theory*, eighth revised and enlarged edition edn. (Springer, 2000)

Analysis of Vortices Generation Process in Turbulent Boundary Subjected to Pressure Gradient

Artur Dróżdż and Witold Elsner

Abstract The paper is concerned with the experimental study of bursting process in turbulent boundary layer. For this purpose the novel identification process, developed by Dróżdż and Elsner, J Phys: Conf Ser 318(6):062007, 2011, [3] based on VITA technique combined with quadrant analysis was applied. By the detection of four possible combinations of instantaneous gradients of u and v phase-averaged velocity traces this method allows to demonstrate such properties of vortices motion as: swirling direction, ascending or descending direction, the trajectory inclination. The analysis gives an evidence of four types of vortical structures present in the TBL which are responsible for the production of Q-type events, namely prograde and retrograde vortices, with the ascending and descending direction of motion. It was found that detected coherent structures have dominant share of the overall energy of velocity fluctuations.

1 Introduction

It is already known that the near-wall region is characterized by the presence of low-speed streaks and hairpin vortices that used to be assembled into large-scale coherent groups termed as vortex packets. In the vortex packet the bursting process occurs which induces high gradients of velocity in time and in space. Kim et al. [6] revealed that the bursting process, which produces roughly 70 % of total turbulence, is a result of break-up of shear layer in the buffer layer caused mainly by ejection event in a buffer layer followed by sweep event. The effect of those phenomena is the appearance of fine-scale structures in the flow. Study of bursting process in pressure gradient flows and especially adverse pressure gradient is important from the viewpoint of complexity of the physical phenomenon as well as practical appli-

A. Dróżdż (✉) · W. Elsner
Institute of Thermal Machinery, Czestochowa University of Technology,
Armii Krajowej 21, 42-201 Czestochowa, Poland
e-mail: arturdr@imc.pcz.czest.pl

W. Elsner
e-mail: welsner@imc.pcz.czest.pl

M. Stanislas et al. (eds.), *Progress in Wall Turbulence 2*,
ERCOFTAC Series 23, DOI 10.1007/978-3-319-20388-1_23

cations. The structure of turbulent boundary layer (TBL) developed on a flat plate in the presence of favourable (FPG) and/or adverse pressure gradients (APG) were the subject of many studies. Especially interesting is the formation of the outer peak of fluctuation under the adverse pressure gradient condition. The reason of such phenomenon is poorly understood and cause problems in the CFD modeling. Harun et al. [5] show that turbulence of the outer layer, for both FPG and APG conditions, is mainly the result of the large-scale motion. In the paper the attention is drawn to the role of the small-scale structures.

The paper aims to apply the novel approach of bursting structure identification process, developed by Dróżdż and Elsner [3] based on VITA technique and combined with quadrant analysis for TBL subjected to APG conditions. By the detection of four possible combinations of instantaneous gradients of u and v phase-averaged velocity traces method allows to demonstrate such properties of vortices motion as: swirling direction, ascending or descending direction, the angle of the motion, as well as the relative speed of vortex propagation in the flow.

2 Experimental Apparatus and Conditions

The experiment was performed in an open-circuit wind tunnel, where the turbulent boundary layer developed along the flat plate, which was 2807 mm long and 250 mm wide. The test section is located in the rear part of the wind tunnel. The upper wall of test section was shaped according to the assumed distribution of the pressure gradient corresponding to the conditions encountered in axial compressor blading.

The facility was equipped with a computer-controlled, 2D traversing system (in streamwise and wall-normal direction). The traverse carriage was driven over a maximum displacement of 180 mm by a servomotor. The uncertainty of the driver step was 0.001 mm with the smallest step equal to 0.01 mm.

Static pressure measurements were done using 70 pressure holes and the results of measurements are shown in Fig. 1. The pressure distribution is typical of a turbomachinery case, where after short region of zero pressure gradient flow accelerates (from $x_s = 197$ mm) and then (from $x_s = 427$ mm) decelerates. It is seen that pressure gradient values varies within the range of $-0.27 \div 0.28$ Pa/mm.

Velocity profiles were measured with single hot-wire anemometry probe of a diameter $d = 3\,\mu$m and length $l = 0.4$ mm (Dantec Dynamics 55P31). Those measurements were supplemented with X-wire probe of wire diameter $d = 5\,\mu$m and length $l = 1.25$ mm (Dantec Dynamics 55P61). The probes were combined with the DISA 55M hot-wire bridge connected to a 14 bit PC card. Acquisition was maintained at frequency 50kHz with 10 s sampling records. For the assumed sampling frequency the nondimensional inner scale representation was $f^+ \approx 1$. During the measurements ambient conditions were carefully controlled. The scatter of ambient temperature at the end of the test section did not exceed 0.2°. At the same time the free-stream velocity was monitored by means of a Prandtl tube. The scatter of free-stream velocity was found to be around 0.2 % of the mean value.

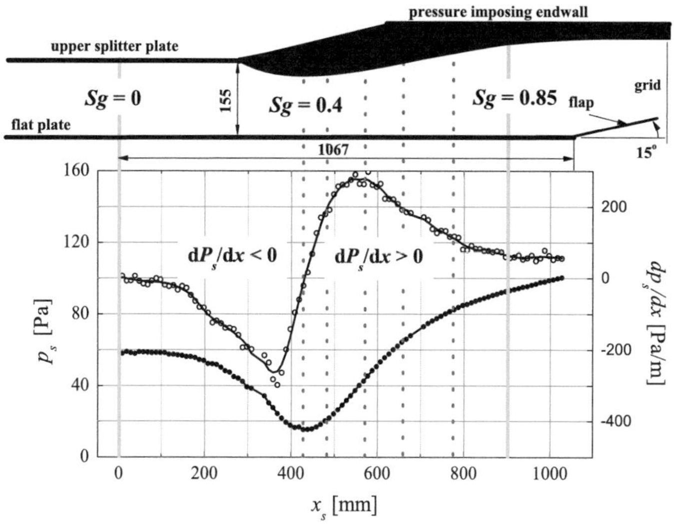

Fig. 1 The shape of the channel *upper wall* and the corresponding static pressure and pressure gradient distributions along the flat plate

Table 1 Location and parameters of the analyzed profiles

PG	Traverse number	x_s (mm)	Sg (−)	Single wire	X-wire	U_∞ (m/s)	u_τ (m/s)	τ_w (Pa)
ZPG	8	427	0.4	x	x	17.48	0.776	0.686
APG	10	487	0.456	x	x	16.97	0.718	0.587
APG	13	577	0.541	x	x	15.64	0.582	0.388
APG	16	667	0.625	x	x	14.45	0.473	0.258
APG	20	787	0.738	x	x	13.21	0.370	0.155

The position at the wall closest point of the hot-wire probe was determined using the mirrored image. The positions of five measuring traverses, marked by dotted lines, are shown in Fig. 1. The distances of traverses from the inlet plane, the corresponding dimensionless distances $Sg = x_s/L$, where L is the length of the test section ($L = 1067$ mm) and the most important flow parameters are given in Table 1. The conditions determined in inlet plane ($Sg = 0$), located in zero pressure gradient area are the mean velocity in core flow $U_\infty \approx 15$ m/s and the turbulent intensity $Tu = 0.4\,\%$.

3 Modified VITA Scheme and Optimization of Detection Parameters

One of the most widely used method to detect bursting process is a VITA (Variable Interval Time Averaging) method (Blackwelder and Kaplan [2]). This method was improved by [3], who extended the procedure by the quadrant decomposition using additional condition which segregate detections with respect to time derivative of u and v velocity components of the signal. The quadrant decomposition is commonly used to describe a relation of u and v velocity components in four quadrants of streamwise—wall-normal plane, where four events commonly named as Q1 ($u > 0, v > 0$), Q2 ($u < 0, v > 0$), Q3 ($u < 0, v < 0$), and Q4 ($u > 0, v < 0$) exist. It means that quadrant analysis allows to identify the bursting process, i.e., sweep (Q4) and ejection (Q2) events. The VITA detection scheme is based on the analysis of a running variance $var(t, T)$ of a velocity signal $a(t)$ given by equation:

$$var(t, T) = \frac{1}{T}\int_{t-T/2}^{t+T/2} a(t')^2 dt' + \frac{1}{T}\left(\int_{t-T/2}^{t+T/2} a(t')dt'\right)^2 \tag{1}$$

where t is a time of a signal and T is a time-averaging window of running variance. In case of X-wire measurement the running variance was calculated for both $u(t)$ and $v(t)$ velocity signals and in both signals the detection process was applied. The reason for that is the fact that bursting vortical structures induce the events in both signals because of the mean shear created by the wall.

Parameters of the detection process were properly tuned in order to obtain the best possible efficiency of the procedure. One of the most important parameter is time-averaging window T, which should be related to the scale of dominant structure. In order to select the appropriate T, the maximum number of positive du/dt detection $N(+)$ was used. The number of these structures was significantly larger in comparison with negative $N(-)$. Constant T^+ value in the whole boundary layer was kept. The second important parameter is the detection threshold value k of detection function $D(t)$. For both signal components the detection level applied on running variance was equal to $k(u')^2$ because of the higher value of $(u')^2$ than $(v')^2$ for all positions across boundary layer. The detection function $D(t)$ is used to detect structures in $var(t, T)$ calculated separately for u and v velocity signals. To determine the signs of the gradients of u and v velocity traces a slope detection function $D(t)$ was used which takes following values: $\{-2, -1, 1, 2\}$. Sign of the detection function describes the slope of gradients of u velocity component during detection, values 1 and 2 correspond to ascending and descending vortices respectively. Finally, for each type of the four possible structures the phase-averaging procedure of u and v signals was applied. During the preliminary analysis it was found that the original VITA method does not give the accurate time location of the structure center, especially when two structures are very close to each other. To improve structure alignment the calculation of center point between maximum and minimum of smoothed velocity time traces was introduced. This additional procedure substantially improves the phase-averaging process. The sample results of the procedure is shown in Fig. 2, where the detection with negative slope is significantly stronger.

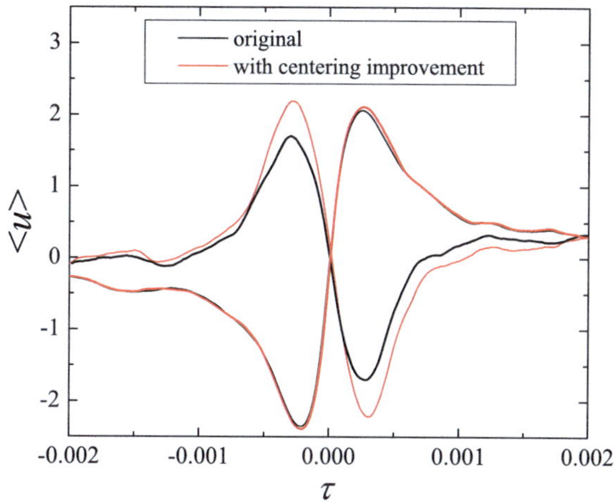

Fig. 2 The effect of centering procedure on phase-averaged signal

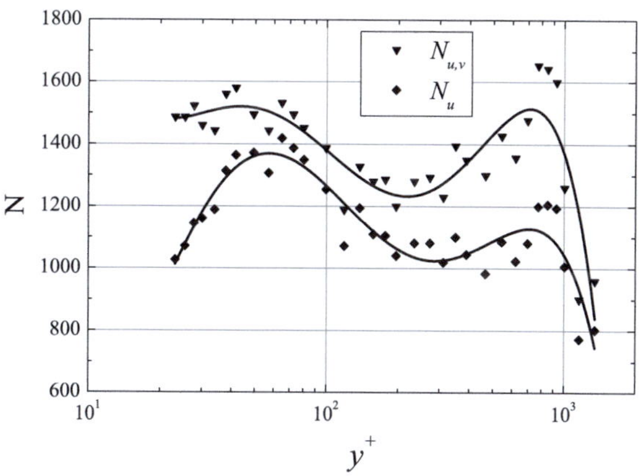

Fig. 3 Comparison of VITA detection numbers for detection based on one component (u) and both components (u and v)

Presented above detection technique is different to those, which can be found in the available literature. In most of the study the detection is preformed usually on single streamwise component, which is insensitive to detect vortices in case when the $\mathrm{d}U/\mathrm{d}y$

is minor. In order to obtain more precise results, there is a need to apply detections algorithms on both u and v components separately. Figure 3 shows the difference in number of detections obtained using only u and both u and v components. The significant increase of number of detected structures is observed especially close to the wall and at the edge of boundary layer. The reason for this is that in those regions the predominant direction of vortex motion is the streamwise direction and therefore only v signal is sufficiently strong to reach the detection level.

4 Interpretation of Bursting Process Based on VITA Structures

The physical interpretation of bursting phenomena is possible based on the analysis of VITA events. Four different pairs of quadrant events can be identified according to four combination of u and v gradients, namely: Q2/Q4, Q4/Q2, Q1/Q3, and Q3/Q1 [3]. It is assumed that those pairs are the result of spanwise vortex passing through the sensor. For particular quadrant events pair the trajectory of the vortex and swirling direction can be identified. Those vortices which move outward the wall (ascending direction) produce negative uv correlations (even quadrant pairs) and those which move toward the wall (descending direction) produce positive uv correlations (odd quadrant pairs). Based on above assumptions the vortical structures can be identified using single-point measurement of u and v components. The types of detected structures and the resulting phase-averaged u and v velocity components of four types of VITA structures for $y^+ = 22.6$ of the first cross-sections are presented in Fig. 4.

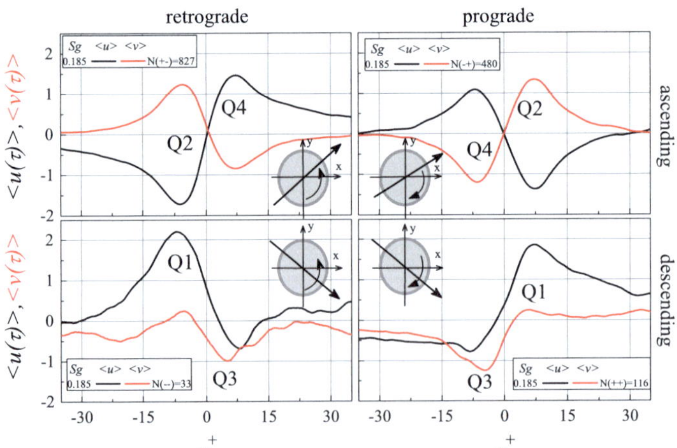

Fig. 4 Distributions of $< u >$ and $< v >$ of four VITA structures types for $y^+ = 22.6$: retrograde in *left column*, prograde in *right column*, ascending in *upper row*, descending in *lower row* (*N*—number of detection for particular structures)

The bursting process is a result of break-up of shear layer caused mainly by ejection event Q2, which is followed by sweep event Q4. The most typical burst-like structure characterized by the positive gradient (+) of u and the negative gradient (−) of v velocity components, which is marked as (+−) on the figure, can be a footprint of so-called *retrograde* (positive vorticity) vortex passing through the sensor in the ascending direction. The VITA method also detects the negative gradients (−) of u and the positive gradient (+) of v velocity components, marked as (−+), which are the effect of *prograde* vortex (negative vorticity) passes through the sensor. For both vortical structures two convection directions exist: ascending (moving away from the wall) and descending (moving toward the wall). For the letter case the both gradients u and v are positive for prograde (++) and negative for retrograde (−−). The rotation direction in the schemes indicates type of vortex, while the black arrow shows the vortex passage direction through the sensor.

It can be observed that for the ascending vortices their velocity components are in opposite phase and give Q2 and Q4 events. Particularly, while retrograde ascending vortex passing through the measuring point then it gives Q2 event before and Q4 event after the vortex center. For the prograde ascending vortex the order of the events is inversed. While descending vortices passing through the measuring point the velocity components are in phase and give less common Q1 and Q3 events. It is widely accepted that majority of vortices move outward the wall, e.g., Adrian [1], what is confirmed by the number of particular structures marked as N in the legend in Fig. 4.

The signs of ascending or descending direction of motion are revealed also by asymmetry of the signal. It can be observed (Fig. 4) that for ascending vortices the distributions of $< u >$ velocity component are slightly shifted toward negative values, while $< v >$ distributions are shifted toward positive values. If zero level corresponds to the mean values of U and V velocity components then any shift determine the direction of motion. Ascending structure arrives from a lower momentum zone, thus should have a lower convection velocity (negative shift) than the mean velocity U at the measuring point. The opposite situation is visible for descending structures.

5 Statistical Analysis of the Bursting Process

Valuable information about the bursting process is provided by mean bursting interval (*MBI*). The *MBI* is a measure of average time between detections. It means that if *MBI* decreases the level of burst energy contribution increases. If the process is influenced by generation of phenomena that occur in the inner or outer layer zone, then the appropriate timescale adoption should normalize *MBI* values in such a way, that the result will be independent of other conditions. Normalisation by the inner viscous scale was adopted and the resulting parameter was marked as MBI^+. Figure 5 presents MBI^+ as a function of y^+ determined for single-wire measurement cross-sections shown in Fig. 1 and described in Table 1. It is seen that along the flow the viscous scaled MBI^+ is almost constant in the inner region of turbulent boundary layer, while decrease of this parameter in outer zone is noticed. It suggests that in the outer region of pressure gradient flows the bursting process is driven by different mechanism than in the ZPG. For X-wire probe instead of *MBI* the number of detected vortical structures is analyzed. X-wire measurements make it possible to detect four types of vorices, i.e., retrograde ascending ($N(+-)$), prograde ascending ($N(-+)$), retrograde descending ($N(--)$), and prograde descending ($N(++)$) vortices number. The results presented in Fig. 6 are reduced by total number of detected vortices N from X-wire probe.

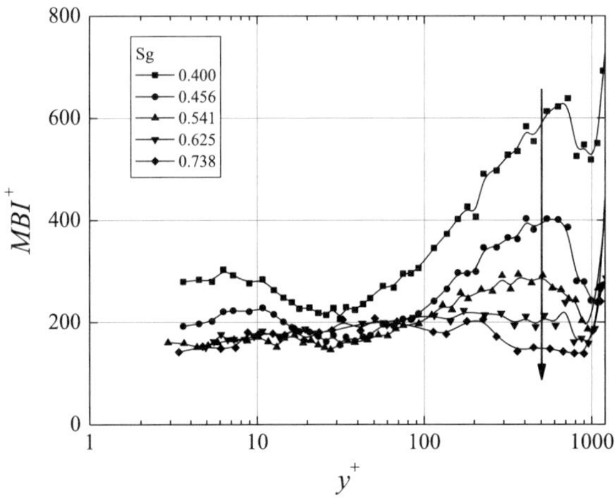

Fig. 5 Influence of APG on viscous scaled mean bursting interval MBI^+

It is seen that number of detections is different for the each type of structures. The dominant type of the structure is retrograde ascending one, which fraction is almost two times higher the fraction for prograde ascending. Significantly smaller fraction is of prograde descending structures.

It can be noticed that the percentage share of detections for each type of vortices varies across y^+. The percentage share of the dominant $N(+-)/N$ structures strongly decrease toward the edge of boundary layer, whereas the trend is reversed for outer structures. However, the increase in the wake region is observed for both prograde vortices. It is worth to note that at the edge of boundary layer ascending and descending structures are more or less equally frequent. It supports observation of Adrian [1] that at the edge of boundary layer the direction of prograde vortices trajectory is changing from ascending to descending due to bulge structures related to the large-scale motions. In turn, the impact of the pressure gradient is minor. Some changes are observed mainly close to the wall, where fraction of ascending structures decrease while descending structures increase along the flow. It means that under the influence of APG the change of the vortex motion from ascending to descending occurs. The drop of the inner scaled bursting process in the near-wall region (see

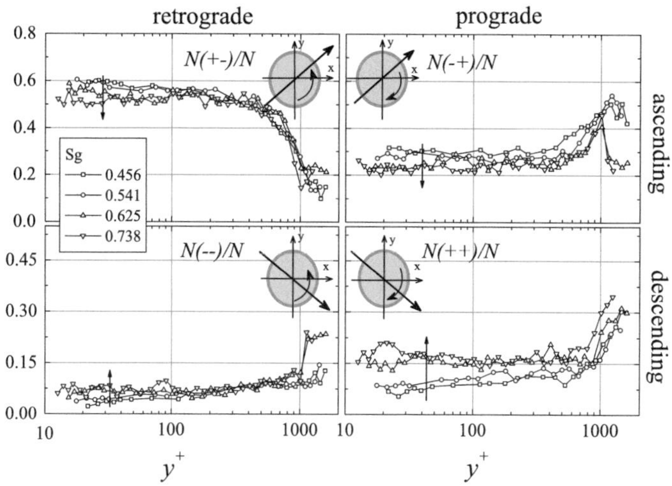

Fig. 6 Influence of APG on number of particular vortical structures

MBI^+ distribution in Fig. 5) could be due to the increased fraction of descending vortices (Fig. 6) coming from outer zone. The increased energy of small scales in the near-wall region was also observed by Harun et al. [5] for higher Reynolds number with APG conditions.

Interesting conclusions can be drawn from analysis of phase-averaged fluctuation components $< u' >$ and $< v' >$. Analysis of the $< u' >$ and $< v' >$ distributions (Fig. 7b) reveals substantial reduction of $< v' >$ value in the inner layer (below $y^+ \approx 100$), while in the outer region the strong increase of $< u' >$ value (above $y^+ \approx 100$). So strong separation of u and v components suggests that the trajectory of vortices is more inclined under the APG conditions. The increased population of descending structures as well as with more inclined descending trajectory supports the idea of Dróżdż and Elsner [4], where it was shown that large-scale structures sweeps the bursts toward the wall. It can be also concluded that those distributions are similar to distribution of u' and v' presented in Fig. 7a. Another confirmation of this conclusion are results shown in Fig. 8, where contribution of coherent structures to overall energy of velocity fluctuations are presented. Figure 8 shows that the ratio $< u' > /u'$ has almost constant value at the level about 45 % except the near-wall region and the boundary layer edge, where it takes higher values. This picture does not change along the flow from the first to the last cross-section, which means that bursting process governs the production of turbulence not only near the wall but also as high as to the end of logarithmic zone.

The presented analysis confirms that the turbulent kinetic energy production is closely related to the presence of coherent structures and that APG conditions have a distinct impact upon the intensity of bursting process. In particular one may conclude that the bursting process is damped near the wall (where the first maximum of

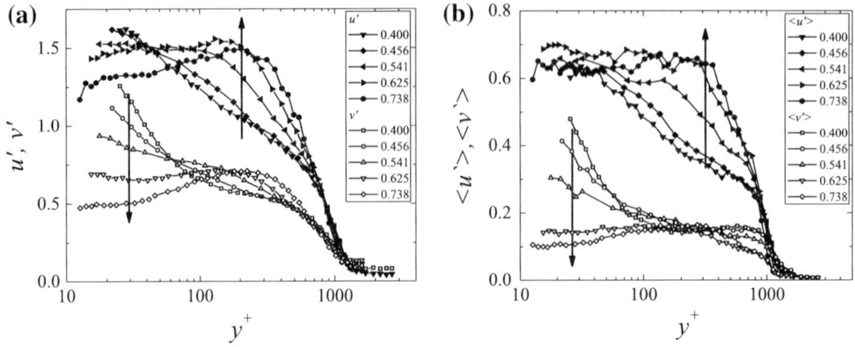

Fig. 7 Influence of APG on u' and v' (**a**) and phase-averaged $< u' >$ and $< v' >$ (**b**) fluctuations

Fig. 8 Influence of FPG and APG on phase-averaged $< u' >^+$ single-wire fluctuations (**a**) and ratio $< u' > /u'$ (**b**). Arrows indicate changes under FPG and APG conditions

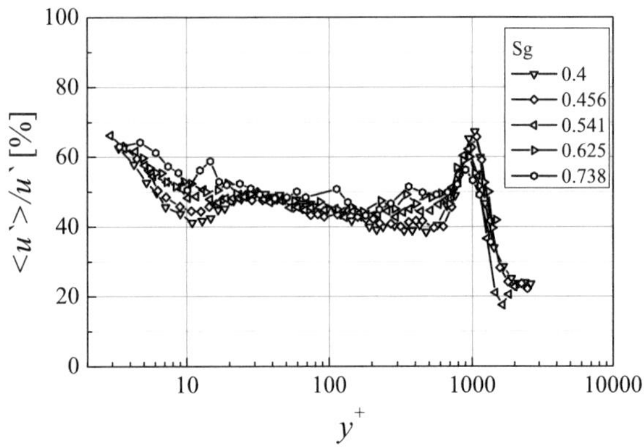

fluctuations occurs) and enhanced in the logarithmic zone of boundary layer where the second maximum of fluctuations is located.

6 Conclusions

The novel approach of bursting structure identification process, developed by Drozdz and Elsner [3] based on VITA technique combined with quadrant analysis was presented. By the detection of four possible combinations of instantaneous gradients of u and v phase-averaged velocity traces, this method allows to demonstrate such properties of vortices motion as: swirling direction, ascending or descending direction, and the trajectory inclination. The analysis gives an evidence of four types of vortical structures present in the TBL, which are responsible for the production of Q-type events, namely prograde and retrograde vortices, with ascending and descending direction of motion. The analysis shows that mean bursting interval scaled on wall friction velocity decreases downstream the flow, especially in the outer region, where outer maximum of velocity fluctuations is observed. The analysis of four types of vortices confirms that in the adverse pressure gradient the fraction of descending vortices is increasing, while at the same time the ascending vortices decreases, what explains the change in distributions of mean velocity fluctuations and confirms a less pronounced contribution of near-wall region to the downstream development of TBL. The presented analysis confirms additionally that the turbulent kinetic energy production is closely related to the presence of coherent structures and that APG condition has a distinct impact upon the intensity of bursting process.

Acknowledgments The investigation was supported by National Science Centre under Grant no. DEC-2012/07/B/ST8/03791.

References

1. R.J. Adrian, Hairpin vortex organization in wall turbulence. Phys. Fluids **19**(4), 041301 (2007). doi:10.1063/1.2717527
2. R.F. Blackwelder, R.E. Kaplan, On the wall structure of the turbulent boundary layer. J. Fluid Mech. **76**, 89–112 (1976)
3. A. Dróżdż, W. Elsner, Detection of coherent structures in a turbulent boundary layer with zero, favourable and adverse pressure gradients. J. Phys.: Conf. Ser. **318**(6), 062007 (2011). doi:10.1088/1742-6596/318/6/062007
4. A. Dróżdż, W. Elsner, Amplitude modulated near-wall cycle in a turbulent boundary layer under an adverse pressure gradient. Arch. Mech. **65**(6), 511–525 (2013)
5. Z. Harun, J.P. Monty, R. Mathis, I. Marusic, Pressure gradient effects on the large-scale structure of turbulent boundary layers. J. Fluid Mech. **715**, 477–498 (2013)
6. H. Kim, S.J. Kline, W.C. Reynolds, The production of turbulence near a smooth wall in a turbulent boundary layer. J. Fluid Mech. **50**, 133–160 (1971)

Experimental Investigation of a Turbulent Boundary Layer Subject to an Adverse Pressure Gradient at Re_θ up to 10000 Using Large-Scale and Long-Range Microscopic Particle Imaging

Tobias Knopp, Nicolas A. Buchmann, Daniel Schanz, Christian Cierpka, Rainer Hain, Andreas Schröder and Christian J. Kähler

Abstract We present an experimental investigation and data analysis of a turbulent boundary layer flow at a significant adverse pressure gradient for two Reynolds numbers $Re_\theta = 6200$ and $Re_\theta = 8000$. We perform detailed multi-resolution measurements by combining large-scale and long-range microscopic particle imaging. We investigate scaling laws for the mean velocity and for the total shear stress in the inner layer. In the inner part of the inner layer the mean velocity can be fitted by a log-law. In the outer part a modified log-law provides a good fit, which depends on the pressure gradient parameter and on a parameter for the mean inertial effects. Emphasis is on the Reynolds number effects on the mean velocity and shear stress.

1 Introduction

The mean behaviour of a turbulent boundary layer subject to a significant adverse pressure gradient (APG) is still an open question [1, 8]. In the present work we consider the inner part of the boundary layer $y < 0.15\delta_{99}$ called the inner layer. As described in [1], several authors report that in case of a positive pressure gradient still a region of a log-linear mean velocity law $u^+ = \log(y^+)/K_i + B_i$ can be observed when plotted in viscous units $u^+ = U/u_\tau$ and $y^+ = yu_\tau/\nu$, where u_τ is the friction velocity and ν is the kinematic viscosity. It is open in the literature if K_i and B_i have

T. Knopp (✉) · D. Schanz · A. Schröder
Institute of Aerodynamics and Flow Technology, DLR (German Aerospace Center), Bunsenstr. 10, 37073 Gottingen, Germany
e-mail: Tobias.Knopp@dlr.de

N.A. Buchmann · C. Cierpka · R. Hain · C.J. Kähler
Institute for Fluid Mechanics and Aerodynamics, Universität der Bundeswehr München, Werner-Heisenberg-Weg 39, 85577 Neubiberg, Germany

M. Stanislas et al. (eds.), *Progress in Wall Turbulence 2*,
ERCOFTAC Series 23, DOI 10.1007/978-3-319-20388-1_24

the same value as in the log-law for zero-pressure gradient (ZPG) flows. Some recent publications propose a functional dependence of K_i and B_i on the so-called pressure gradient parameter $\Delta p_x^+ = \nu/(\rho u_\tau^3)\mathrm{d}p/\mathrm{d}x$, see [6, 16]. Additionally, some authors [2] propose that, beyond the log-layer, a linear stress region forms, where the mean velocity follows a half-power law [2] or a modified log-law [13, 18].

In order to investigate these questions, a database of high-quality experiments and direct numerical simulation (DNS) is needed. A test case is suitable for such a data base if effects of surface curvature, strong non-equilibrium and flow history are small, and if the Reynolds number is sufficiently large. In order to increase the database available from the literature, we designed a new flow experiment which matches these requirements. We performed detailed multi-resolution measurements by combining large-scale and long-range microscopic particle imaging [5, 9, 11, 12].

2 Classical Theory

We assume a two-dimensional, incompressible turbulent boundary layer flow without external forces throughout this work. The equation for the wall-parallel mean velocity component U integrated from the wall to the wall-distance y becomes

$$\tau^+(y^+) \equiv u_\tau^{-2}\left(\nu\frac{\partial U}{\partial y} - \overline{u'v'}\right) = 1 + \Delta p_x^+ y^+ + I_{cu}^+(y^+) + I_{cv}^+(y^+) \qquad (1)$$

where the contribution of the Reynolds normal stresses has been neglected. Therein $\tau^+(y^+)$ is the total shear stress, and $I_{cu}^+(y^+) = u_\tau^{-2} I_{cu}$ and $I_{cv}^+(y^+) = u_\tau^{-2} I_{cv}$ are the mean inertial terms in viscous units, where I_{cu} and I_{cv} are defined by

$$I_{cu}(y) = \int_{y'=0}^{y} U\frac{\partial U}{\partial x}\mathrm{d}y'\,, \quad I_{cv}(y) = \int_{y'=0}^{y} V\frac{\partial U}{\partial y'}\mathrm{d}y'\,.$$

A wall law for u^+ can be derived and assessed by scaling the mean velocity gradient using τ^+ and y^+, leading to the mean velocity slope diagnostic function [7]

$$\Xi = \frac{y^+}{\sqrt{\tau^+(y^+)}}\frac{\mathrm{d}u^+}{\mathrm{d}y^+}. \qquad (2)$$

The diagnostic function (2) should show a plateau in case that the underlying wall law for u^+ is satisfied. In this argument τ^+ is often approximated using a linear stress model $\tau^+(y^+) \approx 1 + \lambda \Delta p_x^+ y^+$, see e.g. [13]. Therein, [13] approximates the effects of the mean inertial term using $\lambda = 0.7$ for flows at mild pressure gradient near equilibrium. The integral of (2) then gives a modified log-law, see e.g. [13],

$$u^+ = \frac{1}{K_o}\log(y^+) + \frac{2}{K_o}\left[\sqrt{1+\lambda\Delta p_x^+ y^+} - 1 + \log\left(\frac{2}{\sqrt{1+\lambda\Delta p_x^+ y^+}+1}\right)\right] + B_o. \tag{3}$$

The theoretical mean velocity gradient is $1/(\kappa y^+)$ for the classical log-law and $\sqrt{1+\lambda\Delta p_x^+ y^+}(K_o y^+)^{-1}$ for the modified log-law (3).

3 Experimental Setup

We designed a new flow experiment of a turbulent boundary layer over a flat plate followed by an s-shaped deflection for producing an adverse pressure gradient, see Fig. 1. The aim was to enable PIV measurements in the inner part of the boundary layer at high Reynolds numbers. The contour geometry was designed using RANS simulations with the DLR TAU code. The experiment has been carried out in the Eiffel type atmospheric wind tunnel of the Universität der Bundeswehr in Munich which has a 22m-long test section with a rectangular cross-section of $2 \times 2\,\text{m}^2$. The freestream velocities are $U_\infty = 9\,\text{ms}^{-1}$ and $U_\infty = 12\,\text{ms}^{-1}$. The flow passes along both the upper side (u) and the lower side (l), but only the upper side flow is of interest. Downstream of the nose (a) at $x = 0\,\text{m}$ the flow is tripped and develops to a fully turbulent flow over a long-flat plate of 6 m length (b). At the end of the ZPG inlet section, we obtain $Re_\tau = 2400$ and $Re_\theta = 6200$ resp. $Re_\tau = 3030$ and $Re_\theta = 8000$. The flow is then deflected (c) from $x = 6.38\,\text{m}$ to $x = 7.88\,\text{m}$ at low curvature. Then, the flow passes an inclined flat plate (d) of length 0.8 m from $x = 7.88\,\text{m}$ to $x = 8.66\,\text{m}$ at an opening angle of 13°. In the middle of this focus region we reach $Re_\theta = 18000$ for $U_\infty = 12\,\text{ms}^{-1}$. Downstream of a second deflection (e) and a second flat plate (f), a flap (g) is attached. This flap controls the circulation around the model. The streamwise distribution of u_τ and Δp_x^+ is shown in Fig. 2 to illustrate the flow.

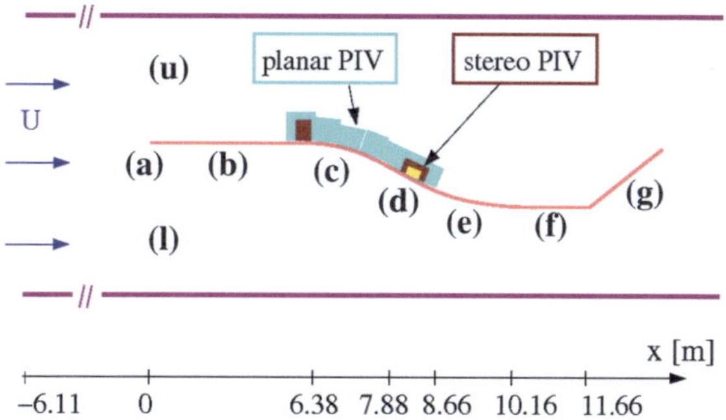

Fig. 1 Sketch of the flow experiment and the PIV systems used (*top view*)

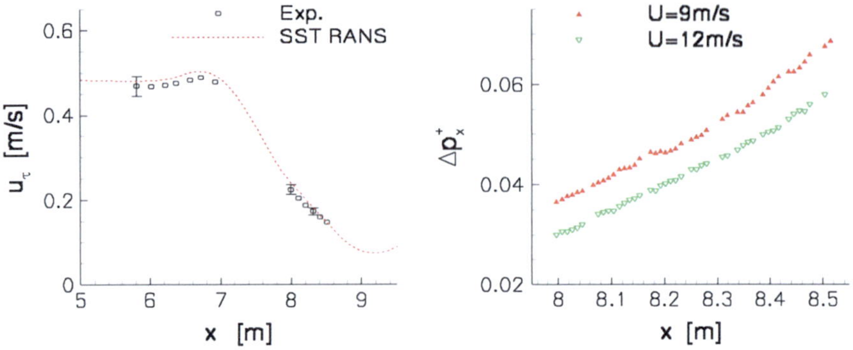

Fig. 2 *Left* Comparison of u_τ between SST $k-\omega$ RANS and experiment for $U = 12\,\text{ms}^{-1}$. *Right* Distribution of Δp_x^+ on the inclined flat plate *(d)* in Fig. 1 for $U_\infty = 9\,\text{ms}^{-1}$ and $U = 12\,\text{ms}^{-1}$

4 Measuring Technique

For particle imaging we applied a multi-resolution approach [5, 9]. We used large-scale PIV for $60 < y^+ < 0.9\delta_{99}^+$ (planar and stereo PIV), and long-range microscopic PIV with a particle tracking velocimetry (PTV) algorithm (LR-μPTV) for $y^+ < 200$, see [4, 10]. LR-μPTV was applied at three distinct locations on the inclined flat plate, see (d) in Fig. 1, where 3000 instantaneous image pairs were evaluated. The

Table 1 Summary of the experimental parameters for the PIV and PTV measurements

	Planar PIV	Stereo PIV (lower res.)	Stereo PIV (higher res.)	LR-μPTV
Centre of field of view x_{ref}	7.61 m	8.28 m	8.28 m	8.28 m
Field of view	0.60 m × 0.20 m	0.55 m × 0.21 m	0.16 m × 0.21 m	11 mm × 17 mm
Resolution	6 $\text{px}_{\text{obj}}/\text{mm}$	8.5 $\text{px}_{\text{obj}}/\text{mm}$	18 $\text{px}_{\text{obj}}/\text{mm}$	45.6 $\text{px}_{\text{obj}}/\text{mm}$
Interrogation window size	16 × 16 px	24 × 24 px	24 × 24 px	–
Interrogation step size	6 px	8 px	8 px	–
ν/u_τ for $U_\infty = 12\,\text{ms}^{-1}$	47 µm	89 µm	89 µm	89 µm
Interrogation window size (l^+)	30	39	18	–
Wall normal extent (y^+_{max})	$3770 = \delta_{99}^+$	$2300 \approx 0.9\delta_{99}^+$	$2300 \approx 0.9\delta_{99}^+$	200
Wall-normal resolution (Δy^+)	10	13	6	0.3

The flow was seeded with DEHS droplets with a diameter of approx. 1 µm. The flow parameters are for the streamwise reference position x_{ref} and correspond to $U_\infty = 12\,\text{ms}^{-1}$

instantaneous velocity fields were averaged in streamwise direction over an extent of $\Delta x = 10.5$ mm ($\Delta x^+ = 120$) and sorted into bins of 2 pixel with 50 % overlap in wall-normal direction. The number of velocity samples per bin ranges between 5000 and 8000. For LR-μPTV, the first accurate data point above the wall is at $y^+ = 1.5$ for $U_\infty = 12\,\mathrm{ms}^{-1}$. The experimental parameters are summarized in Table 1.

The wall-shear stress τ_w was determined directly from the mean velocity profiles in the viscous sublayer using the LR-μPTV data. For the planar and stereo PIV data we used different indirect methods based on the Clauser chart. The indirect method was assessed by reference with the direct method for the LR-μPTV data in [11].

5 Results

The mean velocity profiles in the inner layer $y < 0.15\delta_{99}$ can be described by a composite velocity profile similar to the proposal in [2], see [11, 12]. The inner part can be described by a log-linear fit and the outer part can be described using the modified log-law (3). This is shown in Fig. 3 for the two Reynolds numbers.

5.1 Scaling of the Inner Part of the Inner Layer

In the inner part, the mean velocity profiles can be fitted using a log-linear relation

$$u^+ = \frac{1}{K_\mathrm{i}} \log(y^+) + B_\mathrm{i} \tag{4}$$

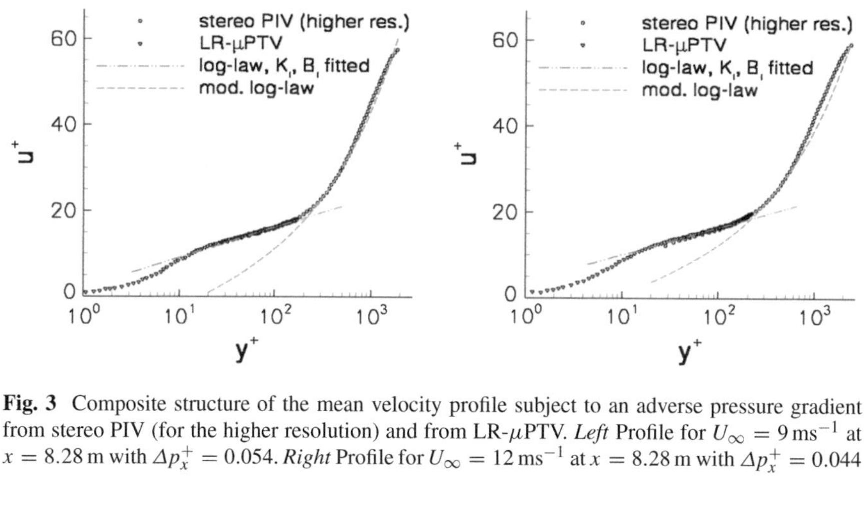

Fig. 3 Composite structure of the mean velocity profile subject to an adverse pressure gradient from stereo PIV (for the higher resolution) and from LR-μPTV. *Left* Profile for $U_\infty = 9\,\mathrm{ms}^{-1}$ at $x = 8.28$ m with $\Delta p_x^+ = 0.054$. *Right* Profile for $U_\infty = 12\,\mathrm{ms}^{-1}$ at $x = 8.28$ m with $\Delta p_x^+ = 0.044$

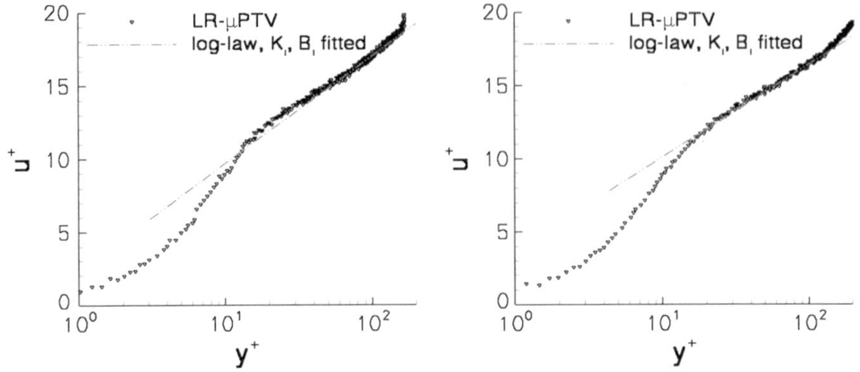

Fig. 4 Mean velocity profiles at $x = 8.28$ m in the viscous sublayer and in the log-law fit region at $\Delta p_x^+ = 0.054$ for $U = 9\,\mathrm{ms}^{-1}$ (*left*) and at $\Delta p_x^+ = 0.044$ for $U = 12\,\mathrm{ms}^{-1}$ (*right*)

in a thin region $y^+_{\log,\min} < y^+ < y^+_{\log,\max}$, see Figs. 3 and 4. From Fig. 4 we find $y^+_{\log,\min} \approx 60$ and $y^+_{\log,\max} \approx 130$ by visual inspection. This region is called the log-law fit region. We study the influence of *Re* on its extent, as suggested by [14]. For this purpose, we consider the mean velocity slope diagnostic function for the log-law $\Xi_{\log} = y^+ \mathrm{d}u^+/\mathrm{d}y^+$. Figure 5 shows Ξ^{-1} for the LR-μPTV data after smoothing and for the stereo PIV data (higher res.). We can observe a first inflection point of Ξ^{-1} at $y^+ \approx 80$ and a second inflection point at $y^+ \approx 115$ for $U = 9\,\mathrm{ms}^{-1}$ resp. at $y^+ \approx 120$ for $U = 12\,\mathrm{ms}^{-1}$. As we do not have a clear plateau in Ξ the log-law is not an exact description for this region. It is worthwhile to study if the second inflection point shifts toward larger y^+-values for higher *Re*. In Fig. 4 we can see an overshoot of u^+ over the log-law for y^+ around 38. This can be also seen in the corresponding plot for Ξ^{-1} in Fig. 5. Such an overshoot is found also for ZPG flows as described by [3]. We point out that these findings need to be studied for additional test cases, in particular at higher Reynolds numbers. Regarding the measurement

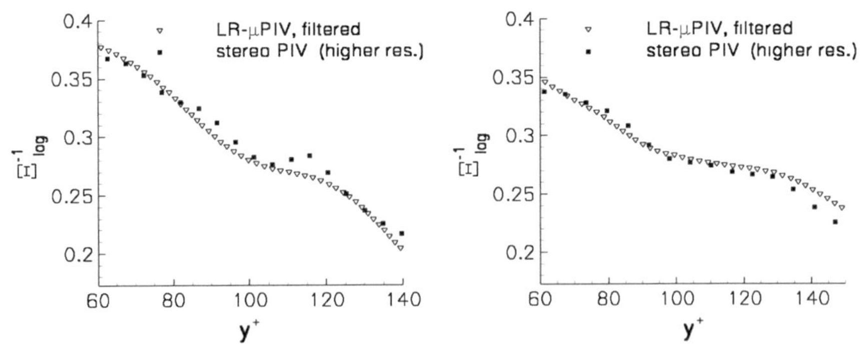

Fig. 5 Log-law diagnostic function $\Xi_{\log} = y^+ \mathrm{d}u^+/\mathrm{d}y^+$ at $x = 7.895$ m with $\Delta p_x^+ = 0.052$ for $U = 9\,\mathrm{ms}^{-1}$ (*left*) and at $x = 8.28$ m with $\Delta p_x^+ = 0.044$ for $U = 12\,\mathrm{ms}^{-1}$ (*right*)

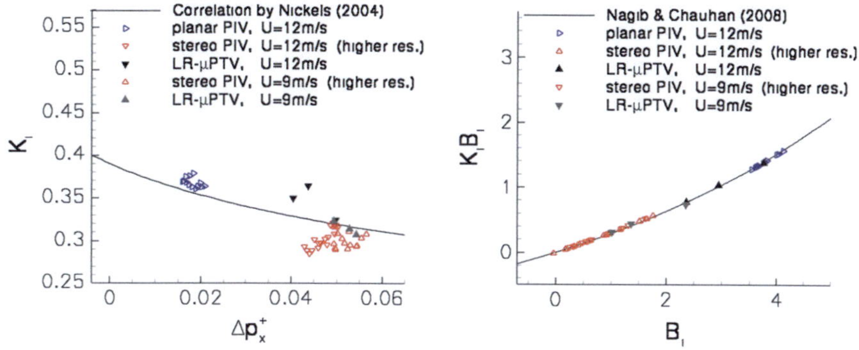

Fig. 6 *Left* Variation of K_i versus. Δp_x^+ in the log-law fit region for the present experiment and correlation (5) by Nickels [16] (denoted by Nickels (2004) in the legend). *Right* Variation of $K_\text{i} B_\text{i}$ vs B_i and correlation (6) by Nagib and Chauhan [15] (denoted by Nagib and Chauhan (2008) in the legend)

technique, a larger number of samples could help to avoid smoothing of the LR-μPTV data. We finally remark that the log-law fit region occupies a y^+-range, which for zero-pressure gradient flows could be influenced by low-Re effects, see [19].

Then, we consider K_i and B_i in (4). We study the idea that K_i and B_i could change w.r.t. Δp_x^+. Nickels [16] examined data for $-0.02 < \Delta p_x^+ < 0.06$ and proposed that

$$K_\text{i} = \frac{\kappa_0}{\sqrt{1 + \Delta p_x^+ y_c^+}}, \quad \text{with} \quad \Delta p_x^+ \left(y_c^+\right)^3 + \left(y_c^+\right)^2 - Re_c^2 = 0 \tag{5}$$

with $\kappa_0 = 0.39$, $Re_c = 12$, and y_c being the smallest positive root of the second equation. A variation of B_i in conjunction with K_i is proposed by Nagib and Chauhan in [15] for positive and negative pressure gradients

$$K_\text{i} B_\text{i} = 1.6\,(\exp(0.1663 B_\text{i}) - 1)\,. \tag{6}$$

For the present data, the variation of K_i w.r.t. Δp_x^+ is shown in Fig. 6 (left) and the variation of $K_\text{i} B_\text{i}$ versus B_i is plotted in Fig. 6 (right). The values for K_i and B_i are obtained by a least-squares fit of (4) to the experimental mean velocity profiles. The correlation (5) shows a good agreement with the experimental results. This supports the results shown in [16]. The correlation (6) almost collapses with the experimental results, which supports the results in [15] for adverse pressure gradients.

5.2 *Scaling of the Outer Part of the Inner Layer*

Then, we consider the outer part of the inner layer. The scaling law argument (2) involves the total shear stress $\tau^+(y^+)$, which is shown in Fig. 7. The total shear

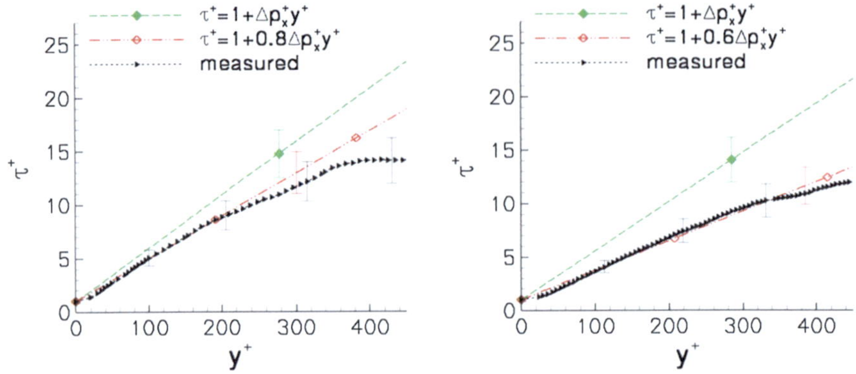

Fig. 7 Total shear stress at $x = 8.19\,\mathrm{m}$ with $\Delta p_x^+ = 0.050$ for $U_\infty = 9\,\mathrm{ms}^{-1}$ (*left*) and at $x = 8.21\,\mathrm{m}$ with $\Delta p_x^+ = 0.046$ for $U_\infty = 12\,\mathrm{ms}^{-1}$ (*right*)

stress is computed from the stereo PIV data (higher res.) as the right hand side of (1) where I_{cu} and I_{cv} are determined by numerical integration. The total shear stress is approximated using the linear stress model for $y^+ < 220$ for $U_\infty = 9\,\mathrm{ms}^{-1}$ resp. for $y^+ < 330$ for $U_\infty = 12\,\mathrm{ms}^{-1}$. We determine λ by a least-square fit and obtain $\lambda = 0.8$ for $U_\infty = 9\,\mathrm{ms}^{-1}$ and $\lambda = 0.6$ for $U_\infty = 12\,\mathrm{ms}^{-1}$ with a relative uncertainty of 10 %.

Then, we consider the mean velocity slope diagnostic function (2). For τ^+ we substitute the measured total shear stress and alternatively the linear stress model. Note that we compute Ξ for the spatially filtered mean velocity field by applying a Gaussian filter in wall-normal and in wall-parallel direction to reduce the wiggles in the gradients compared to the unfiltered data. We seek a plateau in Ξ^{-1} whose extent gives the region of the modified log-law. As we are interested in the inner layer, we consider the region $y^+ < 0.2\delta_{99}^+$, corresponding to $y^+ < 450$ for $U_\infty = 9\,\mathrm{ms}^{-1}$ and $y^+ < 500$ for $U_\infty = 12\,\mathrm{ms}^{-1}$. The results are shown in Fig. 8 (left).

The behaviour for Ξ^{-1} is similar for both choices for τ^+ for $150 < y^+ < 300$, where we find that Ξ^{-1} is clearly decreasing with increasing y^+ for both Reynolds numbers. For $y^+ > 300$ the behaviour of Ξ^{-1} becomes little different depending on the choice for τ^+. Substitution of the linear stress model gives a slower decrease in Ξ^{-1} than for the measured τ^+. Albeit the slope of Ξ^{-1} is becoming smaller for $y^+ > 300$, a clear plateau in Ξ^{-1} cannot be observed. Moreover, we cannot find a region where both the linear stress model and the modified log-law (3) hold. One point could be that the Reynolds number is too small. Albeit we reach $Re_\theta \approx 18000$, the requirements for a sufficiently large Reynolds number are probably even larger for APG turbulent boundary layers than for ZPG flows.

Then, we determine K_o and B_o by a least squares fit of (3) to the data in the region of the modified log-law ($y_{\log,\mathrm{mod,min}}$, $y_{\log,\mathrm{mod,max}}$). The result for K_o is sensitive w.r.t. u_τ and depends on the value for λ and on the lower and upper bound of the modified log-law region. The sensitivity of K_o w.r.t. u_τ is found to be small for the present

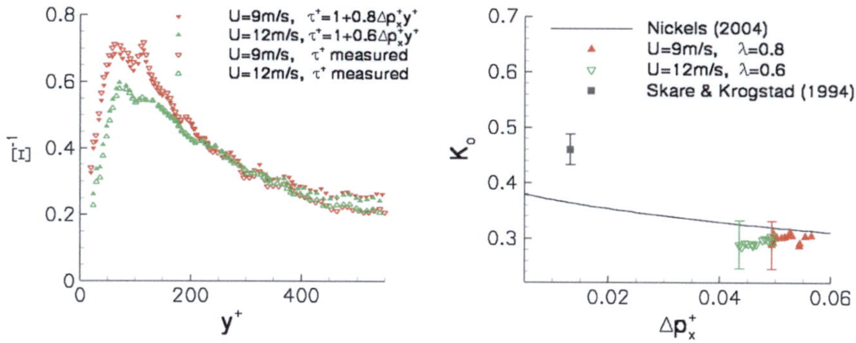

Fig. 8 *Left* Diagnostic function at $x = 8.19\,\text{m}$ with $\Delta p_x^+ = 0.050$ for $U_\infty = 9\,\text{ms}^{-1}$ and at $x = 8.21\,\text{m}$ with $\Delta p_x^+ = 0.046$ for $U_\infty = 12\,\text{ms}^{-1}$. *Right* Slope coefficient K_o of the modified log-law (3)

stereo PIV (higher res.) data. To understand this, we recall that $\Delta p_x^+ > 0.04$ and hence $\Delta p_x^+ y^+ \gg 1$ for $y^+ > 300$. Then the dominant term on the right-hand side of (3) is $(1 + \lambda \Delta p_x^+ y^+)^{1/2}$ which behaves like $1/u_\tau(\lambda y/\nu \mathrm{d}p/\mathrm{d}x)^{1/2}$ and has the same dependence on u_τ as u^+. We estimate that the relative uncertainty for K_o is 15 %, using a relative uncertainty of 10 % for λ and a relative uncertainty of 5 % due to a variation of $y_{\text{log,mod,max}}$. We note that K_o is decreasing with increasing $y_{\text{log,mod,max}}$ and that K_o is decreasing with decreasing λ.

Finally, it is of interest to describe K_o and B_o quantitatively, either as a constant value or using a functional dependence like (5). The behaviour of K_o versus Δp_x^+ is shown in Fig. 8 (right) as a first attempt. The values for K_o at the two Reynolds numbers and for the different values of Δp_x^+ are of a similar magnitude. This is obtained thanks to the flow dependent choice for λ instead of using a constant value. The value of K_o for the equilibrium boundary layer [17] at $\Delta p_x^+ = 0.013$ is significantly larger, where we use $\lambda = 0.9$ found in [11]. From this observation we cannot expect that K_o has a constant value. Interestingly, the values for K_o are close to the correlation (5) for K_i within the estimated error range for the present flow experiment. However, this needs to be studied for additional data sets including different streamwise distributions of the pressure gradient and higher Reynolds numbers.

6 Conclusions

We applied particle imaging with multi-resolution [5, 9] in a turbulent boundary layer at a strong adverse pressure gradient (APG) by combining large-scale PIV and long-range microscopic particle tracking velocimetry (LR-μPTV) for resolving the near-wall region [4, 10]. The mean velocity profiles in the inner layer $y < 0.15\delta_{99}$ can be described by a composite structure [2]. The inner part can be described as a "log-law fit" region. The slope $1/K_\text{i}$ and the intercept B_i of the log-linear fit are found to change under the effect of the pressure gradient, supporting the proposal by [15, 16]. The outer part of the inner layer can be fitted using the modified log-law (3) by [13, 18]. Its main parameters are the pressure gradient parameter Δp_x^+ and the shear stress gradient parameter λ, which describes the effect of the mean inertial terms on the wall-normal gradient of the total shear stress. The classical scaling law argument (2) is based on the existence of a region where both the linear stress model and the modified log-law (3) can be observed. For the present experiment, such a region cannot be found for the Reynolds numbers achieved. Regarding the composite structure of the mean velocity profile, the present experimental data are seen only as a very first step toward a law-of-the-wall for the inner part of the boundary layer for APG flows. The extent of the log-law fit region and of the modified log-law region need to be characterized in terms of suitable flow parameters [14]. Additionally the modified log-law itself needs to be assessed for other data sets and its coefficients K_o and B_o need to be studied. We admit and emphasize that the findings for the present flow could depend on the special distribution of the streamwise pressure gradient and on the Reynolds number.

Acknowledgments The authors are grateful to Prof. Skare for providing his data. The authors are also grateful to Profs. Rossow, Radespiel, Nagib and Herwig and to Drs. B. Eisfeld, A. Krumbein and D. Schwamborn for valuable discussions and suggestions.

References

1. A.E. Alving, H.H. Fernholz, Mean-velocity scaling in and around a mild, turbulent separation bubble. Phys. Fluids **7**, 1956–1969 (1995)
2. K.C. Brown, P.N. Joubert, The measurement of skin friction in turbulent boundary layers with adverse pressure gradients. J. Fluid Mech. **35**, 737–757 (1969)
3. K.A. Chauhan, P.A. Monkewitz, H.M. Nagib, Criteria for assessing experiments in zero pressure gradient boundary layers. Fluid Dyn. Res. **41**, 021404 (2009)
4. C. Cierpka, S. Scharnowski, C.J. Kähler, Parallax correction for precise near-wall flow investigations using particle imaging. Appl. Opt. **52**, 2923–2931 (2013)
5. C.M. de Silva, E.G. Gnanamanickam, C. Atkinson, N.A. Buchmann, N. Hutchins, J. Soria, I. Marusic, High spatial range velocity measurements in a high Reynolds number turbulent boundary layer. Phys. Fluids **26**:025117–1–19 (2014)
6. S.A. Dixit, O.N. Ramesh, Determination of skin friction in strong pressure-gradient equilibrium and near-equilibrium turbulent boundary layers. Exp. Fluids **47**, 1045–1058 (2009)
7. K. Gersten, H. Herwig, *Strömungsmechanik. Grundlagen der Impuls-, Wärme- und Stoffübertragung aus asymptotischer Sicht*. Vieweg, 1992
8. A.G. Gungor, Y. Maciel, M.P. Simens, J. Soria, Analysis of a turbulent boundary layer subjected to a strong adverse pressure gradient. J. Phys.: Conf. Ser. **506**, 012007 (2014)
9. S. Herpin, C.Y. Wong, M. Stanislas, J. Soria, Stereoscopic PIV measurements of a turbulent boundary layer with a large spatial dynamic range. Exp. Fluids **45**, 745–763 (2008)
10. C.J. Kähler, S. Scharnowski, C. Cierpka, On the uncertainty of digital PIV and PTV near walls. Exp. Fluids **52**, 1641–1656 (2012)
11. T. Knopp, N.A. Buchmann, D. Schanz, B. Eisfeld, C. Cierpka, R. Hain, A. Schröder,C.J. Kähler, Investigation of scaling laws in a turbulent boundary layer flow with adverse pressure gradient using PIV. J Turbul **16**, 250–272 (2015)

12. T. Knopp, D. Schanz, A. Schröder, M. Dumitra, R. Hain, C.J. Kähler, Experimental investigation of the log-law for an adverse pressure gradient turbulent boundary layer flow at Re_θ up to 10000. Flow Turbul. Combust. **92**, 451–471 (2014)
13. H. McDonald, The effect of pressure gradient on the law of the wall in turbulent flow. J. Fluid Mech. **35**, 311–336 (1969)
14. H. Nagib. Personal communication. 2013
15. H.M. Nagib and K.A. Chauhan. Variations of von Karman coefficient in canonical flows. Phys. Fluids, 20:101518–1–10, 2008
16. T.B. Nickels, Inner scaling for wall-bounded flows subject to large pressure gradients. J. Fluid Mech. **521**, 217–239 (2004)
17. P.E. Skare, P.A. Krogstad, A turbulent equilibrium boundary layer near separation. J. Fluid Mech. **272**, 319–348 (1994)
18. W. Szablewski, Turbulente Strömungen in divergenten Kanälen. Ing.-Arch. **22**, 268–281 (1954)
19. T. Wei, R. Schmidt, P. McMurtry, Comment on the clauser chart method for determining the friction velocity. Exp. Fluids **38**, 695–699 (2005)

The Structure of APG Turbulent Boundary Layers

Ayse G. Gungor, Yvan Maciel and Mark P. Simens

Abstract The characteristics of three-dimensional intense uv-structures (Qs) in a strongly decelerated large-velocity-defect boundary layer are analyzed. The Q2 and Q4 structures are found to be different from those of turbulent channel flows studied by Lozano-Durán et al. (J Fluid Mech 694:100–130, 2012). They are less streamwise elongated, less present near the wall and wall-detached structures are more numerous. Moreover, contrary to channel flows, wall-detached Q2, and Q4 structures contribute significantly to the Reynolds shear stress.

1 Introduction

In canonical wall-bounded flows, a production peak of turbulent kinetic energy is found near the wall and can be related to a large extent to the coherent streaky near-wall structures and a large mean shear [12]. In contrast, the near-wall production peak is absent or very small when the boundary layer has a large mean velocity defect due to a strong or prolonged adverse pressure gradient (APG) [15]. This absence or attenuation has been related to the lack of strong streaky structures and a lower mean shear than in canonical wall-bounded flows [9]. The main production peak is found in the outer region of the flow, which affirms that the pressure gradient causes a change in

A.G. Gungor (✉)
Faculty of Aeronautics and Astronautics, Istanbul Technical University, 34469 Istanbul, Turkey
e-mail: ayse.gungor@itu.edu.tr

Y. Maciel
Department of Mechanical Engineering, Laval University, Quebec City, QC G1V 0A6, Canada
e-mail: Yvan.Maciel@gmc.ulaval.ca

M.P. Simens
School of Aeronautics, Universidad Politécnica de Madrid, 28040 Madrid, Spain
e-mail: mark@torroja.dmt.upm.es

M. Stanislas et al. (eds.), *Progress in Wall Turbulence 2*,
ERCOFTAC Series 23, DOI 10.1007/978-3-319-20388-1_25

the distribution of energy in a turbulent boundary layer (TBL). By analyzing several large-velocity-defect TBLs, Gungor et al. [2] have concluded that these boundary layers are globally less efficient in extracting turbulent energy from the mean flow than the zero-pressure gradient (ZPG) TBL. The Reynolds stresses and the production of turbulent kinetic energy were found to be weaker in the lower half of the large-velocity-defect boundary layers than in the ZPG TBL. Furthermore, the outer-region turbulent statistics of TBLs close to detachment were found to resemble those of single-stream mixing layers. These various observations suggest that the physical mechanisms and coherent structures responsible for the production and transport of turbulence might be different.

Unfortunately, detailed analyses of the coherent structures found in APG TBLs are rare. By analyzing the DNS data of a turbulent separation bubble of Na and Moin [8], Chong et al. [1] suggested that in the APG zone prior to detachment more of the eddies which contribute to the Reynolds shear stress are eddies which are not connected to the wall. In the case of an equilibrium APG TBL, Krogstad and Skare [4] found that the lower part of the boundary layer is strongly dominated by Q4 motions which are more frequent and last longer than other quadrant motions, while in a ZPG TBL second and fourth quadrant events are equally important. The streamwise correlation length of u was also found to be considerably shorter in the APG case throughout the boundary layer, a result also obtained later by [2, 11]. In the outer region of a large-velocity- defect TBL that eventually separates, Rahgozar and Maciel [10] observed that the predominance of streaky u-structures is less than in the ZPG case and that this predominance even disappears near detachment. Although the pressure gradient reduced the frequency of appearance of the u-structures, it did not significantly affect their dimensions and arrangement in the upper half of the boundary layer. By analyzing the same flow, Rahgozar and Maciel [11] found that large-scale u-structures are less elongated than those of ZPG TBLs, especially in the lower part of the boundary layer.

Lozano-Durán et al. [7] (hereafter referred to as LFJ) recently generalized the quadrant analysis of Reynolds shear stress events to three-dimensional structures that they called Qs. They showed that intense Q structures are important since they are responsible for most of the wall-normal flux of momentum and production of turbulent energy. In the case of turbulent channel flows, LFJ found that wall-detached Qs are background stress fluctuations while wall-attached Qs are bigger and carry most of the mean Reynolds shear stress. The number of wall-attached Qs decreases away from the wall, but the fraction of Reynolds shear stress that they carry is independent of their size and location.

In the present work, the characteristics and temporal evolution of Q structures in a strongly decelerated boundary layer is investigated to advance our knowledge and understanding of the APG boundary layer flows. The focus is on Q structures because of their dominant role in terms of momentum flux and production of turbulent energy.

1.1 Database and Structure Identification Method

The database of an APG TBL obtained via direct numerical simulation (DNS) by [2] is used in the present study. The parameters of the DNS are summarized in Table 1. The desired constant streamwise velocity gradient is controlled by imposing a constant uniform suction at the top simulation boundary. The streamwise and spanwise velocities at the top boundary satisfy free-slip conditions. The laminar Hiemenz profile is prescribed at the inflow, and the velocities at the outflow are estimated by a convective boundary condition, with small corrections to enforce global mass conservation [13]. The turbulent transition is triggered by a disturbance strip located close to the inflow and modeled using the immersed boundary method [14]. The presence of a small transitional separation bubble near the inflow helps achieve the fully turbulent regime within a shorter distance than what would be obtained with attached transition through Tollmien-Schlichting waves. The Reynolds number close to the exit of the numerical domain is $Re_\theta = 2175$ and the shape factor $H = 2.5$.

Profiles of the Reynolds shear stress at two streamwise locations, corresponding to $H = 2, Re_\theta = 1755$ and $H = 2.5, Re_\theta = 2175$ are shown in Fig. 1a. They are also compared to a profile of the ZPG TBL of Simens et al. [13] at a comparable Reynolds number ($Re_\theta = 1975$). The Reynolds shear stress is normalized with the

Table 1 Parameters of the simulation

$(L_x, L_y, L_z)/\theta_0$	$\Delta x^+, \Delta y^+_{min}, \Delta z^+$	N_x, N_y, N_z	$(B_x, B_y, B_z)/\delta_a$	N_Q	N_{Q^-}
2380 × 450 × 1100	2.2 × 0.2 × 2.0	1537×201×768	4.5 × 2 × 8.0	1106735	676657

L_x, L_y and L_z are the box dimensions along the three axes. N_x, N_y, and N_z are the corresponding grid sizes, and the Δ values are the resolutions in wall units at $Re_\theta = 2175$. The momentum thickness, θ_0, is measured at the inflow. B_x, B_y, and B_z are the dimensions of the box used for the detection of the Qs. N_Q and N_{Q^-} are the number of Qs and Q^-s extracted

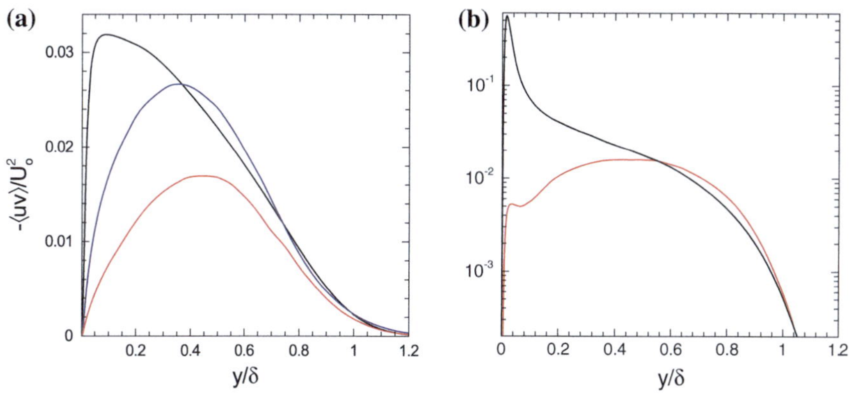

Fig. 1 **a** Reynolds shear stress. **b** Production of turbulent kinetic energy normalized with U_o and δ. Present flow: $H = 2$, *blue line*; $H = 2.5$, *red line*. ZPG TBL [13], *black line*

outer velocity scale U_o introduced by Gungor et al. [2], who have shown that it can properly scale the mean velocity defect profiles and various one-point velocity statistics of turbulence in the high velocity region of a wide range of flows: ZPG TBLs, small and large-defect APG TBLs, and mixing layers. U_o is defined as twice the velocity defect at the middle of the shear layer: $U_o = 2(U_e - U(y = 0.5\delta))$. It is similar to free-shear-layer velocity scales and approximately equivalent to the Zagarola-Smits velocity scale in the sense that both velocity scales are proportional to the mean velocity deficit of the outer flow. Figure 1a shows that in the present large-defect TBL the maximum of the Reynolds shear stress is not near the wall like in canonical wall flows, but rather in the middle of the boundary layer. The Reynolds shear stress decreases in the lower half of the boundary layer as the velocity defect increases. All these characteristics are commonly found in large-defect TBLs [2].

The difference between ZPG and large-defect APG TBLs is even more pronounced for the production of turbulent kinetic energy, as shown in Fig. 1b. In large-defect APG TBLs, the maximum production is in the outer region and production in the upper half of the boundary layer is comparable to that of ZPG TBLs when scaled properly [2]. A near-wall production peak exists in the present flow even at the position corresponding to $H = 2.5$ but it is two orders of magnitude smaller than in the ZPG TBL and four times smaller than the outer peak.

As was detailed in [2] the influence of the upstream bubble shear layer instability only becomes small beyond $x/\theta_0 \approx 1800$, corresponding to $Re_\theta \approx 1700$, where θ_0 is the momentum thickness measured at the inflow. Since we would like to consider Q structures in an APG boundary layer in the absence of large-scale perturbations and at a sufficiently high Reynolds number, the region of the flow used for the extraction of the Qs starts at $Re_\theta = 1314$ and ends at $Re_\theta = 2207$. In this region, the boundary layer is in strong nonequilibrium state and possesses a large mean velocity defect ($H = 1.72 - 2.56$). The box covers $\delta_{0.995}/\theta_0 \approx 97 - 184$, with an average of $\delta_a/\theta_0 = 140.7$. The box dimensions are given in Table 1.

The procedure adopted to identify the Qs follows the method used by [7] with some small changes to make the method applicable to boundary layers. The Qs are defined as regions of connected points that satisfy $|\tau| > \mathrm{H}u'v' + 10^{-6}$, where $\tau(\mathbf{x}) = u(\mathbf{x})v(\mathbf{x})$, is the pointwise instantaneous Reynolds shear stress, u' and v' are, respectively, the streamwise and wall-normal root-mean-squared intensities of fluctuations and H is the threshold constant, also called hyperbolic-hole size. The value 10^{-6} is added to filter out the freestream to prevent spurious clusters to be detected. Following, the notation of LFJ, the Q2 events (ejections) and Q4 events (sweeps) will be referred as Q^-, and the Q1 events (outward interactions) and the Q3 events (inward interactions) as Q^+, hereafter.

As in [7] a percolation analysis has been performed to determine what would be the value for H that gives an equilibrium between detecting only a few very big objects and detecting only a few small and very intense Qs. Figure 2a shows the ratio of the number of detected Qs to the maximum number of detected Qs as a function of the H. This ratio, together with the ratio of the volume of the largest identified structure to the total volume (not shown), shows that an hyperbolic-hole size in the approximate range $0.5 \leq \mathrm{H} \leq 3$ is acceptable. The value $\mathrm{H} = 1.75$ has been chosen

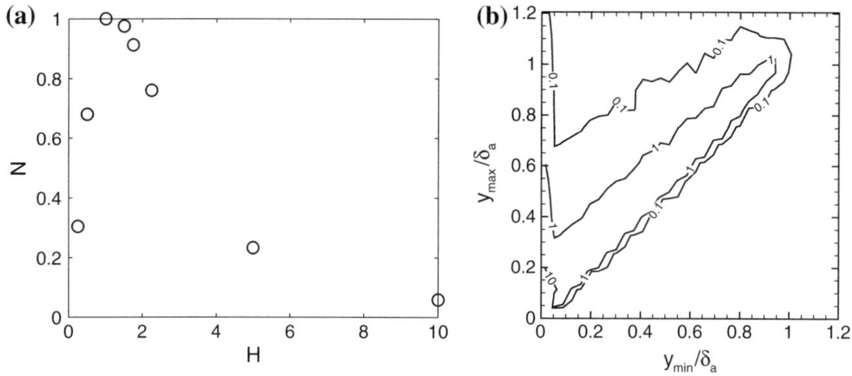

Fig. 2 **a** Ratio of the number of detected Qs to the maximum number of detected Qs as a function of the threshold. **b** Joint pdf of y_{min}/δ_a and y_{max}/δ_a for Q^-s. Contour levels are 0.1, 1 and 10

because it is the same value as used in [7] and it is in the middle of the acceptable range. The Qs are characterized by the volume V_c of a rectangular box circumscribing them, the sides of this box being denoted as Δx_c, Δy_c and Δz_c and the mid-height position of the box y_c. Structures that have $\Delta x_c/\delta_a = 4.5$ and $V_c < (\Delta x)^3$, where Δx is the streamwise grid resolution, are disregarded because either the length of these structures is undetermined or their sizes are not well resolved on the numerical grid. Furthermore, only Qs whose center is below the boundary layer thickness, $y_c/\delta_a \leq 1$, are considered to facilitate a comparison with channel flow. It has been checked that if $y_c/\delta_a \leq 1$ only detached clusters are disregarded.

2 Results and Discussion

With the above procedure, 1106735 Qs are identified in 460 statistically independent velocity fields in the extraction zone defined in the previous section. Of those, 62 % are Q^- structures. Figure 2b shows the joint probability density function (pdf) of the minimum and maximum wall distances for the Q^-s. The structures separate into two groups: wall-attached and wall-detached structures. The wall-attached Q^-s form the narrow vertical band with $y_{min} < 0.05\delta_a$ of the joint pdf, while wall-detached structures form the wide diagonal band. Figure 2b shows that the height of wall-attached Q^-s can exceed the boundary layer thickness. Wall-attached Q^-s as tall as approximately $2\delta_a$ are found. When comparing with similar joint pdfs of the Q^-s of LFJ for channel flows, it is found that in the present APG flow the probabilities are higher everywhere, except near the origin. This implies that, in proportion, there are definitely less small Q^-s close to the wall in the APG TBL than in channel flows. Such a result is expected since turbulence activity is significantly reduced near the wall in a large-defect boundary layer.

Table 2 Number fraction with respect to total number of Qs, and volume fraction with respect to extraction zone volume ($B_x \times B_y \times B_z$) for the four different types of Qs. Channel at $Re_\tau = 2003$ [7]

Case	N_1	N_2	N_3	N_4	V_1	V_2	V_3	V_4
Present (all)	0.20	0.31	0.19	0.30	0.002	0.026	0.003	0.013
Present (attached)	0.03	0.09	0.02	0.10	0.001	0.018	0.001	0.009
Channel (all)	0.19	0.33	0.21	0.28	0.004	0.059	0.008	0.022
Channel (attached)	0.02	0.14	0.007	0.11	0.000	0.053	0.000	0.014

Table 2 summarizes the number and volume fractions for the various types of Qs. Like in channel flow, Q^+s are less frequent than Q^-s, and they occupy a very small fraction of the space, 0.5 % against 4 % for Q^-s. Although the number fraction of Q2s and Q4s is comparable between the channel flow and the present flow, the volume occupied by these structures is much less for the large-defect TBL. In particular, Q2s in the TBL occupy less than half the space of Q2s in channel flow. In both flows Q4s occupy less space than Q2s, but the difference is less pronounced for the large-defect TBL.

Wall-attached Q2s and Q4s represent 32 % of the total number of Q^-s and 69 % of their volume. In channel flow, these percentages are respectively, about 40 and 80 %. The number and size proportions of attached Q^-s are therefore reduced in a large-velocity-defect boundary layer. Q structures whose center is in the range $0.2 < y_c/\delta_a < 0.6$ were also analyzed separately. They are important because $0.2 < y/\delta < 0.6$ corresponds to the region of maximum turbulence intensity, Reynolds shear stress and turbulent kinetic energy production (see Fig. 1). In that range, 94 % of the Q^-s are actually detached structures.

Figure 3 presents joint pdfs of the logarithms of the sizes, normalized with δ_a, of the boxes circumscribing Q2s and Q4s for attached (c) and detached structures (a,b), and for detached structures whose center is in the range $0.2 < y_c/\delta_a < 0.6$ (d). Like in LFJ for turbulent channel flows, the Q2 and the Q4 structures have similar sizes, with Q2s slightly bigger. The structures are approximately as long as they are tall and wide, $\Delta_x \approx \Delta_y \approx \Delta_z$, except for attached Q^-s which tend to be more elongated in the streamwise direction, by a factor of almost 2. This elongation is however, not as pronounced as in channel flows, where LFJ found the attached Q^-s to be three times longer than tall and wide for all sizes. In [4, 5, 11], it was also found that above a certain height the length of the u-velocity structures becomes shorter in an APG flow in comparison with a ZPG flow. For both attached and detached structures, the Q2s reach bigger dimensions than the Q4 structures. This is similar to what is found in channel flow. The largest structures are attached Q2s that can reach $\Delta_x \approx 3\delta_a$ and $\Delta_y \approx \Delta_z \approx 2\delta_a$. For channel flows, LFJ found very long attached Q2s reaching $\Delta_x \approx 20h$ and $\Delta_y \approx \Delta_z \approx 2h$. The rapid streamwise variations and strong nonequilibrium state of the present APG TBL probably prohibit the existence of such long motions. The detached Q^-s are found to be globally smaller than their

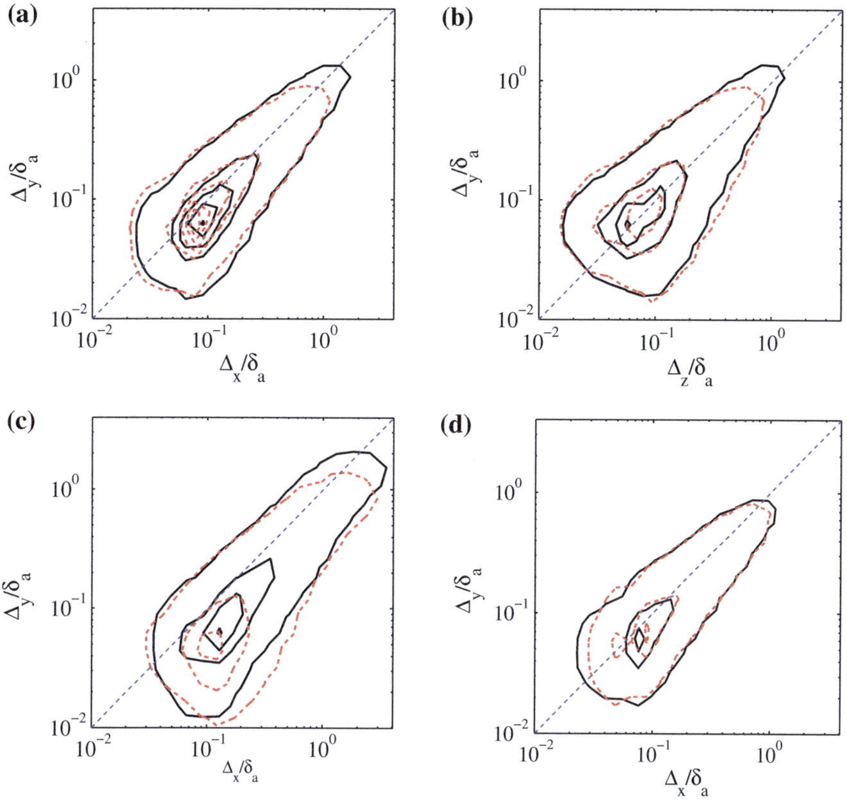

Fig. 3 Joint pdfs of the logarithms of the sizes of the boxes circumscribing Q^-s. Q2s in *black* and Q4s in *red*. Wall-detached Q^-s: **a** $p(\Delta_x/\delta_a, \Delta_y/\delta_a)$, **b** $p(\Delta_z/\delta_a, \Delta_y/\delta_a)$. **c** $p(\Delta_x/\delta_a, \Delta_y/\delta_a)$ for wall-attached Q^-s. **d** $p(\Delta_x/\delta_a, \Delta_y/\delta_a)$ for wall-detached Q^-s whose center is in the range $0.2 < y_c/\delta_a < 0.6$. The *straight dashed lines* are $\Delta_y = \Delta_x$ or $\Delta_y = \Delta_z$

attached counterparts, but not significantly. The detached Q^-s whose center is in the range $0.2 < y_c/\delta_a < 0.6$ are found to have a distribution of dimensions similar to that of the ensemble of all detached Q^-s.

Figure 4 shows joint pdfs of the logarithm of Δ_y/δ_a and τ_m/U_{oa}^2 for Q2s and Q4s, where τ_m is the volumetric average of $-uv$ computed over the volume of the Q structure. The joint pdfs are separated in four plots depending if the Q^-s are attached (a, b) or detached (c, d), and depending on the position y_c of the center of the structure: $y_c/\delta_a < 0.2$ for a and c, $0.2 < y_c/\delta_a < 0.6$ for b and d. Wall-attached Q^-s are definitely different in terms of both size and intensity between the wall region and the middle of the boundary layer. In the latter region, wall-attached Q^-s are more intense and Q2s and Q4s are relatively similar in terms of size and intensity. For $y_c/\delta_a < 0.2$, the attached Q2s are bigger and more intense than attached Q4s. In contrast, wall-detached Q^-s are fairly similar in both regions and between Q2s and

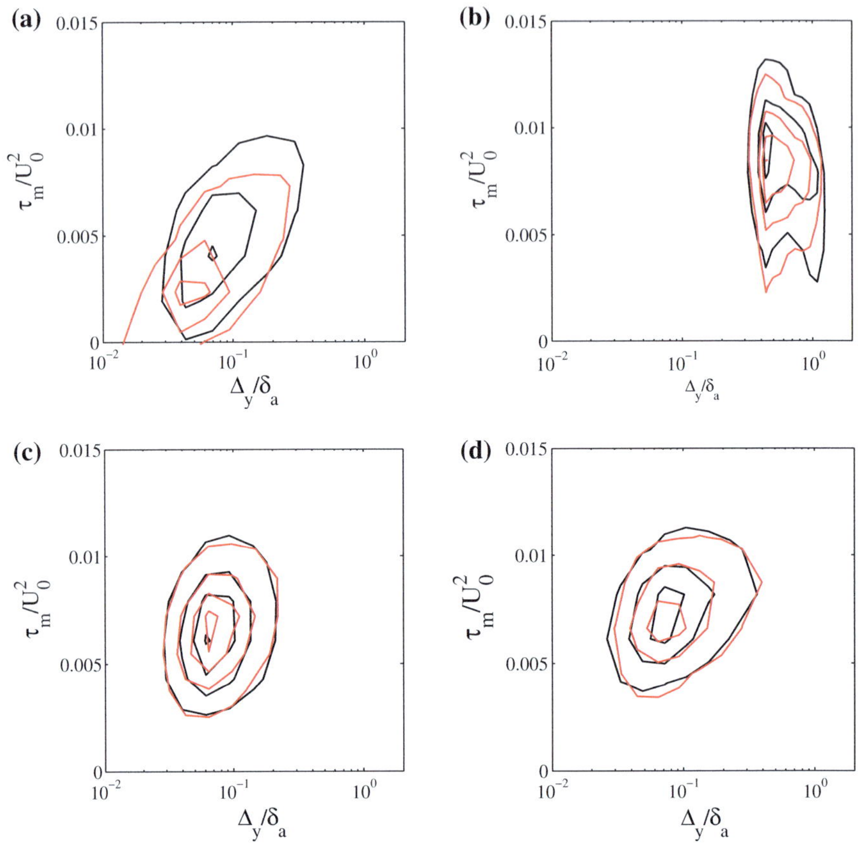

Fig. 4 Joint pdf of the logarithm of Δ_y/δ_a and τ_m/U_{oa}^2 for Q2s in *black* and Q4s in *red*. **a** and **c**: $y_c/\delta_a < 0.2$. **b** and **d**: $0.2 < y_c/\delta_a < 0.6$. **a** and **b**: Wall-attached Q$^-$s. **c** and **d**: Wall-detached Q$^-$s. Contour levels are 2, 12 and 22, except for **b**: 2, 22, 42

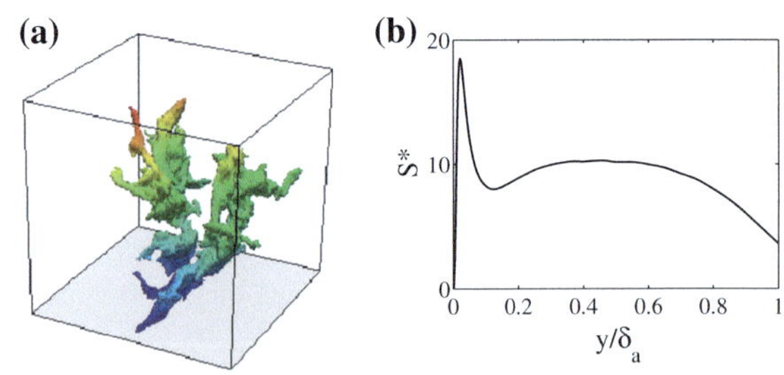

Fig. 5 **a** Instantaneous visualization of an attached structure. The box is $1\delta \times 1\delta \times 1\delta$. The flow is from left to right. **b** The averaged shear parameter as a function of the wall distance

Q4s. Because the detached Q$^-$s are numerous in the middle region (94 % of the total number of Q$^-$s) and because their values of τ_m are of the same order as those of attached Q$^-$s, these joint pdfs suggest that their contribution to the Reynolds shear

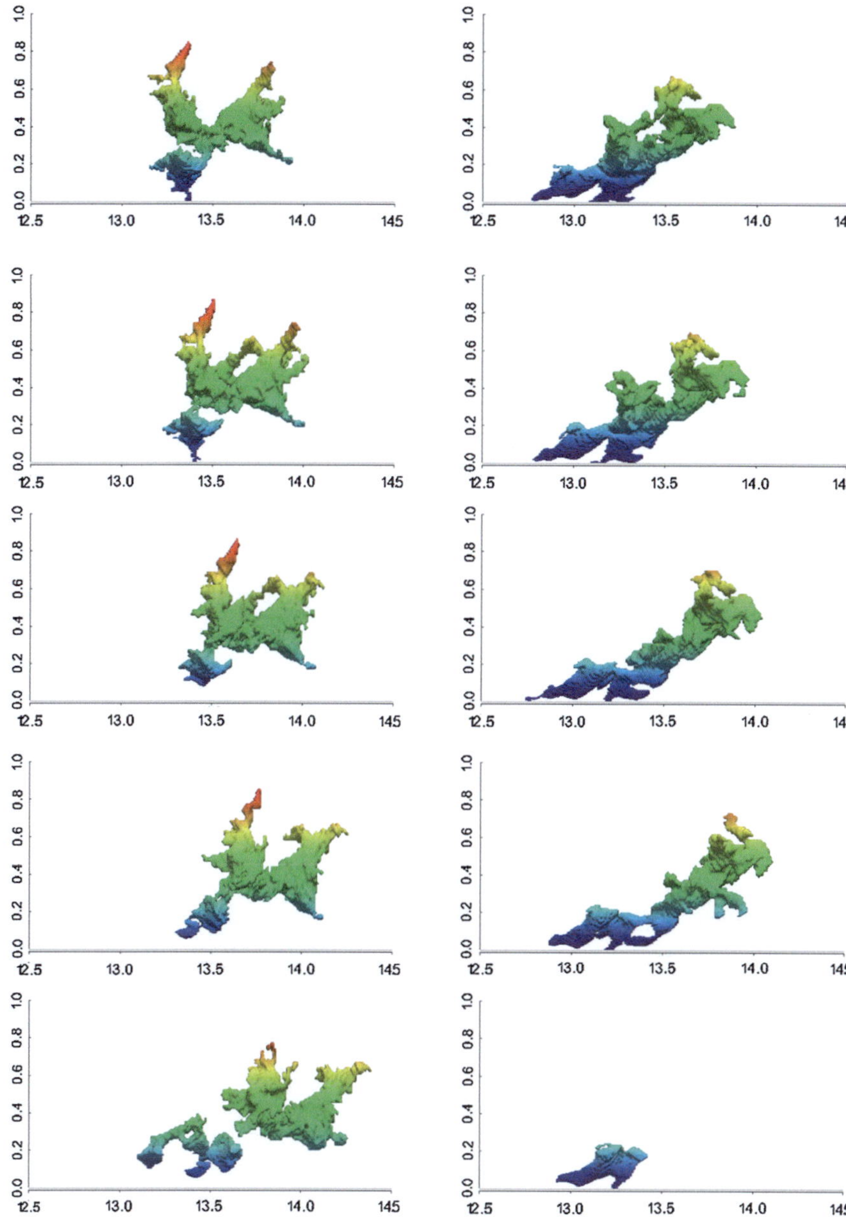

Fig. 6 Temporal evolution of a tall-attached structure. *Left column* Q4; *right column* Q2, with the axes scaled with δ_a. The structure is colored with the distance to the wall. Flow is from left to right. Time is from top to bottom

stress is more important than that of attached Q^-s, contrary to what was found by LFJ for channel flows.

It is recently becoming possible to study the multiscale character of turbulence through space and time-resolved evolution of turbulent flows [6]. The spatiotemporal evolution of an attached structure is studied using DNS sequences. The temporal separation among fields, $t^* = tU_{ea}/\delta_a \approx 0.15$ is short enough for individual objects to be tracked. There are few tall wall-attached structures that penetrate in the outer region and carry significant amount of tangential Reynolds stress, $-uv$, as shown in Fig. 5a. Figure 5b displays the average shear parameter, $S^* = Sq^2/\varepsilon$, which measures the ratio between the energy decay time and the shear deformation time, acting on this turbulent structure. The mean shear dominates, and controls the evolution of the turbulent scales if $S^* \gg 1$ [3].

The attached structure starts as side-by-side Q2 and Q4 structures as shown in Fig. 6. These initially wall-attached structures stay next to each other in the cross-stream direction in their lifetime. Note how different those structures look. The temporal evolution of the Q4 is shown in the left column. The Q4 structure detaches from the wall and splits into two. These new Q4s move toward the wall. The large coherent Q2 (shown in the right column) convects with the local mean streamwise velocity, is elongated in the streamwise direction and inclined with the wall. As seen from the temporal evolution, the Q2 merges with smaller detached objects in front of it, looses structures from behind, and moves upward from the wall. The split occurs at the end of the life of the structure. The lifetime of the Q2 is roughly equal to the mean flow deformation time.

3 Conclusions

Wall-attached and wall-detached Q^- structures are found everywhere in the present large-defect boundary layer. Wall-attached Q^-s are encountered less frequently, are smaller and are much less streamwise elongated that those found in turbulent channel flows by LFJ. Again in contrast to turbulent channel flows, detached Q^- structures are comparable in size and aspect ratio to attached Q^-s, and they carry a significant amount of the Reynolds shear stress. In the zone of maximum turbulence intensity, Reynolds shear stress and turbulent kinetic energy production of the present flow, approximately $0.2 < y/\delta < 0.6$, 94 % of the Q^-s are detached Q^-s. Moreover, in that region the detached Q^-s are as intense as the attached ones. The type of Q structures that contribute mostly to turbulence in the present large-defect boundary layer are therefore very different from those of canonical wall flows.

Acknowledgments Funded in part by the Multiflow program of the European Research Council. YM was supported in part by NSERC of Canada. We are grateful to A. Lozano-Durán for providing the data processing programs for the analysis of the Q structures.

References

1. M.S. Chong, J. Soria, A.E. Perry, J. Chacin, B.J. Cantwell, Y. Na, Turbulence structures of wall-bounded shear flows found using DNS data. J. Fluid Mech. **357**, 225–247 (1998)
2. A.G. Gungor, Y. Maciel, M.P. Simens, J. Soria, Analysis of a turbulent boundary layer subjected to a strong adverse pressure gradient. J. Phys.: Conf. Ser. **506**, 012007 (2014)
3. J. Jiménez, How linear is wall-bounded turbulence? Phys. Fluids **25**, 110814 (2013)
4. P.A.A. Krogstad, P.E. Skåre, Influence of a strong adverse pressure gradient on the turbulent structure in a boundary layer. Phys. Fluids **7**, 2014–2024 (1995)
5. J.H. Lee, H.J. Sung, Structures in turbulent boundary layers subjected to adverse pressure gradients. J. Fluid Mech. **639**, 101–131 (2009)
6. A. Lozano-Durán, J. Jiménez, Time-resolved evolution of the wall-bounded vorticity cascade. J. Phys.: Conf. Ser. **318**, 062016 (2011)
7. A. Lozano-Durán, O. Flores, J. Jiménez, The three-dimensional structure of momentum transfer in turbulent channels. J. Fluid Mech. **694**, 100–130 (2012)
8. Y. Na, P. Moin, Direct numerical simulation of a separated turbulent boundary layer. J. Fluid Mech. **374**, 379–405 (1998)
9. Y. Nagano, M. Tagawa, T. Tsuji, in *Turbulent Shear Flows 8*, ed. by F. Durst, R. Friedrich, B. Launder, F. Schmidt, U. Schumann, J. Whitelaw (Springer, Berlin, 1993), pp. 7–21
10. S. Rahgozar, Y. Maciel, Low- and high-speed structures in the outer region of an adverse-pressure-gradient turbulent boundary layer. Exp. Therm. Fluid Sci. **35**, 1575–1587 (2011)
11. S. Rahgozar, Y. Maciel, Statistical analysis of low- and high-speed large-scale structures in the outer region of an adverse pressure gradient turbulent boundary layer. J. Turbul. **13**, 1–24 (2012)
12. S.K. Robinson, Coherent motions in the turbulent boundary layer. Annu. Rev. Fluid Mech. **23**, 601–639 (1991)
13. M. Simens, J. Jiménez, S. Hoyas, Y. Mizuno, A high resolution code for turbulent boundary layers. J. Comput. Phys. **228**, 4218–4231 (2009)
14. M.P. Simens, A.G. Gungor, The effect of surface roughness on laminar separated boundary layers. J. Turbomach. **136**, 031014 (2013)
15. P.E. Skåre, P.A.A. Krogstad, A turbulent equilibrium boundary layer near separation. J. Fluid Mech. **272**, 319–348 (1994)

Adverse Pressure Gradients and Curvature Effects in Turbulent Channel Flows

A.B. de Jesus, L.A.C.A. Schiavo, J.L. Azevedo and J.-P. Laval

Abstract LES results are presented for the flow over different bump profiles installed on the bottom wall of a channel. Three bump geometries are considered in order to investigate the effects of adverse pressure gradients and changes in curvature on the mean flow and Reynolds stresses. The first bump geometry corresponds to a profile for which DNS results are available at a Reynolds number $Re_\tau = 617$ based on the channel inlet friction velocity. The other two geometries are generated by modifying the rear portion of the initial profile to produce a longer APG region and promote smoother curvature changes as compared to the original profile. LES computations are performed at the same Reynolds numbers than previous DNS, which is also used to validate LES results in the original bump. Finally, RANS results are also presented for the three bump profiles. RANS computations are performed using a two-equation eddy-viscosity Shear Stress Transport (SST) model and one Reynolds stresses transport model.

1 Introduction

Modeling of near wall turbulence in the presence of adverse pressure gradients (APG) is of crucial importance for many practical problems in aerodynamics. Direct numerical simulations (DNS) of turbulent flows around a two-dimensional bump located

A.B. de Jesus (✉) · L.A.C.A. Schiavo
Instituto Tecnológico de Aeronáutica, São José dos Campos, SP, Brazil
e-mail: antoniobjota@gmail.com

L.A.C.A. Schiavo
e-mail: augustoschiavo@gmail.com

J.L. Azevedo
Instituto de Aeronáutica e Espaço, São José dos Campos, SP, Brazil
e-mail: joaoluiz.azevedo@gmail.com

J.-P. Laval
CNRS, UMR 8107, F-59650 Villeneuve dAscq, France
e-mail: jean-philippe.laval@univ-lille1.fr

M. Stanislas et al. (eds.), *Progress in Wall Turbulence 2*,
ERCOFTAC Series 23, DOI 10.1007/978-3-319-20388-1_26

in the bottom wall of a channel were performed at Laboratoire Mécanique de Lille (LML) during the recent WALLTURB European project [8]. Those DNS were executed using the MFLOPS3D semi-spectral code for two Reynolds numbers, $Re_\tau = 395$ and 617, based on the friction velocity at the channel inlet. The results obtained showed a mild separation of the mean flow over the APG region at the rear portion of the bump. Strong peaks of turbulent kinetic energy which caused a sharp recovery of the friction coefficient and, consequently, flow re-attachment was identified in the separated flow region. A similar peak was found at the top wall in the APG region, although flow separation has not been observed. These peaks in turbulent kinetic energy at both walls, were found to be linked to instability of the near wall streaks which seemed to be promoted by adverse pressure gradients [6].

Large Eddy Simulations (LES) on the WALLTURB bump profile were also performed, with the same MFLOPS3D code, for Reynolds numbers starting at $Re_\tau = 617$, for which DNS results are available, and including also computations at $Re_\tau = 950$ and $Re_\tau = 2000$ [2–4]. LES results at $Re_\tau = 617$ are very consistent with DNS in terms of separation and reattachment locations, overall skin-friction coefficient distribution along the bump and Reynolds stresses evolution near the wall. Particularly, it was found that the strong peaks of turbulent kinetic energy identified in the APG region of the original DNS are also present in LES computations. Computations at $Re_\tau = 950$ also presented a mild separated flow region at the bottom wall, although shorter than in the previous $Re_\tau = 617$ simulation, while at $Re_\tau = 2000$ no mean flow separation was identified. Similar Reynolds stresses evolution is observed in both $Re_\tau = 950$ and 2000 simulations as compared with the $Re_\tau = 617$ calculations, including similar strongs peaks of turbulent kinetic energy in the APG region. These results show that the streaks instability modes found in the original DNS at $Re_\tau = 617$ are also present at higher Reynolds numbers and even in fully attached flow conditions ($Re_\tau = 2000$).

RANS computations on the original WALLTURB bump profile were also performed for the same range of Reynolds numbers, $Re_\tau = 617$, 950 and 2000 [2, 3]. A commercial CFD code (Ansys Fluent) was used with two-equation eddy-viscosity turbulence models, namely a realizable k-ε model [10] and the SST model [9], and one Reynolds stresses transport model [5]. Overall it was found that none of the RANS models considered were able to reproduce the friction coefficient along the rear portion of the bump neither the peaks of turbulent kinetic energy in the APG region as observed in DNS and LES results. It was noticed that RANS models formulation did not include the streaks instability mechanisms in their original formulation and therefore could not reproduce the results from DNS or LES.

Current work aims at continuing the previous investigations based on LES computations using the MFLOPS3D code. In the present study, the effects of modifications applied to the geometry of the rear portion of the bump on the velocity profiles and Reynolds stresses evolution are investigated. Two new bump profiles are considered based on modifications of the original WALLTURB bump geometry. The rear portion of the original bump is elongated to produce a longer APG region and promote smoother curvature changes as compared to the original profile. LES results from the MFLOPS3D code are presented for the original and the two modified bump

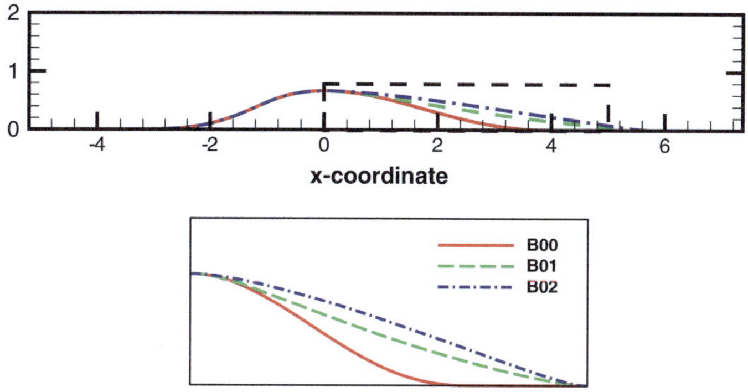

Fig. 1 Computational domain with the three bump profiles (*inset*: zoom of the divergent part)

geometries in order to investigate the resulting effects in the velocity profiles over the bump and the evolution of the Reynolds stresses in the APG region. RANS results for one eddy-viscosity model and one Reynolds stresses transport model are also presented.

2 Geometric and Physical Models

The different bump geometries used in this work are presented in Fig. 1. As it can be noticed three bump profiles are identical up to the summit. The first modified bump, named as B01, consists in elongating the rear portion of the Wallturb bump, B00, maintaining curvature continuity in both ends. The second modified profile, B02, also has an elongated rear portion, as compared to B00, but in that case only tangency continuity is maintained at the summit. In B02 the point of curvature changes to accommodate the flat surface is moved downstream.

3 Large Eddy Simulations

Large Eddy Simulations are performed using the MFLOPS3D semi-spectral code. The Navier-Stokes equations for an incompressible flow are discretized using fourth-order finite differences in the streamwise direction, while Chebyshev polynomials are employed in the normal direction. A Fourier transform is performed in the spanwise direction. Time integration is performed using an implicit second-order backward Euler method for the terms containing the Cartesian components of the Laplacian operator, while an explicit second-order Adams-Bashforth method is used for all other terms, including the sub-filter stresses.

Table 1 Mesh Parameters for LES as compared to the original DNS computations

...	Re_τ	Nx	Ny	Nz	Δx^+	Δy^+	Δz^+
LES	617	512	129	128	21.8	[0.17–123.5]	21.8
DNS	617	2304	385	576	5.7	[0.01–3.8]	3.8

Note: Values of Δx^+, Δy^+ and Δz^+ are computed at the summit of the bumps

Further details of the transformed equations and numerical methodology can be found in previous work which uses the MFLOPS3D code [7]. Boundary conditions for the inlet flow are obtained from a precursory periodic channel flow computation. No-slip conditions are imposed at the top and bottom walls, and the spanwise direction is assumed to be 2π periodic with a symmetry at π. The outflow boundary of the computational domain is treated as a convective boundary with uniform velocity. The Wall Adapting Local Eddy viscosity (WALE) subgrid scale model is used in all computations [2].

Computations are performed for a $Re_\tau = 617$ Reynolds number based on the friction velocity at the channel inlet. Table 1 brings the mesh parameters used, including the number of points in each direction and the grid spacing in wall-units at the top of the bumps. The same mesh resolution is used in all LES computations.

3.1 Wallturb Bump—LES and DNS Results

Figure 2 brings comparisons between DNS [8] and LES results for B00 profile in terms of friction coefficient and pressure coefficient along the bump. A very good agreement between LES and DNS is observed, particularly in capturing separation and reattachment locations. Figure 3, present the evolution of the tangential Reynolds

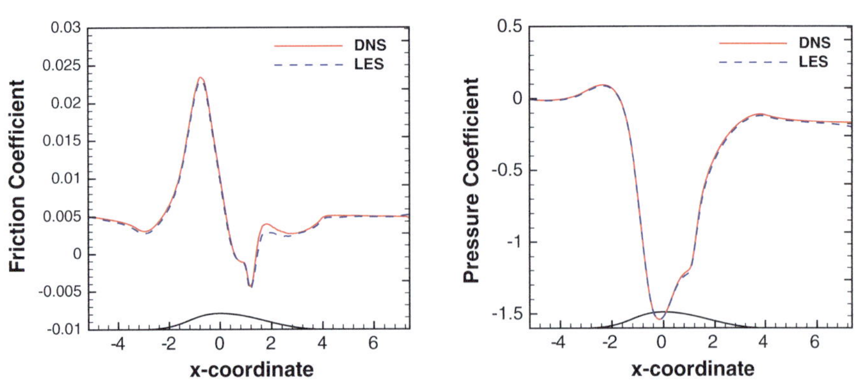

Fig. 2 B00—friction coefficient and pressure coefficient—DNS versus LES

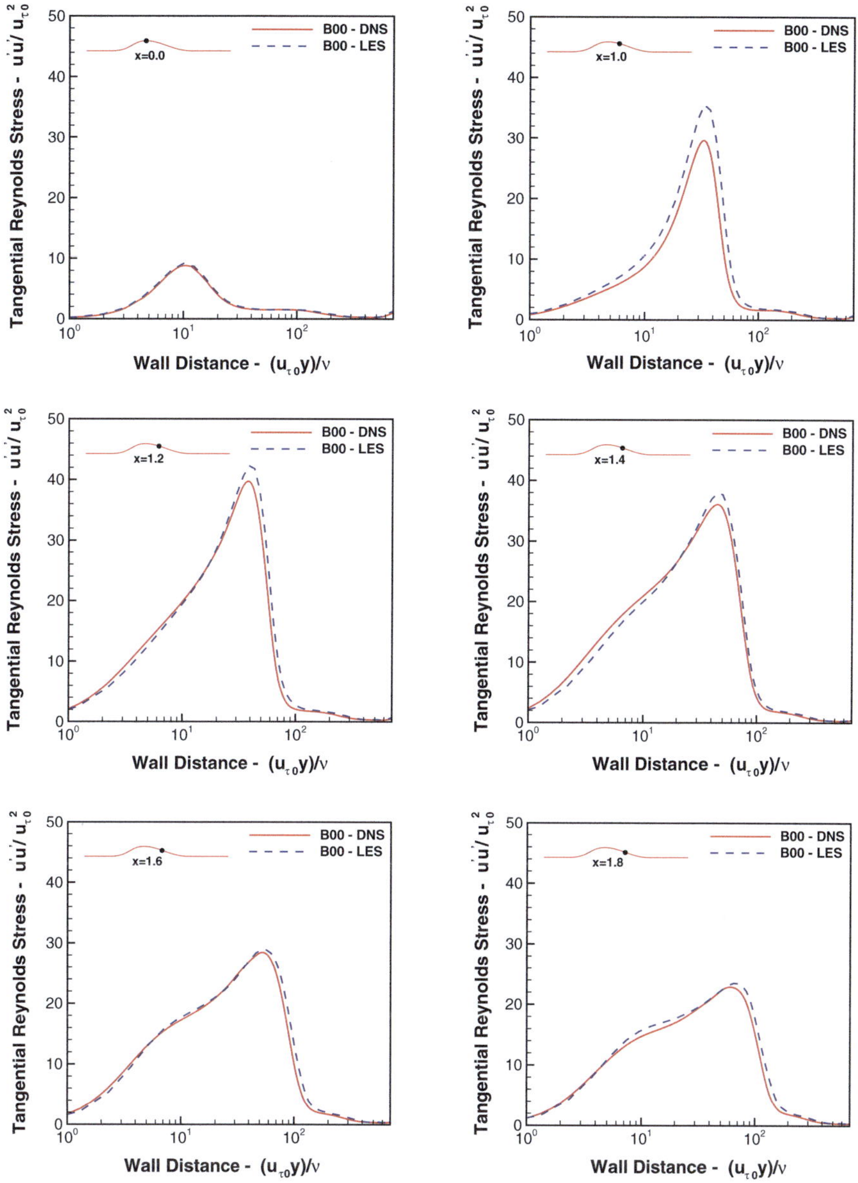

Fig. 3 Wallturb bump—tangential reynolds dtress—DNS versus LES

stress in the adverse pressure gradient region. The Reynolds stress values are presented along lines normal to the bump profile as a function of a normalized distance to the wall. Both DNS and LES results clearly show that the initial near wall Reynolds stress peak tends to move away from the wall and strongly increases in magnitude in the APG region (see stations $x = 1.0$ and $x = 1.2$) and turns to a two peaks

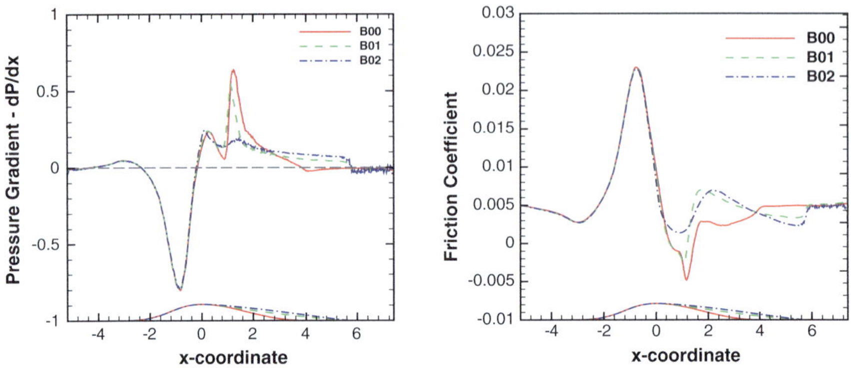

Fig. 4 LES—pressure gradient and friction coefficient—B00, B01 and B02

profile further downstream, as observed in stations $x = 1.4$, $x = 1.6$, and $x = 1.8$. As already noticed in previous reference [2–4], LES results are very similar to the DNS ones.

3.2 Modified Bumps—LES Results

Pressure gradient and friction coefficient obtained from LES with three bump configurations are shown in Fig. 4. The favorable pressure gradient in the converging part are very similar for the three configurations. In the adverse pressure gradient region, B00 shows one initial peak followed by a steep increase and a second peak which decreases until the flat surface region. B01 shows an adverse pressure gradient distribution with an initial peak similar to the one with B00 but with a smaller and shorter second peak and a slower decrease to zero-pressure gradient. This is a consequence of the elongation of the rear portion of the original bump maintaining curvature continuity at the top. The APG region of B02 shows a similar initial peak but slightly displaced upstream when compared to the two other cases. The rear portion of B02 is longer than in the other two bumps and tangent continuity is maintained a the top of the bump. As a consequence, the second peak of adverse pressure gradient is much smaller than in the other configurations with a relatively higher APG level compared to B01 when moving toward the flat wall region. The friction coefficient for the three bumps can be directly correlated to the pressure gradients being very similar in the favorable pressure gradient region and in the beginning of the APG region up to station $x = 0.6$. Further downstream, both the B00 and B01 show flow separation followed by a strong recovery of the friction coefficient. Due to the smaller and shorter APG second peak in B01, the separated flow region is smaller than in B00 and the recovery value of the friction coefficient is higher. The evolution of its level toward the flat region is, of course, slower in B01 due to the longer rear portion. In B02 the second APG peak is delayed and much smaller than

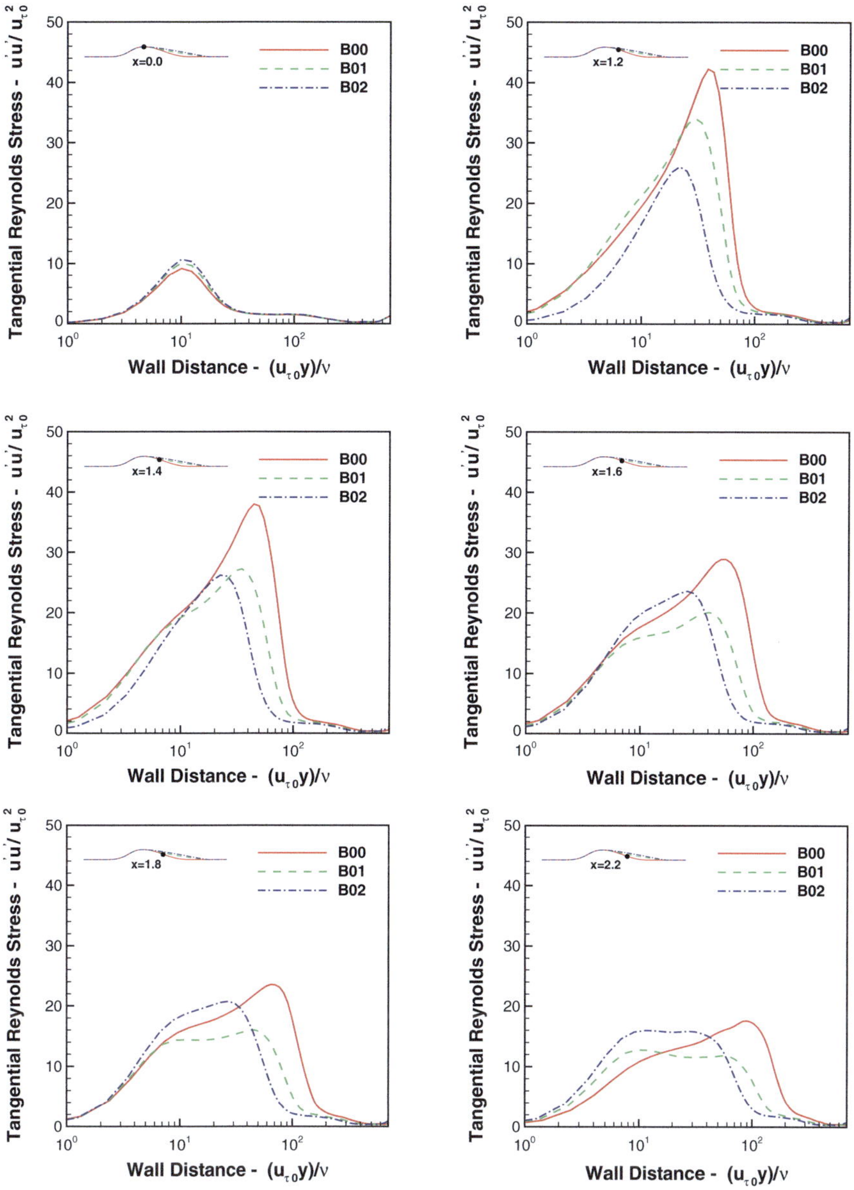

Fig. 5 Wallturb bump—tangential reynolds stress—LES—wallturb bump, Bump B01 and Bump B02

in the other two bumps. Therefore, flow does not separate and the recovery of the friction coefficient takes longer than for the other cases. As the pressure gradient is higher in the back region of B02, the friction coefficient evolves to a relatively lower values than for B02 before reaching the flat plate.

The wall-normal evolution of the tangential Reynolds stresses in the APG region of the three bumps are presented in Fig. 5. Similar behaviors are observed in all three configurations. A near wall peak is noticed at the top of the three bumps. This peak strongly increases in magnitude and is displaced away from the wall when moving downstream in the APG region. Further downstream, the single peak evolves to a double peak profile. The magnitude of the first peak and the extension along the bump until the development of the double peak are related to the adverse pressure gradient. The B01 case shows a similar evolution of the Reynolds stresses as compared to B00 but with a smaller increase of the first peak which also remains closer to the wall. This seems to be related to the smaller pressure gradient peak observed in B01. The formation and evolution of the second peak is very similar in both bumps up to station $x = 1.8$. Further downstream, the second peak is larger and closer to the wall in B01 than in B00. This can be associated to the fact that in B01 the adverse pressure gradient is sustained in a longer region when compared to the original bump. In B02, as the APG values are much lower than in the other bumps, the increase of the first tangential Reynolds stress peak is smaller and the formation of the second peak occurs earlier.

Overall it can be seen that the three bump configurations show very similar patterns. It is concluded that the streaks instability mechanism which was identified in the DNS results is also present in these new bumps with different intensities depending on the APG levels of each case. Particularly, it is observed that in both separated or mildly separated flows, as in B00 and B01, or in fully attached flows, as in B02, the same behavior is identified.

4 RANS Simulations

RANS calculations are performed using the Ansys Fluent v.14.0 commercial CFD package. A 2-equation Shear Stress Transport (SST) [9] turbulence model and one Reynolds stress transport model (RSM) [5] are compared. The steady-state, two-dimensional, incompressible pressure-based SIMPLE solver is used with the default settings of the Fluent package. At the walls, the RSM calculations use the so-called enhanced wall treatment option, which employs a two-layer formulation for the near wall region. Hence, in this region, the length scales used to compute the turbulent viscosity and the turbulent dissipation (ε) are adjusted to take into account the effects of the wall proximity [1]. In addition, the RSM model specifies values for the Reynolds stresses, in the first cell off the wall, through the use of the log-law and the assumption of equilibrium [1].

Similarly to the LES computations, a precursory periodic channel flow is run to generate inlet conditions that are used for the calculations with the bump in the channel. No-slip conditions are imposed at the walls and a constant pressure is assumed at the outlet. A two-dimensional mesh with 400×100 quadrilaterals in the streamwise and wall-normal directions, respectively, is used for the RANS calculations over the bump under study. The centroid of the first off-the-wall cell is placed at 0.25 mm

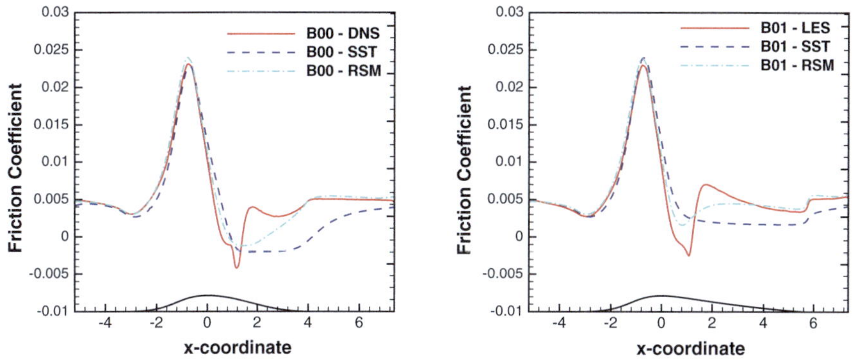

Fig. 6 Wallturb bump—friction coefficient and pressure gradient—RANS, DNS/LES—B00 and B01

from the wall, which approximately corresponds to a maximum $y^+ = 0.3$ for the present calculations.

Figure 6 brings comparisons between DNS or LES and RANS computations of the friction coefficient along the bottom wall of B00 and B01 profiles. Up to the summit it can be argued that both RANS models under study present similar results with good comparison to reference DNS or LES. Further downstream in the APG region a different picture is observed. None of the RANS models reproduces the sharp decrease followed by a rapid increase of the skin-friction coefficient in the APG region. In B00, both models predict flow separation although the separation point is located more downstream than what is predicted by DNS and no model predicts the lower peak of skin friction. The SST model shows a large separation region with a constant negative C_f which starts recovering only downstream of the bump, at the flat surface. The RSM model shows a smaller separation region, when compared to SST, but still much larger than DNS prediction. The recovery of C_f is more rapid than with SST model and just downstream of the bump the RSM model predicts a skin friction level similar to the DNS one. In B01 none of the RANS models predict flow separation but an overall behavior similar to B00 is observed. SST leads to a constant low, but positive, skin friction in the rear portion of the bump followed by a small increase downstream. RSM model shows a faster response of the skin friction in the rear portion of the bump, with a better comparison to LES than the SST model. Overall it can be concluded that, although none of the RANS models correctly predicts the skin-friction evolution in APG region, the RSM model presents a better performance when compared to SST.

Figure 7 brings a comparison of DNS and RSM evolution of the tangential component of the Reynolds stresses along the APG region for B00. It is noticed that initially, in the APG region, the RSM model is able to capture the evolution of the first Reynolds stress peak at stations x = 0.0 and x = 1.0. Further downstream, however, the RSM model does not predict the formation of the second near wall peak, associated to the instability of near wall streaks and which can be associated with the

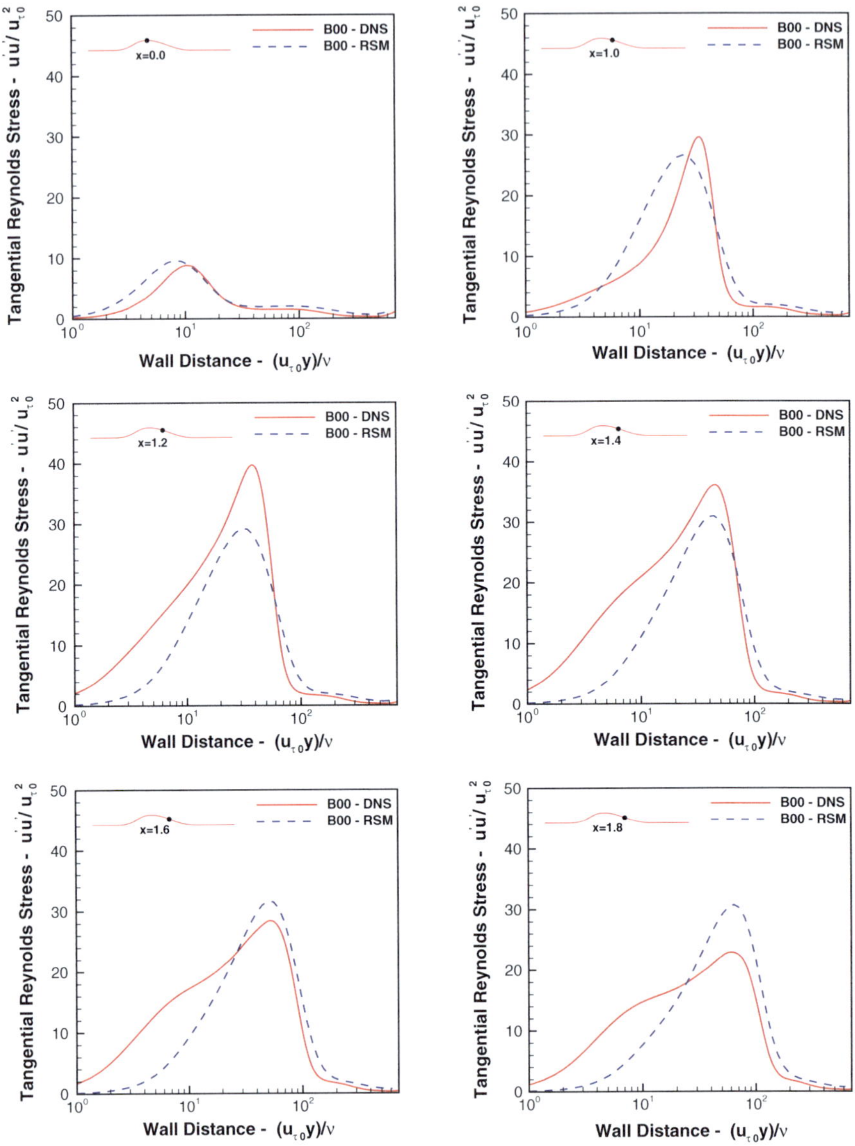

Fig. 7 Wallturb bump—tangential reynolds stress—DNS, RANS—wallturb bump

slower recovery of skin friction observed in RSM results when compared to DNS (see Fig. 6). It can be concluded that the streaks instability mechanism and the formation of a second near wall peak of tangential Reynolds stresses which is not taken into account in the formulation neither of standard two-equation RANS models nor the more sophisticate RSM model maybe responsible of the low performance of the models in the APG region.

5 Conclusions

Large eddy simulations were performed to study the effects of modifications applied to the geometry of the rear portion of two-dimensional bumps on the velocity profiles and Reynolds stresses evolution along the adverse pressure gradient region. Simulations performed for a bump corresponding to available DNS results were complemented by simulations with two modified bump geometries. LES results with initial geometry were found to produce accurate predictions when compared to DNS, particularly, the formation of a second near wall peak in the tangential component of Reynolds stresses which had been associated with instability of the near wall streaks and caused a sharp increase of friction coefficient in the APG region. Simulations of the modified bump profiles presented similar results to the original bump even though the modified configurations provided smoother profiles, with lower pressure gradients and either mild or no flow separation. These results indicate that the streaks instability mechanism is present in a broader class of APG flows. Then, RANS simulations were also performed for the three cases using both standard two-equation model (SST) and RSM models. It is found that none of these models are able to predict the flow characteristics in the near wall APG region. This lack of performance may be associated to the disability of the models to take into account in their formulation the physics associated to the streaks instability.

Acknowledgments The authors gratefully acknowledge the partial support for this research provided by Conselho Nacional de Desenvolvimento Científico e Tecnológico, CNPq, under the Research Grant No. 309985/2013-7. This work is also supported by Fundação de Amparo à Pesquisa do Estado de São Paulo, FAPESP, through a M.S. scholarship for the second author under the Research Grant No. 2013/12015-2. J-P Laval gratefully acknowledge the support of International Campus on Safety and Intermodality in Transportation, the European Community, the Regional Delegation for Research and Technology, the Ministry of Higher Education and Research, and the National Center for Scientific Research. The computational resources provided by Centro Nacional de Supercomputação, CESUP, of Universidade Federal do Rio Grande do Sul, in Porto Alegre is also gratefully acknowledged.

References

1. ANSYS, Inc., ANSYS FLUENT Theory Guide Release 14.0 (2011)
2. A.B. Jesus, J.L.F. Azevedo, J.P. Laval, Large eddy simulations and RANS computations of adverse pressure gradient flows, in *AIAA Paper No. AIAA 2013-0267, 51st AIAA Aerospace Sciences Meeting including the New Horizons Forum and Aerospace Exposition*, Grapevine, TX (2013)
3. A.B. Jesus, L.A.C.A. Schiavo, J.L.F. Azevedo, J.P. Laval, An assessment of attached and mildly separated flows in adverse pressure gradient regions, in *AIAA Paper No. AIAA 2014-0583, 52nd AIAA Aerospace Sciences Meeting including the New Horizons Forum and Aerospace Exposition*, National Harbor, Maryland (2014)
4. L. Kuban, J.P. Laval, W. Elsner, A. Tyliszcak, M. Marquillie, Les modeling of converging-diverging turbulent channel flow. J. Turbul. **16**(11), 1–19 (2012)

5. B.E. Launder, N. Shima, Second-moment closure for the near-wall sublayer: development and application. AIAA J. **27**(10), 1319–1325 (1989)
6. J.P. Laval, M. Marquillie, U. Ehrenstein, On the relation between kinetic energy production in adverse-pressure gradient wall turbulence and streak instability. J. Turbul. **13**, 1–19 (2012)
7. M. Marquillie, J.P. Laval, R. Dolganov, Direct numerical simulation of separated channel flows with a smooth profile. J. Turbul. **9**(1), 1–23 (2008)
8. M. Marquillie, U. Ehrenstein, J.P. Laval, Instability of streaks in wall turbulence with adverse pressure gradient. J. Fluid Mech. **681**, 205–240 (2011)
9. F.R. Menter, M. Kuntz, R. Langtry, Ten years of industrial experience with the SST turbulence model, *Turbulence, Heat and Mass Transfer*, vol. 4 (Begell House, Inc., Redding, 2003)
10. T.H. Shih, W.W. Liou, A. Shabbir, Z. Yang, J. Zhu, A new $k-\varepsilon$ eddy-viscosity model for high Reynolds number turbulent flows—model development and validation. Comput. Fluids **24**(3), 227–238 (1995)

On the Response of a Separating Turbulent Boundary Layer to High Amplitude Excitation

Vitali Palei and Avi Seifert

Abstract Experimental results of turbulent separating boundary layer, subjected to nominally 2D pulsed excitation, are presented and discussed. The effect of the adverse pressure gradient on the vortices circulation and convection speed has been documented. A search for instability mechanism did not result in any that were amplified. Therefore, pulsed excitation that intermittently enhances the skin friction with optimal time lag should be explored.

1 Introduction

Separation of turbulent boundary layers is an important engineering subject due to its mostly negative effect on fluid related system and a lasting scientific challenge [1, 2]. Control of turbulent boundary layers is, therefore, of great practical importance and is extensively studied on airfoils and generic configurations [3, 4]. While turbulent, free-shear flows were found to possess unstable modes [5], turbulent wall-bounded flows are known to be stable [6]. The current study was initiated with an attempt to identify flow instability in search of an efficient AFC methodology, assuming that instability would amplify the introduced excitation modes. However, not unexpectedly, no such instability mechanism was identified, so the question remains: What are the appropriate time scales for efficient excitation in such situations? In this paper, evaluation of temporal and spatial scales are presented and discussed. Following excitation of the boundary layer, subjected to local adverse pressure gradient, with high frequency and amplitude modulated excitation were used.

Many similar experiments were performed on airfoils and on generic configurations, dealing with the response to excitation [7–9] and more. However, no study known to us linked the change in global response to the local boundary layer features

V. Palei (✉) · A. Seifert
School of Mechanical Engineering, Tel-Aviv University, Tel-Aviv, Israel
e-mail: vitali.palei@gmail.com

A. Seifert
e-mail: seifert@eng.tau.ac.il

M. Stanislas et al. (eds.), *Progress in Wall Turbulence 2*,
ERCOFTAC Series 23, DOI 10.1007/978-3-319-20388-1_27

and the excitation parameters, especially, the actuator created vorticity and circulation. Moreover, most studies where performed in deep stall, where a free-shear layer was naturally present and controlled. In our case, a rather short region of incipiently separated flow was controlled and detailed PIV study was performed on the interaction of the excitation with the turbulent boundary layer. Detailed data analysis was performed to quantify effects. The structure of the paper is as follows. Next we present the experimental details. It is followed by presentation of the experimental results and modeling effort. Finally a discussion and conclusions are offered.

2 Experiment Setup

The experiments were conducted in a small size, open-loop wind tunnel [10]. The test section was 50 (mm) high at the entrance and 150 (mm) wide throughout. Its total length is about 1 (m). See Fig. 1.

The measurements were conducted near the flat, lower surface of the test section. To eliminate pressure gradient due to boundary layer growth, the test section height was increased from 50 (mm) at the entrance to 78 (mm) at the outlet over a distance of about 970 (mm) (See Fig. 1).

To regulate transition the boundary layer was tripped at $X = 140 - 175$ (mm) from the entrance to the test section. Velocity profiles were measured by a hot-wire mounted on a 3D traverse system. The sensor was placed on a long holder that penetrated from the outlet. Velocity profiles measured along the test section centerline enabled the calculation of the displacement of momentum thicknesses as well as the shape factor, as shown in Fig. 2a.

At the entrance to the test section the shape factor is about 2.4, corresponding to laminar boundary layer with a slight acceleration. At X close to 600 (mm), both natural and tripped boundary layers reach $H = 1.4$, corresponding to low Reynolds number turbulent state. Roughness accelerates and regulates transition such that turbulent state prevails for $X > 400$ (mm). Turbulent velocity profiles are plotted in wall coordinates in Fig. 2b. A clear logarithmic region can be seen. Even the viscous sublayer could be identified at the proximity of the wall. In order to impose adverse pressure gradient without curvature, suction was applied (see Fig. 1) opposite to the test wall. A porous section was positioned between $540 < X < 580$ (mm),

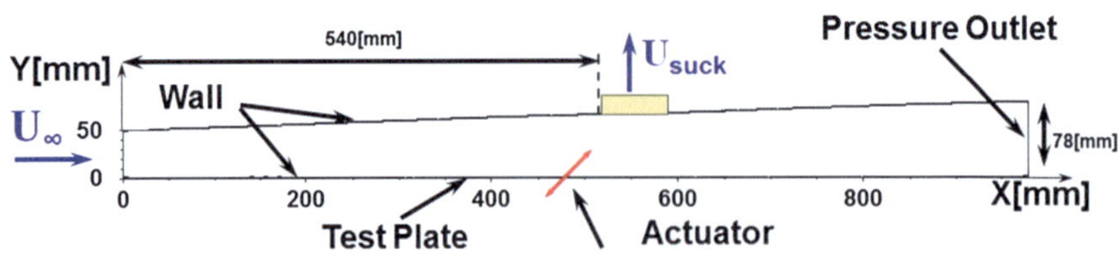

Fig. 1 Test section dimensions, width is 150 (mm), $X_p = 530$ or 560 (mm)

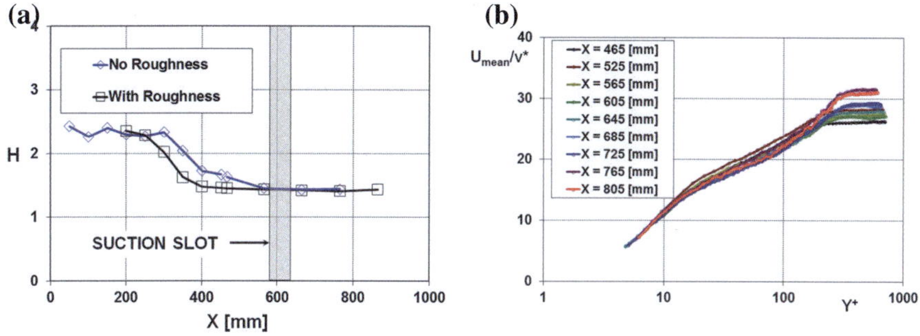

Fig. 2 **a** Velocity profiles shape factor along the test plate. Roughness placed at $X = 140 - 175$ (mm), $U_\infty = 7$ (m/s), $U_s = 0$. **b** Turbulent velocity profiles plotted in wall coordinates, with roughness, $U_\infty = 7$ (m/s), $U_s = 0$

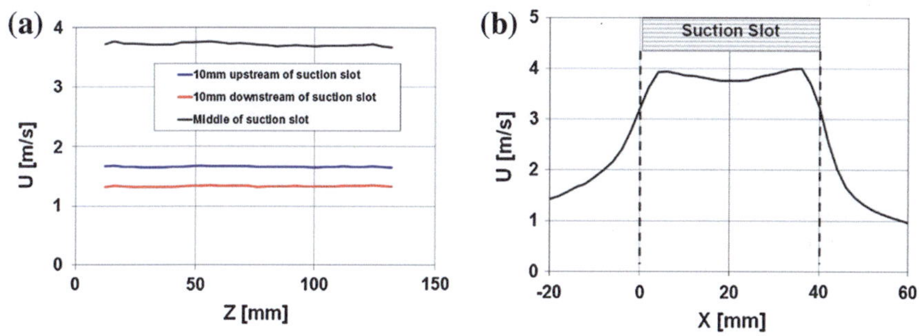

Fig. 3 Suction velocity as measured by a hot-wire placed 1[mm] *below* the suction panel, $U_\infty = 0$ (m/s), $U_s = 4$ (m/s). Suction velocity increased to reduce measurement uncertainty

and an ejector created various degrees of suction in the cavity. Currently, we only consider suction level of $U_s = 1.9$ (m/s), with a free-stream velocity of $U_\infty = 7$ (m/s) and the no suction case. Figure 3a presents the Z and X dependence of the suction velocity measured with a hot-wire 1 mm below the porous plate with $U_\infty = 0$ and $U_s = 4$ (m/s). It can be seen that the suction velocity is Z independent, and is clearly localized in terms of X. Figure 4a present the static pressure and its gradient along the test section (X). It can be seen that the open suction panel has an effect on the pressure gradient even without the ejector operating, but a weak one. With $U_s = 1.9$ (m/s) (Fig. 4b), a positive pressure gradient with a peak of 60 (Pa/m) is created around $X = 560$ (mm). Note that the stagnation pressure at 7 (m/s) is only 30 (pa).

Mean and turbulence velocity profiles and Preston tubes data (referenced to the wall static pressure) can be seen in Figs. 5 and 6. A conventional velocity profiles can be seen for the $U_s = 0$ case, with a clear turbulent activity peak at $Y < 1$ (mm) along the entire test section (X of each profile is indicated below each profile. Above the profile, we plotted the difference between the Preston and the static pressure tubes measured at the same X for each pair. For the $U_s = 0$ case we can see about 2 (Pa)

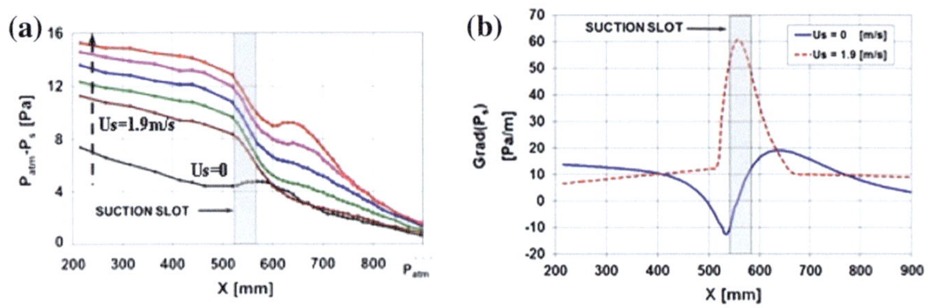

Fig. 4 The static pressure along the test section with increasing opposite wall suction velocity (*left*), and the pressure gradient along the test section for $U_s = 0$ and 1.9 (m/s). $U_\infty = 7$ (m/s). The *Gray* regions indicate $540 < X < 580$ (mm) range

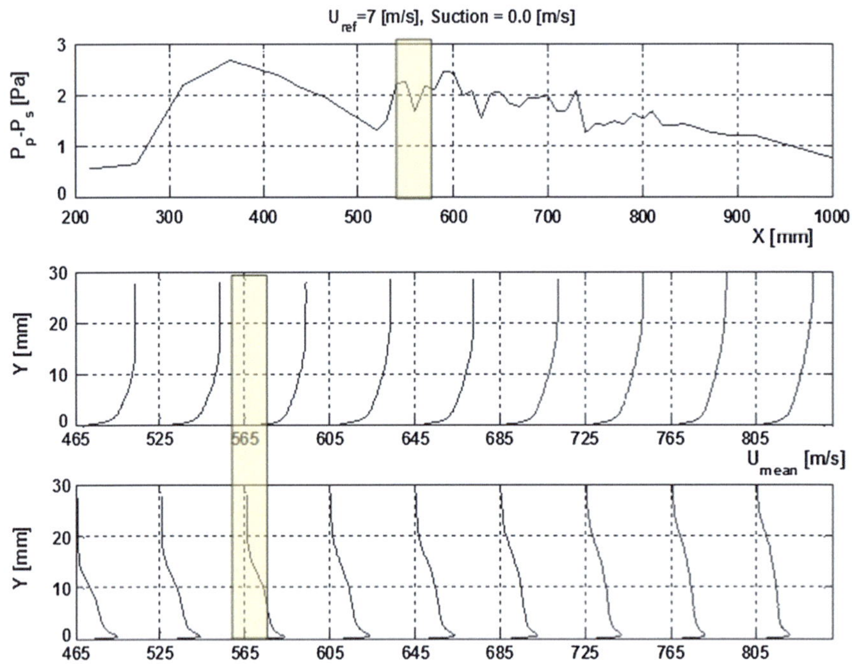

Fig. 5 Pressure difference between Preston and Static pressure ports along the test wall, mean, and fluctuating velocity profiles versus X, $U_s = 0$. Suction region marked by the lightly *colored rectangle*

higher Preston tube pressure, than static pressure, indicating attached flow. Figure 6 presents similar data as Fig. 5, only for $U_s = 1.9$ (m/s) test condition. It is evident that the near wall velocity gradient decreased, the turbulence peak moved away from the wall, and a secondary peak develops at Y 10 (mm). The pressure difference between the Preston and static pressure ports decreased to below 1 (Pa) from the suction panel

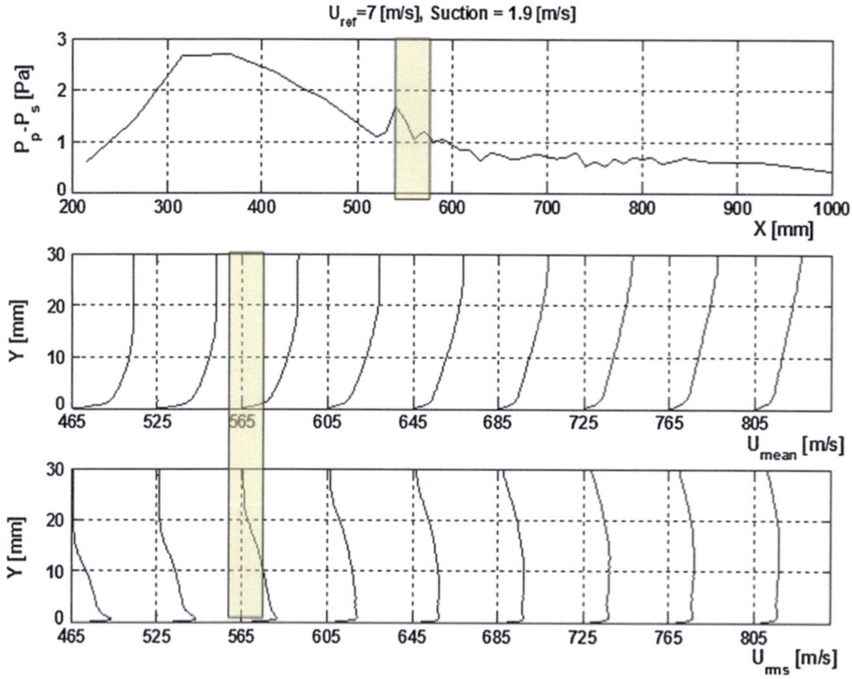

Fig. 6 Pressure difference between Preston and Static pressure ports along the test wall, mean and fluctuating velocity profiles versus X, $U_s = 1.9\,(\text{m/s})$. Suction region marked by the lightly *colored rectangle*

and downstream, indicating separating boundary layer. It should be mentioned that no mean reverse flow was measured by the PIV for this U_s magnitude.

In Fig. 7a we present velocity spectra at one X station with and without opposite wall suction. Both spectra were measured at Y corresponding to the peak turbulent activity close to the wall. Both spectra contain a region with $-5/3$ slope and well-behaved turbulent activities at all scales with no dominant peaks.

Actuation was imparted on the turbulent boundary layer by a Piezo-fluidic zero-mass-flux actuator, spanning the entire 150 (mm). The slot of the test section width was 1.0 (mm) and it ejected the periodic excitation at an angle of 30° with respect to the test plate, pointing downstream. The actuator is identical to the one used in Refs. [10] and [11]. A uniform cavity pressure oscillation was achieved by 6 Piezo actuators. An unsteady pressure sensor was installed inside the cavity, on its floor opposite the slot. A calibration of the peak exit velocity versus the cavity fluctuating pressures Fig. 7b was performed using a single calibrated hot-wire, positioned at the slot exit at three span locations. Figure 7b shows the peak exit velocity versus the cavity pressure RMS. Clearly, the excitation is quite 2D and shows the typical non-linearity of the slot [12].

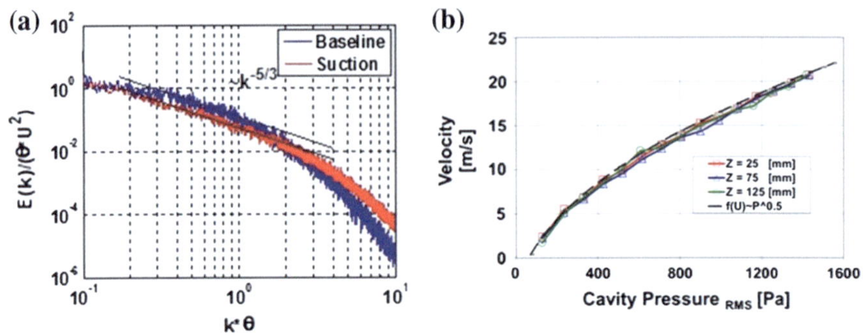

Fig. 7 **a** Turbulent boundary layer velocity spectra at Y of peak velocity, $X = 665\,(\mathrm{mm})$, $U_\infty = 7\,(\mathrm{m/s})$, $U_s = 0$, and $1.9\,(\mathrm{m/s})$. **b** Actuator calibration, slot peak velocity versus cavity pressure fluctuations level

Two types of excitation signals were used during the current study. A pure sine excitation created a steady-state high frequency actuation at the actuator's Helmholtz frequency, 1060 (Hz). In order to create low excitation frequency amplitude modulation (AM discussed later) actuations were used.

3 Discussion of Results

The effect of the pure sine excitation of the separating TBL can clearly be seen in Fig. 8a. The chart presents phase-locked velocity vectors and vorticity contours. The mean boundary layer vorticity is negative and it is significantly enhanced by the periodic excitation. The "positive" vortices, with rotation opposite to the mean boundary layer shear direction, extend well outside the boundary layer. It should be noted that the interrogation region extends from about 4 boundary layer displacement

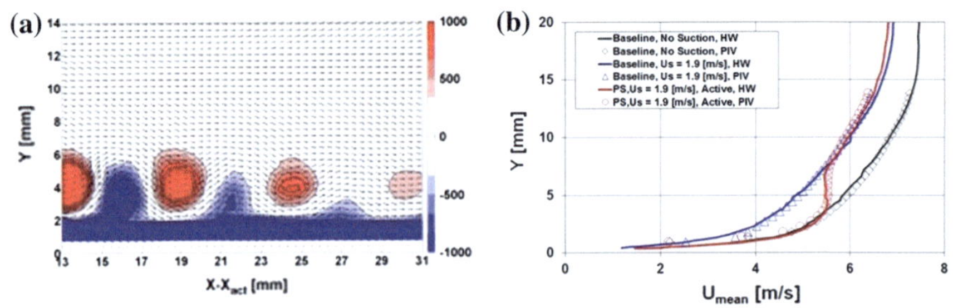

Fig. 8 **a** Velocity vectors and vorticity contours for PS excitation, $U_\infty = 7\,(\mathrm{m/s})$, $U_s = 0$, *PS* excitation, $U_p = 20\,(\mathrm{m/s})$. **b** A comparison of mean velocity profiles measured by either hot-wire (HW) or by 2D-PIV (PIV). Three conditions as indicated in the legend, Active: $U_p = 15.9\,(\mathrm{m/s})$

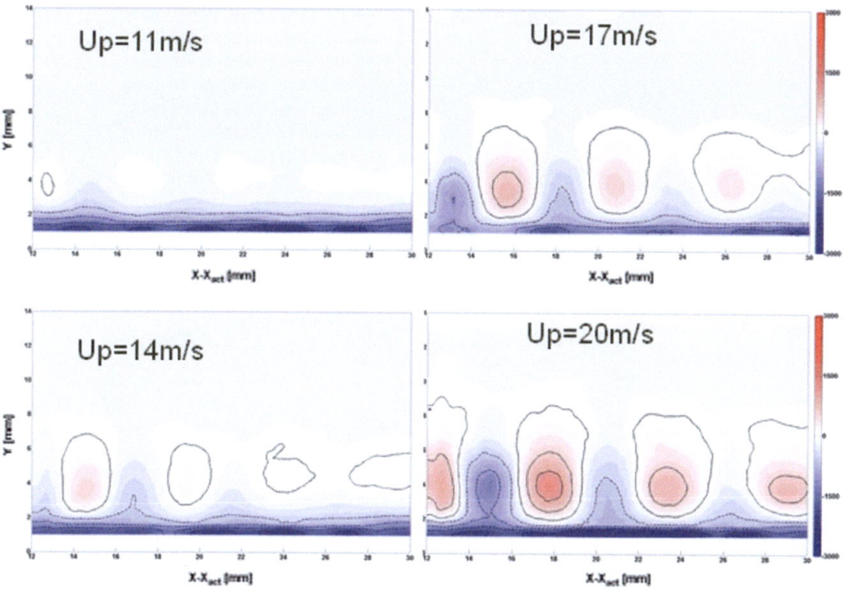

Fig. 9 Vorticity contours of four PS excitation cases, U_p indicated on charts. $X_{act} = 530\,(\text{mm})$, $U_\infty = 7\,(\text{m/s})$, $U_s = 1.9\,(\text{m/s})$, $f = 1060\,(\text{Hz})$

thicknesses downstream of the slot, to a distance of about 14 boundary layer displacement thicknesses.

In Fig. 8b we compare hot-wire and PIV data taken at the same flow conditions. It could be seen that the opposite wall suction significantly reduces the skin friction and the PS excitation "fills" the near wall velocity deficit. The agreement between the PIV and HW data is very good, slightly deteriorating at the first 1 (mm) from the wall. The effect of increasing the magnitude of the PS excitation can be seen in Fig. 9. The U_p is indicated on each plot. It can clearly be seen that as the excitation magnitude decreases, the phase-locked vorticity increases and stronger "positive" vortices form away from the wall.

Further analysis of the vortices' features was conducted. Vortices' centers and borders were identified, its streetwise location identified, from which convection speeds and circulation were calculated. Figure 10 presents the "positive" vortices' circulation from the entrance to the exit from the interrogation region. Data is presented for four amplitude of PS excitation and the two pressure gradient conditions tested, i.e., $U_s = 0$ and $U_s = 1.9\,(\text{m/s})$. The data for $X < 14\,(\text{mm})$, indicating increase in circulation results from entering of the vortices into the interrogation region and can be discarded. It is evident that the vortices accumulate more circulation in the adverse pressure gradient TBL, for all excitation levels. It is also evident that the circulation decays faster in the ZPG case, especially at the highest excitation magnitude. This is hypothesized to be linked to the nonlinear interaction with the TBL.

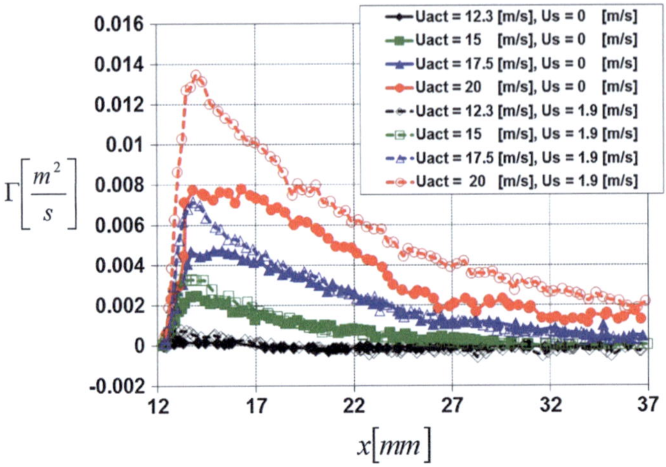

Fig. 10 The circulation of the PS generated vortices, with magnitude and U_s as indicated in the legend. $X_{act} = 530\,(\text{mm})$, $U_s = 1.9\,(\text{m/s})$, $U_\infty = 7\,(\text{m/s})$

The time evolution of the vortices' X locations was also performed. Following the entry stage, the vortices locations create a straight line, indicating a constant convection speed. Vortices convection speeds were calculated by the slope of a liner fit to the data of the X locations of the vortex centers versus X. Figure 11 presents the normalized, by the free-stream velocity of 7 (m/s) throughout, convection speeds for the PS created vortices in the two boundary layers. It is evident that the convection speeds increase with Up and decrease with the adverse dP/dX. This must be a result of the nonlinear interaction with the finite magnitude vortices and the TBL and the changing near wall velocity gradient, where the excitation has its strongest effect. For reference and stability considerations, an Amplitude modulated excitation was also

Fig. 11 Vortex convection speed for PS and AM excitation versus excitation magnitude for U_s as indicated in legend. $U_\infty = 7\,(\text{m/s})$, $f_{AM} = 18\,(\text{Hz})$

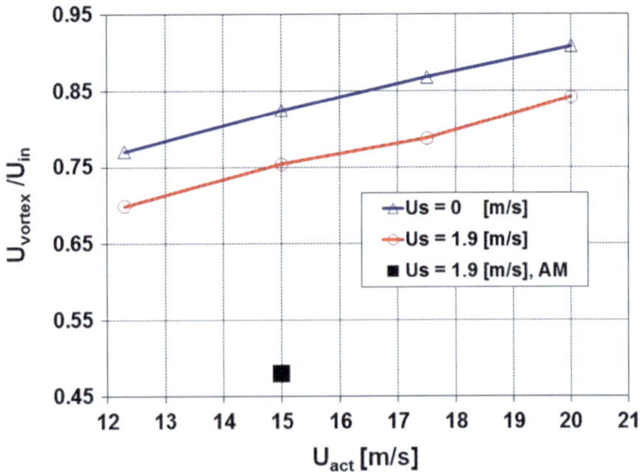

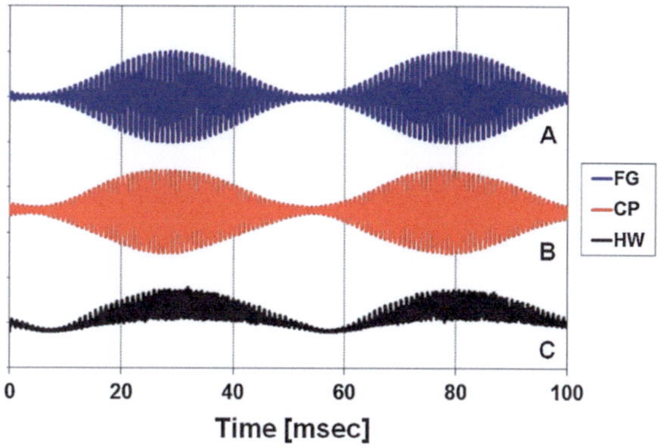

Fig. 12 Amplitude Modulated (AM) excitation

used. With this excitation (Fig. 12), the envelope of the PS excitation is harmonically time varying at low frequency. In Fig. 12 we present the excitation signal (A) the resulting cavity pressure oscillations (B) and the resulting slot exit velocity (C) (when $U_\infty = 0$). These vortices also created harmonic boundary layer velocity fluctuations from which convection speed can be calculated. The convection speed for $U_p = 15$ (m/s) and AM frequency of 18 (Hz) is also presented in Fig. 11. It is evident that this type of excitation generated vortices are convected significantly slower that the PS excitation, resulting from the significant lower momentum and vorticity imparted by the vortices on the separating TBL. The main purpose of using AM excitation was to identify an unstable 2D excitation that will be amplified by the separating TBL. Effective low amplitude excitation is creating by the AM excitation resulting from the sum and difference of the side lobes created by such excitation around the carrier frequency [13, 14].

Finally, Fig. 13 presents the velocity spectra resulting from AM excitation effects on the separating TBL at two X stations. The 1st X station is positioned 25 (mm) ($\sim 9\delta^*$) downstream of the slot, while the second X station is positioned 30 (mm) ($\sim 12\delta^*$) downstream. It is clear that the AM excitation is capable of creating low frequency excitation at $X_s = 25$ (mm). It is also evident that all frequencies tested

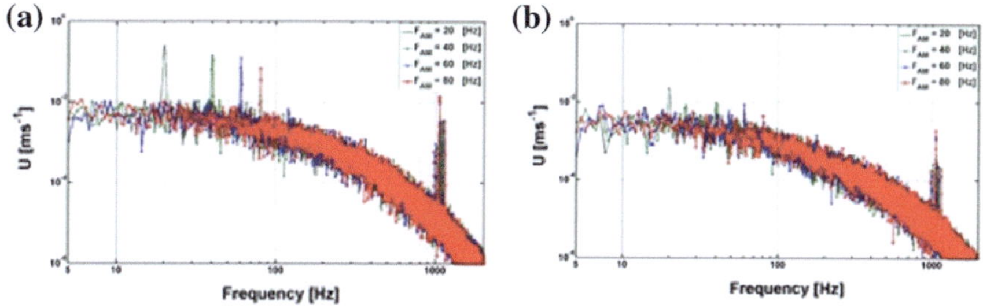

Fig. 13 Velocity spectra with AM excitation at two X stations, **a** $X = 555$ (mm), $X_s = 25$ (mm), **b** $X = 685$ (mm), $X_s = 55$ (mm)

$(10 - 100\,(\mathrm{Hz}))$ decay further downstream. These results are expected since there is no inflexion point in the TBL velocity profiles, at least not in the mean sense (Figs. 5 and 6). The next stage of the investigation, to be reported elsewhere, would describe the effect of pulsed excitation aimed at identifying optimal repetition rates based on the natural TBL recovery times.

References

1. H. Schlichting, *Boundary-Layer Theory*, Chap. 5 (McGraw-Hill, New York, 1968)
2. R.L. Simpson, Turbulent boundary layer separation. Annu. Rev. Fluid Mech. **21**, 34–205 (1989)
3. A. Seifert, A. Darabi, I. Wygnanski, Delay of airfoil stall by periodic excitation. J. Aircr. **33**(4), 691–698 (1996)
4. A. Seifert, L.G. Pack, Active control of separated flow on a wall-mounted hump at high Reynolds numbers. AIAA J. **40**(7), 1363–1372 (2002). (Part of AIAA paper 99-3430)
5. E. Kit, M. Gaster, I. Wygnanski, Large-scale structures in a forced turbulent mixing layer. J. Fluid Mech. **150**, 23–39 (1985)
6. K.M. Butler, B.F. Farrell, Three-dimensional optimal perturbations in viscous shear flow. Phys. Fluids A **4**, 1637 (1992)
7. A. Darabi, I. Wygnanski, Active management of naturally separated flow over a solid surface. Part 1. The forced reattachment process. J. Fluid Mech. **510**, 105–129 (2004)
8. M. Amitay, A. Glezer, Controlled transient of flow reattachment over stalled airfoils surface. Fluids A **4**, 19–699 (1637)
9. J.E. Williams, D.L. Hawkings, Sound generation by turbulence and surfaces in arbitrary motion. Philos. Trans. R. Soc. Lond. A **264**, 321–342 (1969)
10. T. Yehoshua, A. Seifert, Boundary condition effects on oscillatory momentum generators, in *33rd AIAA Fluid Dynamics Conference*, 23–26 June 2003
11. T. Yehoshua, A. Seifert, Boundary condition effects on the evolution of a train of vortex pairs in still air. Aeronaut. J. **110**(1109), 397–417 (2006)
12. U. Ingard, H. Ising, Acoustic nonlinearity of an orifice. J. Acoust. Soc. Am. **42**, 6–17 (1967)
13. J.M. Wiltze, A. Glezer, Manipulation of free shear flows using piezoelectric actuators. J. Fluid Mech. **249**, 261–285 (1993)
14. T. Yehoshua, A. Seifert, Empirical model for the evolution of a vortex-pair introduced into a boundary layer. J. Aerosp. Lab (6) (2013)

Statistical and Temporal Characterization of Turbulent Rayleigh-Bénard Convection Boundary Layers Using Time-Resolved PIV Measurements

Christian E. Willert, Ronald du Puits and Christian Resagk

Abstract This contribution reports on near-wall flow field measurements in turbulent Rayleigh-Bénard convection (RBC) in air at a fixed Prandtl number $Pr = 0.7$ and Rayleigh number $Ra = 1.45 \times 10^{10}$. For the experiment, the large-scale convection (LSC) was confined to a rectangular box of $2.5 \times 2.5 \times 0.65\,m^3$ made of transparent acrylic sheets. Prior video-graphic visualizations of the bottom boundary layer flow by means of laser light sheet illumination of small particles indicated the presence of highly dynamic flow behaviour at flow conditions that classical stability analysis predicts to still be in the laminar regime. While theory predicts a transition to turbulence at Reynolds numbers $Re_\delta \approx 420$, the present investigation exhibits highly unsteady flow at a much lower Reynolds number of $Re_\delta \approx 260$ based on boundary layer thickness. With the help of the PIV data, it can be demonstrated that the entrainment of turbulent structures from the mean wind into the boundary layer acts, alongside with the destabilization due to inner shear, as a second mechanism on its path to turbulence. Both contributions must be considered when predicting the critical bound towards the *ultimate regime* of thermal convection. The measurements rely on the acquisition of long, continuous sequences of particle image velocimetry (PIV) data from which both statistical and spectral information can be retrieved. Contrary to conventional implementation of the PIV technique the field of view is restricted to a narrow strip, generally extending in wall-normal direction. In this way, both the acquisition frequency and the total number images of the employed high-speed camera are proportionally increased. The temporally oversampled data allows the use of multi-frame PIV processing algorithms which reduce measurement uncertainties with respect to standard dual-frame analysis.

C.E. Willert (✉)
DLR Institute of Propulsion Technology, German Aerospace Center,
51170 Köln, Germany
e-mail: chris.willert@dlr.de

R. du Puits · C. Resagk
Department of Mechanical Engineering, Technische Universität Ilmenau,
P.O. Box 100565, 98684 Ilmenau, Germany
e-mail: ronald.dupuits@tu-ilmenau.de

C. Resagk
e-mail: christian.resagk@tu-ilmenau.de

M. Stanislas et al. (eds.), *Progress in Wall Turbulence 2*,
ERCOFTAC Series 23, DOI 10.1007/978-3-319-20388-1_28

1 Introduction

Many natural or technical flows are associated with a heat transfer from hot or cold surfaces to the surrounding fluid. Because of its rather controllable boundary conditions—a confined fluid cooled from above and heated from below—Rayleigh-Bénard convection (RBC) has been the subject of numerous studies in the past decades, both numerically and experimentally (for a comprehensive review, see Ahlers et al. [1]). However, particularly in the case of turbulent RBC, the knowledge about the temperature and velocity field inside the convective boundary layer is still rather limited, which in turn affects the predictability of the local heat transfer coefficient. This process can be investigated experimentally in the "Barrel of Ilmenau" [5], a large facility which offers both high Rayleigh numbers and large flow scales with boundary layer thicknesses in the tens of millimetre range. Especially, the latter property permits a maximum of spatial resolution for established optical and probe-based measurement techniques.

Previous measurements in the facility recovered both the temperature and velocity profiles on the top and bottom walls using glass-encapsulated microthermistors [2] and laser Doppler anemometry (LDA) [3, 10]. While these single-point measurements provide valuable flow statistics, they only give limited insight to the topology of the flow. In this regard, planar techniques such as particle image velocimetry (PIV) are methods of choice yet so far have found little application in high-Ra Rayleigh-Bénard convection in air ($Pr = 0.7$), and to our knowledge have not been applied for the detailed investigation of the transitional boundary layers.

Extensive PIV measurements of both the global flow field as well as the boundary layer structure have been performed Zhou and Xia [16] and Sun et al. [12] in smaller scale facilities for similar Ra numbers with water as working fluid ($Pr = 4.3$). In these works, continuous PIV sequences in excess of 4 h were recorded at sample rates of 2.2 Hz. Among the findings was that the intermittent emissions of coherent structures caused by thermal plumes do not modify the Blasius-type laminar velocity profile which indicates that heat transfer across the thermal boundary layer is mainly by conduction. Modification of this heat transfer is expected with the onset of turbulence. Based on the Prandtl-Blasius theory of a flat-plate boundary layer and the stability criterion $Re = U\,\delta/\nu = 420$ derived by Tollmien [13], the transition in heat transfer in turbulent RBC was firstly predicted by Kraichnan [9] and later on defined more precisely by Grossmann and Lohse [8]. The latter authors predict a critical Ra number of $Ra_c \approx 10^{13} \ldots 10^{14}$ as the stability criterion in turbulent RBC at $Pr = 0.7$.

Video-graphic visualizations of the bottom boundary layer in the "Barrel of Ilmenau" facility by means of laser light sheet illumination of small particles indicated the presence of highly dynamic flow behaviour at flow conditions that classical stability analysis predict to still be in the laminar regime [4, 6]. The Reynolds number of the present investigation is $Re_\delta \approx 260$ which is considerably lower than the predicted transition Reynolds number of $Re_\delta \approx 420$. Also the corresponding Rayleigh number of $Ra = 1.45 \times 10^{10}$ is well below the extrapolated value of $Ra \approx 2 \times 10^{13}$ reported by Sun et al. [12]. In their measurements, they observed a clear departure of the time-averaged data from the Prandtl-Blasius laminar boundary layer profile which would be indicative of a transitional behaviour. However, Zhou and Xia [16] pointed out that a normalization of the velocity profile by the instantaneous viscous boundary layer thickness δ_ν results in a good agreement with the predicted Prandtl-Blasius profile.

In an effort to elucidate the transient behaviour of the boundary layer and to augment existing time-averaged data [10], the following article presents results obtained with long, time-resolved sequences of the bottom boundary layer in the "Barrel of Ilmenau" large-scale RBC facility. The first part of this contribution describes the experimental methods using time-resolved PIV for investigation of turbulent flows. These methods form the basis for the characterization of the boundary layer of turbulent Rayleigh-Bénard convection provided in the second and main part of the article.

2 Time-Resolved PIV for Near-Wall Flow Measurement

The investigation of turbulent flows in general strongly depends on gain information on the statistics of the flow. For the experimental investigation of wall bounded flow, this requirement is fulfilled by hot wire anemometry (HWA) and laser Doppler anemometry (LDA) as these are capable of providing measurement data with both high temporal and spatial resolution. On the other hand, particle image velocimetry (PIV) captures snapshots of the flow field—even volumetric—and thereby provides valuable information on the topology of the flow. PIV has been used extensively for the investigation of turbulent flows essentially ever since the time it has been developed. Due to the spatial averaging nature and rather low sampling rate (typ < 1 kHz), PIV generally has significant limitations in providing statistical information comparable to HWA and LDA. Nonetheless, current technology allows the technique to be pushed to higher sampling rates by reducing the spatial resolution of the sensor and focussing on a particular area of interest with increased magnification. Rather than capturing 2-D or 3-D maps of the flow field, the implementation of PIV utilized for the present investigation is aimed at obtaining measurement data along a narrow strip with a high number of samples at acquisition rates matching the time scales of the flow. In effect, this approach has the potential of bridging the gap between single-point measurement techniques (e.g. HWA, LDA) and conventional PIV, and shall be the main subject of this contribution.

2.1 Near-Wall Measurement of Turbulent Wall Flow

Reliable near-wall measurements with high frame rate PIV become feasible when the magnification of imaging system exceeds resolutions in the range of one pixel per wall unit (z^+) and the frame-to-frame displacements are limited to a few tens of pixels. While this becomes increasingly challenging with higher Reynolds number, the measurements presented here have viscous sublayer thicknesses in the millimetre range such that conventional imaging optics (macro lenses) can be used. At the same time, the outer velocities typically are below 1 m/s which requires camera frame rates of 100–200 Hz to keep frame-to-frame displacements within bounds.

From a hardware point of view, the measurement setup is rather straightforward as it can be reduced to a medium-speed camera and a continuous wave laser of 2–5 W radiant power. The laser light is formed into a narrow, thin light sheet and is directed into the facility at a wall-normal direction, ideally through a clean glass surface. This narrow strip is then imaged by the high-speed camera, whose electronic shutter prevents excessive particle streaking on the sensor.

The technique was first applied for the investigation of the developing turbulent boundary layer of the air flow inside a square rectangular duct with $\mathrm{Re}_D = 20,000$ and $\mathrm{Re}_\delta = 4,900$ using image sequences with lengths up to $N \approx 180,000$ at frame rates of 10–50 kHz [15]. While the high sample count ensures statistical convergence, the time-resolved nature of the data allows the estimation of both velocity spectra and space-time correlations.

2.2 *Estimation of the Wall Shear Rate*

Wall shear stress is directly linked to the wall shear rate, that is, the velocity gradient $\dot{\gamma} = du/dz$ at the wall:

$$\tau_\mathrm{w} = \mu \frac{du}{dz}\bigg|_{y=0}. \tag{1}$$

To recover estimates of the wall, shear stress using particle-based imaging methods requires that the linear portion of the sublayer is sufficiently well resolved such that the gradient can be obtained through finite differing of velocity estimates. Within the linear region, the fluid motion and along with it the motion of the particles is wall parallel. Therefore, it is possible to limit the frame-to-frame displacement estimation to recover only the wall-parallel motion. With the camera properly aligned, this can be achieved by cross-correlating single rows of wall-parallel pixels.

Figure 1 shows the mean wall-parallel velocity of particles within 30 pixels from the wall. Aside from deviations really close to the wall, caused by stationary particles or dust on the wall, the profile is linear throughout owing to the nearly 5-mm-thick viscous sublayer of the RBC boundary layer. The slope of a linear fit to this linear portion provides an estimate of the wall shear rate $\dot{\gamma}_w = du/dz$ and can be used for the straightforward calculation of the friction velocity u_τ and associated normalization factors. For a value of $\dot{\gamma}_w \approx 17.95\,\mathrm{s}^{-1}$ deduced from Fig. 1, this yields

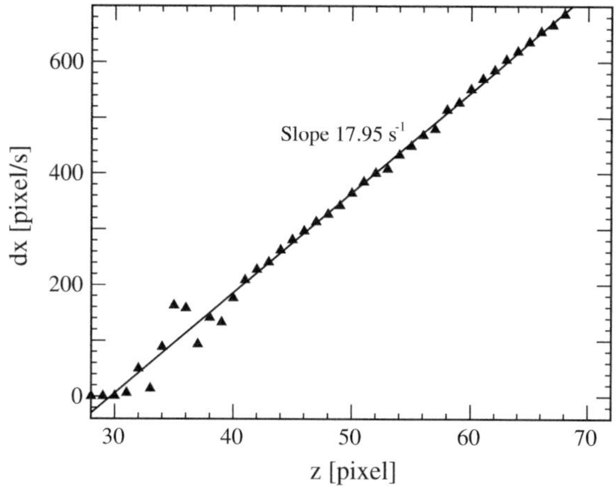

Fig. 1 Mean near-wall particle velocity obtained by single-line image cross-correlation. The wall is approximately located at 29 pixel and magnification is 45.7 μm/pixel

$$u_\tau = \sqrt{\frac{\tau_w}{\rho}} = \sqrt{\nu \frac{du}{dz}\bigg|_{y=0}} \approx 17.1\,\text{mm/s} \tag{2}$$

$$\frac{x_i^+}{x_i} = \frac{u_\tau}{\nu} \approx 1.05\text{mm}^{-1}. \tag{3}$$

While Fig. 1 provides mean values, the single-line cross-correlation technique can also be used to estimate the instantaneous wall shear rate and with it the time varying wall shear stress τ_w [15].

3 Near-Wall Measurements of Rayleigh-Bénard Convection

Turbulent Rayleigh-Bénard convection (RBC) is initiated in a confined fluid in the presence of a sufficiently strong temperature difference between a cooler top surface and a warmer bottom surface. The present measurements are aimed at gaining further insight on the transient behaviour of the boundary layer on the bottom wall previously observed through visualization by du Puits et al. [6]. The acquisition of long PIV image sequences of up to $N \approx 59{,}000$ samples are motivated by achieving both statistical convergence as well as providing continuous time records of the transient flow.

3.1 Experimental Facility and Operating Conditions

For the experiment, the large-scale convection is confined to a rectangular box of $2.5 \times 2.5 \times 0.65\,\text{m}^3$ made of transparent acrylic sheets Fig. 2. The temperature difference between the bottom and top wall is $\Delta T = 10\,\text{K}$. With air as the working

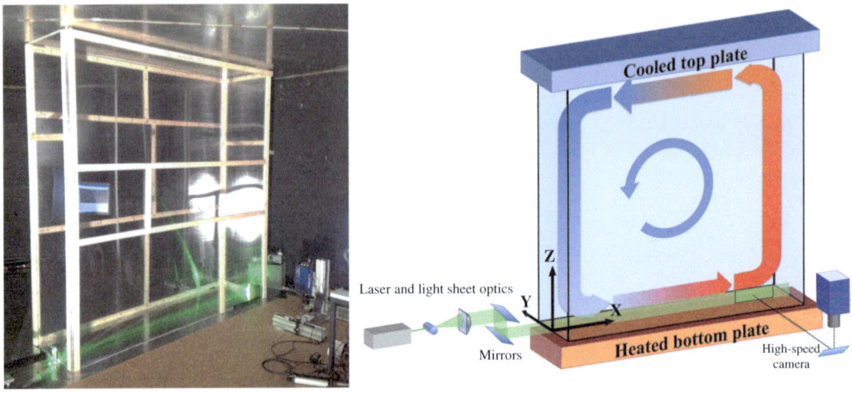

Fig. 2 PIV imaging setup for the investigation of RBC within a confined rectangular cell inside the *Barrel of Ilmenau*. *Right* schematic of the global flow pattern within the RBC cell, PIV measurement setup and coordinate system

fluid (Pr = 0.7) the Rayleigh number amounts to Ra = 1.45×10^{10}. Further details of the facility itself are given in [5].

3.2 Data Acquisition in the RBC Facility

To obtain more detailed information on the statistics and the temporal evolution of the boundary layer flow, long PIV image sequences were recorded at various positions near the centerline of the heated bottom plate (Fig. 2). Laser light sheet illumination was realized with a 2 W continuous wave laser. The light sheet thickness was approximately 1–2 mm with a uniform height of approximately 70 mm across the horizontal length of 2.5-m-wide cell. A smoke generator, based on an evaporation–condensation principle, was used to seed the flow with 1–2 μm oil droplets whose life time exceeded 1 h.

Images of the illuminated particles were acquired with a high-speed camera (*PCO GmbH, Dimax-S4*) at a frame rate of 200 Hz and a spatial resolution of 2016 × 600 pixels. With the camera positioned upright, a 90° mirror in front of the macrolens (*Zeiss Macro-Planar 100/2*) aligned the optical axis with the normal to the light sheet. Aside from its high light sensitivity, the main benefits of this camera are its large dynamic range of 12 bits and large internal storage to capture long sequences of more than 20,000 frames at the chosen resolution. Particle streaking by the CW illumination could be reduced by limiting the sensor exposure to 2 ms.

As indicated in Table 1, various image sequences at two magnifications were captured, one to obtain the global features of the boundary layer and a second, higher resolution to retrieve selected profiles with improved statistical convergence. A high magnification of up to 22.1 pixel/mm was chosen in order to resolve the viscous sublayer (a wall unit corresponds to roughly $z^+/z = 1$ mm, see Table 2). To increase

Table 1 Imaging parameters for the acquired PIV sequences of near-wall RBC

	Symbol	Overview	Detailed view A, B, C, D	Unit
Magnification	m	9.1	22.1	pixel/mm
Field of view	$[W \times H]$	220 × 65	13 × 68	mm^2
	$[W \times H]$	2016 × 600	288 × 1500	pixel
Camera frame rate	f_{acq}	200	100/200	Hz
Camera exposure	t_{exp}	2.0	2.0	ms
Duration	t_{seq}	106	592/296	s
Number of frames	N	21161	59235	

Table 2 Boundary layer data obtained for measurement positions A through D

		A	B	C	D	Unit
Position	x/L	0.09	0.48	0.64	0.92	–
Boundary layer thickness	δ	22.8	32.8	35.8	62.9	(mm)
Boundary layer thickness	$\delta_{99\,\%}$	18.7	25.3	27.7	51.3	(mm)
Kinematic boundary layer thickness	δ_ν	8.05	9.82	10.9	21.6	(mm)
Displacement thickness	δ_*	5.32	6.78	7.54	14.3	(mm)
Momentum thickness	θ	2.49	3.06	3.45	6.13	(mm)
Shape factor	$S = \delta_*/\theta$	2.14	2.22	2.18	2.34	–
Maximum velocity	U_{max}	−37.8	169	155	83.2	$(mm\ s^{-1})$
Wall shear rate	$\dot{\gamma}_W = \frac{du}{dy}$	3.83	16.7	13.7	2.85	(s^{-1})
Friction velocity	u_τ	7.88	16.4	14.9	6.79	$(mm\ s^{-1})$
Wall unit	z^+/z	0.486	1.015	0.920	0.419	(mm^{-1})
Reynolds number	Re_δ	(53.2)	343	343	323	–
Reynolds number	$Re_{\delta_{99}}$	(43.8)	264	265	263	–
Reynolds number	Re_{δ_ν}	(18.8)	103	104	111	–

the duration of the acquisition, one sequence was acquired at 100 Hz which also doubled the particle displacements to a level comparable to the light sheet thickness. As a consequence, the loss of signal due to out-of-plane particle motion increased and required careful adjustment of PIV processing parameters.

3.3 PIV Analysis and Data Post Processing

The acquired image sequences were processed pair-wise with conventional PIV processing software (PIVTEC GmbH, PIVview2C, v3.5) using sample sizes of 24×16 pixels ($W \times H$) and 64×8 pixels at 50 % overlap, where the latter sample size was chosen to improve the resolution in the presence of strong, near-wall shearing motions. The recovered velocity data was additionally temporally filtered using a Gaussian weighted averaging kernel to reduce noise in the individual velocity estimates. The chosen kernel width of 6 frames (e^{-2}) corresponds to a temporal filter width of 30 ms, which is well below the smallest time scales present in this flow. For example, at $Ra = 1.45 \times 10^{10}$, the Kolmogorov length scale is $\eta \approx 4.4$ mm with a time scale $\tau \approx 1.2$ s.

Figure 3 provides an overview of the acquired measurement domains. The four narrow (red) stripes indicate positions that were imaged with higher resolution and an increased number of samples (nearly 60,000 images each). Each of the twelve sampling areas were acquired with a pause of about 30 min in between, which to a large extent is due to the time required to download images from the camera as well as camera repositioning and length calibration. Figure 4 provides the mean velocity distribution at three relevant positions, namely the impingement area of the downwelling flow near the right corner (A), a position near the middle where the boundary layer is considered to be well established (C), and finally an area near the right corner where the mean flow is directed upward (D).

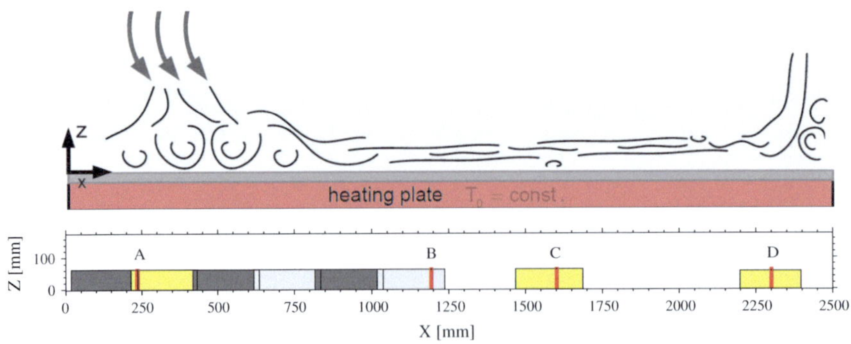

Fig. 3 Schematic of flow field (*top*) and overview of PIV measurement areas. The *narrow vertical strips* labelled A,B,C and D indicate areas used for profile measurements. Flow fields for the *yellow* areas are provided in Fig. 4

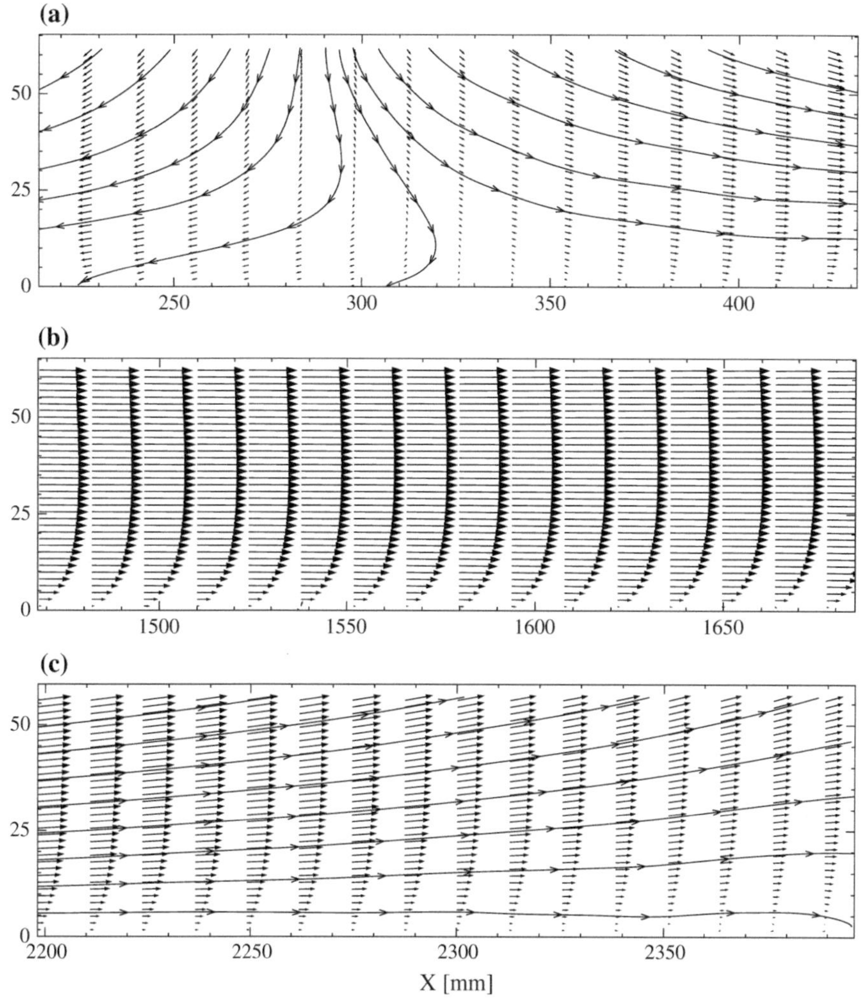

Fig. 4 Overview of mean velocity field at three positions along centerline: downwelling region (**a**), middle area (**b**), and upwelling region (**c**); vectors desampled 16× horizontally, 2× vertically

The flow near the left corner (Fig. 4) is clearly dominated by downward motion and even reverses with a clearly visible stagnation point at $x \approx 330$ mm. The flow then accelerates with increasing horizontal distance establishing a boundary layer with a maximum velocity of $U_{\max} \approx 160$–170 mm/s at a wall distances of $z = 30$–35 mm. Towards the right corner, the mean streamlines indicate the deflection of the horizontal shear flow towards the upper plate by the vertical sidewall.

In order to estimate the mean wall shear rate $\dot{\gamma}_w$, a modified PIV processing scheme is used that employs cross-correlation on single line of pixels (wall parallel). The procedure is described in Sect. 2.2 and provides velocity estimates at

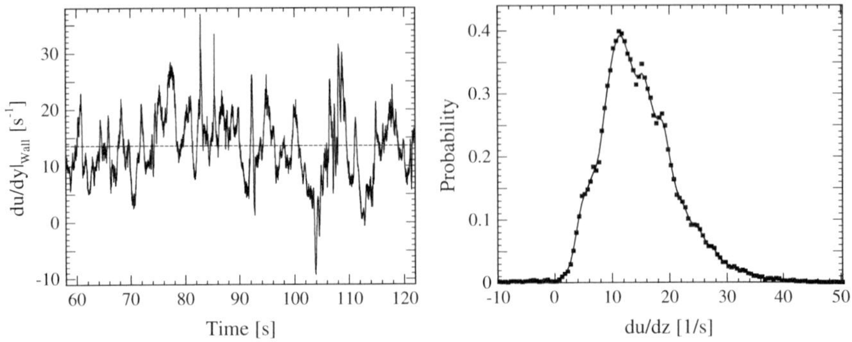

Fig. 5 Temporal evolution (12,000 samples) and probability density function (*right*) of the wall shear rate $\dot{\gamma}_w = du/dz$ at $x/L = 0.64$ (mean: $13.7\,\mathrm{s}^{-1}$, $\sigma : 6.8\,\mathrm{s}^{-1}$, skew: 0.51)

single-pixel increments with a wall-normal resolution of one pixel (45.2 μm). A linear least squares fit over these estimates provides the near-wall velocity gradient in units of [pixel/pixel], and hence is dimensionless. Division by the time difference between image recordings provides the wall shear rate. This gradient can be obtained for each image pair such that the statistics and temporal evolution can be further analysed. A two-minute portion of the highly fluctuating wall shear rate is shown in Fig. 5, left. The probability density function for the entire record is given in Fig. 5, right. A positive skew is indicative of the intermittent behaviour as shown by the high-amplitude bursts in the time trace.

4 Results and discussion

The characteristic values for the four detailed boundary layer measurement areas are summarized in Table 2. Profiles of the normalized mean velocity profile at two positions are provided in Fig. 6. At position $x/L = 0.48$, the maximum velocity $U_{\max} =$ 169 mm/s is reached at a wall distance of $\delta = 32.8$ mm. At greater wall distances, the mean velocity slowly decays. Further downstream ($x/L = 0.64$, $x/L = 0.92$) the boundary layer thickens, while the maximum velocity reduces; the Reynolds number stays approximately constant. This indicates that the boundary layer does not receive additional momentum from the outer flow aside from the momentum added in the downwelling region at $x/L \lesssim 0.3$. At $x/L = 0.92$, the mean velocity profile exhibits a reduced velocity gradient at the wall which is caused by the separating flow in this area (Fig. 8).

The velocity profiles also clearly deviate from the laminar Prandtl-Blasius profile which corresponds to results reported for DNS [11, 14] as well as previous LDA measurements [2, 10, 16]. In Fig. 6, the viscous sublayer is present where the profile follows the Prandtl-Blasius profile and begins to depart at $z/\delta \gtrsim 0.1$ corresponding

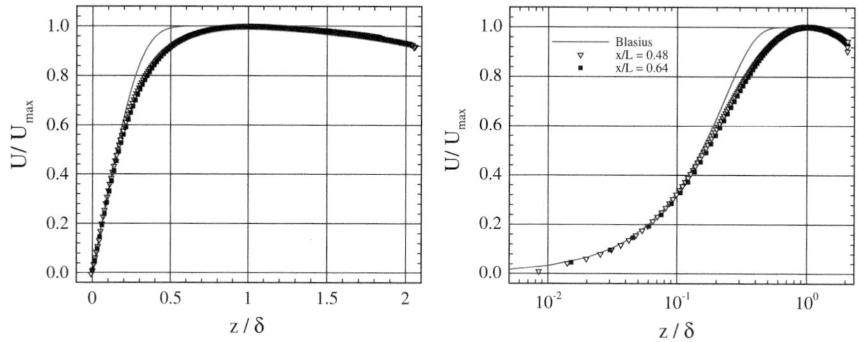

Fig. 6 Wall-normal distribution of the mean horizontal velocity at $x/L = 0.48$ (▽) and $x/L = 0.64$ with mean velocities of $U_{max} = 169$ mm/s and $U_{max} = 155$ mm/s at wall distances $\delta = 32.3$ mm and $\delta = 35.8$ mm, respectively

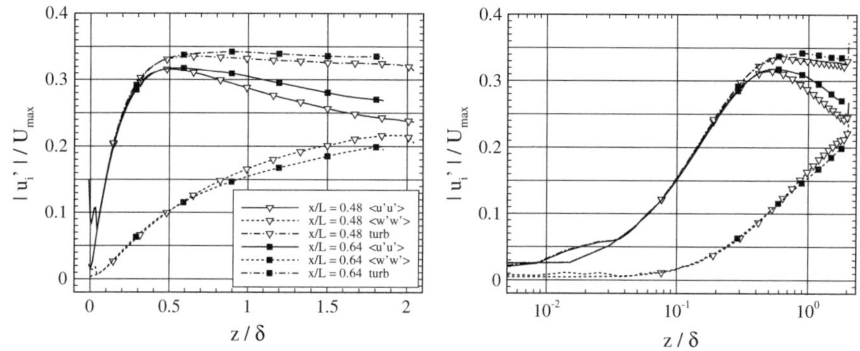

Fig. 7 Wall-normal distribution of fluctuating components at $x/L = 0.48$ (▽) and $x/L = 0.64$

to $z^+ \approx 4$. However, a fully developed logarithmic region does not exist in the mean velocity profile (Fig. 6, right) due to the transitional character of the boundary layer.

The wall-normal distribution of the time-averaged fluctuating velocity components $<u'_i u'_j>$ is shown in Fig. 7. While the horizontal fluctuations u' reach a maximum at a normalized wall distance $z/\delta \approx 0.5$ corresponding roughly to $16\,z^+$ and decay after that, the vertical fluctuations w' continue to increase asymptotically approaching the value of the horizontal fluctuations. This indicates increasing isotropy of the fluid motion (or turbulence) with increased wall distance. The combined quantity—the square root of their sum—reaches at constant value near $z/\delta \approx 0.5$ and only slightly decays thereafter which suggests near constant turbulent kinetic energy in the outer regions of the boundary layer.

The planar nature of PIV data allows the retrieval of differential quantities that are difficult to obtain with single-point techniques such as laser Doppler anemometry (LDA) and hotwire anemometry (HWA). As an example, Fig. 9 shows the wall-normal distribution of mean and fluctuating values of the out-of-plane vorticity

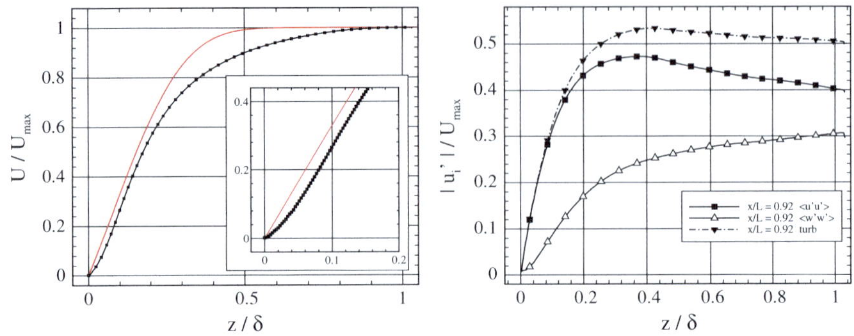

Fig. 8 Wall-normal distribution of mean horizontal velocity (*left*) and fluctuations (*right*) at $x/L = 0.92$

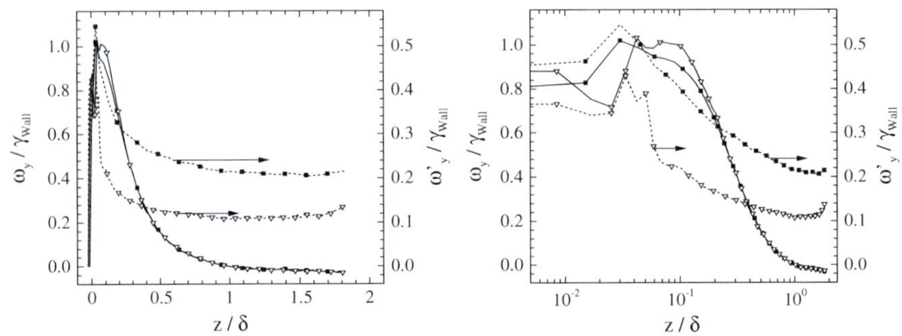

Fig. 9 Wall-normal distribution of mean and fluctuations of vorticity component ω_y at $x/L = 0.48$ ($\triangledown$) and $x/L = 0.64$

component ω_y. Normalized by the wall shear rate, its value peaks near the wall at about $\omega/\dot{\gamma}_w = 0.4$. Closer to the wall, the measurement uncertainty increases due to light scattering at the surface, which "locks" the displacement to zero. Correspondingly, the RMS value of vorticity decays in the immediate vicinity of the wall for $z/\delta < 0.1$, which is not representative. Compared to the velocity fluctuations (Fig. 6, right), the strongest vorticity fluctuations occur much closer to the wall and have nearly decayed to levels close of the outer flow.

While the previous results focussed on the statistics of the boundary layer flow, we will now highlight some of temporal characteristics of the flow. Figure 11 shows a good example of the intermittent behaviour of the flow by exhibiting two very different flow states at position $x/L \approx 0.45$ separated in time by only 1.7 s. A highly dynamic flow, most likely associated with a previously occurred thermal pluming event (Fig. 10), is carried downstream with the convection velocity of the boundary layer returning the boundary layer flow to a quieter state that closely resembles that of a laminar boundary layer.

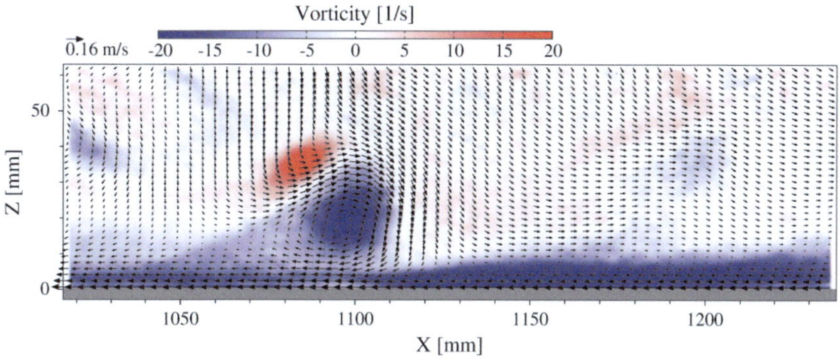

Fig. 10 Thermal plume at $x/L = 0.5$. For better visibility, mean horizontal velocity of 0.16 m/s has been subtracted; vectors are downsampled 4× horizontally and 2× vertically

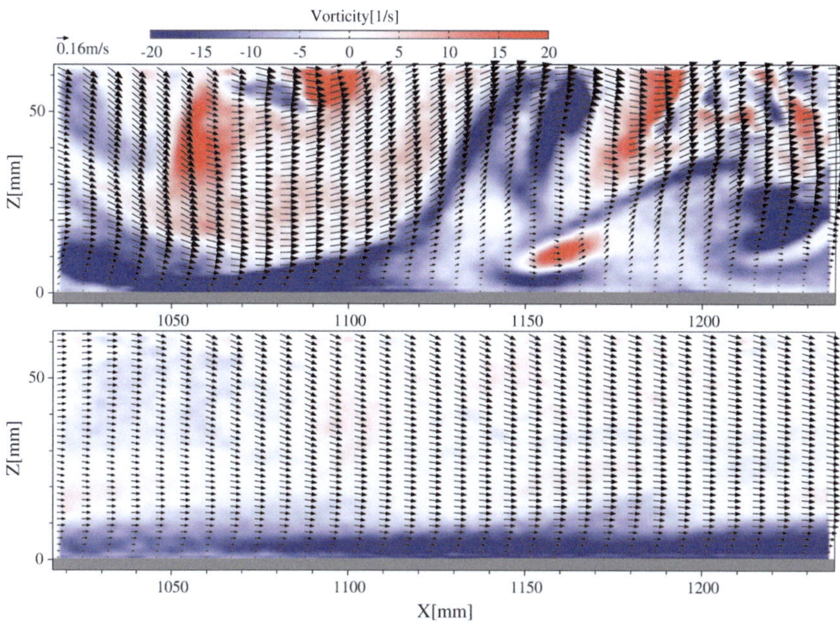

Fig. 11 Two flow conditions at position $x/L = 0.5$ separated by 1.7 s in time. Vectors are downsampled 8× horizontally and 2× vertically

The intermittent behaviour of the boundary layer flow is clearly visible in the time trace of normalized velocity $u/U_{\max}$ and vorticity $\omega/\dot{\gamma}_w$ provided in Fig. 12. The time trace was obtained at a wall distance of $z/\delta = 1$ at sampling position B, $x/L = 0.48$. The red and green vertical lines in Fig. 12, right, correspond to the time instances for the flow maps presented in Fig. 11. In particular, the vorticity $z/\delta = 1$

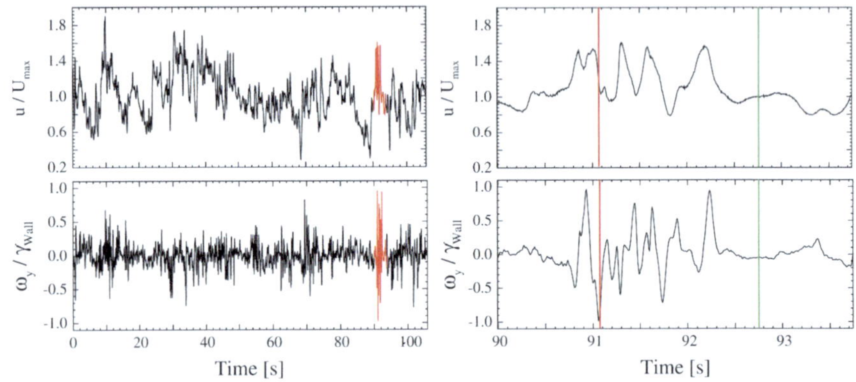

Fig. 12 Time traces of normalized velocity $u/U_{\max}$ (*top*) and normalized vorticity $\omega/\dot{\gamma}_w$ (*bottom*) at position $x/L = 0.48$ and $z/\delta = 1$. Portion highlighted in *red* on left plot is magnified in right plot. *Red* and *green vertical lines* correspond to Fig. 11 *top* and *bottom*, respectively

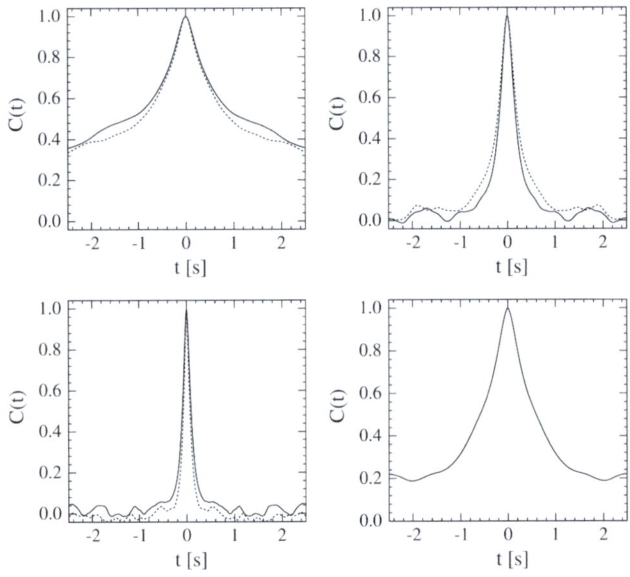

Fig. 13 Autocorrelation functions at position $x/L = 0.64$ for velocity components u and w (*top* row), vorticity ω_y (*bottom left*) and wall shear rate $\dot{\gamma}_w$ (*bottom right*) for wall distances $z/\delta = 1$ (——) and $z/\delta = 1.8$ (- - - -)

exhibits a period of strong bursts with amplitudes reaching that of the mean wall shear rate.

The temporal correlation of selected flow components at selected locations provides some information on the time scales within the boundary layer and the interaction between different quantities. The autocorrelation of both velocity components

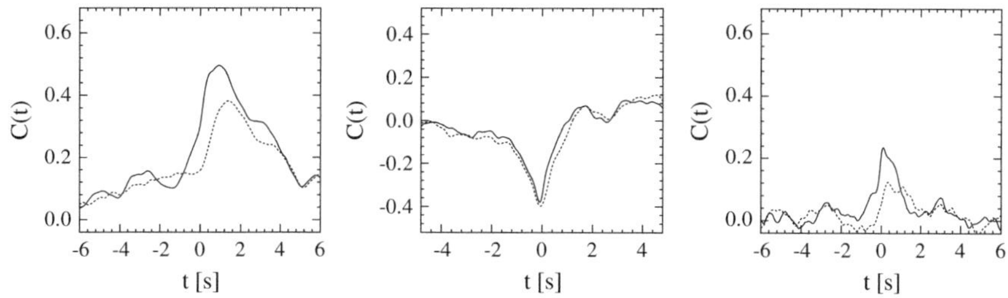

Fig. 14 Cross-correlation functions at position $x/L = 0.64$ between wall shear rate $\dot{\gamma}_w$ and velocity components u (*left*) and w (*middle*), and vorticity ω_y (*right*) for wall distances $z/\delta = 1$ (——) and $z/\delta = 1.8$ (- - - -)

u and w, the out-of-plane vorticity ω_y and the wall shear rate $\dot{\gamma}_w$ are shown in Fig. 13. Among these, the vorticity has the shortest time scale, which is due to their convection with the mean flow. The horizontal velocity component has the highest self-correlation of several seconds which is related to the rather steady horizontal motion of the flow. The self-correlation of the wall shear rate has a duration of roughly 1 s. The cross-correlations of velocity and vorticity with wall shear are depicted in Fig. 14. Here, the vortex structures convected with the bulk of the boundary layer have limited influence on the wall shear rate. On the other hand, both velocity components of the bulk flow have a pronounced influence on the wall shear. While the correlation with horizontal velocity component shows temporal lag of 1–2 s, the response of the wall shear to vertical motions is essentially without delay. This interaction seems independent of the velocity sampling position from the wall, which suggests that the wall shear rate is forced from the outside by the bulk flow of the RBC (e.g. falling plumes, impinging jets).

4.1 Statistical Convergence of the Data

When plotted side-by-side, the time-averaged data domains exhibit discontinuities at the borders. This indicates that the mean flow is not stationary between the acquisitions of the individual sequences (30 min between individual sequences). Variations in the fluctuations further indicate that the statistics are not fully converged, in spite of the rather considerable length of the sequences. For the given operating conditions and geometry of the RB cell, a free-fall velocity $U_f = (g\,\alpha\,\Delta T\;H)^{0.5}$ can be estimated at $U_f \approx 68.3$ mm/s with a corresponding free-fall time of $T_f = H/U_f \approx 36$ s. The mean data for the local overviews covers about 3 T_f. On the other hand, Emran and Schumacher [7] estimate the average loop time of Lagrangian tracers at 20 T_f, corresponding to a duration of 12 min. This is nearly achieved for the high-resolution profile PIV measurements at position B ($x/L = 0.48$). Laser Doppler measurements for similar configurations reported by du Puits et al. [2] and Li et al. [10] sampled each point for up to 1 h and longer.[1] A comparable PIV measurement using the imaging configuration reported herein would require 360,000 images if sampled at 100 Hz corresponding to 400 GB of 8bit raw image data, if temporal coherence is to be preserved.

5 Summary and Conclusions

The velocity field close to the bottom horizontal plate in turbulent Rayleigh Bénard convection in air has been studied experimentally using long sequences of time-resolved PIV data. To the knowledge of the authors, these are the first PIV boundary layer measurements of RBC in air.

Whether the RBC boundary layer flow is turbulent or not cannot be answered conclusively. On the one hand, the Reynolds number of $\mathrm{Re} = 265$ is below the critical threshold of $\mathrm{Re} = 420$. The lack of a log-law region in the velocity profile is another indication that the boundary layer is not turbulent. Furthermore, a shape factor of $S = 2.2$–2.35 is closer to that of the laminar Prandtl-Blasius solution with $S \approx 2.5$ rather than that of a turbulent boundary layer or wall jet with $S = 1.3$–1.4. On the other hand, the flow is highly intermittent with significant velocity variations ($>30\,\%$ at half the mean kinematic boundary layer thickness).

As shown in Fig. 6, right, the peak horizontal fluctuations are observed at $\approx 16\,z^+$ ($z/\delta = 0.5$) which is significantly more than for RBC measurements in water at similar Rayleigh number [12] and is closer to the value of $12\,z^+$ for classical flat-plate boundary layers. The difference to the measurements in water can be related to the much thicker thermal boundary layer in air, which is of similar magnitude as the kinematic boundary layer [11] while it is confined to the viscous sublayer in water [12]. Therefore, the interaction of turbulent structures from the mean wind with the boundary layer will not only destabilize the inner shear of the kinematic boundary layer, but will also destabilize the thermal boundary layer. This is substantiated by the cross-correlation functions between wall shear rate and outer flow (Fig. 14) which show that in particular, vertical motions of the outer flow directly act all the way into the viscous sublayer. This is in agreement with similar conjectures reported in the literature that the wall-normal forcing is more dominant than wall-parallel direction forcing [14]. This vertical forcing results from buoyancy effects that manifest themselves in the form of plume impingement and detachment. In essence, both contributions, inner shear and external forcing of the boundary layer, must be considered when predicting the critical bound towards the "ultimate regime" of thermal convection [9].

[1] du Puits et al. [2] reported on 48 h hot-film measurements in the RB cell and observed a 30 % variation of the typical mean velocity over a duration of about 5 h.

References

1. G. Ahlers, S. Grossmann, D. Lohse, Heat transfer and large scale dynamics in turbulent Rayleigh-Bénard convection. Rev. Mod. Phys. **81**, 503–537 (2009). doi:10.1103/RevModPhys.81.503
2. R. du Puits, C. Resagk, A. Thess, Mean velocity profile in confined turbulent convection. Phys. Rev. Lett. **99**, 234504 (2007). doi:10.1103/PhysRevLett.99.234504
3. R. du Puits, C. Resagk, A. Thess, Structure of viscous boundary layers in turbulent Rayleigh-Bénard convection. Phys. Rev. E **80**, 036318 (2009). doi:10.1103/PhysRevE.80.036318
4. R. du Puits, J. Rilk, C. Resagk, A. Thess: Boundary layers in turbulent Rayleigh-Bénard convection in air. (2012). arXiv:1209.6201v1[physics.fluspsdyn]
5. R. du Puits, C. Resagk, A. Thess, Thermal boundary layers in turbulent Rayleigh-Bénard convection at aspect ratios between 1 and 9. New J. Phys. **15**(1), 013040 (2013)
6. R. du Puits, L. Li, C. Resagk, A. Thess, C. Willert, Turbulent boundary layer in high Rayleigh number convection in air. Phys. Rev. Lett. **112**, 124301 (2014). doi:10.1103/PhysRevLett.112.124301
7. M.S. Emran, J. Schumacher, Lagrangian tracer dynamics in a closed cylindrical turbulent convection cell. Phys. Rev. E **82**, 016303 (2010). doi:10.1103/PhysRevE.82.016303
8. S. Grossmann, D. Lohse, Scaling in thermal convection: a unifying theory. J. Fluid Mech. **407**, 27–56 (2000). doi:10.1017/S0022112099007545
9. R.H. Kraichnan, Turbulent thermal convection at arbitrary prandtl number. Phys. Fluid (1958-1988) **5**(11), 1374–1389 (1962). doi:10.1063/1.1706533
10. L. Li, C. Resagk, R. du Puits, Viscous boundary layers in turbulent Rayleigh-Bénard convection. J. Phys.: Conf. Ser. **318**(8), 082004 (2011). doi:10.1088/1742-6596/318/8/082004
11. N. Shi, M.S. Emran, J. Schumacher, Boundary layer structure in turbulent Rayleigh-Bénard convection. J. Fluid Mech. **706**, 5–33 (2012). doi:10.1017/jfm.2012.207
12. C. Sun, Y.H. Cheung, K.Q. Xia, Experimental studies of the viscous boundary layer properties in turbulent Rayleigh-Bénard convection. J. Fluid Mech. **605**, 79–113 (2008). doi:10.1017/S0022112008001365
13. W. Tollmien, Über die Entstehung der Turbulenz. 1. Mitteilung. Nachrichten von der Gesellschaft der Wissenschaften zu Göttingen. Mathematisch-Physikalische Klasse **1929**, 21–44 (1929)
14. M. van Reeuwijk, H.J.J. Jonker, K. Hanjalić, Wind and boundary layers in Rayleigh-Bénard convection. ii. Boundary layer character and scaling. Phys. Rev. E **77**, 036312 (2008). doi:10.1103/PhysRevE.77.036312
15. C. Willert, High-speed particle image velocimetry for the efficient measurement of turbulence statistics. Exp. Fluids (submitted)
16. Q. Zhou, K.Q. Xia, Measured instantaneous viscous boundary layer in turbulent Rayleigh-Bénard convection. Phys. Rev. Lett. **104**, 104301 (2010). doi:10.1103/PhysRevLett.104.104301

Large-Scale Organization of a Near-Wall Turbulent Boundary Layer

R. Dekou, J.-M. Foucaut, S. Roux and M. Stanislas

Abstract Large-scale streaky structures play an important role in the turbulence production process of a boundary layer. Adrian has proposed a model at very large scales which could explain the organization of the boundary layer, but at high Reynolds number, their main characteristics (size, intensity and life time) and the way they interact with the near-wall structures is still not fully understood. To tackle these points, an experimental database at a Reynolds number based on momentum thickness Re_θ close to 9800 was recorded in the Laboratoire de Mécanique de Lille wind tunnel with stereo-PIV (SPIV) and hot-wire anemometry (HWA). With a Linear Stochastic Estimation (LSE) procedure based on correlations computation, a three-component velocity field was reconstructed at high frequency from stereo-PIV at 4 Hz and hot-wire data at 30 kHz. To extract large streaky structures, a threshold is applied to normalized streamwise velocity fluctuations from the reconstructed PIV field, and then 3D morphological operations (erosion and dilatation) are combined with a volume-size-based cleaning procedure to remove the noise and smooth the object boundaries. Some statistical characteristics of the large streaks are obtained.

R. Dekou (✉) · J.-M. Foucaut · M. Stanislas
Laboratoire de Mécanique de Lille, Boulevard Paul Langevin,
59655 Villeneuve d'ascq Cédex, France
e-mail: raoulflorent.dekoutiomajou@ec-lille.fr

J.-M. Foucaut
e-mail: jean-marc.foucaut@ec-lille.fr

M. Stanislas
e-mail: michel.stanislas@ec-lille.fr

S. Roux
Laboratoire de Thermocinétique de Nantes, Avenue du Professeur Jean Rouxel BP 539, 44475
Carquefou cedex, France
e-mail: Stephane.Roux@univ-nantes.fr

M. Stanislas et al. (eds.), *Progress in Wall Turbulence 2*,
ERCOFTAC Series 23, DOI 10.1007/978-3-319-20388-1_29

1 Introduction

Investigation on superstructures in the log and outer regions of a turbulent boundary layer is an active field of research in turbulence. Because they play an important role in turbulence production across the boundary layer [6–10] and are thought to be responsible for small-scale amplitude modulation near the wall [15], many authors have investigated them in turbulent wall layers for various flow configurations (zero pressure-gradient turbulent boundary layer [12, 16], pipe flows [6] and channel flows [18]). From these studies, general conclusions can be drawn. First, streamwise elongated regions of alternatively high- and low-speed fluid populate the logarithmic region [1, 6, 7, 12, 14], and their streamwise length scale on δ increases with the wall-normal distance in the log region and decreases beyond [6, 8]. Together with a detailed analysis of large-scale vortices, a very Large-Scale Motion Model (VLSM) was provided by Adrian. This model suggests that hairpin-type vortices are bounding regions of low-speed fluid with ejections between their legs and sweeps outside [6–10, 12]. Investigation of the energetic contribution of hairpins packets within the flow at moderate Reynolds number $Re_\tau = 1060$ shows that the Reynolds shear stress mean value inside low uniform momentum zones is $20U_\tau^2$, $Q2$ and $Q4$ events' maximum contribution are $35U_\tau^2$ and $20U_\tau^2$, respectively, [8]. Finally, it is thought that the meandering behaviour of streaks can generate a swirling activity [12]. When we switch from low and moderate Reynolds numbers to higher ones, the structures size increases. Studies at Reynolds $Re_\tau \approx 1100$ reveal that their streamwise extent is 2δ [8, 12], and further studies at Mach 2 [6] and at $Re_\tau = 6.6 \times 10^5$ [12] reveal structures which can go up to 8δ and 20δ, respectively. This long streamwise extent, combined with the three-dimensional aspect of the structures, and their meandering behaviour complexify their extraction and analysis. Thus, experiments and DNS need to be performed at high Reynolds number to complete the existing model. It is for this purpose that an experimental database at high Reynolds number ($Re_\tau = 3610$) was built in the frame of WALLTURB project. Measurements were made in a zero pressure-gradient turbulent boundary layer over a flat plate using Stereo PIV at 4 Hz and hot-wire anemometry at 30 kHz. At high Reynolds number, the PIV sampling rate is not sufficient to follow superstructures evolution in time and the field of view is strongly limited by the available laser power. In contrast, HWA allows time-resolved measurements but limited to one velocity component. Another option is to combine experimental tools and use recent mathematical methods such as LSE in order to reconstruct a fully time-resolved field with three components. The first part of this paper describes the experimental setup, and then the LSE procedure used for reconstruction is described and a statistical validation is performed on the reconstructed field. Then, a streaks extraction algorithm is described. The last part is dedicated to the characterization of the structures: a statistical analysis of relevant quantities (diameter, life time and Reynolds stress) is carried out on coherent structures and conclusions provided.

2 Experimental Setup

The experiment was carried out in the LML wind tunnel during a WALLTURB test campaign. The test section of the wind tunnel is 20 m long with a cross section of $2 \times 1\,m^2$. The free stream velocity is 10 m/s, and the turbulence level is 0.3 %. A full description of this wind tunned can be found in [3], and the programme is described in [21].

The present experiment was carried out with a free stream velocity $U_\infty = 5\,m/s$ and a Reynolds number based on momentum thickness $Re_\theta = 9830$. A Clauser chart fit was used to estimate the friction velocity $u_\tau = 0.188$ corresponding to a Reynolds number based on friction velocity $Re_\tau = 3610$. Table 1 summarizes the main characteristics of the boundary layer. The hot-wire rake (HWR) displayed in Fig. 1a was positioned streamwise at $x = 18\,m$ from the boundary layer starting point. This hot-wire rake is made of 143 single hot-wire probes grouped in 13 vertical combs along the spanwise direction z with 11 probes on each of them. The probes are logarithmically distributed as shown in Fig. 1b. The two extreme rows of probes are located at 0.4 ± 0.2 mm and 306.9 mm, respectively, from the wall (corresponding to 7 ± 3 and 7365 wall units, respectively). The first two rows are below the PIV measurement plane and were not used in the present study. The sensing wires are 0.5 mm long and 2.5 μm in diameter ($l^+ = 11.8$ and $d^+ = 0.006$, respectively). The

Table 1 Main flow properties with δ the boundary layer thickness, u_τ the friction velocity and θ the momentum thickness

Facility	U_∞ (m/s)	T (K)	δ (m)	u_τ (m/s)	Re_δ	Re_θ	Re_τ
LML	5	288	0.28	0.188	96020	9830	3610

The Reynolds numbers are $Re_\delta = \frac{U_\infty \cdot \delta}{\nu}$, $Re_\tau = \frac{U_\tau \cdot \delta}{\nu}$ and $Re_\theta = \frac{U_\infty \cdot \theta}{\nu}$

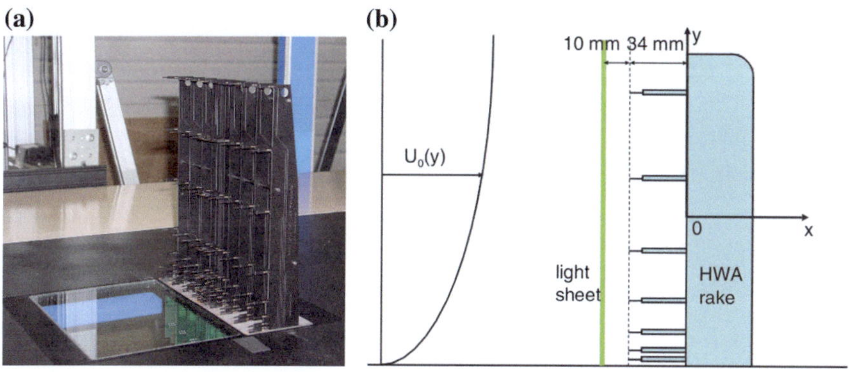

Fig. 1 **a** View of the 143 hot-wire rake built by Institut **PPRIME** installed in the LML wind tunnel. **b** Position of the rake relative to the SPIV measurement plane, probes on the rake are logarithmically distributed in wall-normal direction

acquisition time of the hot-wire signal is 6 s, the sampling frequency is 30 KHz and measurements are repeated over 534 blocks to ensure convergence.

Because hot-wire measurements are limited to one dimension, a stereo-PIV system described in [4] allows measurements at 4 Hz, and the resulting velocity field has 3 components with a spatial resolution of 2 mm in the spanwise and wall-normal directions. The PIV laser is parallel to the hot-wire rake and positioned 1 cm upstream of it as shown in Fig. 1b, and it covers the entire log region over an area of $30 \times 30\,\text{cm}^2$. Note that the PIV measurements were used to calibrate hot-wire probes, and details on the calibration procedure can be found in [20]. The anemometry system used in this experiment is described in [4].

3 Linear Stochastic Estimation

The LSE (see e.g. [11]) is used to reconstruct from the hot wire and PIV measurements of a fully time-resolved three-component velocity field with the same spatial resolution as the PIV. Given a set of observables located in space at x' and in time at t', the LSE allows the linear approximation of the conditional estimate of some quantity at a position x and time t. In our case, the conditional variables to reconstruct at high frequency are the three components of velocity field in $\mathbf{u}'(\mathbf{x}, t) = (u_1', u_2', u_3')(\mathbf{x}, t)$ in the PIV y-z plane. $\mathbf{x} = (x_1, \ldots, x_{N_p})$ and N_p is the number of PIV points. The set of observables includes the streamwise velocity $u_1'(\mathbf{x}', t')$ measured at the N_h hot-wire probes on the two-dimensional rake whose coordinates are given by $\mathbf{x}' = (x_1', \ldots, x_{N_h}')$. A single-time formulation for the linear approximation of the velocity component $\hat{u}_i(\mathbf{x}, t)$ is implemented, such as used in [5]:

$$\hat{u}_i(\mathbf{x}, t') = \sum_{k=1}^{N_h} a_{i,k}(\mathbf{x}) u_1'(x_k', t' + \tau(x_k')) \qquad i = 1, 2, 3, \tag{1}$$

where $a_{i,k}(\mathbf{x})$ are coefficients relating the conditional field to the observers, and $\tau(\mathbf{x}_k')$ is the time delay evaluated between a point $x_{i,i=1-N_p}$ in the PIV plane and the observer $x'_{k,k=1-N_h}$. The $a_{i,k}(\mathbf{x})$ coefficients are solution of a linear system of equations of the form $A\mathbf{y_i} = \mathbf{b_i}$ (i = 1, 2, 3), where $\mathbf{A} \in \mathbb{R}^{N_h \times N_h}$, $\mathbf{b}_i \in \mathbb{R}^{N_h}$, $\mathbf{y}_i \in \mathbb{R}^{N_h}$:

$$A = \begin{bmatrix} \rho_{u_1'u_1'}(x_1', x_1'; 0) & \cdots & \rho_{u_1'u_1'}(x_{N_h}', x_1'; \tau_1 - \tau_{N_h}) \\ \vdots & \ddots & \vdots \\ \rho_{u_1'u_1'}(x_1', x_{N_h}'; \tau_{N_h} - \tau_1) & \cdots & \rho_{u_1'u_1'}(x_{N_h}', x_{N_h}'; 0) \end{bmatrix}, \quad \mathbf{b_i} = \begin{bmatrix} \rho_{u_i'u_1'}(\mathbf{x}, x_1'; \tau_1) \\ \vdots \\ \rho_{u_i'u_1'}(\mathbf{x}, x_{N_h}'; \tau_{N_h}) \end{bmatrix} \tag{2}$$

$$\mathbf{y_i} = \begin{bmatrix} a_{i,1}(\mathbf{x}) \\ \vdots \\ a_{i,N_h}(\mathbf{x}) \end{bmatrix} \quad \text{and} \quad \rho_{q_1q_2}(x, x'; \tau) = \frac{1}{T} \int_{-T/2}^{T/2} q_1(x, t) q_2(x', t - \tau) dt. \tag{3}$$

The time delay $\tau_k = \tau(x'_k)$ is the one that maximizes the correlation between streamwise velocity fluctuations at a PIV point $x_{i,i=1-Np}$ and **HWA** sensor $x_{k,k=1-Nh}$. Note that the same time delay was used here for the streamwise, spanwise and wall-normal velocity components. To obtain $\mathbf{y_i}$, the system $A\mathbf{y_i} = \mathbf{b_i}$ is solved with a Tikhonov regularization procedure as in [13].

4 Statistical Analysis of the Velocity Field

In order to characterize the quality of the reconstructed data, a statistical analysis was performed on the PIV, HWR and reconstructed data and compared to those of [3].

In a previous analysis of the present PIV and HWR data, [22] did show that the HWR generates an obstruction which is maximum in the plane of symmetry and affects mostly the mean velocity field. For this reason, the side rows of the HWR are used here for comparison to minimize the influence of the obstruction. Figure 2 displays profiles of mean U^+ and root mean square (RMS) u^+ velocities. For the mean velocity, a good agreement is observed between the results of the side row of the HWR, the PIV, the results of [3] and the logarithmic law with $\kappa = 0.41$ and

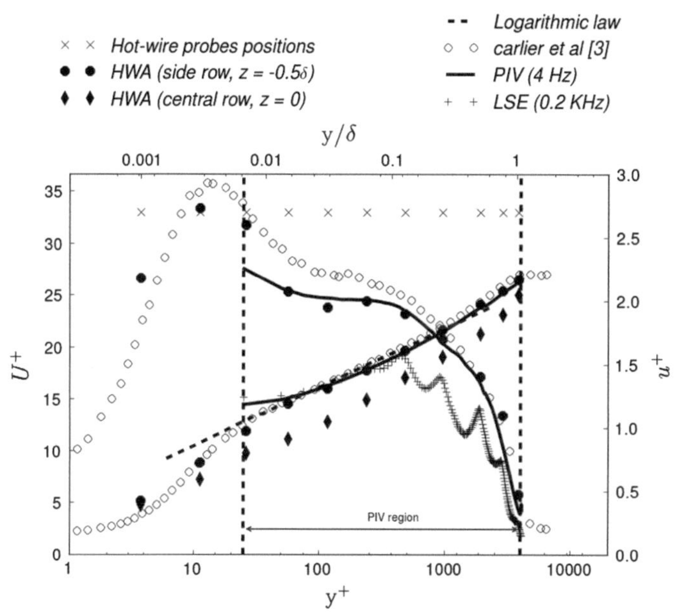

Fig. 2 Mean & RMS velocity profiles obtained here compared with those obtained by [3] using single hot-wire probes ($l^+ = 11$). The *vertical dashed lines* indicate the region of the boundary layer in wall-normal direction covered in the present paper, and the (×) denote the location of hot-wire probes along a given comb

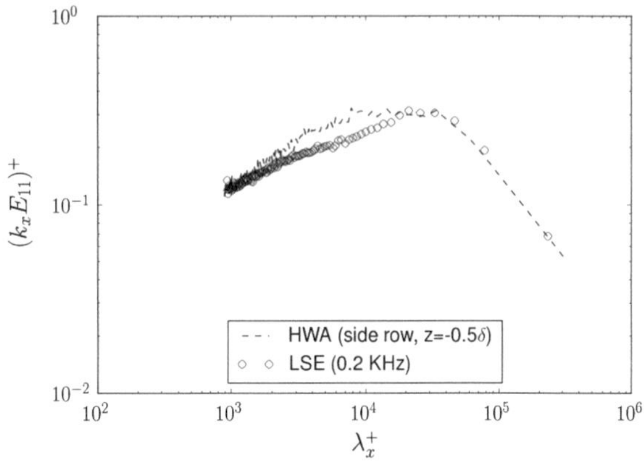

Fig. 3 Premultiplied frequency spectra of the LSE reconstructed streamwise velocity fluctuations (*circle*) compared with hot-wire measurements (*dashed lines*) at $y^+ = 959$

$C = 4.9$. The central row is also plotted to illustrate the blockage effect. Concerning the RMS, a good agreement is observed between all rows of the HWR and the PIV data. A slight underestimation is visible compared to the data of [3]. The RMS reconstructed by LSE at 0.2 KHz displays an underestimation and wavy patterns. Each crest corresponds to a hot-wire probe position highlighted by the crosses on the top part of the figure. Because the LSE is based on spatio-temporal correlations between the hot-wire signal and the original PIV field, points in the PIV domain which are between two hot-wire probes are less correlated and thus lead to lower amplitudes. The RMS of the reconstructed field follows the same trend as the one obtained by [3], but its amplitude is lower denoting a loss of energy. Part of it is also due to the filtering of the small scales by LSE reconstruction. Also the PIV plane and the HWA are separated streamwise by $\Delta x = 1\,\text{cm}$ ($\Delta x^+ = 125$), and therefore the correlation between the PIV and hot-wire measurements reaches values closed to 0.8 and not 1 as might be expected. Using Taylor hypothesis, the premultiplied spectrum of the streamwise velocity reconstructed by LSE can be plotted at different wall-normal positions and compared to the HWR data. Figure 3, for $y^+ = 959$, illustrates that the amplitude of the LSE reconstructed field is underestimated at wavelengths below 2.10^4 which correspond to about 5δ down to 3.10^3 (about δ). This loss of energy should be viewed as a broadband filtering by LSE affecting most scales but keeping the large scales of interest in the present study.

5 Detection of Streaks

5.1 Detection Algorithm

To detect streaks, a suitably chosen threshold is applied directly on the fluctuating part of the velocity field. In order to ensure normalization of the fluctuations, the fluctuating part was divided here by its standard deviation (averaged in time and

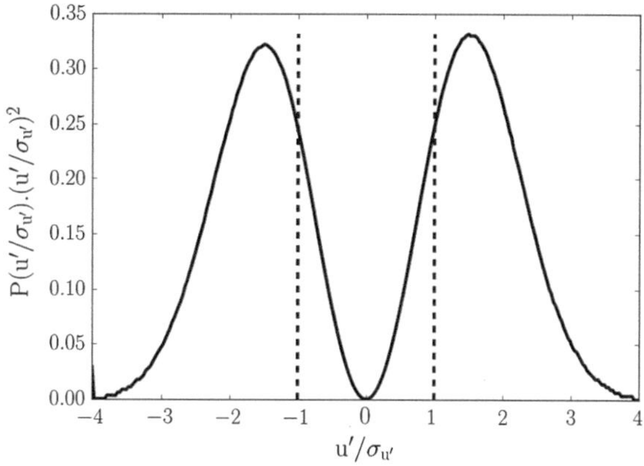

Fig. 4 Energetic contribution of $\frac{u'}{\sigma_{u'}}$ at $y^+ = 237$, the *dashed lines* provide the threshold values chosen for low- and high-speed streaks

in the spanwise direction as the flow is homogenous in these two directions). The detection function is

$$Fd_u = \frac{u'(t, y, z)}{\sigma_{u'}(y)}. \tag{4}$$

The threshold was chosen to keep the most energetic structures and to minimize the noise. The probability density function of the detection function times the energy $P(\frac{u'}{\sigma_{u'}}).(\frac{u'}{\sigma_{u'}})^2$ is displayed in Fig. 4. For low-speed streaks, a threshold value of -1 was used as it is the standard value in the literature and applied in previous studies of the group [3]. Similarly, a threshold of $+1$ was used for high-speed streaks. The thresholding of low-speed streaks and high-speed streaks, respectively, can be summarized with the following two formulas:

$$F_i = \begin{cases} 1 \ if \ Fd_x < C_t \\ 0 \ if \ Fd_x \geq C_t \end{cases} \quad and \quad F_i = \begin{cases} 1 \ if \ Fd_x > C_t \\ 0 \ if \ Fd_x \leq C_t \end{cases}. \tag{5}$$

Erosion and dilatation are classical morphological operations (Serra [19]). The dilatation (respectively erosion) of a set X by a structuring element B is the Minkowski addition (respectively substraction) of X and the transposed set $\hat{B}$ of B. The resulting set $\delta_B(X)$ (respectively ε_B) is given by

$$\delta_B(X) = X \oplus \hat{B} = \bigcup_{\substack{x \in X \\ y \in \hat{B}}} \{x + y\} \quad \left(\varepsilon_B(X) = X \ominus \hat{B} = \bigcup_{\substack{x - y \in X \\ y \in \hat{B}}} \{x\} \right). \tag{6}$$

These two operations are generally combined together in closing and opening operations. An opening operation of a set X by a structuring element B is an erosion followed by a dilatation, the resulting set X_B is written as

$$X_B = (X \ominus \hat{B}) \oplus B. \tag{7}$$

A closing operation of a set X by B is a dilatation followed by an erosion, and the resulting set X^B is written as

$$X^B = (X \oplus \hat{B}) \ominus B. \tag{8}$$

These mathematical morphology tools were used in an interactive way on the reconstructed data in order to smooth the noise existing in the data generated by the above thresholding procedure.

A cleaning procedure is generally used to remove high-frequency noise left by morphological operations. Given a threshold C_1, the cleaning step can be summarized as follows (Lin et al. [17]):

$$F_i = \begin{cases} 1 \; if \; volume > C_1 \\ 0 \; if \; volume \leq C_1 \end{cases}. \tag{9}$$

The threshold is selected in order to keep only the elongated streaky structures both low and high speed.

5.2 Results

The main objective of using morphological operations and a cleaning algorithm is to remove the noise from the original data in order to have a processable field as an output. The constraint is that these operations should not alter the size and shape of the original objects, which contain the physical information. While a rough erosion can create big holes in the field which can be hard to recover, a rough dilatation can increase the size of small objects thus increasing the noise. This is the reason why erosions, dilatations and the volume-based cleaning procedure were combined all together and tested step by step in an interactive way with caution on the size of the structuring elements and the threshold used for the cleaning. Therefore, the method used here is certainly not universal but follows the previous constraints in such a way to stay close to the raw field. In Fig. 5, an example of low-speed streaks is displayed and compared with the raw field. Values in z and y are scaled by δ. And the time is scaled with the free stream velocity $U_e = 5$ m/s and the boundary layer thickness δ so that $t^* = \frac{t \cdot U_e}{\delta}$. All the small objects are eliminated, and the main features from the raw field are smoothed. This was evident in all our reconstructed $3D$ fields of size $1000 \times 163 \times 141$ grid points. The same algorithm with the same parameters (but $C_t = +1$) was used to also extract the high-speed streaks. An example of cleaned

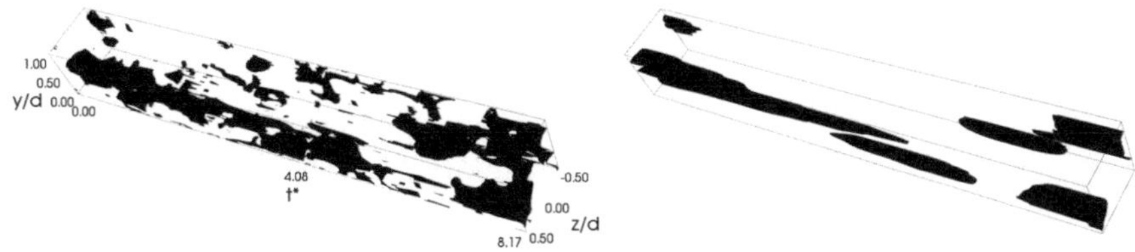

Fig. 5 Comparison between the raw field (*left*) and the cleaned velocity fluctuations (*right*)

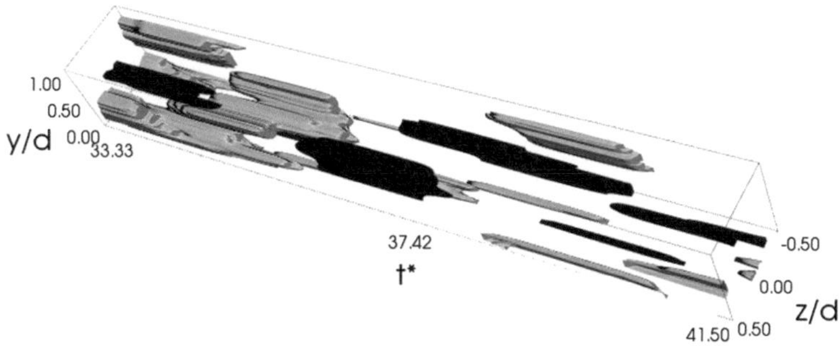

Fig. 6 Indicative functions of the low-speed streaks (*black*) and high-speed streaks (*grey*)

low- and high-speed streaks in the same field is displayed in Fig. 6 for a time interval $t^* = [33.33, 41.50]$. Alternate regions of low and high speed are evident. Based on this careful validation, this filtering procedure was applied in the whole fields where t^* range is [0, 83.25], y/δ ranges [0, 1] and z/δ ranges $[-0.5, 0.5]$ giving results similar to Fig. 6.

6 Characterization of Streaks

Once the streaks are extracted, it is possible to characterize them with their hydraulic diameter d_h in the (y,z) plane, their life time t_l and the $<u'u'>$ and $<u'v'>$ Reynolds stresses computed inside the streaks. Figure 7a, b, respectively, displays the histogram of the mean hydraulic diameter $\overline{d}_h/\delta$ of low- and high-speed streaks (dashed lines and plain lines, respectively) normalized with the boundary layer thickness and the histogram of their life time t_l^*. From these plots, it is obvious that the averaged mean hydraulic diameters and life times of low- and high-speed streaks follow the same trend with a peak at $\overline{d}_h = 0.06\delta$ and $t^* = 1.16$ for both. Figure 7c, d displays the streamwise and shear Reynolds stresses contribution to the flow from low-speed streaks (dashed lines) and high-speed streaks (plain lines) compared with the total stresses (empty circles), and the same contribution from low- and high-speed streaks together is also plotted (filled circle). The contributions from both types of streaks

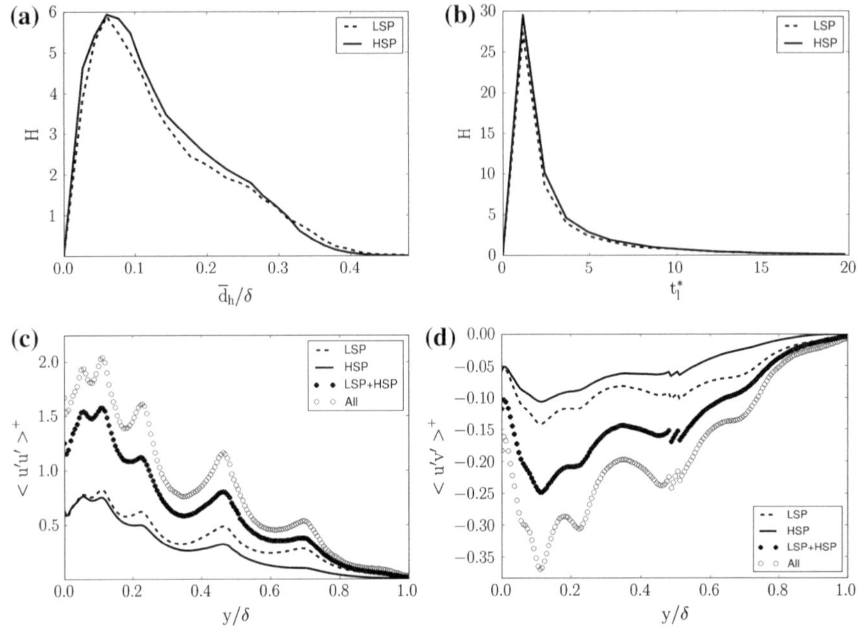

Fig. 7 **a** Histogram of the mean hydraulic diameter of streaks normalized with the boundary layer thickness, **b** Histogram of the normalized life time of streaks, **c** Streamwise Reynolds stress contribution of low- and high-speed streaks, compared with the total stress and **d** Reynolds shear stress contribution of low- and high-speed streaks compared with the total stress

are comparable for both types of stresses. Globally, the contribution from low- and high-speed streaks to $<u'u'>$ is of order of 40 and 30 %, respectively, which means that most of this Reynolds stress is generated inside these two types of structures. For $<u'v'>$, the contribution from low-speed streaks remains 40 %, and the one from high-speed streaks is a bit less (25 %), especially close to the wall with a higher contribution of high-speed streaks indicating that a significant part of this shear is generated between the streaks. The percentage of contribution from low-speed streaks peak at $y/\delta \approx 0.86$ and in high-speed streaks, this percentage is always decreasing as we move away from the wall.

7 Conclusion

Time-resolved hot-wire rake measurements at 30 KHz were combined with SPIV measurements at 4 Hz to reconstruct via LSE a 3 component velocity field resolved in space and time. A statistical analysis performed on the outcoming field shows that LSE leads to a loss of energy affecting most scales but keeping the large scales of interest. Streaks extraction reveals the existence of altering zone of low-speed fluid and high-speed fluid throughout the log region. Low-speed regions are found to be slightly more energetic than their high-speed twins but in average, their cross section and life span are nearly equal. It is also found that Reynolds stress contributions of streaks to the total stresses are important, about 40 % for low speed and between 25 % and 30 % for high speed. The perspective of this study is to extract and characterize vortical structures in order to provide a complete picture of the log region of a turbulent boundary layer.

References

1. H. Abe, H. Choi, H. Kawamura, Very large-scale structures and their effects on the wall shear-stress fluctuations in a turbulent channel flow up to Re_T= 640. J. Fluids Eng. **126**(5), 835–843 (2004)
2. R. Adrian, C. Meinhart, C. Tomkins, Vortex organization in the outer region of the turbulent boundary layer. J. Fluid Mech. **422**, 1–54 (2000)
3. J. Carlier, M. Stanislas, Experimental study of eddy structures in a turbulent boundary layer using particle image velocimetry. J. Fluid Mech. **535**, 143–188 (2005)
4. J. Delville, P. Braud, S. Coudert, J.-M. Foucaut, C. Fourment, W. George, P.B. Johansson, J. Kostas, F. Mehdi, A. Royer et al., The wallturb joined experiment to assess the large scale structures in a high Reynolds number turbulent boundary layer, *Progress in Wall Turbulence: Understanding and Modeling* (Springer, Berlin, 2011)
5. V. Durgesh, J. Naughton, Multi-time-delay lSE-POD complementary approach applied to unsteady high-Reynolds-number near wake flow. Exp. Fluids **49**(3), 571–583 (2010)
6. B. Ganapathisubramani, N. Clemens, D. Dolling, Large-scale motions in a supersonic turbulent boundary layer. J. Fluid Mech. **556**, 271–282 (2006)
7. B. Ganapathisubramani, N. Hutchins, W. Hambleton, E. Longmire, I. Marusic, Investigation of large-scale coherence in a turbulent boundary layer using two-point correlations. J. Fluid Mech. **524**, 57–80 (2005)
8. B. Ganapathisubramani, E.K. Longmire, I. Marusic, Characteristics of vortex packets in turbulent boundary layers. J. Fluid Mech. **478**, 35–46 (2003)
9. B. Ganapathisubramani, E.K. Longmire, I. Marusic, Experimental investigation of vortex properties in a turbulent boundary layer. Phys. Fluids (1994-present) **18**(5), 055105 (2006)
10. B. Ganapathisubramani, E.K. Longmire, I. Marusic, S. Pothos, Dual-plane PIV technique to determine the complete velocity gradient tensor in a turbulent boundary layer. Exp. Fluids **39**(2), 222–231 (2005)
11. Y. Guezennec, Stochastic estimation of coherent structures in turbulent boundary layers. Phys. Fluids A: Fluid Dyn. (1989–1993) **1**(6): 1054–1060 (1989)
12. W. Hambleton, N. Hutchins, I. Marusic, Simultaneous orthogonal-plane particle image velocimetry measurements in a turbulent boundary layer. J. Fluid Mech. **560**, 53–64 (2006)
13. P.C. Hansen, Regularization tools: A Matlab package for analysis and solution of discrete ill-posed problems. Numer. Algorithms **6**(1), 1–35 (1994)
14. N. Hutchins, I. Marusic, Evidence of very long meandering features in the logarithmic region of turbulent boundary layers. J. Fluid Mech. **579**, 1–28 (2007)
15. N. Hutchins, I. Marusic, Large-scale influences in near-wall turbulence. Philos. Trans. Royal Soc. A: Math. Phys. Eng. Sci. **365**(1852), 647–664 (2007)
16. N. Hutchins, J. Monty, B. Ganapathisubramani, H. Ng, I. Marusic, Three-dimensional conditional structure of a high-Reynolds-number turbulent boundary layer. J. Fluid Mech. **673**, 255–285 (2011)
17. J. Lin, J.-P. Laval, J.-M. Foucaut, M. Stanislas, Quantitative characterization of coherent structures in the buffer layer of near-wall turbulence. Part 1: streaks. Exp. Fluids **45**(6), 999–1013 (2008)

18. J. Monty, N. Hutchins, H. Ng, I. Marusic, M. Chong, A comparison of turbulent pipe, channel and boundary layer flows. J. Fluid Mech. **632**, 431–442 (2009)
19. J. Serra, *Image analysis and mathematical morphology* (Academic Press, London, 1982) [Review by E.B. Fensen in: J. Microsc. **131**, 258 (1983).] Mathematics, Review article General article, Technique Staining Microscopy, Cell size (PMBD, 185707888)
20. M. Stanislas, J.-M. Foucaut, S. Coudert, M. Tutkun, W.K. George, J. Delville, Calibration of the wallturb experiment hot wire rake with help of PIV, *Progress in Wall Turbulence: Understanding and Modeling* (Springer, Berlin, 2011)
21. M. Stanislas, J. Jimenez, I. Marusic, *Progress in Wall Turbulence: Understanding and Modeling Proceedings Of The Wallturb International Workshop Held In Lille*, France, 21–23 April 2009 (2012)
22. M. Tutkun, W.K. George, J.-M. Foucaut, S. Coudert, M. Stanislas, J. Delville, In situ calibration of hot wire probes in turbulent flows. Exp. Fluids **46**(4), 617–629 (2009)

Near-Wall Study of a Turbulent Boundary Layer Using High-Speed Tomo-PIV

Fabio J.W.A. Martins, Jean-Marc Foucaut, Luis F.A. Azevedo and Michel Stanislas

Abstract The fundamental study of the near-wall structure organization in turbulent flows is crucial to understand the self-generation process of turbulence. To investigate such phenomena, an experiment of high-repetition, 6-camera tomo-PIV in a boundary layer was performed. Vector fields generated from BIMART high-quality reconstructed volumes resulted in low measurement uncertainties. The comparison of turbulence statistics from tomographic PIV and hot-wire anemometer data shows an excellent agreement. Preliminary vortex detection from Q-criterion is presented and allows the identification of dispersed vortices around the low-speed streaks in the boundary layer. Nevertheless an accurate identification of turbulent structures is not yet achieved. The postprocessing is being reviewed and the discussion of the interaction and evolution of turbulent structures will be addressed in a future paper.

1 Introduction

The turbulence structure near the wall in a boundary layer has challenged researchers over the last six decades due to its importance as a driver in innumerous practical engineering applications. In this turbulent flow, kinetic energy from the free flow is converted into turbulent fluctuations and then dissipated into internal energy by the viscosity. A population of many eddies of different scales interact with each other in a complex phenomenon of a continuous self-sustaining process. The presence of quasi-periodic patterns of coherent motion in the flow seems to be responsible for the maintenance of turbulence in a boundary layer [1, 5, 23, 27]. Nonetheless, the near-wall turbulence structures, their evolution, and their interplay are still not fully understood [9, 28].

F.J.W.A. Martins · J.-M. Foucaut · M. Stanislas (✉)
Laboratoire de Mécanique de Lille (LML), Villeneuve-d'Ascq, France
e-mail: michel.stanislas@ec-lille.fr

L.F.A. Azevedo
Mechanical Engineering Department, PUC-Rio, Rio de Janeiro, Brazil

M. Stanislas et al. (eds.), *Progress in Wall Turbulence 2*,
ERCOFTAC Series 23, DOI 10.1007/978-3-319-20388-1_30

Turbulence measurements achieved a notorious improvement with the recent advances in Particle Image Velocimery (PIV). Nowadays, variations of this nonintrusive technique are able to capture the full three-dimensionality of unsteady flows by means of multi-plane PIV [15], three-dimensional particle tracking velocimetry (3D PTV) [16], defocusing PIV [21], holographic and digital holographic PIV (HPIV and DHPIV) [26], scanning-PIV [13], and tomo-PIV [8]. This work explores the latter, which has a high potential of 3D-3C velocity measurement [24].

The tomographic PIV approach is based on the reconstruction of the particle distribution inside an illuminated volume of a seeded flow. The light scattered by the particles is recorded by multiple, simultaneous camera views, from which the images are used to reconstruct the volume. The instantaneous flow is estimated by the particle displacement between two light pulses, through a 3D cross-correlation over a pair of reconstructed volumes calculated from these views [8]. The reconstruction of the particles from a limited number of views is an indeterminated system solved by algebraic techniques, which are very time consuming [2, 29, 33]. Fortunately, alternative methods for optimizing the reconstruction process were created in order to improve the quality and to decrease the computational time [7, 20, 22, 33].

Recently, Thomas and co-workers [29] proposed the block-iterative multiplicative algebraic reconstruction technique (BIMART) [4], which, in their synthetic data, obtained an equivalent accuracy as the classic MART [8] spending half of its processing time.

In the present study, a 6-camera high-repetition tomo-PIV was used to study the unsteady character of the near-wall boundary layer flow over a flat plate using a spanwise-wall-normal thin volume. The final purpose of the work is to access the full velocity gradient tensor and to reconstruct the time history of the turbulent structures.

2 Experimental Setup

The turbulent boundary layer tomo-PIV experiment took place in the Laboratoire de Mécanique de Lille (LML) large-sized wind tunnel, whose boundary layer can reach up to 300 mm thickness for a Reynolds number based on the momentum thickness, R_θ, of 8000. The tunnel test section is 1 m high, 2 m wide and 20 m long to allow for the development of the boundary layer [5]. The closed-loop wind tunnel is able to maintain the air temperature to within ± 0.2 K and presents a free streamwise velocity stabilization of $\pm 0.5\,\%$.

The tomographic PIV arrangement was composed of six high-speed CMOS cameras: three Phantom V9, one Phantom V10, and two Photron Fastcam APX [11]. All of them recorded a spanwise-wall-normal volume produced by a Quantronix laser (2×30 mJ@1kHz) as can be seen in Fig. 1. The cameras 1–6 were mounted on Scheimpflug adapters positioned under the wind tunnel in a forward scattering configuration with θ_z angles of about $-35°$, $-145°$, $-145°$, $-35°$, $-135°$ and $-45°$, subsequently, and θ_y of 35°, 145°, $-145°$, $-35°$, 180° and 0° as depicted in Fig. 2. The Phantom cameras (1600×1200 pixels@1kHz and pixel size of 11.5 μm)

Fig. 1 Cameras imaging an illuminated volume (upper-center position in the picture) assembled in a tomo-PIV arrangement under the wind tunnel

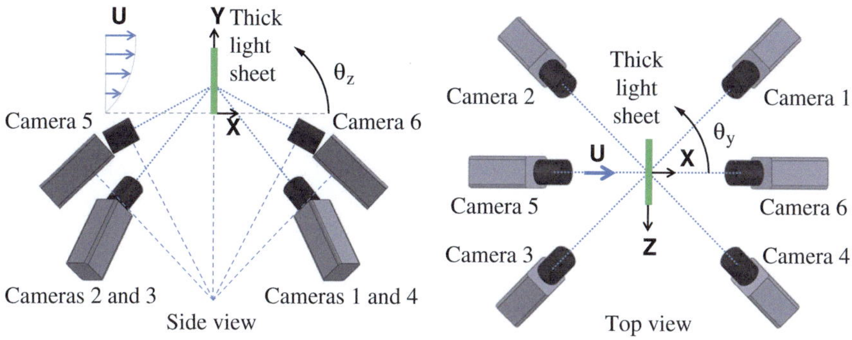

Fig. 2 Sketch of the cameras arrangement in the boundary layer experimental setup

were equipped with micro Nikkor 200 mm lenses at f# 5.6, while the Photron cameras (1024 × 1024 pixels@2kHz and pixel size of 17 μm) were equipped with micro Nikkor 105 mm lens at f# 5.6 combined with a doubler. The laser volume thickness was carefully limited to 5 mm by a knife-edge filter to create an investigation volume of $5 \times 45 \times 45\,mm^3$ (i.e. 40 × 360 × 360 wall units). This thin-volume was chosen in order to overcome some difficulties with low light energy [24] and to improve the measurement accuracy [3].

The whole flow was seeded with polyethylene glycol smoke, which generates particles with sizes of about 1 μm. The PIV images were recorded at 1 kHz with a time delay between pulses of 300 μs. The maximum gray level, not including reflections at the wall, was 155 in 8-bit images and the particle-image size was about 1 to 2 pixels. A total of 5 runs of 1725 tomo-PIV fields from an average particle per pixel of 0.05 ppp were recorded.

The cameras were calibrated using a pinhole model [32]. The calibration target was transparent in order to allow the visualization of the markers from both sides.

It was made by printing a crosses pattern on a transparent plastic sheet, which was sandwiched between two 2 mm-thick glass plates. The yz-target plane was scanned in seven equally spaced, out-of-plane positions from $x = -3$ to $x = 3$ mm. A micrometer with 5 μm precision was used as a translation stage. The crosses pattern presented a distance of 5 mm between two consecutive markers in both y and z directions.

Due to the form that the target was manufactured, the refraction through the 2 mm-thick glass plates created significant triangulation errors. After the removal of the target, the error between the position of the markers along x from the cameras placed downstream and the ones upstream were greater than 2 mm. So, a large number of spurious particle matchings appeared and spread the disparity peak, making the correction of the mapping function by the standard self-calibration method [31] very difficult. To circumvent this problem, the target planes along x were artificially translated by 1.1 mm for the downstream cameras and -1.1 mm for the upstream ones. These values were computed taking into account the refraction properties. An initial self-calibration was applied using the pinhole model with a longitudinal translation on low particle density images followed by three more self-calibration steps for each experimental run. The final pinhole mapping functions had projection errors of about 0.01 pixel and maximum triangulation errors lower than 0.18 pixel, which is enough to produce good quality reconstruction volumes [8, 31]. It is noteworthy that this problem is not faced in the stereoscopic PIV [9].

3 Volume Reconstruction and PIV Processing

The reconstruction was performed using the BIMART [29], given by the equation

$$E(x_j, y_j, z_j)^{n+1} = E(x_j, y_j, z_j)^n \prod_{i,k \in B_Q \cap N_j} \left(\frac{I_k(x_i, y_i)}{\sum\limits_{l \in L_{i,k}} w_{i,l,k} E(x_l, y_l, z_l)^n} \right)^{\mu w_{i,j,k}}, \tag{1}$$

where $I(x, y)$ are the image pixel intensity distribution, $E(x, y, z)$ are the voxel intensity distribution inside the reconstructed volume, $L_{i,k}$ represents the voxels intercepted by the line of sight (LOS) of the ith pixel in the kth camera, $w_{i,j,k}$ is a weight function related to the contribution of the jth voxel (x_j, y_j, z_j) in these pixels, N_j represents the pixels that have a LOS crossing a given voxel j, B_Q contains the pixels inside the block Q, n indicates the iteration and μ is a relaxation parameter.

In order to improve the volume reconstruction quality and to reduce the time requirements, a meticulous analysis was performed. The optimization of the tomo-PIV parameters together with several image preprocessing methods were studied on synthetic and experimental data [17, 29]. As a result, a conservative 8-iteration BIMART with a block size equal to four, with an initialization based on the minimum line of sight (MinLOS) [18] and with 2-iteration volume filter of 0.004 was applied to reconstruct the volumes with a voxel size of 71 μm. The 6-camera images were preprocessed using a time-average background subtraction followed by a 3×3 Gaussian filtering before the reconstruction procedure [17].

From a pair of volumes comprising the particle distribution, the velocity fields were computed by means of a 3D multipass cross-correlation with subpixel shift. Two initial steps with an interrogation window size of $36 \times 72 \times 72$ voxels were performed followed by three final steps with a window size of $36 \times 36 \times 36$ voxels and 75 % overlap. The resulting 8625 vector fields of $5 \times 67 \times 67$ grid were validated by a normalized median filtering [30] and the spurious vectors, amounting to about 1.5 %, were successfully replaced by the interpolation of their adjacent neighbors.

The tomo-PIV software was developed in C^{++} as a result of the partnership between Pprime (Poitiers), Coria (Rouen), and LML (Lille) laboratories. The post-processing is based on the Matlab platform.

4 Results and Discussion

The comparison of the velocity statistics of tomographic PIV against the standard hot-wire anemometer data (HWA) [5] are presented in the Fig. 3. From Fig. 3a, b, it can be verified that the tomo-PIV and HWA velocity profiles virtually collapse and that they perfectly follow the theoretical curves. Figure 3c shows an excellent agreement between the turbulent–velocity fluctuation profiles from tomo-PIV and HWA. However, the tomo-PIV data is slightly under the HWA curve, which insinuates that the size of interrogation window is filtering the data. A small difference is observed at points near the wall where the reflections of the laser light were stronger. It is also important to mention that the $\sqrt{<v'^2>}/u_{\tau_w}$ near the wall for the hot-wire anemometer is overestimated due to the size of the probe. From Fig. 3d, a perfect concordance with the Van Driest model is observed. Finally, the agreement between the tomo-PIV and HWA is an indication that the record time of 8.62 s is sufficient to reach converged statistics.

Figure 4 compares the probability densities of the three velocity components calculated at 50 wall units by tomo-PIV and HWA. A small discrepancy is observed for $v'/\sqrt{<v'^2>}$ and $w'/\sqrt{<w'^2>}$ measured by the two techniques probably due to a slight difference in the HWA and tomo-PIV experimental conditions as the data were collected at distinct moments. In Fig. 4c, a marginal asymmetry is observed in the tomo-PIV that is seemingly caused by a little reminiscence of the peak-locking effect, since the deformation method was not applied to the interrogation window during the correlation process [24].

From the 3C-3D velocity field, the velocity gradients in x, y and z directions were computed by a second-order central difference scheme [10] to access both the velocity gradient tensor and the divergence of the fluctuating velocity.

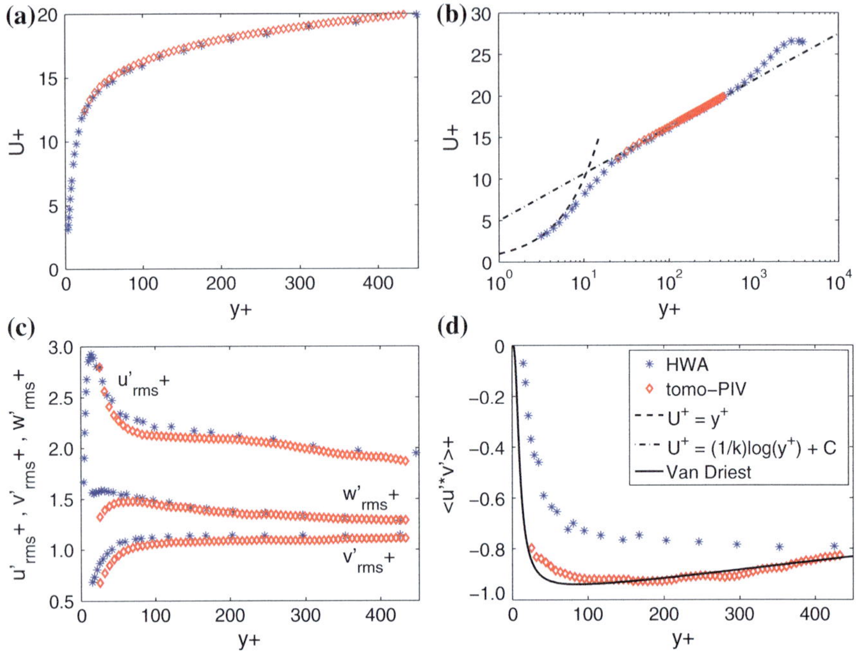

Fig. 3 Streamwise velocity profile in **a** linear and **b** semi-log scales and turbulent velocity fluctuations **c** $\sqrt{<u'^2>}/u_{\tau_w}$, $\sqrt{<v'^2>}/u_{\tau_w}$ and $\sqrt{<w'^2>}/u_{\tau_w}$ and **d** $\sqrt{<u'v'>}/u_{\tau_w}$

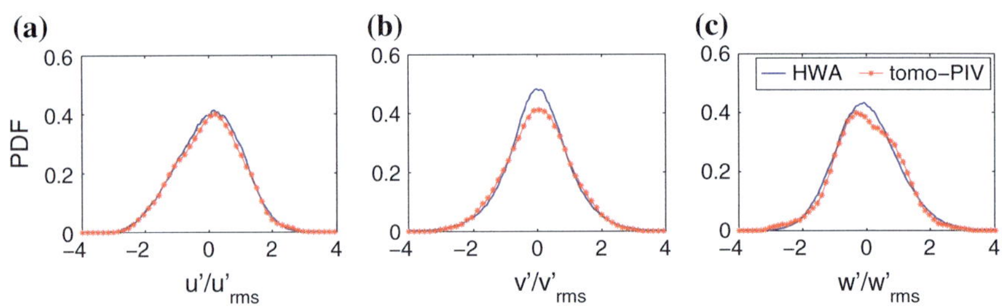

Fig. 4 Probability density functions at $y^+ = 50$ for **a** $u'/\sqrt{<u'^2>}$, **b** $v'/\sqrt{<v'^2>}$ and **c** $w'/\sqrt{<w'^2>}$

The analysis of the local velocity-fluctuation divergence allows us to estimate the uncertainty on the velocity field [3, 19]. Since, for an incompressible flow without error measurements, the divergence of the velocity must be zero, the uncertainty in the gradient $\delta(\partial u'_i/\partial x_i)$ can be computed as the root-mean-square of the divergence. Assuming that the vector spacing and the uncertainty $\delta(u)$ in each direction are uniform, the uncertainty on the velocity can be given as

$$\delta(u) = \sqrt{\frac{2\Delta^2}{3}}\delta\left(\frac{\partial u'_i}{\partial x_i}\right), \tag{2}$$

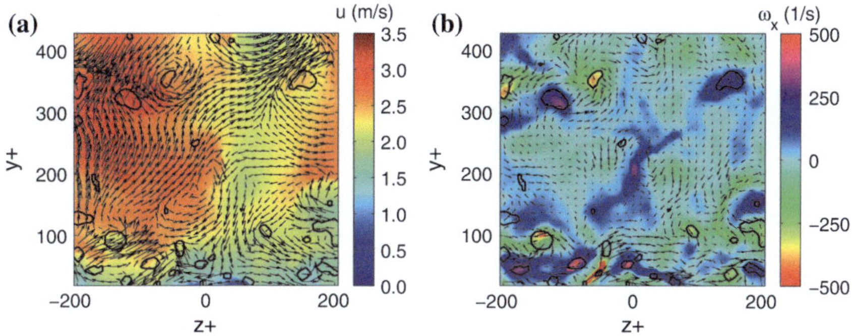

Fig. 5 Comparison between an instantaneous **a** velocity field and its correspondent **b** vorticity. The color represent the third component and the black isolines are the Q-criterion for both plots

where Δ is the physical spacing between velocity vectors, in this work equal to 635 μm (i.e. 6.5 wall units).

The root-mean-square of the divergence through the volume was 49.3 s^{-1}, which corresponds to a velocity uncertainty of 0.0256 m/s (equivalent to 0.12 voxels). This good value for a tomo-PIV air experiment was probably achieved due to the measurement in a thin volume and the use of six cameras. The uncertainty level obtained is comparable to the literature [3, 25].

Preliminary results for the detection of turbulent structures are presented in Fig. 5. The middle plane of an instantaneous 3D-velocity field from tomographic PIV is plotted together with the corresponding 3D-vorticity field. The vortical detection was made by the Q-criterion, whose isolines are embedded in the plots. To enhance the visualization, only one over two vectors are displayed. As can be seen in the Fig. 5b, the isolines appear in the expected regions of high vorticity magnitude.

The Q-criterion [14] identifies vortices as flow regions with positive second invariant of the velocity gradient tensor, i.e., $Q > 0$, with the requirement of pressure in the eddy region to be lower than the ambient pressure. In this work, the additional pressure condition was not used as it is done by other researchers [6]. The Q-criterion for an incompressible flow is defined as

$$Q = -\frac{1}{2}\frac{\partial u_i}{\partial x_j}\frac{\partial u_j}{\partial x_i}. \tag{3}$$

Since the tomo-PIV data from the present experiment are time resolved, 3D plots, with the third dimension being time, complement the 2D analysis, helping in the inference of turbulent structure dynamics and evolution. In favor of the vortical-structure clarification, the elimination of noise in the vortex detection was made by a robust discrete-cosine-transform filter [12] followed by a volume filtering in the labelized field. The visualization of low-speed streaks together with vortical structures, before and after filtering, are displayed in the Fig. 6. From the sample results presented from this ongoing work, it is possible to identify vortices neighboring the

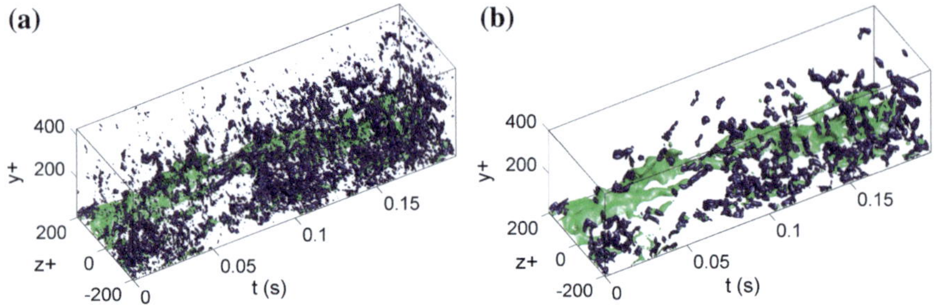

Fig. 6 Interaction between low-speed streaks (in *green*) and turbulent vortical structures (in *blue*) inside a turbulent boundary layer from **a** raw and **b** filtered data

low-speed streaks. The results, however, still do not present sufficient resolution to allow the distinction of boundary layer turbulence structures.

The preliminary results exposed in the present paper clearly indicate the potential of the experiment to allow a comprehensive analysis of boundary layer turbulence structures.

5 Conclusions

A tomo-PIV technique with six high-speed cameras was applied to study the unsteady character of the near-wall flow in a wind tunnel boundary layer for a momentum-thickness Reynolds number of 8000.

The cameras were carefully calibrated using a pinhole model and a modified self-calibration procedure. Due to refraction in the target, a procedure of artificial translation of target planes was necessary to minimize the errors.

The optimization of the image preprocessing and the volume reconstruction was successfully performed. The vector fields generated from BIMART reconstructed volumes resulted in low measurement errors, which were estimated by the velocity-fluctuation divergence.

Turbulence statistics from over 8600 vector fields were presented. An excellent agreement between tomographic PIV and hot-wire anemometer data is observed in terms of velocity profiles and turbulent fluctuations. Also, the experimental data match theoretical curves, which reveals that the recording time was sufficient to obtain converged statistics.

The probability density functions of the velocity fluctuations from the tomo-PIV method were also compared against the hot-wire anemometer showing good agreement.

Preliminary vortex detection results were demonstrated in terms of vorticity and Q-criterion, which allow the identification of some vortices interacting with low-speed streaks in the wall boundary layer. Nevertheless, an accurate identification of turbulent structures was not yet achieved and is an ongoing work.

Acknowledgments This work was carried out in the frame of the joint supervision of PhD of Fabio Martins held at both PUC-Rio (Brazil) and EC-Lille (France). It was funded by the PUC-Rio and the Brazilian scholarship CAPES grant no. BEX 9249/12-5. The experiment had the financial support of AFDAR European project, ANR Vive3D contract, and CISIT. The tomo-PIV software was developed as a result of the partnership between Pprime (Poitiers), Coria (Rouen), and LML (Lille) laboratories in the frame of the VIV3D ANR project. L. David, B. Tremblais, and P. Braud—from Pprime—and B. Lecordier, G. Godard and C. Gobin—from Coria—are acknowledged for the cooperation in the tomo-PIV software and S. Coudert and A.C. Avelar for the participation in the experiment.

References

1. R. Adrian, C. Meinhart, C. Tomkins, Vortex organization in the outer region of the turbulent boundary layer. J. Fluid Mech. **422**, 1–54 (2000)
2. C. Atkinson, J. Soria, An efficient simultaneous reconstruction technique for tomographic particle image velocimetry. Exp. Fluids **47**(4–5), 553–568 (2009)
3. C. Atkinson, S. Coudert, J.-M. Foucaut, M. Stanislas, J. Soria, The accuracy of tomographic particle image velocimetry for measurements of a turbulent boundary layer. Exp. Fluids **50**(4), 1031–1056 (2011)
4. C. Byrne, Block-iterative algorithms. Int. Trans. Oper. Res. **16**(4), 427–463 (2009)
5. J. Carlier, M. Stanislas, Experimental study of eddy structures in a turbulent boundary layer using particle image velocimetry. J. Fluid Mech. **535**(36), 143–188 (2005)
6. P. Chakraborty, S. Balachandar, R. Adrian, On the relationships between local vortex identification schemes. J. Fluid Mech. **535**(2005), 189–214 (2005)
7. S. Discetti, T. Astarita, A fast multi-resolution approach to tomographic PIV. Exp. Fluids **52**(3), 765–777 (2012)
8. G. Elsinga, F. Scarano, B. Wieneke, B.W. van Oudheusden, Tomographic particle image velocimetry. Exp. Fluids **41**(6), 933–947 (2006)
9. J.-M. Foucaut, S. Coudert, M. Stanislas, J. Delville, Full 3D correlation tensor computed from double field stereoscopic PIV in a high Reynolds number turbulent boundary layer. Exp. Fluids **50**(4), 839–846 (2011)
10. J.-M. Foucaut, M. Stanislas, Some considerations on the accuracy and frequency response of some derivative filters applied to PIV vector fields. Meas. Sci. Technol. **13**(7), 1058–1071 (2002)
11. J.-M. Foucaut, S. Coudert, A. Avelar, B. Lecordier, G. Godard, C. Gobin, L. Thomas, P. Braud, L. David, Experiment of high repetition tomographic PIV in a high Reynolds number turbulent boundary layer wind tunnel, in *PIV'11—Ninth International Symposium on Particle Image Velocimetry*. Kobe, Japan (2011)
12. D. Garcia, Robust smoothing of gridded data in one and higher dimensions with missing values. Comput. Stat. Data Anal. **54**(4), 1167–1178 (2010)
13. T. Hori, J. Sakakibara, High-speed scanning stereoscopic PIV for 3D vorticity measurement in liquids. Meas. Sci. Technol. **15**(6), 1067 (2004)
14. J. Hunt, A. Wray, P. Moin, Eddies, streams, and convergence zones in turbulent flows, in *Studying Turbulence Using Numerical Simulation Databases*, vols. 1, 2, (1988), pp. 193–208
15. C.J. Kähler, J. Kompenhans, Fundamentals of multiple plane stereo particle image velocimetry. Exp. Fluids **29**(1), S070–S077 (2000)
16. H.G. Maas, A. Gruen, D. Papantoniou, Particle tracking velocimetry in three-dimensional flows. Exp. Fluids **15**(2), 133–146 (1993)
17. F. Martins, J.-M. Foucaut, L. Thomas, L. Azevedo, M. Stanislas, Volume reconstruction optimization for tomo-PIV experimental data, in *Final International Workshop on Advanced Flow Diagnostics for Aeronautical Research—AFDAR*, Lille, France (2014)

18. D. Michaelis, M. Novara, F. Scarano, B. Wieneke, Comparison of volume reconstruction techniques at different particle densities, in *15th International Symposium on Applications of Laser Techniques to Fluid Mechanics*. Lisbon, Portugal (2010)
19. R. Moffat, Describing the uncertainties in experimental results. Exp. Therm. Fluid Sci. **1**(1), 3–17 (1988)
20. M. Novara, K. Batenburg, F. Scarano, Motion tracking-enhanced MART for tomographic PIV. Meas. Sci. Technol. **21**(3), 35401 (2010)
21. F. Pereira, M. Gharib, D. Dabiri, D. Modarress, Defocusing digital particle image velocimetry: a 3-component 3-dimensional DPIV measurement technique. Application to bubbly flows. Exp. Fluids **29**(1), S078–S084 (2000)
22. S. Petra, C. Schnörr, A. Schröder, B. Wieneke, Tomographic image reconstruction in experimental fluid dynamics: synopsis and problems, in *Mathematical Modelling of Environmental and Life Sciences Problems* (2007)
23. S. Robinson, Coherent motions in the turbulent boundary layer. Annu. Rev. Fluid Mech. **23**(1), 601–639 (1991)
24. F. Scarano, Tomographic PIV: principles and practice. Meas. Sci. Technol. **24**(1), 012001 (2013)
25. F. Scarano, C. Poelma, Three-dimensional vorticity patterns of cylinder wakes. Exp. Fluids **47**(1), 69–83 (2009)
26. U. Schnars, W. Jüptner, Direct recording of holograms by a CCD target and numerical reconstruction. Appl. Opt. **33**(2), 179–181 (1994)
27. W. Schoppa, F. Hussain, Coherent structure generation in near-wall turbulence. J. Fluid Mech. **453**(1), 57–108 (2002)
28. M. Stanislas, L. Perret, J.-M. Foucaut, Vortical structures in the turbulent boundary layer: a possible route to a universal representation. J. Fluid Mech. **602**, 327–382 (2008)
29. L. Thomas, B. Tremblais, L. David, Optimisation of the volume reconstruction for classical tomo-PIV algorithms (MART, BIMART and SMART): synthetic and experimental studies. Meas. Sci. Technol. **25**(3), 035303 (2014)
30. J. Westerweel, F. Scarano, Universal outlier detection for PIV data. Exp. Fluids **39**(6), 1096–1100 (2005)
31. B. Wieneke, Volume self-calibration for 3D particle image velocimetry. Exp. Fluids **45**(4), 549–556 (2008)
32. C. Willert, Stereoscopic digital particle image velocimetry for application in wind tunnel flows. Meas. Sci. Technol. **8**(12), 1465–1479 (1997)
33. N. Worth, T. Nickels, Acceleration of Tomo-PIV by estimating the initial volume intensity distribution. Exp. Fluids **45**(5), 847–856 (2008)

The Effects of Superhydrophobic Surfaces on Skin Friction Drag

Hyunwook Park and John Kim

Abstract Direct numerical simulation of a turbulent boundary layer developing over a superhydrophobic surface (SHS) was performed in order to investigate the effects of SHS on skin friction drag. Significant modifications of near-wall turbulence structures were observed, which resulted in large skin friction drag reduction. For the considered Reynolds number ranges and SHS geometries, it was found that the effective slip length normalized by viscous wall units was a key parameter. It was shown that the effective slip length should be on the order of the buffer layer in order to have the maximum benefit of drag reduction. It was also found that the width of SHS, relative to the spanwise width of near-wall turbulence structures, was also a key parameter to the total amount of drag reduction. Similarities and differences between the present turbulent boundary layer over SHS and our earlier work of turbulent channel flows with SHS are also discussed.

1 Introduction

Recently, achieving substantial skin friction drag reduction by using superhydrophobic surfaces (SHSs) in high Reynolds number turbulent flows has received much attention. An SHS, a combination of hydrophobicity and surface roughness in micro- and/or nano-scales, can lower the free energy of an air–water interface. In certain situations, air pockets are stably entrapped between the surface features, thereby creating a significant effective slip length on the surface fully submerged in moving water [1, 2]. It is noteworthy that SHSs with an effective slip length on the order of 200–400 μm have been reported [3, 4]. The prospect of reducing skin friction drag without additional energy input, in contrast to other active control schemes (e.g., blowing and suction on the surface and air lubrication through air bubbles), has been the main reason for interest in SHSs. The UCLA group has demonstrated that

H. Park (✉) · J. Kim
Department of Mechanical and Aerospace Engineering, University of California, Los Angeles, 420 Westwood Plaza, Los Angeles 90095-1597, CA, USA
e-mail: hwpark@ucla.edu

J. Kim
e-mail: jkim@seas.ucla.edu

M. Stanislas et al. (eds.), *Progress in Wall Turbulence 2*,
ERCOFTAC Series 23, DOI 10.1007/978-3-319-20388-1_31

superhydrophobic state under water in the presence of adverse conditions could be maintained through an electrolysis [5]. Samaha et al. [6] reported that the stability of underwater superhydrophobic states was improved by considering the distribution of SHS geometries.

Although some key aspects of SHSs relevant to skin friction drag reduction have been investigated, experimental observations of noticeable drag reduction on SHSs have been elusive and largely limited to laminar flows [2]. In turbulent flows, from a noticeable amount of drag reduction [7–10] to large drag reduction [11] have been reported. Meanwhile, some researchers [12, 13] reported that the effects of SHSs on drag reduction were negligible. In numerical studies, where superhydrophobicity was assumed to prevail and SHSs were modeled through a boundary condition, several investigators have reported significant drag reduction in both laminar and turbulent channel flows. The drag reduction in laminar flows is a direct consequence of the effective slip on SHS [14], whereas the mechanism by which drag reduction was achieved in turbulent flows is not well understood [15–17]. Recently, Busse and Sandham [18] found that the changes in streamwise velocity and wall shear–stress fluctuations were bounded together to the variation of turbulent skin friction drag on SHS. Park et al. [19] reported the effects of SHSs on turbulence structures and the resulting skin friction drag reduction in turbulent channel flows. They also showed that the effective slip length normalized by viscous wall units was the key parameter to characterize the performance of SHSs in turbulent channel flows.

All numerical studies to date were conducted in a turbulent channel flow, whereas most experiments were conducted in a turbulent boundary layer, thus preventing direct comparisons. In the present study, direct numerical simulations (DNS) of a turbulent boundary layer (TBL) developing over SHS were carried out in order to gain further insights into the mechanism by which skin friction drag in a TBL was affected by SHS. In Sect. 2, details of numerical experiments are described. The effects of SHS on turbulence structures and skin friction drag in a TBL are presented in Sect. 3, followed by a summary and concluding remarks in Sect. 4.

In this paper, u, v, w denote, respectively, streamwise (x), wall-normal (y) and spanwise (z) velocity, δ^* and u_τ denote the displacement thickness and a wall shear velocity, and superscipt $^+$ denotes a variable normalized by the kinematic viscosity, ν and the wall shear velocity. Subscript $_{\rm in}$ denotes a flow property at the inflow generated through a rescaling process, thus Reynolds numbers (Re) of the inflow are defined as $Re_{\delta^*_{\rm in}} = U_\infty \delta^*_{\rm in}/\nu$ and $Re_{\tau,\rm in} = u_{\tau,\rm in}\delta_{\rm in}/\nu$ using the inlet boundary layer displacement thickness ($\delta^*_{\rm in}$), the inlet boundary layer thickness ($\delta_{\rm in}$), the free stream velocity (U_∞), and the inlet wall-shear velocity ($u_{\tau,\rm in}$).

2 Numerical Details

The effects of SHS, consisting of microgrates aligned with the mean-flow direction, on skin friction drag were investigated through DNS of a TBL. A semi-implicit, modified fraction step method [20, 21] was used. The wall-normal viscous terms were treated implicitly using the Crank–Nicolson method, while all other terms were treated explicitly using a low-storage third-order Runge–Kutta method (RK3). All spatial derivatives were computed using the second-order central difference method on staggered grids, which, in the absence of viscous terms, numerically conserved the kinetic energy. The original channel code for SHS flows [19] was validated by comparing the present results with those by Martell et al. [16] and Min and Kim [15]. Except for the Poisson solver for pressure and the inflow and outflow boundary conditions, other features of the present TBL code were essentially the same as those of the established turbulent channel code. This TBL code was validated by comparing present results with previous TBL results [22, 23].

In order to perform DNS of a spatially developing TBL, an inflow boundary condition is also required. We incorporated the rescale-and-recycle technique developed by Lund et al. [22]. The incoming turbulent boundary layer first developed over a regular smooth wall, followed by an SHS downstream, as shown in Fig. 1. The recycling station was located within a regular smooth wall region, and it was located at $40\delta^*_{\text{in}}$ ahead of the SHS so that the flow field at the recycling station was not affected by the upstream influence of the SHS. A convective boundary condition, satisfying the global mass conservation, was used at the exit.

SHSs were modeled through a shear-free boundary condition on the air–water interface and no-slip boundary condition on the top surface of microridges, aligned in the direction of flow [16, 17, 19]. That is,

$$\left.\frac{\partial u}{\partial y}\right|_{\text{wall}} = 0, \quad \left.\frac{\partial w}{\partial y}\right|_{\text{wall}} = 0, \tag{1}$$

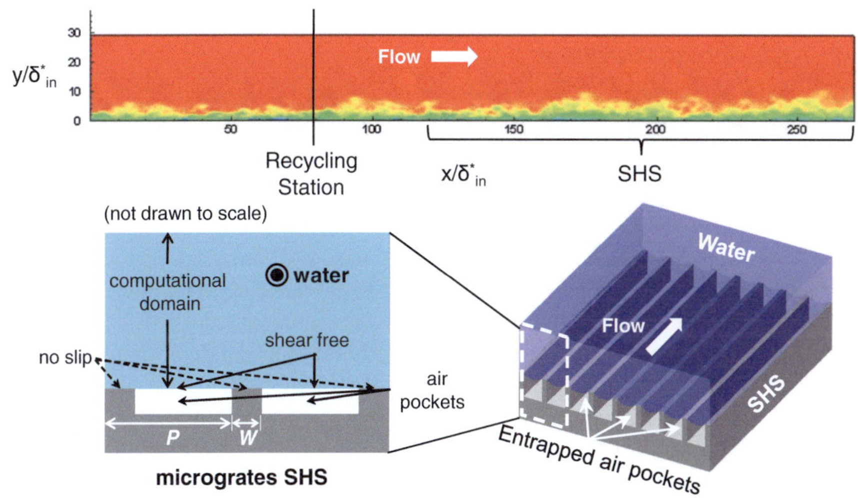

Fig. 1 A schematic of a TBL over SHS with micrograts aligned along the mean-flow direction

on shear-free interfaces and no-slip boundary conditions on solid surfaces since this is more convenient for investigating the effects of SHS geometries than Navier's slip model with a prescribed slip length used by Min and Kim [15]. Assuming flat and nondeformable air–water (i.e., shear-free) interfaces, the shear-free boundary condition was imposed implicitly using the staggered grids, i.e., without a finite-difference approximation for the boundary condition. The effective slip length (b) is defined and nondimensionalized as follows:

$$U_{\text{slip}} = b \left. \frac{\partial u}{\partial y} \right|_{\text{wall}}, \quad b^+ = \frac{b u_\tau}{\nu}. \tag{2}$$

In order to investigate the effects of Reynolds number and SHS geometry variables such as gas fraction ratio (*GF*) and the pitch (P), parametric studies were performed. Considered Reynolds numbers were relatively low as $Re_{\tau,\text{in}} = 140$ ($Re_{\delta^*_{\text{in}}} = 500$) and 200 ($Re_{\delta^*_{\text{in}}} = 720$). These Reynolds numbers were chosen such that they roughly match with those of the water tunnel experiment at UCLA [11]. *GF* is defined as $GF = (P - W)/P = d/P$, where P, W, and d are, respectively, the pitch of microgrates, the width of the no-slip top surface of the micrograte and the air–water interface (see Fig. 1). Three different *GF* cases were considered as 0.5, 0.75, and 0.917 for $Re_{\tau,\text{in}} = 140$ and 0.5, 0.75, and 0.875 for $Re_{\tau,\text{in}} = 200$. P/δ^*_{in} varied as 35, 17.5, 8.75, 4.375, 2.1875, and 1.09375. Thus, a total of 36 ($2 \times 3 \times 6$) cases was considered.

3 Results and Discussions

3.1 Turbulence Structures

In all TBL simulations, a TBL first develops over a regular wall followed by an SHS. Figure 2 shows contours of iso-surface of λ_2, which is commonly used to depict organized vortical structures in TBLs [24], and contours of streamwise slip velocity (note that only a partial computational domain including SHS is shown in the figure). As expected, significant modification of near-wall turbulence structures was observed over the SHS, which in turn led to significant skin fraction drag reduction. It is evident that the significant reduction of drag was accompanied by the suppression of near-wall turbulence structures, which supports the mechanism of drag reduction proposed previously by Min and Kim [15] and Park et al. [19]. This is in contrast to the observation by Martell et al. [17], who reported that turbulence structures were simply shifted toward SHS.

In general, organized near-wall turbulence structures, such as streamwise vortices, increase skin friction drag in TBLs. However, as shown in Fig. 2b–d, SHSs reduced the strength of the organized near-wall turbulence structures. Compared to a TBL flow over a smooth surface, near-wall turbulence structures over SHSs became

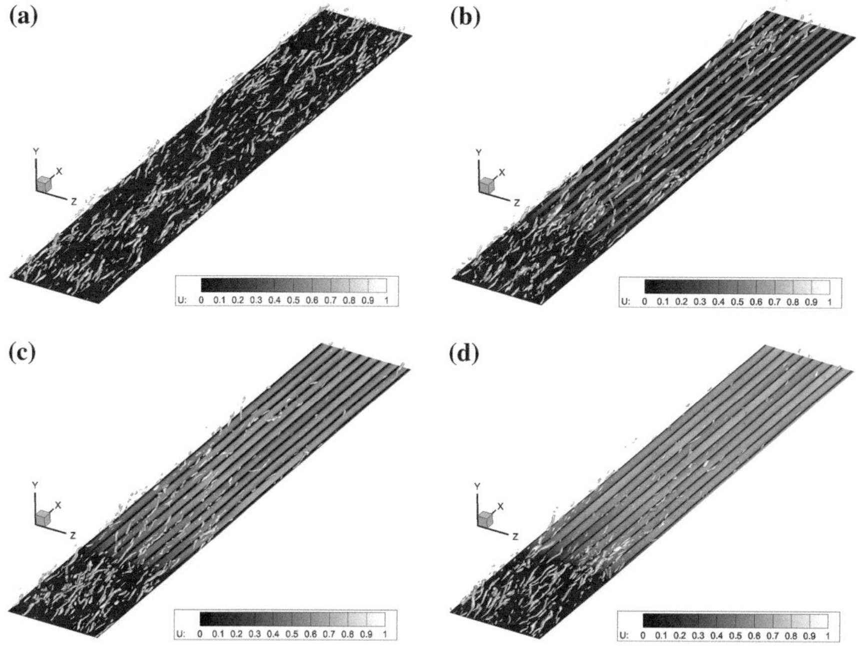

Fig. 2 Isosurface of $\lambda_2 = -0.4$ and contours of streamwise slip velocity over SHS ($Re_{\tau,\text{in}} = 200$): **a** smooth surface, $GF = 0$, **b** SHS, $GF = 0.5$, **c** SHS, $GF = 0.75$, **d** SHS, $GF = 0.875$. Contour levels of slip velocities are shown in the figure

significantly modified with increasing gas fraction ratio, GF. Also, the growth of slip velocities on the air–water interfaces, especially near the front of SHSs, can be observed in Fig. 2b–d from the gray scale of the slip velocity shown on the surface. Compared to the no-slip surface (i.e., smooth surface) shown in Fig. 2a, slip velocities increase as GF and the streamwise location increase. This trend is shown more clearly in Fig. 3a. In order to quantify this trend, proper parameters characterizing this phenomenon need to be introduced and these are discussed in the following subsections.

3.2 The Effective Slip Length

In contrast to fully developed channel flows, where the streamwise direction is homogeneous and SHSs are placed in the entire computational domain in the streamwise (i.e., mean-flow) direction, the streamwise direction in TBLs is inhomogeneous, especially near the beginning and the end of an SHS. As shown in Fig. 3, we observed that the effective slip velocity and hence the effective slip length on an SHS was not

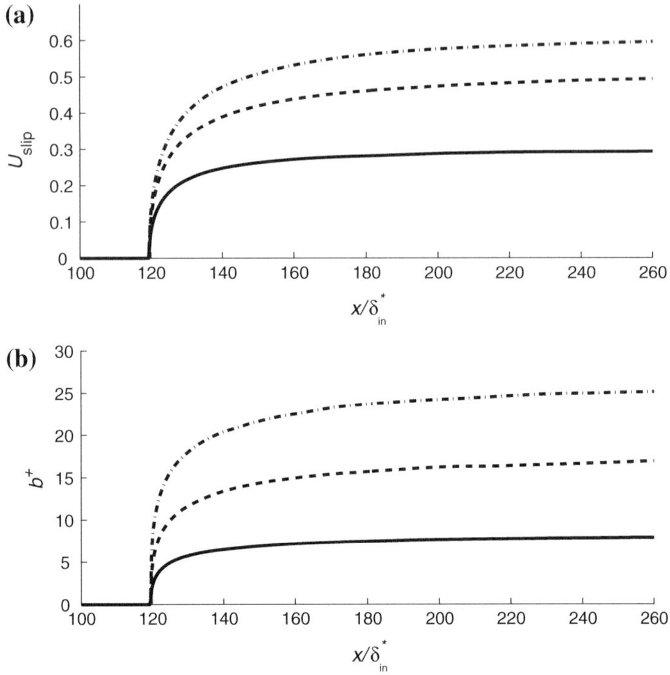

Fig. 3 Growth of **a** slip velocity and **b** the effective slip length in viscous wall units (b^+) in the streamwise direction ($Re_{\tau,\mathrm{in}} = 200$, $P/\delta^*_{\mathrm{in}} = 4.375$). *Solid*, *dashed*, and *dashed dot lines* denote $GF = 0.5, 0.75$, and 0.875, respectively

uniform in the streamwise direction, suggesting that total reduction may depend on the streamwise length of a given SHS.

Similar to the results of turbulent channel flows [19], within the considered Reynolds number range and SHS geometries, it was found that the drag reduction in turbulent boundary layer flows was well-correlated with the effective slip length normalized by viscous wall units, b^+, as shown in Fig. 4. All data were collected at $x/\delta^*_{\mathrm{in}} = 240$, which was shown as nearly converged in Fig. 3. It is apparent that there is a strong correlation between skin friction drag and the effective slip length. It is worth mentioning that the effective surface slip length can be interpreted as a depth of influence into which SHSs affect the flow in the wall-normal direction. As b^+ increases, drag drops rapidly as near-wall turbulence structures are modified. When b^+ becomes larger than 40–50, drag reduction is almost saturated. This implies that the depth of influence must include the buffer layer of wall-bounded turbulent flows, i.e., the effective slip length must be comparable to the buffer layer thickness for maximum drag reduction. It should be noted that turbulence structures responsible for high skin friction drag in turbulent boundary layers are mainly present within the buffer layer.

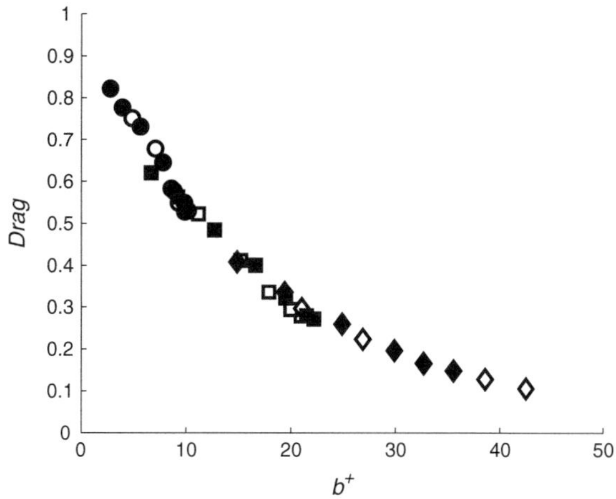

Fig. 4 Variation of normalized drag with the effective slip length in viscous wall units (b^+): *Open symbols* of *circles*, *squares*, and *diamonds* denote $GF = 0.5$, 0.75, and 0.917 of $Re_{\tau,\text{in}} = 140$ flows, respectively. *Closed symbols* of *circles*, *squares*, and *diamonds* denote $GF = 0.5$, 0.75, and 0.875 of $Re_{\tau,\text{in}} = 200$ flows, respectively

The present results from TBL over SHS were essentially the same as those in the turbulent channel with SHS. However, unlike channel flows, it was observed that the effective slip length varied in the streamwise direction, and consequently the drag reduction varied in the streamwise direction. This spatial-variation effect requires further investigation, and it is currently underway.

3.3 The Interface Width

It was also found that the relative size between near-wall turbulence structures and the SHS geometry was a key parameter to the amount of drag reduction. The results shown in Fig. 4, the saturation of drag reduction for $b^+ \succeq 40$–50 in particular, supports the notion that the interaction between an SHS and near-wall turbulence structures is the key to drag reduction by an SHS. More evidence supporting this notion is shown in Fig. 5. A normalized drag is shown as a function of d^+, the width of the shear-free interface, i.e., the width of air–water interface. Note that the maximum drag is limited by $1 - GF$ from the geometric consideration.

What is remarkable is that the normalized drag approaches its limit value when $d^+ \sim 100$, which is the mean spacing of the near-wall streamwise streaky structures. These streamwise streaks are known to be created by near-wall streamwise vortices or vortical structures. When $d^+ > 100$, streamwise vortices cannot be sustained due to the lack of the shear required to sustain those structures. However, when $d^+ < 100$, SHSs can affect these structures partially, thus diminishing its effectiveness in reducing drag. These results indicate that the relative size between near-wall turbulence structures and the SHS geometry also plays an important role for drag reduction. The same trend was observed in turbulent channel flows [19].

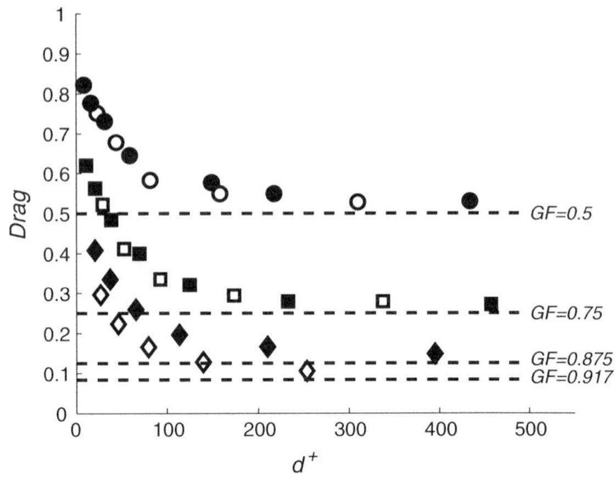

Fig. 5 Variation of normalized drag with the width of shear-free interface in viscous wall units (d^+): *Open symbols* of *circles*, *squares*, and *diamonds* denote $GF = 0.5$, 0.75, and 0.917 of $Re_{\tau,\text{in}} = 140$ flows, respectively. *Closed symbols* of *circles*, *squares*, and *diamonds* denote $GF = 0.5$, 0.75, and 0.875 of $Re_{\tau,\text{in}} = 200$ flows, respectively

4 Summary and Concluding Remarks

DNS of a turbulent boundary layer developing over SHS were performed. The SHS was modeled through a shear-free boundary condition, assuming that the air–water interfaces were flat and it would remain superhydrophobic as Park et al. [19]. It was found that SHS led to substantial drag reduction by weakening near-wall turbulence structures, which were not sustained due to the lack of the shear over SHS. It was also shown that a proper length scale to characterize the flow over SHSs was the effective slip length normalized by viscous wall units, which should be large enough to affect the buffer layer for the maximum drag reduction. Most effects of SHSs can be achieved when its influence depth reaches the buffer layer, indicating that turbulence structures within the buffer region were primarily responsible for high skin friction in wall-bounded turbulent shear flows. Examinations of the effect of SHS geometries (i.e., the pitch and *GF*) revealed that once the width of the air–water region (i.e., shear-free region) became on the order of the width of near-wall streaky structures, a minimum drag state was reached, suggesting that the relative size between near-wall turbulence structures and SHS geometries is also important to the amount of drag reduction. Similarities and differences compared to turbulent channel flows with SHSs were observed.

We plan to continue to conduct the DNS of TBLs, extending to higher Reynolds numbers and a larger set of SHS geometries. This will allow us to compare directly with experimental results, and we will be able to explain similarities and differences observed in numerical and laboratory experiments. In particular, some fundamental differences seem to exist between our results and those observed by Park et al. [11] as to the effect of SHS geometries and resulting drag reduction, and we hope that we can address the origin of these differences. In addition, we will investigate the effect of the streamwise variation of the effective slip length, especially near the beginning and the end of an SHS, on the total drag reduction. These new investigations will ultimately lead to an improved design of SHSs for optimal skin friction drag reduction.

References

1. C.-H. Choi, C.-J. Kim, Large slip of aqueous liquid flow over a nano-engineered superhydrophobic surface. Phys. Rev. Lett. **96**, 66001 (2006)
2. J.P. Rothstein, Slip on superhydrophobic surfaces. Annu. Rev. Fluid Mech. **42**, 89 (2010)
3. C. Lee, C.-H. Choi, C.-J. Kim, Structured surfaces for a giant liquid slip. Phys. Rev. Lett. **101**, 064501 (2008)
4. C. Lee, C.-J. Kim, Maximizing the giant liquid slip on superhydrophobic microstructures by nanostructuring their sidewalls. Langmuir **25**, 12812 (2009)
5. C. Lee, C.-J. Kim, Underwater restoration and retention of gases on superhydrophobic surfaces for drag reduction. Phys. Rev. Lett. **106**, 14502 (2011)
6. M.A. Samaha, H.V. Tafreshi, M. Gad-el, Hak, Influence of flow on longevity of superhydrophobic coatings. Langmuir **28**, 9759 (2012)
7. S. Gogte, P. Vorobieff, R. Truesdell, A. Mammoli, F.V. Swol, P. Shah, C.J. Brinker, Effective slip on textured superhydrophobic surfaces. Phys. Fluids **17**, 051701 (2005)
8. C. Henoch, T.N. Krupenkin, P. Kolodner, M.S. Taylor, J.A. Hodes, A.M. Lyons, C. Peguero, K. Breuer, Turbulent drag reduction using superhydrophobic surfaces. AIAA paper 2006-3192 (2006)
9. R.J. Daniello, N.E. Waterhouse, J.P. Rothstein, Drag reduction in turbulent flows over superhydrophobic surfaces. Phys. Fluids **21**, 085103 (2009)
10. B. Woolford, J. Prince, D. Maynes, B.W. Webb, Particle image velocimetry characterization of turbulent channel flow with rib patterned superhydrophobic walls. Phys. Fluids **21**, 085106 (2009)
11. H. Park, G. Sun, C.-J. Kim, Superhydrophobic turbulent drag reduction as a function of surface grating parameters. J. Fluid Mech. **747**, 722 (2014)
12. J. Zhao, X. Du, X. Shi, Experimental research on friction-reduction with superhydrophobic surfaces. J. Mar. Sci. Appl. **6**, 58 (2007)
13. C. Peguero, K. Breuer, On drag reduction in turbulent channel flow over superhydrophobic surface. in *Advances in Turbulence XII*. Springer Proceedings in Physics (Springer Berlin, 2009)
14. D. Maynes, K. Jeffs, W. Woolford, B.W. Webb, Laminar flow in a microchannel with hydrophobic surface patterned microribs oriented parallel to the flow direction. Phys. Fluids **19**, 093603 (2007)
15. T. Min, J. Kim, Effects of hydrophobic surface on skin-friction drag. Phys. Fluids **16**, L55 (2004)
16. M.B. Martell, B. Perot, J.P. Rhothstein, Direct numerical simulations of turbulent flows over superhydrophobic surfaces. J. Fluid Mech. **620**, 31 (2009)
17. M.B. Martell, J.P. Rhothstein, B. Perot, An analysis of superhydrophobic turbulent drag reduction mechanisms using direct numerical simulation. Phys. Fluids **22**, 065102 (2010)
18. A. Busse, N.D. Sandham, Influence of an anisotropic slip-length boundary condition on turbulent channel flow. Phys. Fluids **24**, 055111 (2012)
19. H. Park, H. Park, J. Kim, A numerical study of the effects of superhydrophobic surface on skin-friction drag in turbulent channel flow. Phys. Fluids **25**, 110815 (2013)
20. J. Kim, P. Moin, Application of a fractional step method to incompressible Navier-Stokes equations. J. Comput. Phys. **59**, 308 (1985)
21. K. Akselvoll, P. Moin, An effcient method for temporal integration of the Navier-Stokes equations in confined axisymmetric geometries. J. Comput. Phys. **125**, 454 (1996)
22. T.S. Lund, X. Wu, K.D. Squires, Generation of turbulent inflow data for spatially-developing boundary layer simulations. J. Comput. Phys. **140**, 233 (1998)
23. P.R. Spalart, Direct numerical simulation of a turbulent boundary layer up to $Re_\theta = 1400$. J. Fluid Mech. **187**, 61 (1988)
24. J. Jeong, F. Hussain, On the identification of a vortex. J. Fluid Mech. **285**, 69 (1995)

Structure and Dynamics of Turbulence in Super-Hydrophobic Channel Flow

Amirreza Rastegari and Rayhaneh Akhavan

Abstract The structure and dynamics of turbulence in turbulent channel flow with super-hydrophobic (SH) walls has been investigated using DNS with Lattice Boltzmann methods. The channel walls consisted of longitudinal arrays of SH microgrooves of width g, separated by distances of w. The liquid/gas interfaces in the SH walls were modeled as idealized, flat, shear-free surfaces. Simulations were performed at a bulk Reynolds number of $Re_b = U_{bulk}h/\nu = 3600$, corresponding to $Re_{\tau 0} = u_{\tau 0}h/\nu \approx 223$. Drag reductions (DR) of 5–47 % and 51–83 % were obtained with $g/w = 1$, and $g/w = 7$ and $g/w = 15$, respectively. DR was found to be primarily due to surface slip. Mathematical analysis shows that the magnitude of DR in both laminar and turbulent flow is given by $DR = U_{slip}/U_{bulk} + O(\varepsilon)$. In laminar flow, where DR is purely due to surface slip, ε is zero. In turbulent flow, ε attains a small nonzero value at high DR, reflecting additional DR effects resulting from modification of the turbulence dynamics in the interior of the flow due to the presence of the SH surface. Analysis of the turbulence statistics and kinetic energy budgets in the drag-reduced flow reveals that the influence of the SH surface remains confined to a surface layer of thickness on the order of the SH microgrooves width, g. Outside of this layer, the 'normalized' turbulence dynamics proceeds as in regular turbulent channel flow. Within the surface layer, the presence of the pattern of longitudinal microgrooves on the SH surfaces gives rise to spanwise variations in all Reynolds-averaged turbulence quantities, leading to development of a mean secondary flow and additional turbulence production and Reynolds shear stresses within the surface layer of the SH channel.

A. Rastegari · R. Akhavan (✉)
Department of Mechanical Engineering, University of Michigan,
Ann Arbor, MI 48109-2125, USA
e-mail: raa@umich.edu

A. Rastegari
e-mail: amirreza@umich.edu

M. Stanislas et al. (eds.), *Progress in Wall Turbulence 2*,
ERCOFTAC Series 23, DOI 10.1007/978-3-319-20388-1_32

1 Introduction

Super-hydrophobic (SH) surfaces are micro-textured surfaces covered with a hydrophobic coating [14]. In the flow of liquids over these surfaces, nano-bubbles form in between the surface micro-textures [14, 17], resulting in slip over the gas/liquid menisci and a concomitant reduction in the friction drag [14, 17]. The magnitude of the slippage is generally quantified by an average slip velocity U_s on the wall, and a corresponding average slip length, l_s, defined as $U_s = l_s \partial U/\partial n$, where n denotes the unit normal to the wall [17].

In laminar flow, drag reductions (DR) of up to 40 % have been reported for SH surfaces with a variety of surface patterns [14]. Flow visualization studies have verified that while the no-slip condition remains valid on the solid/liquid contact areas of these surfaces, an apparent slip velocity develops on the gas/liquid interfaces [9, 16]. For purposes of analytical and computational studies, the SH surface is generally modeled as a pattern of shear-free regions interspersed among no-slip surfaces. Analytical solutions of Stokes flow over these modeled surfaces, with various patterns of no-slip and shear-free boundaries [3, 5, 11, 12], show good agreement with experimental results in laminar channel flow with SH walls [3, 6, 9], justifying the use of flat shear-free boundaries in modeling the gas/liquid interfaces.

In turbulent flow, experiments on DR by SH surfaces have been contradictory. In micro-channels with arrays of longitudinal SH microgrooves on one or both walls, Daniello et al. [2] report DRs of up to 50 %, at $1000 \leq Re_b \leq 4750$ for $0.7 \leq g^{+0} = w^{+0} \leq 4.8$, where g and w denote the widths of the microgrooves and the spacings in between them, respectively, and $+0$ denotes nondimensionalization with respect to the wall friction velocity of a channel with no-slip walls at the same Re_b of the SH channel. Other groups have reported much lower DRs of only up to 11 % at $1350 \leq Re_b \leq 2800$ with $1.3 \leq g^{+0} = 4w^{+0} \leq 2.5$ [19], or no DR at all at $Re_b = 3000$ with $g^{+0} = w^{+0} \approx 0.2$ [10] in micro-channels with SH walls. Experiments in pipe flow with SH random texture on the walls have shown no noticeable DR in turbulent flow, while showing 14 % DR in the same geometry in laminar flow [18].

Furthermore, it is not clear how turbulent DR over SH surfaces scales with the SH geometric properties. Some studies have suggested that DR scales with the slip length in wall units and the Re_τ of the base flow [4, 8], while others have suggested that the width of the SH surface elements in wall units and the fraction of shear-free to no-slip surface area are the proper scaling parameters [7]. It is also not clear whether DR is purely a surface phenomenon, resulting from surface slip, or if additional modifications to the turbulence dynamics are at work. The objective of the present study is to address some of these questions using results from DNS with Lattice Boltzmann Methods (LBM) [15] of turbulent flow in a channel with SH walls.

2 Simulation Parameters and Numerical Methods

All simulations were performed in channels of size $L_x \times L_y \times L_z = 4.6h \times 2.3h \times 2h$ in the streamwise (x), spanwise (y) and wall-normal (z) directions, respectively, where h denotes the channel half-height. A schematic of the channel is shown in Fig. 1a. Both channel walls were covered with a pattern of SH longitudinal microgrooves of width g, separated by distances of w. The gas/liquid interfaces on the SH microgrooves were modeled as flat and shear-free interfaces. Standard D3Q19 single relaxation time Lattice Boltzmann BGK Methods [15] were used for all the simulations. The simulations all were performed at constant flow rate and maintained a bulk Reynolds number of $Re_b = U_b h/\nu = 3600$ in the channel, corresponding to a friction Reynolds number of $Re_{\tau 0} = u_{\tau 0} h/\nu \approx 223$ in the base flow. All simulations were performed with a resolution of $512 \times 256 \times 221$, corresponding to grid spacings of $\Delta^{+0} \approx 2$ in all three directions, where the superscript $+0$ denotes nondimensionalization with respect to the viscosity and friction velocity, $u_{\tau 0}$, of the base flow. A total of 9 cases were studied, of which six had a fraction of shear-free to no-slip surface area of $g/w = 1$, corresponding to $g^{+0} = w^{+0} = 4$, 8, 16, 32, 64 and 128, respectively. Three additional cases investigated the effect of changing the ratio of g/w, and corresponded to $g/w = 7$ with $g^{+0} = 28$, $w^{+0} = 4$ and $g^{+0} = 56$, $w^{+0} = 8$, and $g/w = 15$ with $g^{+0} = 120$, $w^{+0} = 8$. Furthermore, for comparison, all these cases were repeated in the laminar flow regime at $Re_b = 50$.

All SH channel simulations were initialized from fully developed turbulent flow in a base channel with no-slip walls at $tU_b/h = 400$. Figure 1b shows the time histories of the skin-friction coefficient, C_f, in the SH channel. The maximum nominal DR that can be obtained in any SH channel geometry is $g/(g + w)$. This nominal DR is obtained only when the slip velocity on the slip surfaces approaches U_b. For all other

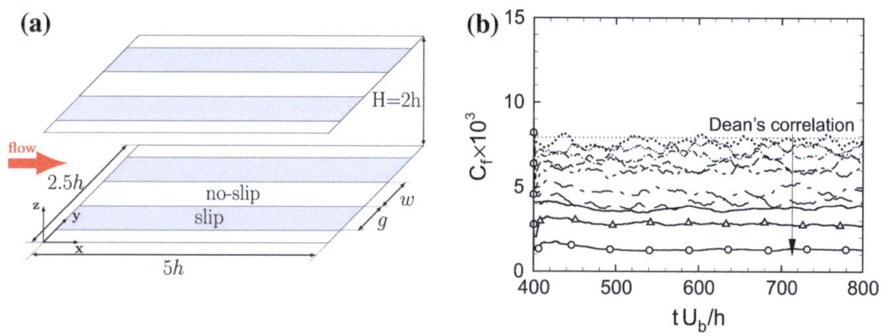

Fig. 1 **a** Schematic of the channel and the coordinate system; **b** Time history of the skin-friction coefficient in DNS of turbulent channel flow with SH walls: · · · · , no-slip channel; – ·· –, $g^{+0} = w^{+0} \approx 4$, 5.0 %*DR*; – · –, $g^{+0} = w^{+0} \approx 8$, 10.9 %*DR*; – – –, $g^{+0} = w^{+0} \approx 16$, 18.2 %*DR*; —— ·· ——, $g^{+0} = w^{+0} \approx 32$, 22.6 %*DR*; —— · ——, $g^{+0} = w^{+0} \approx 64$, 38.2 %*DR*; —— ——, $g^{+0} = w^{+0} \approx 128$, 47.3 %*DR*; ——, $g^{+0} \approx 28$, $w^{+0} \approx 4$, 51.6 %*DR*; —△—, $g^{+0} \approx 56$, $w^{+0} \approx 8$, 63.5 %*DR*; —o—, $g^{+0} \approx 120$, $w^{+0} \approx 8$, 83.2 %*DR*

cases, DR is lower. In the simulations with $g/w = 1$, DRs of 5, 10.9, 18.2, 22.6, 32.8, and 47.3 % were obtained with $g^{+0} = w^{+0} = 4$, 8, 16, 32, 64, and 128, respectively. At higher ratios of g/w, DRs of 51.6, 63.5, and 83.2 % were obtained with $g^{+0} = 28$, $w^{+0} = 4$ ($g/w = 7$), $g^{+0} = 56$, $w^{+0} = 8$ ($g/w = 7$), and $g^{+0} = 120$, $w^{+0} = 8$ ($g/w = 15$), respectively.

3 Scaling of Drag Reduction

Three different scalings of DR with SH surface properties have been proposed in the literature. For laminar flow, the analytical solution of Philip [11, 12] suggests that DR scales with the geometric parameters g/w and $(g + w)/2H$, where $H = 2h$ denotes the full height of the channel. Martel et al. [7] proposed that in turbulent flow, DR scales with g^{+0} instead of $(g + w)/2H$, while Fukugata et al. [4] have proposed a scaling of DR in terms of L_s^{+0} and $Re_{\tau 0}$ for cases of uniform streamwise slip, uniform spanwise slip, and uniform combined slip, based on the assumption of a logarithmic mean velocity profile throughout the cross-section of the channel.

Figure 2 shows the comparison of the present DNS results with these three scalings. For reference, the experimental data of Daniello et al. [2] in turbulent channel flow with longitudinal microgrooves of width $1 \leq g^{+0} = w^{+0} \leq 4$ are also shown. When plotted in Philip's [11, 12] scaling (Fig. 2a), the LB DNS results follow parallel trends in laminar and turbulent flow, both showing an increase in DR with increasing ratios of $(g + w)/2H$ and g/w. However, the DRs in turbulent flow are always higher than in laminar flow. Experiments in turbulent flow, however, show significantly higher DR than both DNS and even theoretical limits for $g/w = 1$. These discrepancies between experiments, DNS, and theoretical limits are not resolved when the results are plotted as a function of g^{+0}, as shown in Fig. 2b. Figure 2c shows the comparison of the DNS results and experimental data [2], with the analytical expressions of Fukagata et al. [4]. The DNS results in turbulent flow, fall within the analytical curves corresponding to uniform streamwise slip and uniform combined slip at the corresponding $Re_{\tau 0}$. The experimental data also show reasonable agreement with the analytical curves, given the scatter of the data. However, the DNS results in laminar flow are not properly described by Fukagata et al.'s expression, which is to be expected, since the expression assumes a logarithmic mean velocity profile.

The trends in experimental data observed in Fig. 2a–c lead us to believe that in the experiments, the surface micro-bubbles may have spread beyond the surface microgrooves, resulting in a larger g/w ratio than the nominal value of $g/w = 1$ reported. A higher g/w ratio can dramatically increase the DR, as observed in both the DNS results and the theoretical limit. Indeed, Cottin-Bizonne et al. [1], have suggested that contamination of the SH surface can act as nucleation sites for formation of nano-bubbles, thus increasing the fraction of shear-free to no-slip surface area, g/w, from its nominal values, and leading to larger slip lengths.

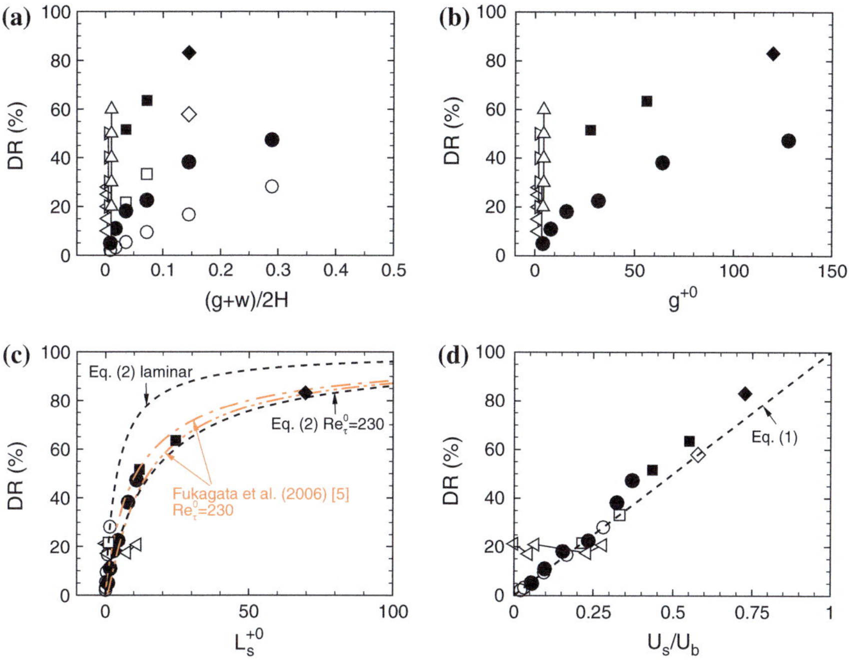

Fig. 2 Scaling of DR with SH surface properties: **a** DR as a function of nondimensional microgroove size in outer scaling; **b** DR as a function of nondimensional microgroove size in inner scaling; **c** DR as a function of nondimensional slip length in inner scaling; **d** DR as a function of nondimensional slip velocity; •, LB DNS at $Re_b = 3600$ ($Re_{\tau_0} \approx 223$) with $g/w = 1$, $4 \leq g^{+0} = w^{+0} \leq 128$; ■, LB DNS at $Re_b = 3600$ ($Re_{\tau_0} \approx 223$) with $g/w = 7$, $g^{+0} \approx 28$, $w^{+0} \approx 4$ or $g^{+0} \approx 56$, $w^{+0} \approx 8$; ♦, LB DNS at $Re_b = 3600$ ($Re_{\tau_0} \approx 223$) with $g/w = 15$, $g^{+0} \approx 120$, $w^{+0} \approx 8$; ○, LB DNS of laminar flow at $Re_b = 50$ with $g = w$; □, LB DNS of laminar flow at $Re_b = 50$ with $g/w = 7$; ◇, LB DNS of laminar flow at $Re_b = 50$ with $g/w = 15$; —◁—, experiments of Daniello et al. [2] at $2450 \leq Re_b \leq 4000$ with $g = w = 30\,\mu$m microgrooves on one wall; —▷—, experiments of Daniello et al. [2] at $1100 \leq Re_b \leq 2900$ with $g = w = 30\,\mu$m microgrooves on both walls; —△—, experiments of Daniello et al. [2] at $1500 \leq Re_b \leq 4000$ with $g = w = 60\,\mu$m microgrooves on both walls; - - -, $DR = U_s/U_b = L_s^{+0}/[1 + (Re_b/Re_{\tau_0})]$; —·—, analytical prediction of Fukagata et al. [4] for uniform streamwise slip on both walls; —··—, analytical prediction of Fukagata et al. [4] for combined slip on both walls

A more general scaling for DR in channels with SH walls, which remains valid in both laminar and turbulent flow regimes, can be obtained from analysis of the Navier–Stokes equations [13]. It can be shown that for both laminar and turbulent flow in a SH channel,

$$DR = \frac{U_s}{U_b} + O(\varepsilon). \quad (1)$$

Where U_s is the average slip velocity on the channel walls, and U_b is the bulk velocity in the channel. When DR is purely due to surface slip, as in laminar flow, $\varepsilon = 0$ and $DR = U_s/U_b$. In turbulent flow, there is an additional contribution to DR beyond surface slip, arising from modification of the turbulence dynamics, in the interior flow due to the presence of the SH surface. In this case, the magnitude of DR is given by $DR = U_s/U_b + O(\varepsilon)$. As such, ε is a measure of any drag reducing effects of the SH surface in turbulent flow beyond surface slip. Figure 2d shows the comparison of the LB DNS results in laminar and turbulent channel flow, along with the experimental results of Daniello et al. [2], in turbulent channel flow, with the line $DR = U_s/U_b$. The DRs observed in DNS of laminar flow are in agreement with the line $DR = U_s/U_b$ for all magnitudes of DR and all ratios of g/w and $(g+w)/2H$. In turbulent flow, the deviations between the line $DR = U_s/U_b$ and the DRs observed in DNS remain negligibly small in the Low Drag Reduction (LDR) regime ($DR < 35\,\%$). In this regime, DR is mainly due to surface slip, ε remains negligible, and the structure and dynamics of turbulence is not significantly affected beyond the viscous sublayer and the buffer layer. In the High Drag Reduction (HDR) regime ($DR > 35\,\%$), the deviations of the DNS data from the line $DR = U_s/U_b$ become more noticeable (see Fig. 2d). However, even in these cases, the deviations remain small and to the first order, DR is still given by $DR \approx U_s/U_b$. Figure 2d also shows the experimental data of Daniello et al. [2] for which the slip velocity was reported. To within the scatter of the experimental data, these results also show good agreement with the scaling $DR = U_s/U_b + O(\varepsilon)$.

Using the relation between the slip velocity and slip length, Eq. (1) can be rewritten in terms of the slip length, L_s^{+0}, as [13]

$$DR = \frac{L_s^{+0}}{L_s^{+0} + \frac{Re_b}{Re_{\tau 0}}} + O(\varepsilon). \tag{2}$$

Because no assumption regarding the shape of the velocity profile were made in the derivation of Eqs. (1) and (2), these equations remain equally valid in both laminar and turbulent flow. Figure 2c shows the comparison of Eq. (2) and the LB DNS data in both laminar and turbulent flow, verifying the good agreement of Eq. (2) with the DR obtained in DNS in both laminar and turbulent flow regimes.

4 Turbulence Statistics

The presence of the longitudinal SH microgrooves on the channel walls give rise to Reynolds-averaged turbulence statistics, $\overline{f}$, which are functions of both the spanwise (y) and wall-normal (z) directions. Failure to account for these spanwise variations results in inaccurate turbulence statistics. To compare these statistics with the corresponding quantities in regular turbulent channel flow, here we report the Reynolds-averaged turbulence statistics after averaging these statistics in the spanwise direction. These spanwise-averaged statistics are denoted by $\langle \overline{f} \rangle$.

Figure 3a–c show the spanwise variations of the mean streamwise velocity for cases of $g^{+0} = w^{+0} = 16$, $g^{+0} = w^{+0} = 64$ and $g^{+0} = 120$, $w^{+0} = 8$. In each

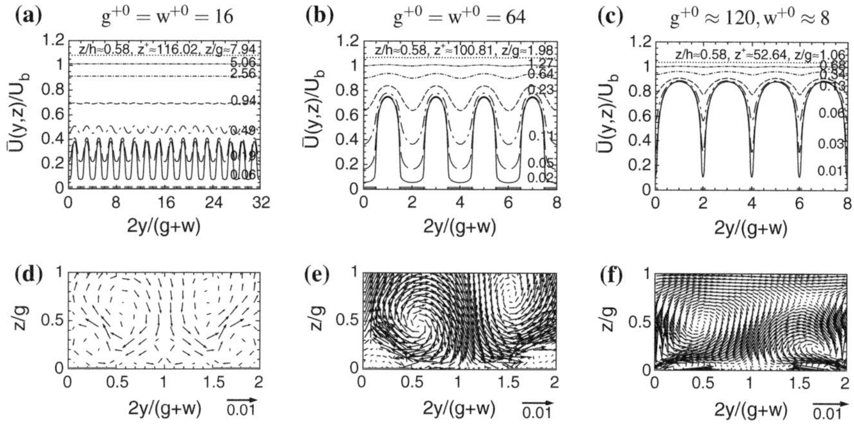

Fig. 3 **a**, **b**, **c** Spanwise variation of the mean streamwise velocity and **d**, **e**, **f** vector plots of the mean secondary flow in turbulent channel flow with SH surfaces: **a**, **d** $g^{+0} = w^{+0} \approx 16, 18.2\,\%DR$; **b**, **e** $g^{+0} = w^{+0} \approx 64, 38.2\,\%DR$; **c**, **f** $g^{+0} \approx 120$, $w^{+0} \approx 8$, $83.2\,\%DR$; ——, $z/h \approx 0.01$; —— ——, $z/h \approx 0.01$; —— · ——, $z/h \approx 0.03$; – – –, $z/h \approx 0.07$; – · –, $z/h \approx 0.19$; – ·· –, $z/h \approx 0.37$; · · · ·, $z/h \approx 0.58$; ▬▬▬ (*blue*), location of the no-slip stripes

case, the spanwise variations are seen to extend to a distance of $z \sim g$ from the SH walls. Beyond this distance, no differences can be observed between the Reynolds averaged turbulence statistics on the slip and no-slip surfaces. Similar spanwise variations are observed in all the Reynolds-averaged turbulence statistics, including Reynolds stresses. The latter gives rise to the development of a mean secondary flow in the channel [13], in the form of pairs of counter-rotating vortices which extend to a distance of $z \sim g$ into the channel, as shown in Fig. 3d–f. Furthermore, the presence of a spanwise gradient in the mean streamwise velocity, $\partial \bar{U}/\partial y$, gives rise to the development of an additional Reynolds stress, $\overline{uv}$, at $z \lesssim g$ [13], and additional production of turbulence kinetic energy over the slip surfaces, through $\overline{uv}\partial \bar{U}/\partial y$.

Figure 4 shows the profiles of the normalized spanwise averaged turbulence statistics in the SH channel compared to the base flow. The profiles of the spanwise-averaged mean streamwise velocity display increasing slip velocity with increasing DR, as seen in Fig. 4a. When the slip velocity is subtracted, the velocity profiles display a narrowing of the buffer layer, along with a downward shift of the logarithmic layer compared to the base flow, as shown in Fig. 4b. These features stand in contrast with other examples of drag-reduced wall-bounded turbulence, for which DR is generally associated with a thickening of the buffer layer and an upward shift of the logarithmic layer. The different behavior arises from the fact that in the present flow DR is primarily due to surface slip, not modification of the turbulence dynamics in the interior flow.

The profiles of the normalized spanwise-averaged turbulence intensities are shown in Fig. 4d–f. There is enhanced production of turbulence in all three components of the velocity fluctuations within a distance of $z \lesssim g$ from the walls. This enhanced

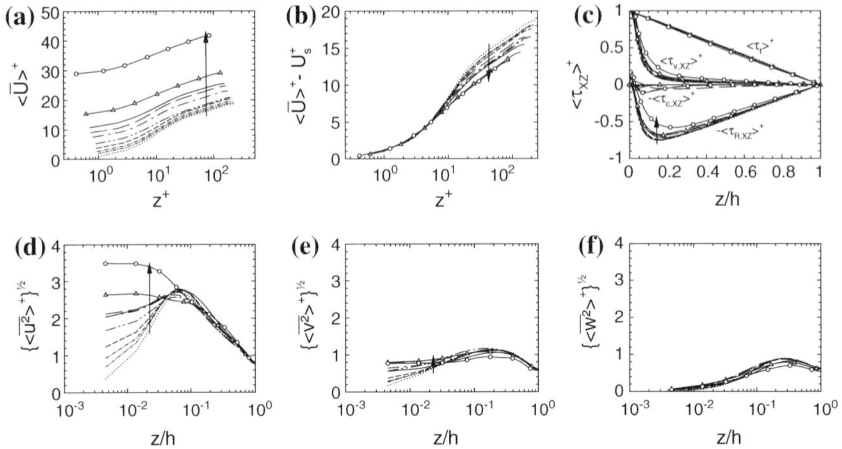

Fig. 4 Spanwise-averaged turbulence statistics in channel flow with super-hydrophobic walls: **a** mean streamwise velocity; **b** mean streamwise velocity with the slip velocity subtracted; **c** Reynolds shear stress $\langle\tau_{R,xz}\rangle^+ = -\langle\overline{uw}\rangle^+$, viscous shear stress $\langle\tau_{v,xz}\rangle^+ = \langle\partial\bar{U}/\partial z\rangle^+$, convective stress $\langle\tau_c\rangle^+ = -\langle\bar{U}\,\bar{W}\rangle^+$, and total shear stress $\langle\tau_t\rangle^+ = \langle\tau_{R,xz}\rangle^+ + \langle\tau_{v,xz}\rangle^+ + \langle\tau_c\rangle^+$; **d**, **e**, **f** streamwise, $\langle\overline{u^2}\rangle^{1/2+}$, spanwise, $\langle\overline{v^2}\rangle^{1/2+}$, and wall-normal, $\langle\overline{w^2}\rangle^{1/2+}$, turbulence intensities; line types as in Fig. 1

production arises from additional production of streamwise turbulent stresses through the $\overline{uv}\partial\overline{U}/\partial y$ term, at the boundary between the slip and no-slip stripes (see Fig. 3a–c), which is transferred to the other components through the pressure strain terms. Beyond $z \gtrsim g$, however, the profiles of normalized turbulence intensities approach those in the base flow, indicating that the dynamics of turbulence has reverted back to that in regular wall-bounded flow and the effect of the SH surface is no longer felt.

The profiles of the normalized spanwise-averaged Reynolds shear stress, $\langle\tau_{R,xz}\rangle^+ = -\langle\overline{uw}\rangle^+$, display a small drop in their peak magnitudes, compared to the base flow, as shown in Fig. 4c. The magnitude of this drop, however, is nowhere as large as the drops observed with other means of DR in wall-bounded flows and is more commensurate with the lower Re_τ of the drag-reduced flow compared to the base flow, once again emphasizing that the main mechanism of DR with SH surfaces is surface slip and not a modification of the turbulence dynamics in the interior of the flow. In the total balance of shear stresses, in addition to contributions from the Reynolds shear stress, $\langle\tau_{R,xz}\rangle$, and the viscous stress, $\langle\tau_{v,xz}\rangle$, there is also a small contribution from the convective stress, $\langle\tau_{c,xz}\rangle^+ = -\langle\overline{UW}\rangle^+$, arising from the mean secondary flow in the channel, such that $\langle\tau_t\rangle^+ = \langle\tau_{R,xz}\rangle^+ + \langle\tau_{v,xz}\rangle^+ + \langle\tau_{c,xz}\rangle^+ = (1 - z/h)$, as shown in Fig. 4c.

5 Turbulence Kinetic Energy Dynamics

The balance of the spanwise-averaged Turbulence Kinetic Energy (TKE) in the SH channel is given by

$$\frac{D}{Dt}\langle\overline{k^2}\rangle = \langle P_{ii}\rangle + \langle t_{ii}^{(\Sigma)}\rangle - \langle\varepsilon_{ii}\rangle, \tag{3}$$

where $\langle P_{ii}\rangle = -\langle\overline{u_i u_j \frac{\partial \bar{U}_i}{\partial x_j}}\rangle$ is the rate of TKE production per unit mass, $\langle\varepsilon_{ii}\rangle = 2\nu\langle\overline{s_{ij}\frac{\partial u_i}{\partial x_j}}\rangle$ is the rate of viscous dissipation and $\langle t_{ii}^{(\Sigma)}\rangle = \frac{\partial}{\partial x_j}\langle\overline{k^2 u_j}\rangle + \frac{1}{\rho}\langle\overline{u_j\frac{\partial p}{\partial x_j}}\rangle + 2\nu\frac{\partial}{\partial x_j}\langle\overline{u_i s_{ij}}\rangle$ is the total rate of TKE transport. Furthermore, $s_{ij} = \frac{1}{2}(\frac{\partial u_i}{\partial x_j} + \frac{\partial u_j}{\partial x_i})$ is the fluctuating strain-rate tensor.

Figure 5 show the profiles of the normalized spanwise-averaged terms in the TKE equation. As with the turbulence statistics, the presence of the SH surfaces affects the dynamics of TKE only within a distance of $z \lesssim g$ from the walls, as shown in Fig. 5a. In addition to the usual production of streamwise TKE due to $-\langle\overline{uw}\partial\bar{U}/\partial z\rangle$, in the SH channel, there is additional production of streamwise TKE due to $-\langle\overline{uv}\partial\bar{U}/\partial y\rangle$,

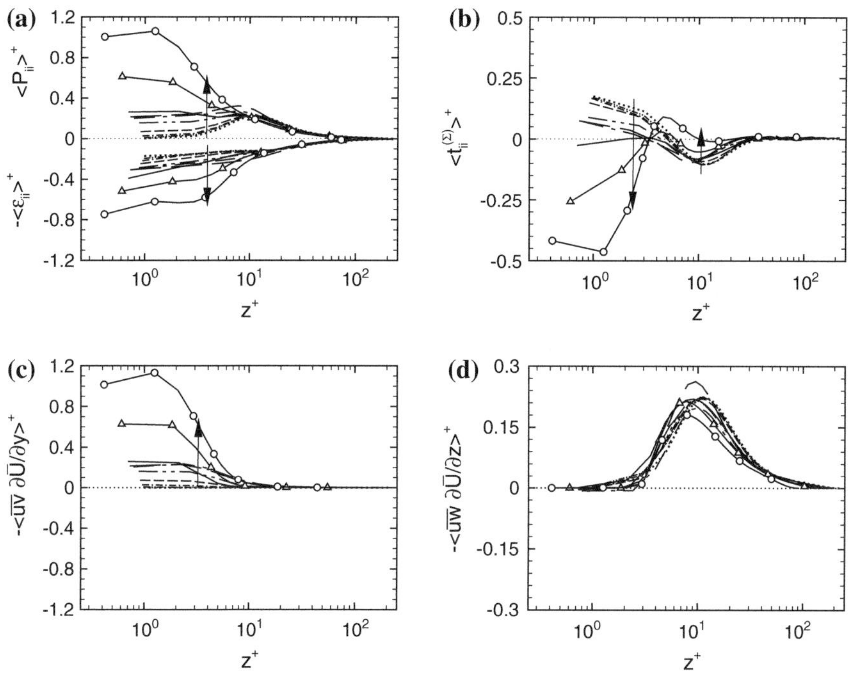

Fig. 5 Spanwise-averaged turbulence kinetic energy budget in channel flow with super-hydrophobic walls: **a** production $\langle P_{ii}\rangle^+$ and dissipation, $\langle\varepsilon_{ii}\rangle^+$; **d** total transport, $\langle t_{ii}^{(\Sigma)}\rangle^+$; **b, c** breakdown of the production term; line types as in Fig. 1

as shown in Fig. 5c, d. The normalized TKE production through $-\langle \overline{uw} \partial \bar{U}/\partial z \rangle$ is not that different from that in the base flow, as shown in Fig. 5c, while the normalized TKE production due to $-\langle \overline{uv} \partial \bar{U}/\partial y \rangle$ is concentrated within a distance of $z \lesssim g$ from the walls, as shown in Fig. 5d. Examination of the components of TKE equation, shows that this production of TKE in the streamwise component is transferred to the other components through the pressure-strain terms. In the LDR regime, the balance between production, dissipation and transport terms are not that different from that in the base flow, as shown in Fig. 5a, b. In the HDR regime, however, the additional production due to the $-\langle \overline{uv} \partial \bar{U}/\partial y \rangle$ term is excessive at $z \lesssim g$ and cannot be fully dissipated locally. In this case, the transport terms act to transfer this production away from the walls to the interior of the flow where it can be dissipated, resulting in a reversal of the sign of the transport terms compared to the base flow at $z \lesssim g$.

6 Summary and Conclusions

Drag reduction in laminar and turbulent channel flow with SH longitudinal microgrooves on both walls has been studied using DNS with Lattice Boltzmann Methods. In both laminar and turbulent flow, mathematical analysis of the Navier–Stokes equations shows that DR in the channel scales as $DR = U_s/U_b + O(\varepsilon)$, where the first term (U_s/U_b) represents the effects of surface slip and the second term $(O(\varepsilon))$ represents additional DR due to modification of the turbulence dynamics in the interior of the flow. In laminar flow, ε is zero and DR is purely due to the surface slip. In turbulent flow, ε is negligible and DR is primarily due to surface slip at LDR. At HDR, ε becomes nonnegligible, but DR is still primarily due to surface slip. The effect of the SH wall is confined to a surface layer of size of the order of the width, g, of the SH stripes. Inside this layer, the presence of the SH longitudinal microgrooves, gives rise to spanwise variations in all Reynolds-averaged turbulence statistics, and development of a secondary flow and additional Reynolds stresses, which lead to additional turbulence production within the surface layer. Outside of this surface layer, the normalized structure and dynamics of turbulence is unaffected. These features are consistent with the observation that the bulk of DR in turbulent flow is due to surface slip.

References

1. C. Cottin-Bizonne, B. Cross, A. Steinberger, E. Charlaix, Boundary slip on smooth hydrophobic surfaces: intrinsic effects and possible artifacts. Phys. Rev. Lett. **94**, 056102 (2005)
2. R.J. Daniello, N.E. Waterhouse, J.P. Rothstein, Drag reduction in turbulent flows over superhydrophobic surfaces. Phys. Fluids **21**, 085103 (2009)
3. A.M.J. Davis, E. Lauga, Effective slip in pressure-driven stokes flow. J. Fluid Mech. **661**, 402–411 (2010)

4. K. Fukagata, N. Kasagi, P. Koumoutsakos, A theoretical prediction of friction drag reduction in turbulent flow by superhydrophobic surfaces. Phys. Fluids **18**, 051703 (2006)
5. E. Lauga, H.A. Stone, Effective slip in pressure-driven stokes flow. J. Fluid Mech. **489**, 55–77 (2003)
6. C. Lee, C. Choi, C. Kim, Structured surfaces for a giant liquid slip. Phys. Rev. Lett. **101**, 064501 (2008)
7. M.B. Martel, J.P. Rothstein, J.B. Perot, An analysis of superhydrophobic turbulent drag reduction mechanisms using direct numerical simulation. Phys. Fluids **22**, 065102 (2010)
8. T. Min, J. Kim, Effect of superhydrophobic surfaces on skin-friction drag. Phys. Fluids **16**, L55 (2004)
9. J. Ou, J.P. Rothstein, Direct velocity measurements of the flow past drag reducing ultrahydrophobic surfaces. Phys. Fluids **17**, 103606 (2005)
10. C. Peguero, K. Bruer, On drag reduction in turbulent channel flow over superhydrophobic surfaces, in *Advances in Turbulence XII Proceedings of the 12th EUROMECH European Turbulence Conference, 7–10 September 2009*, ed. by B. Eckhardt (Springer, Berlin, 2009), pp. 233–236
11. J.R. Philip, Flows satisfying mixed no-slip and no-shear conditions. Z. Angew. Math. Phys. **23**, 353–372 (1972)
12. J.R. Philip, Integral properties of flows satisfying mixed no-slip and no-shear conditions. Z. Angew. Math. Phys. **23**, 960–968 (1972)
13. A. Rastegari, R. Akhavan, On the mechanism of turbulent drag reduction with superhydrophobic surfaces. J. Fluid Mech. **773**, R4 (2015)
14. J.P. Rothstein, Slip on superhydrophobic surfaces. Annu. Rev. Fluid Mech. **42**, 89–109 (2010)
15. S. Succi, *The Lattice Boltzmann Equation for Fluid Dynamics and Beyond* (Oxford University Press, New York, 2001)
16. P. Tsai, A.M. Peters, C. Pirat, M. Wessling, R.G.H. Lammertink, D. Lohse, Quantifying effective slip length over micropatterned hydrophobic surfaces. Phys. Fluids **21**, 112002 (2009)
17. R.S. Voronov, D.V. Papavassiliou, Review of fluid slip over superhydrophobic surfaces and its dependence on the contact angle. Ind. Eng. Chem. Res. **47**, 2455–2477 (2008)
18. K. Watanabe, Y. Ugadawa, H. Ugadawa, Drag reduction of Newtonian fluid in a circular pipe with a highly water-repellent wall. J. Fluid Mech. **381**, 225–238 (1999)
19. B. Woolford, J. Prince, D. Maynes, B.W. Webb, Particle image velocimetry charactrization of turbulent channel flow with rib patterned superhydrophobic walls. Phys. Fluids **21**, 085106 (2009)

Spectral Assessment of the Turbulent Convection Velocity in a Spatially Developing Flat Plate Turbulent Boundary Layer at Reynolds Number $Re_\theta = 13\,000$

Nicolas Renard, Sébastien Deck and Pierre Sagaut

Abstract A method inspired by del Álamo and Jiménez, J Fluid Mech 640:5–26, 2009, [7] is derived to assess the wavelength-dependent convection velocity in a zero pressure gradient spatially developing flat plate turbulent boundary layer at $Re_\theta = 13\,000$ for all wavelengths and all wall distances, using only estimates of the time power spectral density of the streamwise velocity and of its local spatial derivative. The resulting global convection velocity has a least-squares interpretation and is easily related to the wavelength-dependent convection velocity. The method intrinsically provides an estimation of the validity of Taylor's hypothesis by a correlation coefficient identical to the one from del Álamo and Jiménez, J Fluid Mech 640:5–26, 2009, [7]. The results reveal some similarities between the convection of the superstructures, the hairpin packets, and the near-wall structures. The convection velocity of the superstructures is isolated by restricting the global convection velocity to the largest wavelengths. The spatial spectrum is estimated from the temporal spectrum using the frequency-dependent convection velocity. The results are consistent with a classical correlation-based evaluation.

1 Introduction

High Reynolds number wall-bounded turbulence involves complex physics, see e.g. [11, 22], the experimental or numerical study of which is made difficult by the scale separation. At high Reynolds numbers, there appear very large scale motions, also called superstructures, which are much longer than the outer scale unit (see e.g. [10]) and contribute significantly to quantities such as turbulent skin friction (see e. g. [6]). The spatially developing zero pressure gradient flat plate turbulent boundary layer is especially relevant to applications. Because of the differences between the canonical wall-bounded turbulent flows (see e.g. [12]), we focus here on the boundary layer.

N. Renard (✉) · S. Deck
Onera, The French Aerospace Lab, 92190 Meudon, France
e-mail: nicolas.renard@onera.fr

P. Sagaut
Institut Jean le Rond D'Alembert, Université Pierre Et Marie Curie, Paris, France

M. Stanislas et al. (eds.), *Progress in Wall Turbulence 2*,
ERCOFTAC Series 23, DOI 10.1007/978-3-319-20388-1_33

One first insight into the dynamics of turbulent coherent motions is given by their convection velocity. The assumption that the convection term is dominating leads to Taylor's frozen turbulence hypothesis (see [21]). The time and streamwise derivatives are then related by the convection velocity, which is classically set equal to the Reynolds-averaged velocity. Reference [9] emphasize it is not a good approximation in shear flows, unlike homogeneous isotropic turbulence. Mean shear drives time evolution of the turbulent structures, and it is felt differently depending on the size of the coherent motions. This motivates the study of wavelength-dependent convection velocity. The validity of the frozen turbulence hypothesis is all the more marginal as the wavelength is large, as derived theoretically by [17], but it may still be used for the superstructures, see [8].

The assessment of the convection velocity in shear flows was treated in details in [9, 23], and a wealth of slightly varying methods may be found in the literature. A widely used method relies on the space—time correlation. For a given small streamwise separation, the time delay that leads to a maximum correlation is measured. This has been done with a very small spatial separation in [5], and is also used here as a reference, referred to as the correlation-based method. Seeking a maximum correlation for a given time delay (varying the spatial separation), as well as using a spatial separation that is large instead of very small, provide different results. The dependency on the spatial separation is a consequence of the dependency on the length scale of the fluctuation, showing the limitations of a global convection velocity.

In order to describe the dependency on the length scale, time and space signals may be simultaneously used. The convective process is identified in the spatial separation—time delay space, or in the wavenumber—frequency space, as done in an energy-preserving manner by [2]. This finds its limits in a flow that is not homogeneous in the streamwise direction, motivating the use of only time spectra while the spatial information is obtained from the local space derivative only. Using data in only one direction of the space—time plane in order to assess the convection velocity has been successfully done in a channel flow by [7]. When Taylor's hypothesis is not strictly valid, the method of [7] is more easily interpreted than similar methods by [3, 13].

The present study is devoted to the assessment of the length-scale-dependent convection velocity of the streamwise velocity fluctuations in a boundary layer at $Re_\theta = 13\,000$. To the authors' knowledge, this has not been assessed in a boundary layer at such a high Reynolds number from a numerical simulation before, and the available experimental data do not cover the whole boundary layer thickness and suffer from measurement uncertainty or Reynolds number limitations, see e.g. [1, 13–16].

A new method is presented for the estimation of the wavelength-dependent convection velocity, using estimates of the time power spectral density of the streamwise velocity and its local streamwise derivative. It is directly inspired by [7], who make use of the spatial Fourier modes of velocity and its instantaneous time derivative. The suggestion by [7] to deal with the dual problem to theirs (in a time—space meaning) makes the method suitable for a spatially developing flow, because only data very close to the considered streamwise location is needed. Unlike [7] who use the exact Fourier modes available in their spectral channel flow DNS, the present method resorts, with mathematical justification, to statistical estimates of the power spectral densities. The global convection velocity, taking into account all wavelengths, may be recovered in a dual way to [7].

The numerical simulation using the third mode of Zonal Detached Eddy Simulation (ZDES) as a Wall-Resolved LES is first described and visualized, focusing on packets of hairpins and superstructures. The wavelength-dependent convection velocity is then derived, leading to the estimation of the spatial streamwise velocity spectrum from its temporal counterpart and to the assessment of the validity of Taylor's hypothesis. Finally a global convection velocity representing all or only the largest scales of motions is derived and compared to other characteristic velocities.

2 Wall-Resolved Large Eddy Simulation (WRLES) of a High Reynolds Number Boundary Layer

A wall-resolved LES of a zero pressure gradient turbulent boundary layer has been performed up to $Re_\theta = 13\,000$ using the third mode of Zonal Detached Eddy Simulation (ZDES, see [4]), with thorough description and validation in [6], emphasizing that the subgrid scales carry a negligible turbulent kinetic energy in the outer layer. The mesh resolution is $\Delta x^+ = 50$ and $\Delta z^+ = 12$ in the streamwise and spanwise directions. The Onera compressible flow solver FLU3M is used. The low freestream Mach number $M_\infty = 0.21$ entails negligible differences with the incompressible boundary layer. The spatial scheme is a modified low-dissipative AUSM+(P) scheme [19] and the implicit time integration is second-order accurate. A quick overview of the coherent motions near $Re_\theta = 13\,000$ is provided by the numerical Schlieren shown in Fig. 1, and by the streamwise velocity isosurface in Fig. 2. The Schlieren strongly suggests resolved hairpins organizing within packets up to several boundary layer thickness long. At a larger streamwise length scale, one may see in Fig. 2 elongated and meandering areas of low streamwise velocity, of very large scale (exceeding 5δ, clearly longer than the hairpin packets), which are believed to be resolved superstructures.

The time power spectral density of the streamwise velocity is estimated by Welch's method from the time signal stored during a total duration of $1155\delta_0/U_\infty$ (δ_0 being

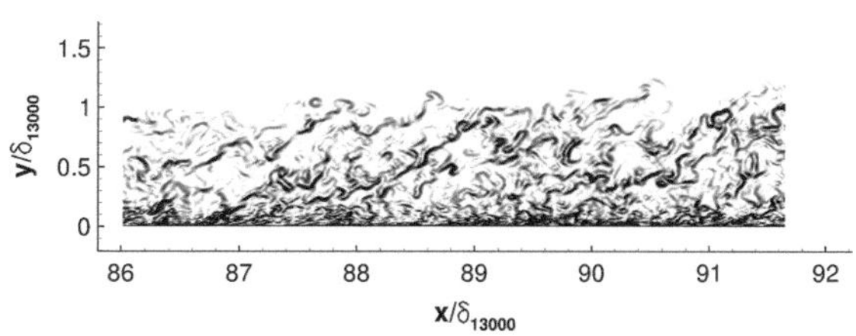

Fig. 1 Numerical Schlieren (magnitude of density gradient in a streamwise-wall normal plane) near the $Re_\theta = 13\,000$ station

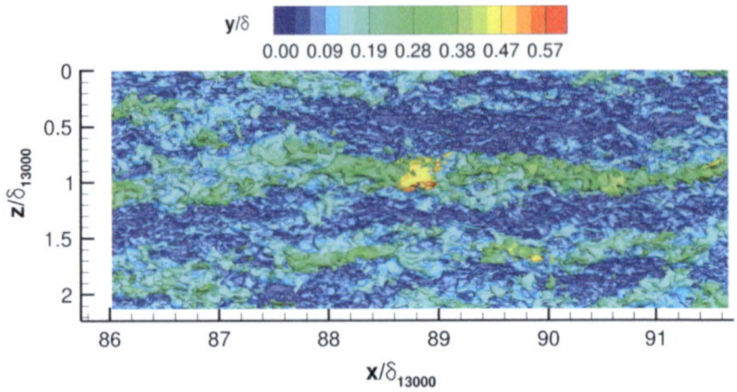

Fig. 2 $u^+ = 20$ isosurface colored by the outer-scaled wall distance near the $Re_\theta = 13\,000$ station

the initial boundary layer thickness and U_∞ the freestream velocity). The frequency is then related to the streamwise wavenumber using Taylor's hypothesis, with the correlation-based convection velocity mentioned in the introduction, so that the streamwise spectrum is estimated. The outer layer spectra in Fig. 3 feature an outer site located near $\lambda_x/\delta = 2-3$ at both $Re_\theta = 5\,200$ and $Re_\theta = 13\,000$, attributed to the hairpin packets. A second outer site appears at $Re_\theta = 13\,000$ near $\lambda_x/\delta = 10$. It is thought to correspond to the superstructures seen in Fig. 2. This is compared with experimental data from [18, 20] in Fig. 4, using the mean velocity as convection velocity for consistency with the experimental spectra. A fair agreement is found in the outer layer, which suggests that the second outer site does correspond to superstructures. This adds to the validation of the present simulation documented in [6].

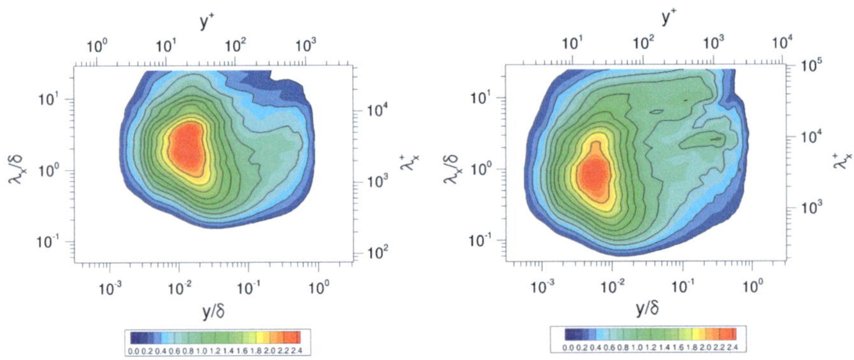

Fig. 3 Reynolds number impact on the premultiplied power spectral density of the streamwise velocity $k_x G_{uu}/u_\tau^2$ (*colors* and *isolines*). *Left* $Re_\theta = 5\,200$. *Right* $Re_\theta = 13\,000$. Dispersion relation given by $U_c = U_{\text{corr}}$ (correlation-based convection velocity)

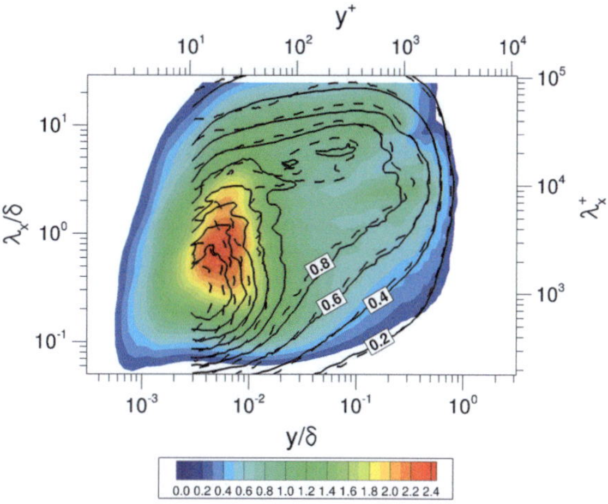

Fig. 4 Premultiplied streamwise power spectral density of streamwise velocity $k_x G_{uu}/u_\tau^2$ (dispersion relation $U_c = \langle U\rangle(y)$) at $Re_\theta = 13\,000$ ($Re_\tau = 3\,600$), compared with experimental data from [20] (*solid isolines*, $Re_\tau = 2\,800$) and from [18] (*dashed isolines*, $Re_\tau = 3\,900$)

3 Spectral Assessment of the Convection Velocity

The convection velocity of the streamwise velocity fluctuations $u(x,t) = U(x,t) - \langle U(x,t)\rangle$, where $\langle\cdot\rangle$ is the Reynolds average, is assessed in the dual manner to [7], as suggested in this reference. Assuming Taylor's hypothesis, the fluctuations are driven by the relation $\frac{1}{U_c}\partial_t u + \partial_x u = 0$, where U_c is the convection velocity. Since the hypothesis is not exactly verified in shear flows, we choose to minimize the quantity $\mathscr{D} = \frac{E\left(\left(\frac{1}{C}\partial_t u + \partial_x u\right)^2\right)}{E\left((\partial_x u)^2\right)}$, where $E(\cdot)$ represents the mathematical expectation. The value of C that leads to the minimum value of $\mathscr{D}(C)$ is defined as the convection velocity C_u, leading to the following least-squares assessment of the convection velocity:

$$C_u = -\frac{E\left((\partial_t u)^2\right)}{E\left(\partial_t u\, \partial_x u\right)} \qquad (1)$$

Only local signals are used, at the streamwise location x. The spatial growth of the boundary layer is better ignored than in a method using data from various streamwise positions. The ultimate goal is indeed to recover a fictitious spatial spectrum at the chosen Reynolds number station. The smaller $\mathscr{D}(C_u)$, the higher the level of validity of Taylor's hypothesis. Consistently with [7], a correlation coefficient is defined by $1 - \gamma_u^2 = \mathscr{D}(C_u)$:

$$\gamma_u = \frac{\left|E\left(\partial_t u\, \partial_x u\right)\right|}{\sqrt{E\left((\partial_x u)^2\right) E\left((\partial_t u)^2\right)}} \qquad (2)$$

γ_u varies between 1 (perfect convection process) and 0. It is identical to the one of the dual case by [7], unlike the convection velocity.

The mathematical expectations are assessed using the estimated power spectral density of the streamwise velocity and of its local space derivative. The time cross power spectral density S^t_{ab} of signals $a(t)$ and $b(t)$ is defined as the Fourier transform of their correlation function $R^t_{ab}(x,\tau) = E\left(a(x,t)b(x,t+\tau)\right)$, so that $S^t_{ab}(x,f) = \int_{-\infty}^{\infty} R^t_{ab}(x,\tau)\exp(-2i\pi f\tau)d\tau$. Equation (1) then becomes

$$C_u = -\frac{\int_0^{\infty}(2\pi f)^2 S^t_{uu}(f)df}{\int_0^{\infty} 2\pi f \mathscr{I}\left(S^t_{u\,\partial_x u}(f)\right)df}, \tag{3}$$

where $\mathscr{I}(\cdot)$ stands for the imaginary part. The correlation coefficient becomes

$$\gamma_u = \frac{\left|\int_{-\infty}^{\infty} -2i\pi f S^t_{u\,\partial_x u}(f)df\right|}{\sqrt{\int_{-\infty}^{\infty} S^t_{\partial_x u\,\partial_x u}(f)df}\sqrt{\int_{-\infty}^{\infty}(2\pi f)^2 S^t_{uu}(f)df}}. \tag{4}$$

If the velocity is narrowly band-pass-filtered around the frequency f, so is its streamwise derivative (because space differentiation and time Fourier transform are commutative). This way, the frequency-dependent convection velocity and the associated correlation coefficient are derived from Eqs. (3)–(4):

$$U_c(x,f) = -\frac{2\pi f \cdot S^t_{uu}(x,f)}{\mathscr{I}\left(S^t_{u\,\partial_x u}(x,f)\right)} \tag{5}$$

$$\gamma_u(f) = \frac{\left|\mathscr{I}\left(S^t_{u\,\partial_x u}(x,f)\right)\right|}{\sqrt{S^t_{uu}(x,f)}\sqrt{S^t_{\partial_x u\,\partial_x u}(x,f)}}. \tag{6}$$

The spatial derivative ∂_x is evaluated using a 4th-order accurate centered finite difference scheme, using the time signal at five successive nodes of the mesh. The resulting convection velocity and the correlation coefficient are plotted in Fig. 5, using the dispersion relation $2\pi f = -k_x U_c(f)$ to recover the streamwise wavelength from Taylor's hypothesis. Consistently with the predictions of [17], the validity of Taylor's hypothesis appears to be all the better as smaller structures are considered further away from the wall. However, even for the footprint of the largest length scales, the correlation coefficient remains reasonably high when compared with values reported by [7]. Of interest, the convection velocity of the superstructures does not vary much through the boundary layer: their footprint is faster and their top is slower than the other structures. In Fig. 6, the convection velocity is divided by the local mean velocity. The convection velocity is globally close to the mean velocity, except near the wall where it is greater than the mean. A consistent pattern may be identified among the inner site, the hairpin packet outer site and the superstructure outer site.

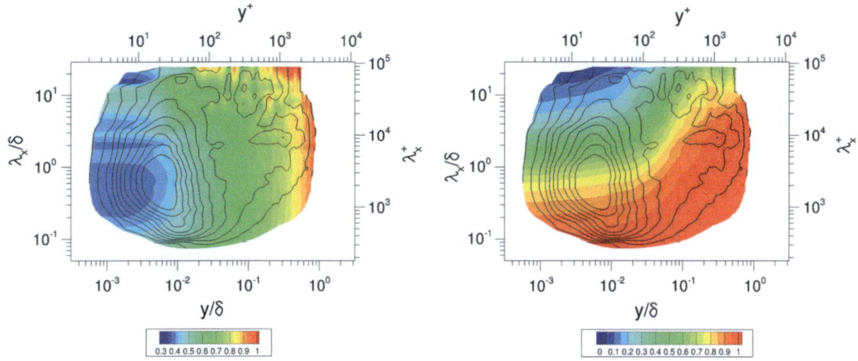

Fig. 5 Convection velocity U_c/U_∞ (colored contours, *left*) and correlation coefficient γ_u (colored contours, *right*) at $Re_\theta = 13\,000$ as a function of the wavelength and the wall distance. Isolines of $k_x G_{uu}/u_\tau^2$ in *solid lines*

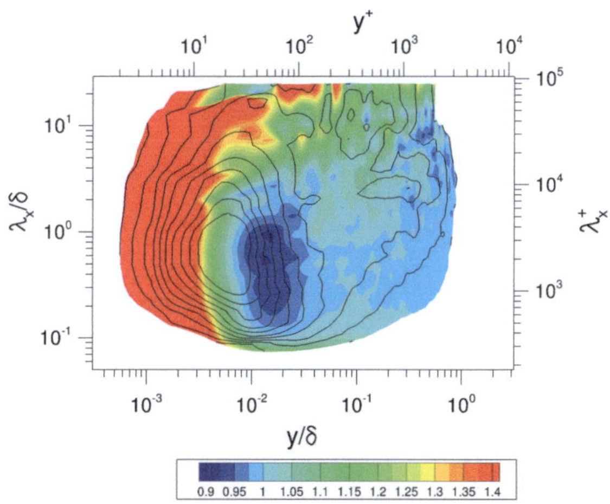

Fig. 6 Convection velocity divided by the mean streamwise velocity $U_c/\langle U\rangle$ (colored contours) at $Re_\theta = 13\,000$ as a function of the wavelength and the wall distance. Isolines of $k_x G_{uu}/u_\tau^2$ in *solid lines*

Indeed, all sites feature a convection velocity higher than the mean velocity near the foot of the site, and lower than the mean near the top of the site.

4 Estimation of the Spatial Spectrum from the Temporal Spectrum

The spatial spectrum may be estimated from the time spectrum, using the frequency-dependent convection velocity. The dispersion relation $2\pi f = -k_x\, U_c(f)$ is used to link the frequency to a streamwise wavelength. The spectra should be matched such

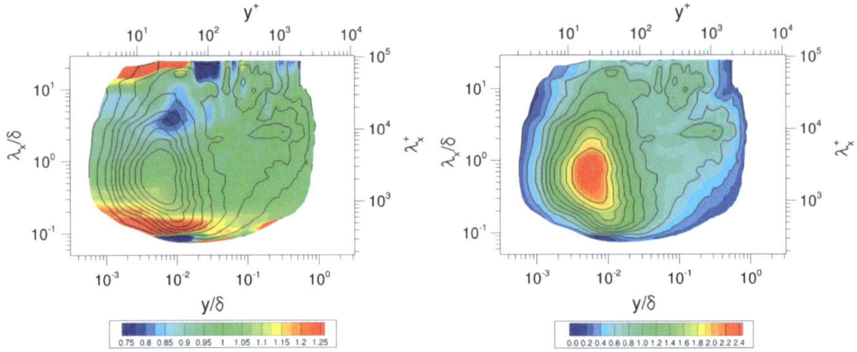

Fig. 7 Dilatation factor (*left*, colored contours) and premultiplied streamwise power spectral density $k_x G_{uu}/u_\tau^2$ estimated from the temporal spectrum at $Re_\theta = 13\,000$ as a function of the wavelength and the wall distance. Isolines of $k_x G_{uu}/u_\tau^2$ in *solid lines*

that $\int_{-\infty}^{\infty} S_{uu}^t(f)df = E(u^2) = \int_{-\infty}^{\infty} S_{uu}^x(k_x)dk_x$, leading to a dilatation factor (see [7]):

$$k_x G_{uu}^x\left(k_x = \frac{2\pi f}{U_c(f)}\right) = f G_{uu}^t(f) \cdot \underbrace{\frac{1}{\left|1 - \frac{f}{U_c(f)}\frac{dU_c(f)}{df}\right|}}_{\text{dilatation factor}}, \tag{7}$$

where G_{uu}^t is the one-sided power spectral density, defined as twice S_{uu}^t for positive frequencies so that $\int_0^{\infty} G_{uu}^t(f)df = \int_{-\infty}^{\infty} S_{uu}^t(f)df$.

The dilatation factor is shown in Fig. 7. The site of values lower than 1 located in between the inner site and the footprint of the superstructures is a direct consequence of the foot of the superstructures being advected faster than the surrounding structures. The values near the very largest and smallest wavelengths are not significant, because the noise tends to dominate the signal in these low-energy regions. The resulting streamwise spectrum is shown in Fig. 7 and may be compared with the spectrum made using the correlation-based convection velocity shown in Fig. 3.

5 Defining a Global Convection Velocity

The global convection velocity of Eqs. (3) and (1) may be written as a function of the local (frequency-dependent) convection velocity of Eq. (5):

$$C_u = \frac{\int_{-\infty}^{\infty}(2\pi f)^2 S_{uu}^t(f)df}{\int_{-\infty}^{\infty}\frac{1}{U_c(x,f)}(2\pi f)^2 S_{uu}^t(f)df} \tag{8}$$

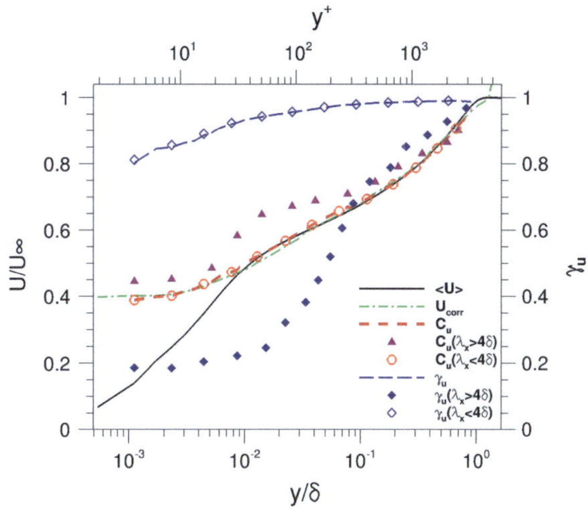

Fig. 8 Outer-scaled streamwise velocities U/U_∞ at $Re_\theta = 13\,000$: mean velocity $\langle U \rangle$, correlation-based convection velocity U_{corr}, global convection velocity C_u, and partially integrated global convection velocity ($C_u(\lambda_x > 4\delta)$ and $C_u(\lambda_x < 4\delta)$). The associated correlation coefficients γ_u are plotted on the *right-hand axis*

and the associated global correlation coefficient of Eq. (2) is

$$\gamma_u = \frac{1}{|C_u|}\sqrt{\frac{\int_{-\infty}^{\infty}(2\pi f)^2 S^t_{uu}(f)df}{\int_{-\infty}^{\infty} S^t_{\partial_x u\,\partial_x u}(f)df}}. \tag{9}$$

The resulting convection velocity profile is shown in Fig. 8. It is close to the mean velocity, except in the inner layer where it is higher. It is also very close to the correlation-based convection velocity at all wall distances. The global correlation coefficient is satisfyingly high, even though it decreases near the wall. The integrals of Eqs. (8) and (9) may be restricted to the largest wavelengths only, namely $\lambda_x/\delta \geq 4$ in Fig. 8. The largest length scales are convected faster than the other scales near the wall, a result that should be taken with caution because of the relatively low value of the associated correlation coefficient.

6 Outlook

The method of [7] has been successfully adapted to a spatially developing flow for which only statistical estimates of the time power spectral densities are available. This enables the assessment of the wavelength-dependent convection velocity in a flat plate turbulent boundary layer at $Re_\theta = 13000$, for which accurate experimental data through the whole boundary layer are missing. The present method enables the intrinsic assessment of the validity of Taylor's hypothesis, the easy interpretation of the least-square-based convection velocity, and the natural definition of a global convection velocity. Some common features of the convection of near-wall structures, hairpin packets, and superstructures are found. The high convection velocity associated to the near-wall footprint of the superstructures is also clearly visible. The evaluation of the streamwise spectrum using the time spectrum and a dilatation factor computed from the frequency-dependent convection velocity provides results that are close to those obtained with the correlation-based convection velocity.

The study of convection velocity may be seen as an additional approach to coherent motions at high Reynolds numbers, which would be enhanced by simulations at even higher Reynolds numbers with clearer scale separation. Using this method to assess the physical validity of resolved motions in wall-modeled LES (WMLES) might also be of interest.

Acknowledgments This work was performed using HPC resources from GENCI-CINES (Project ZDESWALLTURB, Grant 2012-[c2012026817]). The authors wish to thank Pierre-Élie Weiss and Romain Laraufie for fruitful discussions. All the people involved in the evolution of the FLU3M code are warmly acknowledged. The thesis of Nicolas Renard is partly funded by the French defense procurement agency DGA.

References

1. C. Atkinson, N. Buchmann, O. Amili, J. Soria, High-resolution large field-of-view experimental investigation of turbulent convection velocities in a turbulent boundary layer. in *International Symposium on Turbulence and Shear Flow Phenomena (TSFP-8)* (2013)
2. D. Chung, B.J. McKeon, Large-eddy simulation of large-scale structures in long channel flow. J. Fluid Mech. **661**, 341–364 (2010)
3. S. Davoust, L. Jacquin, Taylor's hypothesis convection velocities from mass conservation equation. Phys. Fluids **23**, 051701 (2011)
4. S. Deck, Recent improvements in the Zonal Detached Eddy Simulation (ZDES) formulation. Theor. Comput. Fluid Dyn. **26**, 523–550 (2012)
5. S. Deck, N. Renard, R. Laraufie, P. Sagaut, Zonal Detached Eddy Simulation (ZDES) of a spatially developing flat plate turbulent boundary layer over the Reynolds number range $3\,150 \leq Re_\theta \leq 14\,000$. Phys. Fluids **26**, 025116 (2014)
6. S. Deck, N. Renard, R. Laraufie, P.E. Weiss, Large scale contribution to mean wall shear stress in high Reynolds number flat plate boundary layers up to Re_θ =13650. J. Fluid Mech. **743**, 202–248 (2014). doi:10.1017/jfm.2013.629
7. J.C. del Álamo, J. Jiménez, Estimation of turbulent convection velocities and corrections to Taylor's approximation. J. Fluid Mech. **640**, 5–26 (2009)
8. D.J.C. Dennis, T.B. Nickels, On the limitations of Taylor's hypothesis in constructing long structures in a turbulent boundary layer. J. Fluid Mech. **614**, 197–206 (2008)
9. M.J. Fisher, P.A.L. Davies, Correlation measurements in a non-frozen pattern of turbulence. J. Fluid Mech. **18**, 97–116 (1963)
10. N. Hutchins, I. Marusic, Evidence of very long meandering features in the logarithmic region of turbulent boundary layers. J. Fluid Mech. **579**, 1–28 (2007)
11. J. Jiménez, Cascades in wall-bounded turbulence. Annu. Rev. Fluid Mech. **44**, 27–45 (2012)
12. J. Jiménez, S. Hoyas, M.P. Simens, Y. Mizuno, Turbulent boundary layers and channels at moderate Reynolds numbers. J. Fluid Mech. **657**, 335–360 (2010)
13. R. de Kat, L. Gan, J.R. Dawson, B. Ganapathisubramani, Limitations of estimating turbulent convection velocities from PIV, in *16th International Symposium on Applications of Laser Techniques to Fluid Mechanics*, Lisbon, Portugal (2012)

14. P.A. Krogstad, J.H. Kaspersen, S. Rimestad, Convection velocities in a turbulent boundary layer. Phys. Fluids **10–4**, 949–957 (1998)
15. J.H. Lee, H.J. Sung, Very-large-scale motions in a turbulent boundary layer. J. Fluid Mech. **673**, 80–120 (2011)
16. J. LeHew, M. Guala, B.J. McKeon, A study of the three-dimensional spectral energy distribution in a zero pressure gradient turbulent boundary layer. Exp. Fluids **51**, 997–1012 (2011)
17. C.C. Lin, On Taylor's hypothesis and the acceleration terms in the Navier-Stokes equations. Q. Appl. Math. **X**(4), 154–165 (1953)
18. I. Marusic, R. Mathis, N. Hutchins, High Reynolds number effects in wall turbulence. Int. J. Heat Fluid Flow **31**, 418–428 (2010)
19. I. Mary, P. Sagaut, Large eddy simulation of flow around an airfoil near stall. AIAA J. **40**, 1139–1145 (2002)
20. R. Mathis, N. Hutchins, I. Marusic, Large-scale amplitude modulation of the small-scale structures in turbulent boundary layers. J. Fluid Mech. **628**, 311–337 (2009)
21. P. Moin, Revisiting Taylor's hypothesis. J. Fluid Mech. Focus Fluids **640**, 1–4 (2009)
22. A.J. Smits, B.J. McKeon, I. Marusic, High-Reynolds number wall turbulence. Annu. Rev. Fluid Mech. **43**, 353–375 (2011)
23. J.A.B. Wills, On convection velocities in turbulent shear flows. J. Fluid Mech. **20**(3), 417–432 (1964)

Statistics of Single Self-sustaining Attached Eddy in a Turbulent Channel

Yongyun Hwang

Abstract A Numerical experiment that isolates the motions at a given spanwise length scale is performed based on previous observation on the self-sustaining nature of the eddies in the logarithmic and the wake outer regions [7, 8]. It is shown that the statistics of the isolated self-sustaining motions at a given spanwise length scale are strikingly similar to those of the single attached eddy postulated by Townsend and Perry [1, 2, 5, 6], demonstrating the existence of the attached eddies in turbulent channel flow. Inspecting one-dimensional spectra also leads to build a complete form of the self-similarity of the streamwise length scale and the wall-normal location of all the coherent structures known, including near-wall streaks, quasi-streamwise vortices, very-large-scale motions, and large-scale motions.

1 Introduction

The foundation of the logarithmic dependence in the mean-velocity profile has been understood as a landmark in wall-bounded turbulence research. Apart from the debate on the log versus power law, the only relevant length scale in this region is the distance from the wall 'y' (the wall-normal direction) which fills the gap caused by separation between the inner and the outer length scales at high Reynolds numbers. Townsend [1, 2] pointed out that emergence of this feature would be unimaginable unless the size (or the diameter) of the energy-containing motions (i.e. coherent structures) in this region is proportional to distance of their centres from the wall. Based on this insightful observation, he developed a self-consistent theory which describes the second-order statistics in the logarithmic region by superposing many hypothetical self-similar 'attached eddies' with a probability density distribution yielding constant Reynolds stress, predicting

Y. Hwang (✉)
Department of Aeronautics, Imperial College London,
London SW7 2AZ, UK
e-mail: y.hwang@imperial.ac.uk

M. Stanislas et al. (eds.), *Progress in Wall Turbulence 2*,
ERCOFTAC Series 23, DOI 10.1007/978-3-319-20388-1_34

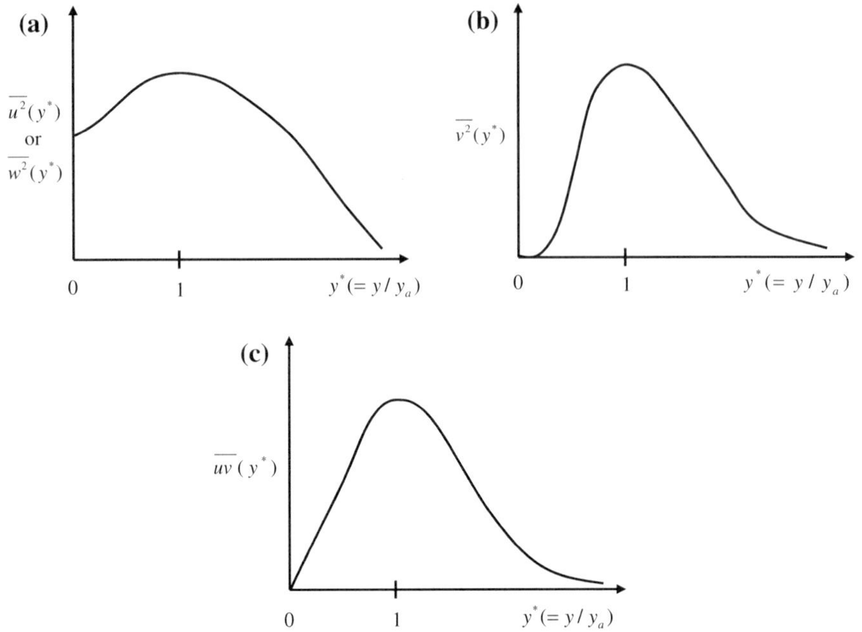

Fig. 1 A sketch of the second-order statistical moments of a single attached eddy according to Townsend [2]: **a** streamwise and spanwise velocities; **b** wall-normal velocity; **c** Reynolds stress. Here, y_a is the location where the single attached eddy shows the maximum intensity, characterising its size

$$\begin{aligned} \overline{u^2} &= A_1 + B_1 \log L_0/y, \\ \overline{v^2} &= A_2, \\ \overline{w^2} &= A_3 + B_1 \log L_0/y, \end{aligned} \tag{1}$$

where A_i $(i = 1, 2, 3)$ and B_j $(j = 1, 2)$ are constants and L_0 is the upper boundary of the logarithmic region. The key of the attached eddy hypothesis is essentially in the assumption of the second-order statistical moments of a 'single' attached eddy (see also Fig. 1). The presence of the wall allows the streamwise and the spanwise components of the single attached eddy to carry non-negligible amounts of energy near the wall (Fig. 1a) whereas it forces the wall-normal velocity component and the Reynolds stress to be very small (Fig. 1b, c). This feature is essential to obtain (1): i.e. the logarithmic dependence in (1) mathematically originates from the non-negligible contribution of the streamwise and the spanwise components to the near-wall region.

Owing to rapid development of advanced experimental techniques and modern computational power, numerical [3] and laboratory experiments [4] have recently confirmed the prediction (1), suggesting that it is very likely that wall-bounded turbulence is composed of the attached eddies. Despite the emerging evidence, the core of the theory addressing the question, 'what really is the single attached eddy?', has not been well addressed for a long time, given the difficulty caused by the presence of coherent structures at scales varying from the inner to the outer unit. In fact, Townsend himself proposed that the single attached eddy might be in the form of a double conical eddy [2], whereas the follow-up studies by Perry and coworkers modelled it as a hairpin or 'Λ' vortex [5, 6].

The purpose of the present study is to address this issue by taking a minimal action to online simulation to isolate the attached eddies at a given length scale. The numerical experiment is designed upon the author's recent observation of the self-sustaining nature of a single attached eddy [7, 8], and calculates the statistics of a single attached eddy. We report, for the first time, the statistics of a single attached eddy given in the form of solution of the Navier-Stokes equation, which will also be shown to be strikingly similar to those given in [1, 2, 5, 6].

2 Numerical Experiment

The key idea for designing the present numerical experiment is from the recent observation on the linear spanwise length-scale growth in the logarithmic region, as also in Fig. 2 where the premultiplied spanwise spectra of the streamwise velocity are shown from DNS at $Re_\tau = 2000$ [9]. This feature directly supports the attached eddy hypothesis as the spanwise length scale of a given eddy would characterise the size of the attached eddy. Based on this observation, in the author's recent study, the near-wall motions at $\lambda_z^+ \leq 100$ have been studied [10]. In the present study, we extend this approach to the region of $100\delta_\nu < \lambda_z < 1.5\,\text{h}$ where the linear growth of the spanwise length scale with the distance from the wall appears.

Let us start by choosing a spanwise length scale, λ_z. Given the scope of the present study aimed to isolate the motion at given λ_z, it is not necessary to resolve the motions wider than λ_z. Therefore, we set the spanwise computational domain to be $L_z = \lambda_z$

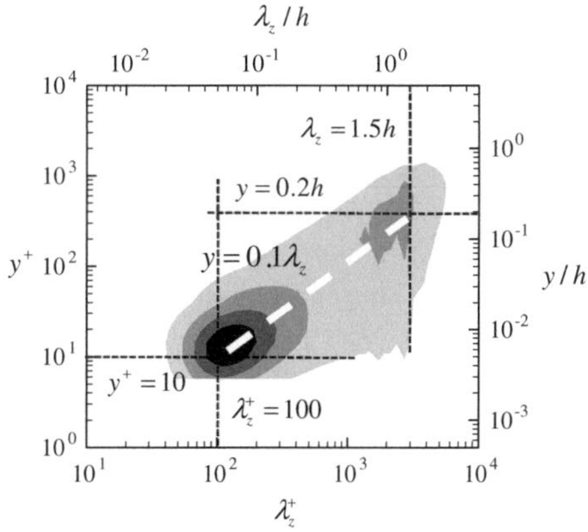

Fig. 2 One-dimensional premultiplied spanwise spectra of the streamwise velocity (from [9])

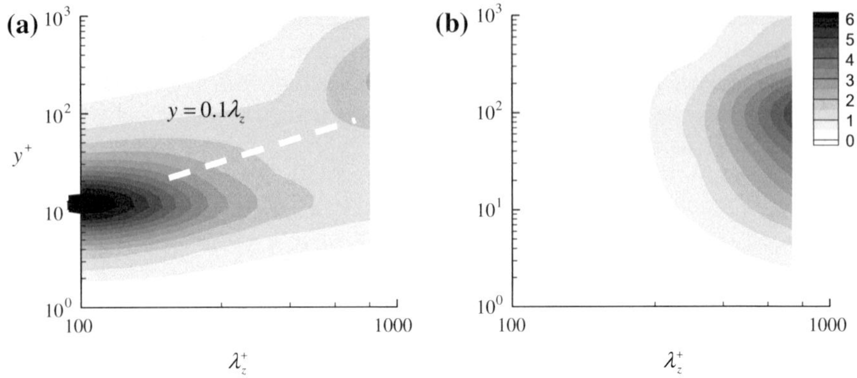

Fig. 3 One-dimensional spanwise spectra of the streamwise velocity ($L1800b$): **a** $C_s = 0.05$; **b** $C_s = 0.20$

Table 1 Parameters of the numerical experiment in the present study

Case	Re_m	Re_τ	L_x/h	L_z/h	L_x^+	L_z^+	$N_x \times N_y \times N_z$	C_s
$F950$	38133	955	12π	6.0	36003	5730	$384 \times 81 \times 128$	0.05
$O950$	38133	1051	12π	6.0	39622	6306	$384 \times 81 \times 128$	0.30
$L950a$	38133	949	12π	0.75	35776	712	$384 \times 81 \times 16$	0.20
$L950b$	38133	1016	12π	1.0	38302	1016	$384 \times 81 \times 21$	0.25
$L1800a$	73333	1519	8π	0.375	38177	570	$512 \times 129 \times 16$	0.10
$L1800b$	73333	1745	8π	0.5	43857	872	$512 \times 129 \times 21$	0.20
$L1800c$	73333	2022	8π	0.75	50818	1517	$512 \times 129 \times 32$	0.25

while we consider a very long computational domain in the streamwise direction ($L_x \geq 8\pi h$). As discussed in [10], the spanwise confined computational domain yields spanwise uniform non-physical motions. A filtering approach to remove such motions has been proposed in [10], and we also adopt this approach here. This then enables us to remove the motions wider than λ_z. An example of this is given in Fig. 3a where the motions wider than $\lambda_z > 0.5h$ are removed by this technique. The removal of the motions smaller than the given spanwise length scale is then performed using the technique described in [7, 8] where a large-eddy simulation with the static Smagorinsky model is used to damp out the smaller motions by increasing the Smagorinsky constant C_s. Using this approach, the motions at the given spanwise length scale can be successfully isolated, as in Fig. 3b where its spanwise spectra are shown. The designed numerical experiment is repeated for different λ_z and Re, and its details are summarised as in Table 1.

3 Results and Discussion

Figure 4 shows the second-order statistics of the isolated motions at a given spanwise length scale. The appearance of non-trivial statistical moments clearly suggests that the motions at a given spanwise length scale are able to be sustained in the absence of the ones at the other length scales. When the length scale and the velocity are respectively normalised by the given spanwise length and the maximum value of the velocity, all the second-order statistical moments are found to be self-similar (Fig. 4). For the statistics of the streamwise velocity and the Reynolds stress, to which the streamwise velocity contributes, the self-similar nature appears for $y < 0.3L_z$, whereas the wall-normal and the spanwise velocities show the self-similarity for $y \simeq 0.5L_z$. It should be mentioned that the statistical moments above those locations are found not to be self-similar, and this is probably because the self-similar motions induce non-negligible fluctuations (for further details on this issue, see also [10]).

The most important finding of the present study is that the second-order statistics of the self-similar part of the motions at a given length scale is strikingly similar to those given in Fig. 1, suggesting that the energy-carrying motion at a given spanwise length scale is the 'attached' eddy that Townsend postulated. The only difference between the original theory given in Fig. 1 and the self-similar part in Fig. 2 appears in the

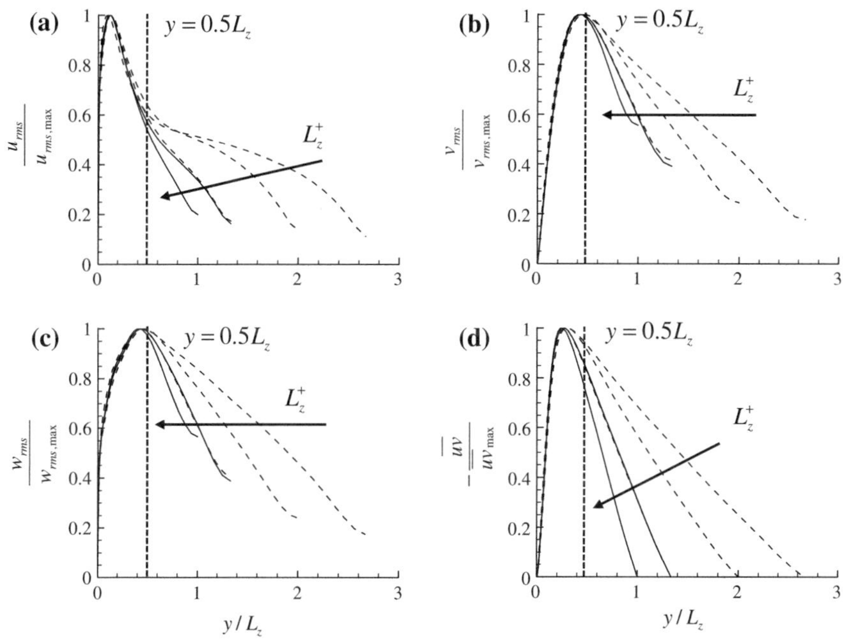

Fig. 4 Normalised second-order statistical moments of the self-sustaining motions at given spanwise length scale: **a** streamwise velocity; **b** wall-normal velocity; **c** spanwise velocity; **d** Reynolds stress

streamwise velocity, which shows the maximum closer to the wall compared to the wall-normal and the spanwise velocities, but this will be further explained (see Fig. 5). This numerical result provides direct evidence of the existence of the 'attached' eddies defined strictly in the sense of Townsend and Perry, and demonstrates that the individual attached eddy can be represented by the energy-carrying motions at a given spanwise length scale between $\lambda_z^+ = 100$ and $\lambda_z = 1.5$. It should also be emphasised that the fact that the attached eddies are isolated by the present numerical experiment indicates their self-sustaining nature, providing direct counter evidence to the visual-inspection-based argument by Adrian [11] that the motions in the logarithmic region are formed by the merger and/or growth of hairpin vortices from the near-wall region.

Figure 5 shows one-dimensional streamwise spectra of the isolated motions at the given spanwise length scale. The self-similarity with respect to the spanwise length scale is also seen. It is interesting to note that the streamwise component of each motion shows the largest energy at $y \simeq 0.1L_z$ and $\lambda_x \simeq 10L_z$ whereas the wall-normal and the spanwise components of the motion contain the largest energy at $y \simeq 0.5L_z$ and $\lambda_x = 2 \sim 3L_z$ where the streamwise velocity component of the spectra also shows a non-negligible amount of energy. This suggests that a single

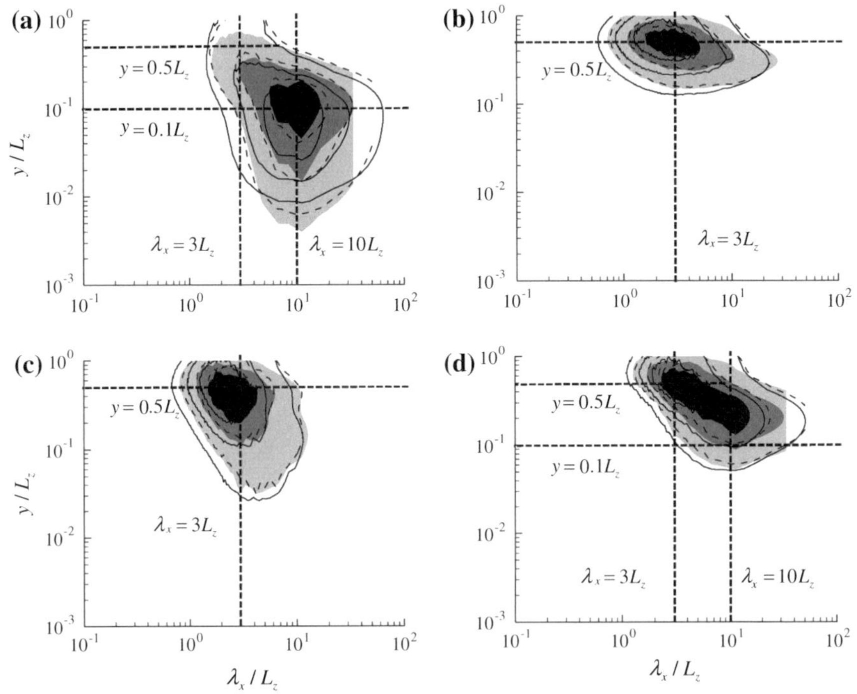

Fig. 5 One-dimensional premultiplied streamwise spectra: **a** streamwise velocity; **b** wall-normal velocity; **c** spanwise velocity; **d** Reynolds stress. Here, the *solid lines*, $L1800a$; the *dashed lines*, $L1800b$; the *shaded*, $L1800c$, respectively. The contour lines represent 0.25, 0.5, 0.75 times of the maximum value of the given contour

attached eddy is composed of two different types of motions, one of which is a long and streamwise velocity dominant motion and the other is short and tall and is rather isotropic in the sense that it carries all the velocity components.

This one-dimensional spectra of the motions at a given spanwise length scale suggest that the energy-carrying motions in the logarithmic region should appear along the two ridges:

$$y \simeq 0.1\lambda_z \text{ and } \lambda_x \simeq 10\lambda_z, \quad (2)$$

$$y \simeq 0.5\lambda_z \text{ and } \lambda_x \simeq 2 \sim 3\lambda_z. \quad (3)$$

It is very helpful to seek what kind of features are expected by (2) and (3) at the bottom and the top ends of the ridges. At the bottom end, the relevant λ_z is the well-known spanwise spacing of the near-wall motions $\lambda_z^+ = 100$. Then, (2) yields $y^+ \simeq 10$ and $\lambda_x^+ \simeq 1000$ while (3) gives $y^+ \simeq 50$ and $\lambda_x^+ \simeq 200 \sim 300$, indicating that the former represents the well-known near-wall streaks whereas the latter represents the near-wall quasi-streamwise vortices. At the top end where the spanwise spacing can be set to be $\lambda_z = 1.5\,h$, (2) leads to $y = 0.2\,h$ and $\lambda_x = 15\,h$, retrieving the wall-normal location and the streamwise length of the VLSMs, whereas (3) gives $y = 0.75\,h$ and $\lambda_x = 3 \sim 4\,h$, yielding those of the LSMs. Therefore, the scaling given by (2) and (3) establishes a complete form of the self-similarity spanning from the inner (streaks and quasi-streamwise vortices) to the outer coherent structures (VLSMs and LSMs).

The one-dimensional spectra also allow us to identify the so-called 'inactive' component of a single attached eddy, the part which does not carry Reynolds stress. Figure 5a suggests that the inactive component in the streamwise velocity appears as the long motions (i.e. the motion given by (2)), the part of which penetrates the region close to the wall. The inactive component in the spanwise velocity appears from the short and tall motions (i.e. the motion given by (3)), and its spanwise component clearly penetrates to the near-wall region without carrying the Reynolds stress. The appearance of such inactive components in the streamwise and the spanwise velocities is essential to explain the logarithmic dependence given in (1), and is also consistent with the recent measurements by [3, 4].

4 Concluding Remarks

So far, the present study has provided direct evidence on the attached eddy hypothesis by extracting the statistics of a single attached eddy. Most of the present results have been found to be consistent with the original theory given by Townsend [2], and complete the self-similarity of all the coherent structures known. More results with a compete discussion are currently in preparation with hope to provide a complete form of statistical description on all the coherent structures in wall-bounded flows.

References

1. A. Townsend, Equilibrium layers and wall turbulence. J. Fluid Mech. **11**(1), 97–120 (1961)
2. A. Townsend, *The Structure of Turbulent Shear Flows* (Cambridge University Press, Cambridge, 1995)
3. J. Jiménez, S. Hoyas, Turbulent fluctuations above the buffer layer of wall-bounded flows. J. Fluid Mech. **611**, 215–236 (2008)
4. I. Marusic, J.P. Monty, M. Hultmark, A.J. Smits, On the logarithmic region in wall turbulence. J. Fluid Mech. **716**, R3 (2013)
5. A.E. Perry, M.S. Chong, On the mechanism of turbulence. J. Fluid Mech. **119**, 173–217 (1982)
6. A.E. Perry, H. Henbest, M. Chong, A theoretical and experimental study of wall turbulence. J. Fluid Mech. **165**, 163–199 (1986)
7. Y. Hwang, C. Cossu, Self-sustained process at large scale in turbulent channel flow. Phys. Rev. Lett. **105**, 044505 (2010)
8. Y. Hwang, C. Cossu, Self-sustained processes in the logarithmic layer of turbulent channel flows. Phys. Fluids **23**, 061702 (2011)
9. S. Hoyas, J. Jiménez, Scaling of the velocity fluctuations in turbulent channels up to Re_τ=2003. Phys. Fluids **18**, 011702 (2006)
10. Y. Hwang, Near-wall turbulent fluctuations in the absence of wide outer motions. J. Fluid Mech. **723**, 264–288 (2013)
11. R.J. Adrian, Hairpin vortex organization in wall turbulence. Phys. Fluids **19**, 041301 (2007)

Scaling the Internal Boundary Layer

Fanxiao Meng, Donald J. Bergstrom and Bing-Chen Wang

Abstract The internal boundary layer (IBL) is an important phenomenon in atmospheric flows which is associated with a step change in the surface roughness. This phenomenon also relates to industrial flows, where a wall may exhibit abrupt changes in surface roughness, perhaps related to corrosion. The present study reports new wind tunnel measurements which consider a Smooth–Rough–Smooth (SRS) configuration. The rough surfaces were created using 40-grit sandpaper glued onto the ground plane of a wind tunnel, and mean velocity measurements were collected using a boundary layer Pitot tube. The development of the IBL is clearly evident in the streamwise evolution of the mean velocity profiles. The results indicate that once the flow encounters the step change in roughness, the flow immediately next to the wall is decelerated due to the enhanced skin friction associated with the surface roughness. The roughness effects propagate further into flow as the IBL grows in the streamwise direction. However, when the surface condition changes back to smooth, the flow begins to accelerate, but does not recover to the initially smooth profile. This implies that some regions of the velocity profile preserve a "memory" of the previous surface condition, and therefore are not in equilibrium with the local surface.

F. Meng · D.J. Bergstrom (✉)
Department of Mechanical Engineering, University of Saskatchewan,
57 Campus Drive, Saskatoon, SK S7N 5A9, Canada
e-mail: don.bergstrom@usask.ca

F. Meng
e-mail: mengf3@myumanitoba.ca

F. Meng · B.-C. Wang
Department of Mechanical Engineering, University of Manitoba,
Winnipeg, MB R3T 5V6, Canada

B.-C. Wang
e-mail: BingChen.Wang@umanitoba.ca

M. Stanislas et al. (eds.), *Progress in Wall Turbulence 2*,
ERCOFTAC Series 23, DOI 10.1007/978-3-319-20388-1_35

1 Introduction

The flow over a step change in surface roughness has both meteorological and industrial applications. For example, in the field of meteorology, the wind flowing over ocean-to-land and land-to-lake boundaries is often modeled as smooth-to-rough and rough-to-smooth transitions, respectively. In industrial duct systems, surface roughness step-changes may occur in localized regions of the wall, due to such effects as corrosion. Therefore, improved knowledge of the behavior of a turbulent boundary layer as it experiences a step change in roughness remains a relevant research topic.

One of the major characteristics of turbulent flow over a step change in surface roughness is the generation of a so-called Internal Boundary Layer (IBL). It is commonly considered that an IBL is the near-wall region where the flow is influenced by the new surface condition. This layer is generated at the beginning of the step change in roughness, and it grows in thickness in the streamwise direction. When a flow encounters a new surface, the fluid close to the surface first experiences the impact of any change in roughness, and then the effect propagates upwards into the surrounding fluid. However, Garratt [5] noted that the definition of the IBL is not always clear, and the criteria used to define the IBL have typically been determined based on practical rather than theoretical considerations.

Over several decades, numerous studies have investigated the characteristics of the IBL. Elliott [4] proposed an empirical formula for describing the growth rate of the IBL thickness, δ_i, which is proportional to the streamwise distance, x, raised to the power n = 0.8. Panofsky and Townsend [7] modified this theory to avoid an abrupt change in the surface shear stress at the top of the IBL. Since these pioneering works, the subject has been studied using analytical, experimental, and numerical methods [3, 6, 8].

The mean velocity and turbulent stress profiles are often used to determine the thickness of an IBL. Shir [9] found that the thickness of an IBL based on a stress criterion was smaller than that based on a velocity criterion since the fluid stress profiles adjust more rapidly than the velocity profiles. Most studies have focused on only a single step change in roughness, either from smooth-to-rough or from rough-to-smooth transition. In contrast, there were only a few investigations on flows subjected to two or more changes in roughness [2]. However, a flow subjected to multiple step changes in roughness has important applications for flows over localized roughness in industrial duct systems.

In this study, the flow over a smooth–rough–smooth transition was investigated experimentally in a wind tunnel based on measurements of the mean velocity field. Mean velocity profiles at different streamwise sections were used to investigate the characteristics of a turbulent boundary layer subjected to consecutive step-changes in roughness. The results show that the flow in the inner layer responds immediately to the change in surface roughness and skin friction. The roughness effects propagate vertically towards the outer edge of the IBL as the flow develops in the streamwise direction. One observation of the consecutive changes in roughness is that some regions of the flow preserve a "memory" of the previous surface condition, and therefore are not in equilibrium with the local surface condition.

2 Experimental Set-Up

The experiments were performed in a subsonic wind tunnel; a detailed description of the experimental facility is available in Akinlade et al. [1]. A trip strip consisting of a metal rod of square cross section (6 mm × 6 mm) was placed across the ground plane 100 mm downstream of the leading edge. The trip strip ensured that the turbulence was fully developed before the flow encounters the rough surface, and was used in all the experiments to ensure a consistent development of the turbulent boundary layer.

A two-dimensional smooth–rough–smooth transition was created by gluing a row of 40-d grit sandpaper to the ground plane in the wind tunnel. The length of the sandpaper section was 230 mm, and the leading edge of the sandpaper was located 1100 mm downstream of the trip strip to ensure the development of a canonical zero-pressure gradient turbulent boundary layer prior to the rough surface. Figure 1 shows a schematic of the experiment. Two origins (x_1 and x_2) were defined based on the two transition locations. Mean velocity profiles were measured using a boundary layer Pitot tube at different streamwise locations within the boundary layer. For each location, the mean streamwise velocity (U) was measured at positions ranging from the wall to the outer edge of the boundary layer. Initially, the study only considered mean velocity profiles. For scaling purposes, the value of the skin friction velocity, u_τ, for the last velocity profile on the smooth surface upstream of the roughness was estimated by profile fitting to the Log Law. For the results presented in the present paper, the free stream velocity was set to $U_e = 30$ m/s. The corresponding Reynolds number based on momentum thickness, Re_θ, ranged from 10,550 to 12,050.

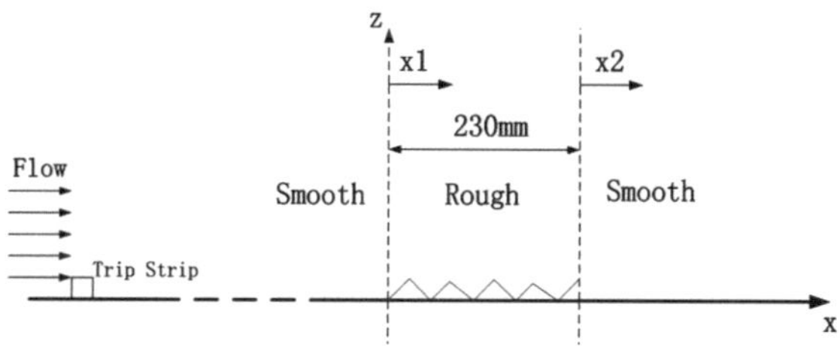

Fig. 1 Schematic of the experimental set-up

3 Results and Discussion

Figure 2a, b shows the mean velocity profiles at different streamwise locations for the smooth–rough and rough–smooth transitions, respectively, when the free-stream wind speed is fixed at $U_e = 30\,\text{m/s}$. The results clearly illustrate the development of the flow over each step-change in roughness. Note that at the first measurement positions, a small change in the vertical position z can result in a significant change to the mean velocity. The irregular behavior of some data points in Fig. 2a, b is partly attributed to the uncertainty in the z measurement.

Figure 2a indicates that when the flow encounters the rough surface, the inner layer is immediately decelerated due to the enhanced skin friction associated with the surface roughness. However, the flow in the outer layer (approximately $z = 3 - 10\,\text{mm}$) is initially not affected by the roughness, so that the mean velocity profiles in this region and above are almost the same as those on the smooth surface.

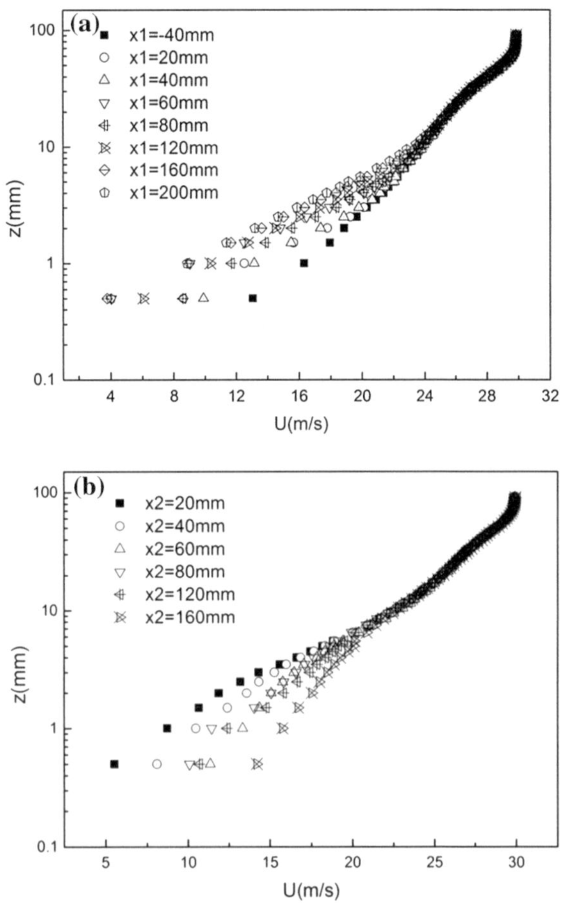

Fig. 2 **a** Mean velocity profiles over smooth–rough transition for $U_e = 30\,\text{m/s}$, **b** Mean velocity profiles over rough–smooth transition for $U_e = 30\,\text{m/s}$

With increasing distance in the x-direction, the roughness effects propagate further into flow, and the profiles in both the inner and outer layers gradually depart from the velocity profile obtained on the smooth surface. When the flow reaches the last section measured on the rough surface ($x_1 = 200\,\mathrm{mm}$), the profile is very close to that at the previous location ($x_1 = 160\,\mathrm{mm}$). This demonstrates that the flow has almost completely adjusted to the new condition created by the rough surface.

Figure 2b illustrates how the flow develops when the surface condition changes back to smooth. In many ways, the process is the reverse of the phenomenon created by the smooth–rough transition. The flow in the inner layer is the first to respond to the new surface condition, and begins to accelerate due to the reduced skin friction at the surface. This effect gradually propagates upwards into the flow. However, unlike the case of the smooth–rough transition, at the last two measurement locations ($x_2 =$ 120 and 160 mm) there is no collapse of the two velocity profiles. This implies that the influence of the smooth surface will penetrate further into the flow and the growth rate of the IBL over the rough–smooth transition is lower than that over the smooth–rough transition. This can be physically explained by the fact that the smooth surface creates a smaller skin friction drag than the rough surface.

Figure 3 compares some select data over the entire test section using inner coordinates, i.e. $u^+ = \frac{U}{u_\tau}$ and $z^+ = \frac{z u_\tau}{\upsilon}$. The friction velocity, u_τ, was obtained by fitting the mean velocity profile on the initial smooth surface to the Log Law. The results indicate the same flow development as shown in Fig. 2a, b. Note that the profile at the last measured location on the second smooth surface exhibits a different behavior from that on the first smooth surface; it seems that the velocity profile does not recover to the original log law profile. This implies that parts of the velocity profile preserve a "memory" of the previous surface condition, and therefore are not in equilibrium with the local surface.

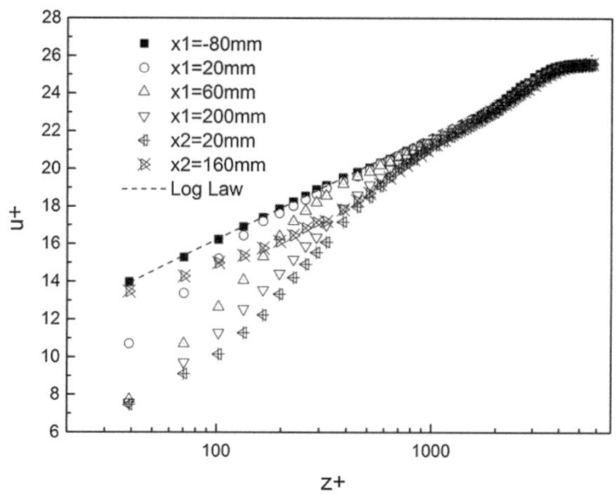

Fig. 3 Mean velocity profiles for the smooth–rough–smooth transition using inner coordinates for $U_e = 30\,\mathrm{m/s}$

4 Conclusions

In this study, the development of a turbulent boundary layer flow over a smooth–rough–smooth transition was investigated experimentally in a wind tunnel. The mean velocity profiles clearly respond to each step-change in roughness, first by decelerating, and then by accelerating. However, the combination of two different transitions results in a part of the flow responding to the previous surface condition. As such, the entire velocity profile is not in equilibrium with the local surface. Further analysis of the turbulent velocity field in terms of the higher-order statistics will be pursued by future studies.

Acknowledgments The authors gratefully acknowledge the financial support of the Natural Science and Engineering Research Council of Canada. The experimental assistance of Prof David Sumner was also much appreciated.

References

1. O.G. Akinlade, D.J. Bergstrom, M.F. Tachie, L. Castillo, Outer flow scaling of smooth and rough wall turbulent boundary layers. Exp. Fluids **37**(4), 604–612 (2004)
2. J. Andreopoulos, D.H. Wood, The response of a turbulent boundary layer to a short length of surface roughness. J. Fluid Mech. **118**, 143–164 (1982)
3. E.F. Bradley, A micrometeorological study of velocity profiles and surface drag in the region modified by a change in surface roughness. Q. J. R. Meteorol. Soc. **94**(401), 361–379 (1968)
4. W.P. Elliott, The growth of the atmospheric internal boundary layer. Trans. Am. Geophys. Union **39**, 1048–1054 (1958)
5. J.R. Garratt, The internal boundary layer–a review. Bound.-Layer Meteorol. **50**(1–4), 171–203 (1990)
6. N.A. Jackson, The propagation of modified flow downstream of a change in roughness. Q. J. R. Meteorol. Soc. **102**(434), 924–933 (1976)
7. H.A. Panofsky, A.A. Townsend, Change of terrain roughness and the wind profile. Q. J. R. Meteorol. Soc. **90**(384), 147–155 (1964)
8. K.S. Rao, J.C. Wyngaard, O.R. Coté, The structure of the two-dimensional internal boundary layer over a sudden change of surface roughness. J. Atmos. Sci. **31**(3), 738–746 (1974)
9. C.C. Shir, A numerical computation of air flow over a sudden change of surface roughness. J. Atmos. Sci. **29**(2), 304–310 (1972)

3D Spatial Correlation Tensor from an L-Shaped SPIV Experiment in the Near Wall Region

Jean-Marc Foucaut, Christophe Cuvier, Sebastien Coudert and Michel Stanislas

Abstract Understanding the turbulence organization near a wall is necessary to help improving turbulence models. From the experimental point of view, many researchers have worked on this subject since the fifties. Recently, Foucaut et al. (Exp. Fluids 50(4, Sp. Iss. SI):839–846, 2011) [16] have proposed a new idea to compute the 3D correlation tensor from two normal velocity fields when there are two homogeneous directions in the flow. The idea of the present contribution is to propose a specific SPIV experiment which allows the computation of the full 3D spatial correlation tensor in the near wall region of the TBL. This experiment composed of two Stereoscopic PIV planes normal to the wall which were simultaneously recorded was performed in the LML wind tunnel. The 3D correlation is then computed from the two velocity planes in order to give some information about the near wall turbulence organization. Conditioning the average by specific events allows us to improve the analysis of the organisation. It can evidence the link between the events.

1 Introduction

Because of its particular importance, near wall turbulence has been examined extensively in the last 50 years [1–3]. The study of its organization is fundamental for the improvement of the physics understanding and for the development of turbulence models. From the experimental point of view, many researchers have worked on this subject since the fifties [4]. A possible approach consists in computing the spatial and/or temporal two-point velocity correlation tensor to investigate the dynamic of

---

J.-M. Foucaut (✉) · C. Cuvier · M. Stanislas
Ecole Centrale de Lille, Laboratoire de Mécanique de Lille(LML), UMR8107 CNRS, Bvd P. Langevin, 59655 Villeneuve d'ascq, France
e-mail: jean-marc.foucaut@ec-lille.fr

S. Coudert
Physico-Chimie des Processus de Combustion et de l'Atmosphère (PC2A), UMR 8522 CNRS/Lille 1, Université Lille 1 Sciences et Technologies, Cité Scientifique, 59655 Villeneuve d'ascq Cedex, France
e-mail: sebastien.coudert@univ-lille1.fr

M. Stanislas et al. (eds.), *Progress in Wall Turbulence 2*,
ERCOFTAC Series 23, DOI 10.1007/978-3-319-20388-1_36

the dominant structures. This approach has been applied with both hot wire rakes data [5] for space-time correlations, and PIV, mainly for the spatial correlations [6]. Favre et al. [7, 8] studied the spatio-temporal structure of the streamwise velocity component by using a pair of spatially separated hot-wire probes. Stanislas et al. [6] showed that double spatial correlations, computed from PIV fields, can help to improve the understanding of the turbulent flow organization. They demonstrated the richness of such a tool when it is coupled with conditional sampling. Kähler [3] characterized the structure size and organization of the buffer region of a turbulent boundary layer using such an approach, based also on PIV data.

Stereoscopic PIV is now a reliable method to study turbulence [9, 10] and in particular the organization of near wall flows [11]. The possibilities of the technique can be enlarged by the use of the light polarization to record velocity fields in two different planes simultaneously. When the planes are parallel, this method is called dual plane stereoscopic PIV [12]. The dual plane technique allows the measurement of 2 velocity fields with an adjustable time delay or spatial separation between them. By varying this delay, the space-time correlation of the velocity field can be computed. By varying the separation, the 3D spatial correlation can be obtained. Ganapathisubramani et al. [13] used the dual plane technique to get the full gradient tensor and to study the near wall flow structures.

If both planes are perpendicular, it can give some information about the spatial properties of the flow [14, 15]. Recently, Foucaut et al. [16] proposed an original method to compute the 3D velocity correlation tensor in a turbulent boundary layer from 2 normal planes. This idea is also proposed here with a specific experiment to study the near wall flow. This experiment is first described. Then the resulting two point correlations in the near wall region are presented.

2 Experimental Setup

The experiment was carried out in the LML turbulent boundary layer wind tunnel. The test section is 1 m high, 2 m wide and 20 m long to allow the development of the boundary layer. The boundary layer thickness can reach 30 cm. The present experiment was carried out at a Reynolds numbers R_θ of 7500 (Re_τ of 2300) which corresponds to a free stream velocity of 3 m/s.

Figure 1 shows a top view of the set-up. The field of view was imaged with two Stereoscopic PIV systems in two normal planes (Fig. 1). Each system is based on Hamamatsu 2k $\times$ 2k pixel cameras and micro Nikkor 105 mm lenses at f# 8. The field of view of each system is about 8 $\times$ 12 cm^2 (corresponding to about 640 $\times$ 960 wall units). Both systems were adjusted so that the two PIV planes share a common wall normal line. In each plane, the grid spacing is 0.5 mm which corresponds to 3.8 wall units. Moreover each SPIV system was adjusted with respect to the Scheimpflug conditions [9]. The seeding was Poly-Ethylene Glycol micron particles. The particle size was of the order of 1 μm. In this configuration, the Airy disk diameter is of the order of 12.3 μm. It gives a particle image size of the order of 1.8 pixel [17]. On

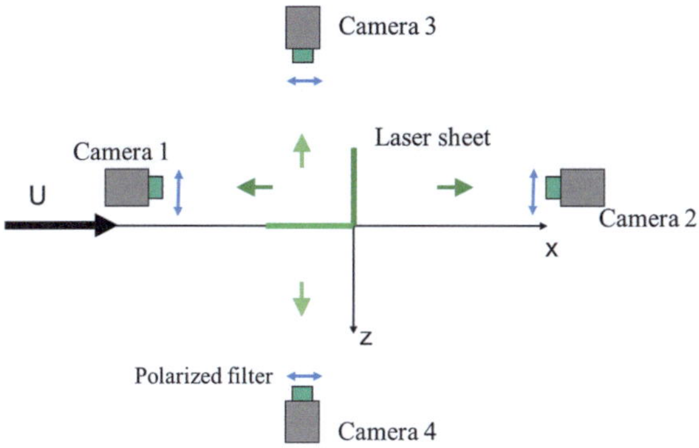

Fig. 1 *Top* view of the experimental setup

the images, the particle image size appears to be effectively of the order of 1.8 pixel. Following [18], such a size should lead to weak peak-locking effect.

The x and z axes are in the streamwise and spanwise direction, respectively. The y axis is normal to the wall. The laser used was a BMI YAG system with 4 cavities, which can deliver 4 beams two by two recombined and orthogonally polarized. Each camera lens was suited with a polarizing filter. Consequently, for each camera, only the corresponding light sheet is recorded on the images as for the dual plane technique. A total of 12000 velocity fields were recorded. Figure 2 gives an example of the instantaneous streamwise velocity component u in the two planes.

The images from both cameras were processed with a standard multi-grid and multi-pass algorithm with image deformation [19]. The analysis was made by the classical FFT-based cross-correlation method with symmetrical integer shift of both

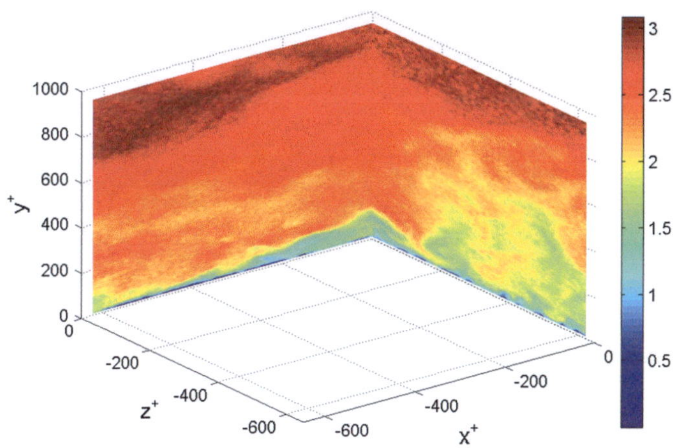

Fig. 2 Example of an instantaneous streamwise component velocity fields

windows. Before the final pass, image deformation was applied according to the displacements estimated from the three previous passes to improve the accuracy [20]. A 1-D Gaussian peak fitting algorithm was used for the sub-pixel displacement determination. The final interrogation window size was 24×32 pixels for each plane. Such a size corresponds to a square window in the physical space of $1.4 \times 1.4\,\text{mm}^2$ (3.8×3.8 wall units). It was optimized close to the wall where the signal is maximum. A mean overlapping of about 66 % was used. The Soloff method using 7 calibration planes was used to reconstruct the three velocity components in the plane of measurement. The analyses were performed using the MatPIV software modified and developed at LML. The calibration was done with different targets using crosses. From the set of recorded calibration planes and SPIV images, the misalignment between the light sheet and the calibration plane was corrected (Coudert and Schön [21]) with a self-calibration method similar to the one proposed by Wieneke [22].

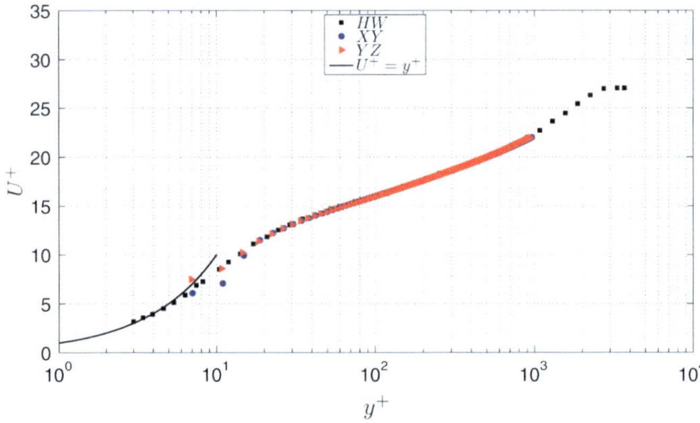

Fig. 3 Mean velocity profiles of the two planes compared to hot-wire data

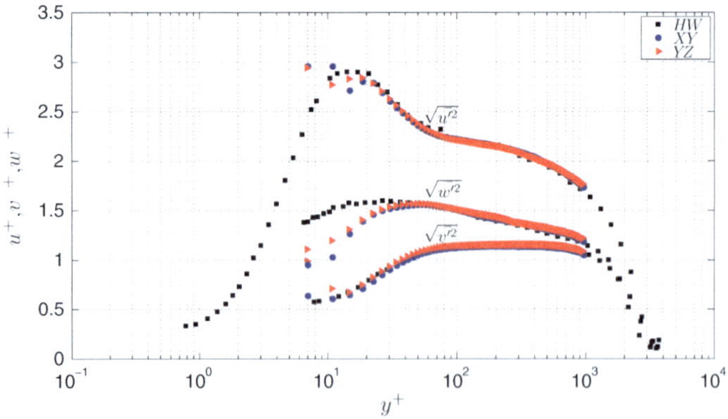

Fig. 4 Turbulent intensity profiles of the two planes compared to hot-wire data

Figures 3 and 4 show the mean streamwise velocity and the turbulent intensity profiles from the two planes, respectively, compared with hot wire anemometry. These results are in good agreement above 15 wall units, except for the spanwise intensity component in Fig. 4 which presents a difference below 40 wall units probably due to the mean velocity gradient at the scale of the X-wire probe which cannot be taken into account during the calibration. Figure 5 presents the streamwise and spanwise spectra computed from the streamwise SPIV plane at a distance of 200 wall units and compared to the HWA ones using a local Taylor hypothesis. The agreement is perfect below $k = 900\,\mathrm{rad/m}$. By comparing the PIV spectrum with the model of Foucaut et al. [23], the noise level of the streamwise component σ_u can be quantified. From this noise level, the uncertainty of each component in the XY plane can be estimated [23]. These values are given in Table 1. In this table, the noise level E_{noise} is deduced by a fit of the model of Foucaut et al. [23] on the PIV spectrum, WS is the interrogation window size in physical units used in the model. Thanks to the similarity of the setup for both planes, the uncertainties in the YZ plane can also

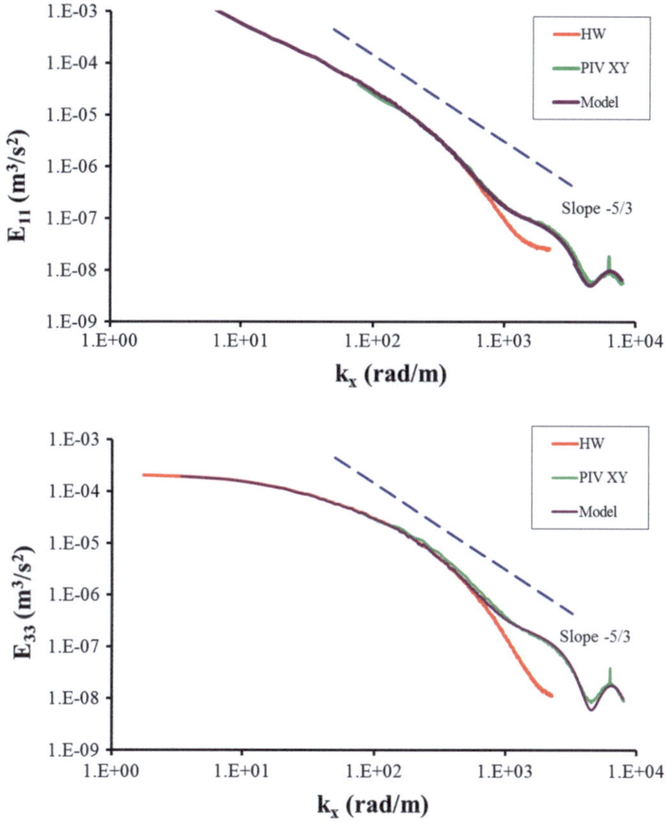

Fig. 5 Comparison of the streamwise and spanwise velocity spectra from PIV and HWA

Table 1 Noise level and uncertainties of the SPIV measurement, E_{noise} is the noise level (see [23]), WS is the PIV window size and σ_u is the PIV measurement uncertainties

Plane XY	$E_{noise}(m^3/s^2)$	WS (mm)	σ_u (m/s)	σ_u (pix)	Plane YZ
u	1.3×10^{-7}	1.4	0.023	0.068	w
v	2.5×10^{-7}	1.4	0.032	0.095	v
w	2.5×10^{-7}	1.4	0.032	0.095	u

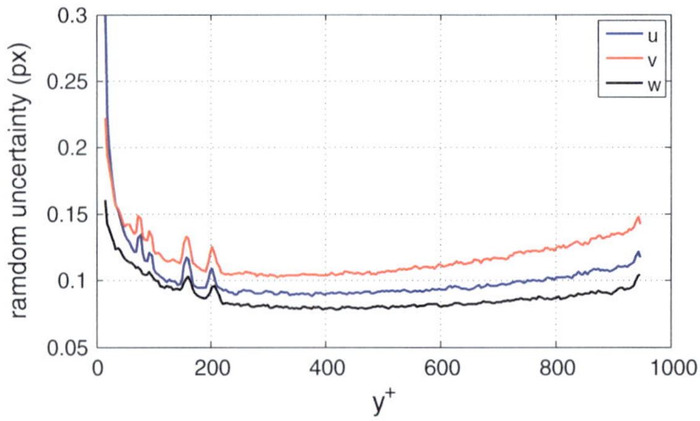

Fig. 6 Estimation of the measurement errors at the intersection line of the two planes

be estimated by a comparison of the different components in the stereoscopic setup. The uncertainties on the components stretched and not stretched by the stereoscopic angle and on the out of plane component are considered respectively the same for the two planes. At the intersection of the two planes, the velocity is estimated by two independent SPIV systems. Along this line the uncertainties can also be estimated by computing the RMS value of the difference between the velocity components given by the two planes (see [24]). Figure 6 shows the profiles of these estimations of the order of the values of Table 1. The errors on the components u and w are a little smaller than v because they are not stretched in one of the two planes. The measurement error is finally of the order of 0.1 pixel which is classical for a good quality SPIV measurement.

Figures 7 show the probability density function of the three fluctuating velocity components from both planes at 250 wall units and compared with hot wire anemometry. They show a very good agreement between the two SPIV systems and the hot wire results. Also they show no filtering of the velocity due to a saturation of the dynamics in the plane normal to the flow.

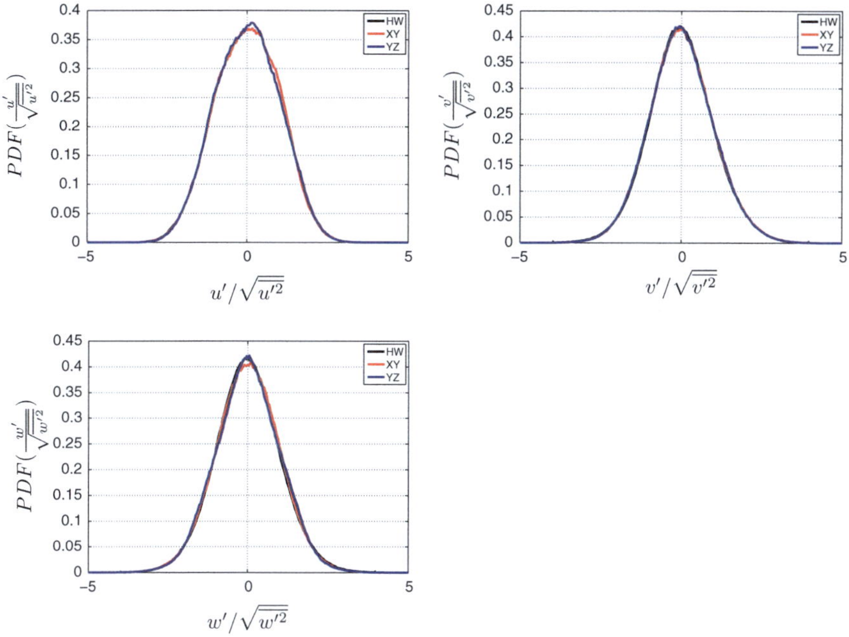

Fig. 7 Comparison of the PDF of the 3 fluctuating velocity components from SPIV and HWA

3 Spatial Correlation

As was detailed in [16], the computation of the 3D correlation tensor needs a volumic measurement as can be obtained for example from Tomographic PIV [25]. However, if the flow is homogeneous along two directions (i.e. x and z), the 3D spatial correlation can be computed from the present experiment using the velocity data of the two perpendicular SPIV planes. As the experiment is L-shaped the correlation is computed in quarters of the final volume. In the first quarter, the fixed point is taken along a line of the XY plane at fixed y and the moving points are in the YZ plane following Eq. 1.

$$R_{ij}(y, \delta x, \delta y, \delta z) = \frac{\overline{u_i^1(x_0 - \delta x, y, 0) u_j^2(x_0, y + \delta y, \delta z)}}{\sqrt{\overline{u_i^1 u_i^1}} \sqrt{\overline{u_j^2 u_j^2}}} \tag{1}$$

In the second quarter, the fixed point is taken along a line of the YZ plane at the same y and the moving points in the XY plane following Eq. 2.

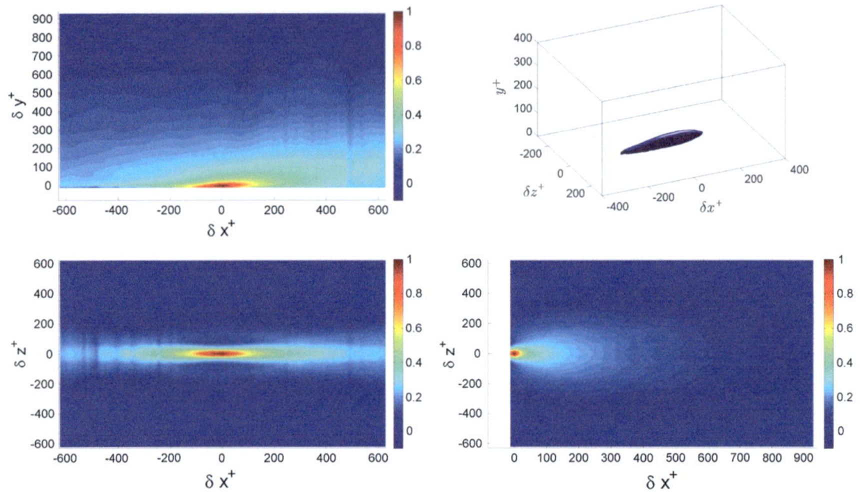

Fig. 8 Maps of the R_{11} correlation with the fixed point at $y^+ = 18$ in planes $\delta_x = 0$ (*top left*), $\delta_y = 0$ (*bottom left*) and $\delta_z = 0$ (*bottom right*) and 3D iso-value at 0.5

$$R_{ij}(y, \delta x, \delta y, \delta z) = \frac{\overline{u_i^2(x_0, y, -\delta z)u_j^1(x_0 + \delta x, y + \delta y, 0)}}{\sqrt{\overline{u_i^1 u_i^1}}\sqrt{\overline{u_j^2 u_j^2}}} \tag{2}$$

In Eqs. 1 and 2, $x = x_0$ and $z = 0$ are the equations of the two planes intersection, u_i^1 and u_i^2 are the fluctuating components in planes $z = 0$ and $x = x_0$ respectively and the over-bar corresponds to an average over the number of independent realizations which is about 12 000 in the present study.

In the present case, as the flow is exactly homogeneous in span, the symmetry can be enforced in order to fill up the two missing quarters and then obtain the full 3D correlation. The correlation R_{ij} must be symmetrical for $i = j$ and also for $i \neq j$ with $i \neq 3$ or $j \neq 3$. In the other cases the correlation must be anti-symmetrical.

Figure 8 shows a 3D isovalue of R_{11} correlation corresponding to a level of 50 % and 3 cuts in the streamwise ($\delta z = 0$) and spanwise ($\delta x = 0$) planes and in a plane parallel to the wall ($\delta y = 0$) with the fixed point at 18 wall units. R_{11} shows a very elongated region of correlation due to the streaky organisation of the streamwise velocity fluctuations. This correlation has a long ovoid shape of about 360 wall units in length with a diameter of about 55 wall units. It is inclined at about 3° to the x axis. Two minima can be observed on each side of the main peak, located at about 60 wall units from this peak. These minima are not negative probably because of the meandering behavior of the streaks. They indicate that the streaks (low or high speed) should be separated by about 120 wall units in span as in Lin et al. [26]. The same correlation computed at different wall distances (not presented here) shows that the size of the ovoid increases with wall distance.

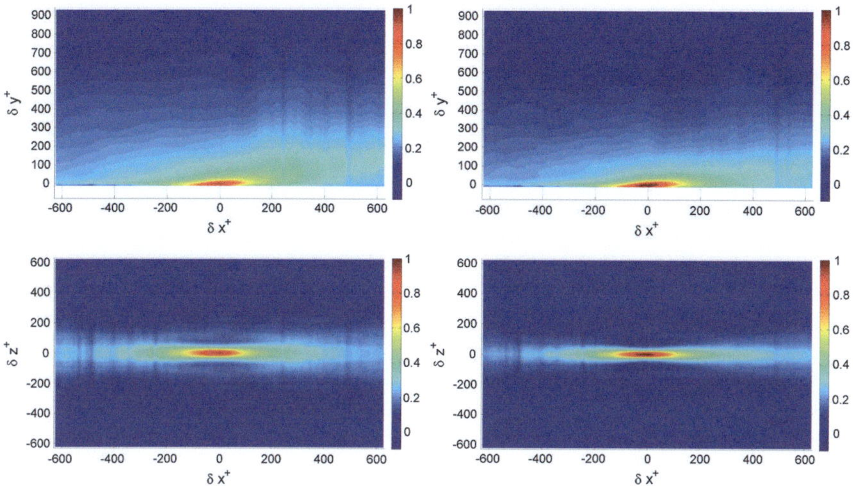

Fig. 9 Maps of the R_{11} correlation with the fixed point at $y^+ = 18$ in XY plane at $\delta_z = 0$ (*top*) and XZ plane at $\delta_y = 0$ (*bottom*) conditioned by high (*left*) and low (*right*) speed streaks

Figure 9 shows two cuts through R_{11} in the steamwise plane ($\delta z = 0$) and in a plane parallel to the wall ($\delta y = 0$), at the same wall distance as in Fig. 8. In this case, the correlation is conditioned by positive and negative velocity fluctuations, corresponding to high (left) and low (right) speed streaks. The width of the correlation increases to 60 wall units for a high speed streaks and decreases to 50 wall units for the low speed one. However the correlation seems more contrasted when it is conditioned by a low speed steaks. It also extends to longer δx. This corresponds to higher coherence on long distance of this structure. Low speed streaks are longer and thinner than high speed ones.

Figure 10 shows the same results as Fig. 8 for the R_{12} cross-correlation at $y^+ = 18$. The 3D plot corresponds to a correlation level of -0.25 (50 % of the minimal value) of $\sqrt{u'^2}\sqrt{v'^2}$. The 3D plot evidences a long region of negative correlation around the plane of symmetry which probably corresponds to a region of negative shear stress. It is about 330 wall units long, 60 wall units large and 90 wall units height based on the chosen threshold. On each side of this negative region, probably linked with the streaks organisation, two positive ones appear almost parallel to it and separated by $\delta z^+ = 90$. The shape of the positive regions is more affected by convergence and by the limited accuracy close to the wall. Although shorter streamwise, the R_{12} correlation has some similarity with R_{11} in Fig. 8.

Figure 11 shows two cuts of R_{12} conditioned by positive and negative streamwise velocity fluctuations, respectively. The negative region conditioned by a high speed streaks shifts clearly toward the negative displacements. High speed streaks seems to appear closer to the leading edge of what can be a sweeping motion ($u > 0$ and $v < 0$) [1]. Low speed streaks appear on their side more centred with respect to what can be an ejection ($u < 0$ and $v > 0$) [1]. It is also interesting to note that the negative

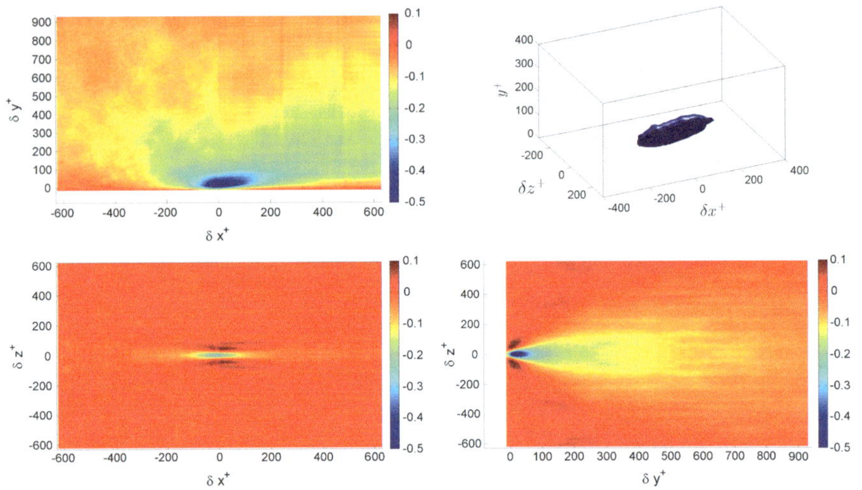

Fig. 10 Maps of the R_{12} correlation with the fixed point at $y^+ = 18$ in planes $\delta_x = 0$ (*top left*), $\delta_y = 0$ (*bottom left*) and $\delta_z = 0$ (*bottom right*) and 3D iso-value at -0.25

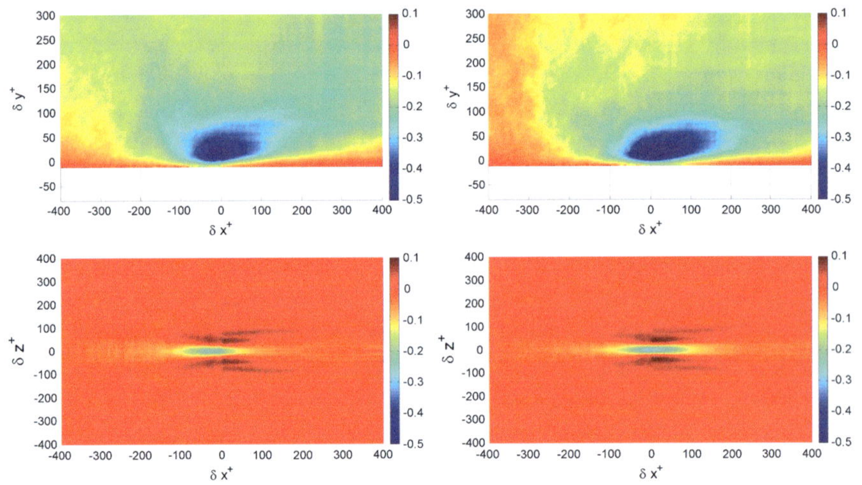

Fig. 11 Maps of the R_{12} correlation with the fixed point at $y^+ = 18$ in XY plane at $\delta_z = 0$ (*top*) and XZ plane at $\delta_y = 0$ (*bottom*) conditioned by high (*left*) and low (*right*) speed streaks

correlation region is spreading far above the fixed point and is tilted with respect to the x axis, which is probably linked with large ejections and sweeps.

Figure 12 presents R_{12} in a plane $\delta_z = 0$ with the fixed point at $y^+ = 18$, 35, 100 and 160. The shape of this correlation varies with the fixed point wall distance. At distances of 100 and 160 wall units, the negative region is inclined upwind with an angle of about 50° to the y axis. It is elongated with a size of about 240 and 350

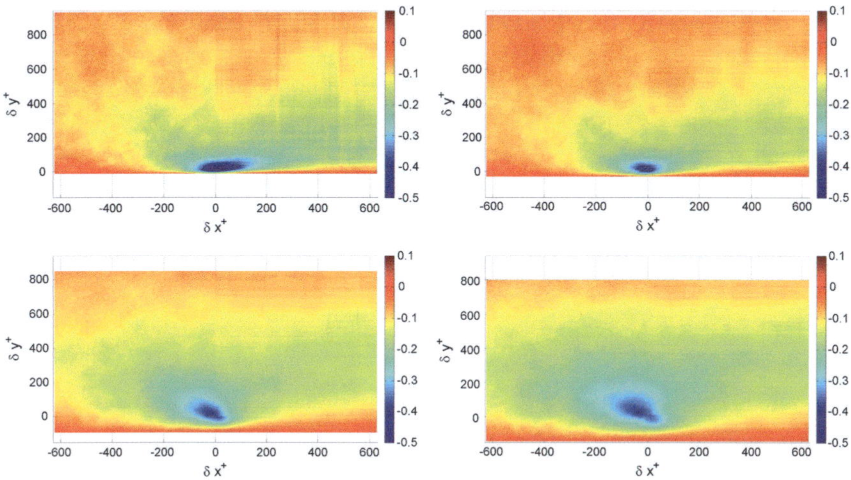

Fig. 12 Maps of the R_{12} correlation with the fixed point at $y^+ = 18$ (*top left*), $y^+ = 35$ (*top right*), $y^+ = 100$ (*bottom left*) and $y^+ = 160$ (*bottom right*) in the plane $\delta_z = 0$

wall units respectively, along the 50° direction. This negative correlation can also be associated to the sweep and ejection motions [1]. The change in shape appears in the buffer layer at a distance of about 35 wall units. At this distance the downwind part of the negative peak disappears. Below this distance the near wall cycle based on streaks and streamwise vortices as proposed by Schoppa and Hussain [27] should dominate. The correlations presented here show that the flow in this region is also strongly influenced by large scale sweeping and ejection motions. Further from the wall ($y^+ > 35$), the upwind part of the negative peak increases. The ejection are more compact and closer to the wall while the sweeps extend more further outside (Figures are not schown here).

4 Conclusion

A method to measure the 3D spatial correlation tensor from a specific PIV experiment in a turbulent boundary layer is detailed. It is based on the fact that there are two homogeneous directions in the flow (one exact and one approximate in the present case). The method is based on two stereoscopic PIV fields recorded in perpendicular planes at the same time. At a given wall distance, the method makes use of all the possible product of velocity in order to fill part of a 3D correlation which is completed by symmetry. The 3D correlations, obtained with the proposed method, allow the study of the turbulence organization of the near wall region. This method needs a large number of samples to obtain converged correlations. Here, a total of 12 000 samples were used. This number is enough close to the fixed point and has to be increased far from this point were the correlation level decreases. From the correlation R_{11} and R_{12}, the size and shape of coherent regions linked to turbulent structures were measured. To complete this study a conditioning of the averaging gives more details about the organization. Close to the wall, for $y^+ < 35$ an organisation with streaks, sweeps and ejections is inferred, in agreement with the model of Schoppa and Hussain [27]. Further from the wall, for $y^+ > 35$, large ejections and sweeps inclined upstream seem to dominate.

Acknowledgments The work was supported through the International Campus on Safety and Inter modality in Transportation (CISIT). S. Coudert is acknowledged for the participation in the experiment.

References

1. E.R. Corino, R.S. Brodkey, A visual investigation of the wall region in turbulent flow. J. Fluid Mech. **37**, 1–30 (1969)
2. R.J. Adrian, C.D. Meinhart, C.D. Tomkins, Vortex organization in the outer layer of the turbulent boundary layer. J. Fluid Mech. **422**, 1–54 (2000)
3. C.J. Kahler, Investigation of the spatio-temporal flow structure in the buffer region of a turbulent boundary layer by means of multiplane stereo PIV. Exp. Fluids **36**, 114–130 (2004)
4. H.T. Kim, S.J. Kline, W.C. Reynolds, The production of turbulence near a smooth wall in a turbulent boundary layers. J. Fluid Mech. **50**, 133–160 (1971)
5. M. Tutkun, W.K. George, J. Delville, M. Stanislas, P.B.V. Johansson, J.M. Foucaut, S. Coudert, Two-point correlations in high Reynolds number flat plate turbulent boundary layers. J. Turbul. **10**(21), 1–23 (2009)
6. M. Stanislas, J. Carlier, J.M. Foucaut, P. Dupont, Double spatial correlations, a new experimental insight in wall turbulence. C. R. Acad. Sci. Paris II B **327**, 55–61 (1999)
7. A. Favre, J. Faviglio, R. Dumas, Space-time double correlations and spectra in a turbulent boundary layer. J. Fluid Mech. **2**, 313–342 (1957)
8. A. Favre, J. Faviglio, R. Dumas, Further space-time correlations of velocity in a turbulent boundary layer. J. Fluid Mech. **3**, 344–356 (1958)
9. C. Willert, Stereoscopic digital particle image velocimetry for applications in wind tunnel flows. Meas. Sci. Technol. **8**, 1465–1479 (1997)
10. S. Soloff, R. Adrian, Z.C. Liu, Distortion compensation for generalized stereoscopic particle image velocimetry. Meas. Sci. Technol. **8**, 1441–1454 (1997)
11. M. Stanislas, L. Perret, J.M. Foucaut, Vortical structures in the turbulent boundary layer: a possible route to a universal representation. J. Fluid Mech. **602**, 327–382 (2008)
12. C. J. Kahler and M. Stanislas. Investigation of wall bounded flows by means of multiple plane stereo PIV, in *Proceeding of 10th International Symposium on Application of Laser Technology to Fluid Mechanics*, Lisbon (2000) (to appear)
13. B. Ganapathisubramani, N. Hutchins, W.T. Hambleton, E.K. Longmire, I. Marusic, Investigation of large-scale coherence in a turbulent boundary layer using two-point correlations. J. Fluid Mech. **524**, 57–80 (2005)
14. W.T. Hambleton, N. Hutchins, I. Marusic, Simultaneous orthogonal plane particle image velocimetry measurements in a turbulent boundary layer. J. Fluid Mech. **560**, 53–64 (2006)
15. B. Ganapathisubramani, E.K. Longmire, I. Marusic, Experimental investigation of vortex properties in a turbulent boundary layer. Phys. Fluids **18**(055105), 1–14 (2006)
16. J.-M. Foucaut, S. Coudert, M. Stanislas, J. Delville, Full 3d correlation tensor computed from double field stereoscopic PIV in a high Reynolds number turbulent boundary layer. Exp. Fluids **50**(4, Sp. Iss. SI), 839–846 (2011)

17. R.J. Adrian, Dynamic ranges of velocity and spatial resolution of particle image velocimetry. Meas. Sci. Technol. **8**(12), 1393–1398 (1997)
18. M. Raffel, C. Willert, J. Kompenhans, *Particle Image Velocimetry* (Springer, Berlin, 1998)
19. F. Scarano, Iterative image deformation methods in PIV. Meas. Sci. Technol. **13**(1), R1–R19 (2002)
20. M. Stanislas, K. Okamoto, C.J. Kahler, J. Westerweel, F. Scarano, Main results of the third international PIV challenge. Exp. Fluids **45**(1), 27–71 (2007)
21. S. Coudert, J.P. Schon, Back projection algorithm with misalignment corrections for 2D3C Stereoscopic PIV. Meas. Sci. Technol. **12**, 1371–1381 (2001)
22. B. Wieneke, Stereo-PIV using self-calibration on particle images. Exp. Fluids **39**(2), 267–280 (2005)
23. J.M. Foucaut, J. Carlier, M. Stanislas, PIV optimization for the study of turbulent flow using spectral analysis. Meas. Sci. Technol. **15**, 1046–1058 (2004)
24. J.M. Foucaut, S. Coudert, C. Braud, C.M. Velte, Influence of light sheet separation on SPIV measurement in a large field spanwise plane. Meas. Sci. Technol. **25**(035304), 1–10 (2014)
25. C. Atkinson, S. Coudert, J.M. Foucaut, M. Stanislas, J. Soria, The accuracy of tomographic particle image velocimetry for measurements of a turbulent boundary layer. Exp. Fluids **50**(4, Sp. Iss. SI), 1031–1056 (2011)
26. J. Lin, J.-P. Laval, J.-M. Foucaut, M. Stanislas, Quantitative characterization of coherent structures in the buffer layer of near-wall turbulence. Part 1: streaks. Exp. Fluids **45**(6), 999–1013 (2008)
27. W. Schoppa, F. Hussain, Coherent structure dynamics in near-wall turbulence. Fluid Dyn. Res. **26**(2), 119–139 (2000)

On Objective and Non-objective Kinematic Flow Classification Criteria

Ramon S. Martins, Anselmo S. Pereira, Gilmar Mompean, Laurent Thais and Roney L. Thompson

Abstract Turbulent flows present several compact and spatially coherent regions generically known as coherent structures. The understanding of these structures is closely related to the concept of vortex, whose definition is still a subject of controversy within the scientific community. In particular, the role of objectivity in the definition of vortex remains a largely open question. The three most usual criteria for vortex identification (Q, Δ and λ_2) are non-objective since they all depend on the fluid's rate-of-rotation, which is not invariant to the reference frame. In the present work, we propose an objective definition of these criteria by using the concept of relative rate-of-rotation with respect to the principal directions of the strain rate tensor. We also explore two novel naturally objective flow classification criteria. All the criteria are applied to instantaneous velocity fields obtained by DNS of both Newtonian and viscoelastic turbulent channel flows. The analysis is carried out here for four friction Reynolds numbers from 180 to 1000, emphasizing the difference between objective and non-objective and classification criteria, as well as between Newtonian and non-Newtonian flows. Moreover, we try to obtain from the results

R.S. Martins (✉) · A.S. Pereira · G. Mompean · L. Thais
Laboratoire de Mécanique de Lille (LML), CNRS, UMR 8107, École Polytechnique Universitaire de Lille, Université Lille Nord de France, F59655 Villeneuve d'ascq, France
e-mail: ramon.martins@polytech-lille.fr

A.S. Pereira
e-mail: anselmo.pereira@polytech-lille.fr

G. Mompean
e-mail: gilmar.mompean@polytech-lille.fr

L. Thais
e-mail: laurent.thais@polytech-lille.fr

R.L. Thompson
Laboratório de Mecânica Teórica Aplicada (LMTA), Department of Mechanical Engineering, Universidade Federal Fluminense, Rua Passo da Pátria 156, Niterói, RJ 24210-240, Brazil
e-mail: rthompson@id.uff.br

M. Stanislas et al. (eds.), *Progress in Wall Turbulence 2*,
ERCOFTAC Series 23, DOI 10.1007/978-3-319-20388-1_37

of flow classification criteria information related to the polymer drag reduction phenomenon.

1 Introduction

Vortices are present in many practical flows and their dynamics dictates several phenomena, such as heat transfer, mixing, combustion, noise generation and hydrodynamic drag. Hence, a solid understanding of vortex dynamics may lead to a better comprehension of these phenomena.

1.1 Previous Work

In fact there is not, among researches, a consensus for the definition of a vortex. Instead, there are several criteria available in the literature. Nevertheless, the majority of these definitions is based on the rate-of-rotation (or vorticity) tensor, **W**. Such tensor is not invariant under general transformations of the reference frame. In other words, tensor **W** does not enjoy the objectivity property. Thus, all criteria which depend on the rate-of-rotation tensor are not objective as well. For instance, the classic Q-criterion by Hunt et al. [10], Δ-criterion by Chong et al. [3], and λ_2-criterion by Jeong and Hussian [11], largely employed in the literature, all depend on the tensor **W** and, therefore, are not objective criteria.

A key concept relevant to the discussion on objectivity is the concept of persistence-of-straining, introduced by Lumley [14], and applied by several authors [2, 7, 17, 22, 23]. In brief, this concept is associated to the capacity of the flow to persistently stretch a material filament. This leads to a physically consistent perspective of the motion of a fluid element, and, consequently, to a new point of view regarding flow classification. In this connection, rigid body motion opposes to maximum persistence-of-straining. In this context, Astarita [2] argued that, because flow classification is mostly used to verify the behavior of constitutive equations, a solid criteria should enjoy (among others) the same invariance properties as those required for the constitutive equations. Besides that, a legitimate flow classification should be an intrinsic character of the flow, and not something that changes depending on the observer. The author proposed a criterion based on the relative rate-of-rotation ($\overline{\mathbf{W}}$), a quantity known to be objective ([5, 6]). The tensor $\overline{\mathbf{W}}$ represents the rate-of-rotation with respect to the principal directions of the strain rate tensor (**D**).

Despite the remarkable work carried out by Astarita, the criterion proposed by him has been proven to present some flaws for certain classes of 3D flows, see [9]. By analyzing these inconsistencies, Thompson and Mendes [23] proposed a criterion based on the concept of persistence-of-straining.

By examining stretching and alignment of material filaments, Tabor and Klapper [20] verified the importance of the use of the relative rate-of-rotation without invoking objectivity. Their analysis reinforces Astarita's criterion which is equivalent to an objective version of the Q-criterion mentioned above.

Haller [8] conducted a remarkable work which defends the importance of objectivity on identifying a vortex and presents a criterion based on the stability analysis of the trajectory of particles immersed in the flow. His criterion uses the covariant strain acceleration tensor, obtained from the covariant convected derivative of the strain rate tensor, in order to quantify the ability of the flow to defy the stretching tendency imposed by the strain rate tensor. The author also presents a simple elucidating example regarding the role played by objectivity. In a subsequent work, Thompson [22] adds more consistent physical meaning to the role played by the covariant strain acceleration tensor, proposing a persistence-of-straining criterion based on this entity.

The present work aims to analyze the behavior of objective and non-objective kinematic flow classification criteria applied to the instantaneous velocity fields of both Newtonian and viscoelastic channel flows obtained by direct numerical simulation (DNS). The interaction between polymer and turbulence, especially near the wall, is also aimed due to its relation to the drag reduction phenomenon.

2 Objective Versions for Classic Flow Classification Criteria

The relative rate-of-rotation, presented by Astarita [2], is the rate-of-rotation "of the fluid" with respect to the principal directions of the strain rate tensor. Mathematically, it takes the following form given by Eq. (1),

$$\overline{\mathbf{W}} = \mathbf{W} - \Omega^D, \tag{1}$$

where Ω^D is the tensor that gives the rate of rotation of the eigenvectors of $\mathbf{D}$. Although both $\mathbf{W}$ and Ω^D are non-objective tensors, Drouot and Lucius [5, 6] have proven that $\overline{\mathbf{W}}$ is objective. If $\overline{\mathbf{W}}$ vanishes, the filaments that are aligned to the eigenvectors of $\mathbf{D}$ have the tendency to continue aligned. In this sense, the stretch is persistent.

An objective redefinition for the Q-criterion is now proposed by replacing the non-objective rate-of-rotation tensor, $\mathbf{W}$, by the objective relative rate-of-rotation tensor, $\overline{\mathbf{W}}$, in its respective original formulation, yielding .

$$Q_s = \frac{1}{2}\left(\|\overline{\mathbf{W}}\|^2 - \|\mathbf{D}\|^2\right) > 0. \tag{2}$$

Applying the same methodology to the Δ and λ_2-criteria, their objective versions take the form of Eqs. (3) and (4), respectively.

$$\Delta_s = \left(\frac{Q_s}{3}\right)^3 + \left(\frac{det(\mathbf{D} + \overline{\mathbf{W}})}{2}\right)^2 > 0 \tag{3}$$

$$\lambda_2^{\mathbf{D}^2 + \overline{\mathbf{W}}^2} < 0 \tag{4}$$

3 Novel Naturally Objective Criteria

We adopt here a decomposition of the covariant strain acceleration tensor, $\mathbf{M} \equiv \overset{\triangle}{\mathbf{D}}$ (where the triangle indicates the covariant convected time derivative) with respect to the strain rate tensor, $\mathbf{D}$, as discussed in [22]. This decomposition splits tensor $\mathbf{M}$ into two additive parts: a part that is *in-phase* with $\mathbf{D}$, ϕ_M^D, and a part that is *out-of-phase* with $\mathbf{D}$, $\widetilde{\phi}_M^D$. These tensors are defined as

$$\phi_M^D = \mathbb{I}^{DD} : \mathbf{M}\ ; \quad \widetilde{\phi}_M^D = \left(\mathbb{I}^{\delta\delta} - \mathbb{I}^{DD}\right) : \mathbf{M}, \tag{5}$$

where the symbol ":" accounts for the double dot product and $\mathbb{I}^{DD}$ is a fourth order tensor given by

$$\mathbb{I}^{DD} = \sum_{i=1}^{3} = \mathbf{e}_i^D \mathbf{e}_i^D \mathbf{e}_i^D \mathbf{e}_i^D, \tag{6}$$

where $\mathbf{e}_i^D$ is an eigenvector of $\mathbf{D}$ and $\mathbb{I}^{\delta\delta}$ is the fourth order identity tensor.

Aligned to the concepts presented by Haller [8], we can define a ratio, IR, that can be interpreted as a measure of how tensor $\mathbf{M}$ corroborates the tendency dictated by $\mathbf{D}$ as

$$IR = 1 - \frac{2}{\pi}\cos^{-1}\left(\frac{||\phi_{\mathrm{M}}^{\mathrm{D}}||}{||\mathbf{M}||}\right). \tag{7}$$

It can be shown that this quantity is the complement with respect to unity to the quantity we would find if we replace ϕ_M^D by $\widetilde{\phi}_M^D$. In this other case this quantity would be a measurer of how tensor $\mathbf{M}$ *defies* the tendency dictated by $\mathbf{D}$.

Because the flow character is, most of the times, anisotropic, we found the necessity to come up with anisotropic ratios that could better represent the competition between the parts that corroborates and defies the $\mathbf{D}$-tendency. Hence, we can define anisotropic ratios depending on the eigendirection of $\mathbf{D}$, as

$$AR(k) = 1 - \frac{2}{\pi}\cos^{-1}\left(\sqrt{\frac{[\mathbf{M}]_{\mathrm{kk}}\,[\mathbf{M}]_{\mathrm{kk}}}{[\mathbf{MM}]_{\mathrm{kk}}}}\right), \tag{8}$$

where $[\mathbf{M}]_{kk}$ is an element of the principal diagonal of tensor $\mathbf{M}$ (terms of $\mathbf{M}$ which are *in-phase* with $\mathbf{D}$), and $[\mathbf{MM}]_{kk}$ is an element of the principal diagonal of tensor $\mathbf{M}^2$.

The anisotropic ratios are reorganized so that $AR1 > AR2 > AR3$. The anisotropic ratios are better interpreted together. While the other criteria aim to give an overall verdict, the anisotropic ratios provide a directional information.

4 Results and Discussions

We discuss here the results obtained with all criteria applied to the instantaneous velocity fields of both Newtonian and viscoelastic channel flows obtained by DNS with the massively parallel algorithm presented by Thais et al. [21]. Viscoelastic effects are taken into account by means of the FENE-P model with maximum polymer chain extensibility L equal to 100 and a friction Weissenberg number We_τ (being $We_\tau = \lambda u_\tau^2/\nu$, where λ is the relaxation time scale, u_τ is the friction velocity and ν is the kinematic viscosity of the solution) equal to 115, which leads to a percentage of drag reduction of 62.3 %.

It is possible to identify the so called *hairpin vortices* in turbulent boundary layers by the application of classic flow classification criteria (such as Q, Δ and λ_2 criteria). The dynamic evolution of these hairpin vortices have been largely explored in the literature (see, for example, Ref. [1] for a remarkable literature review). Regarding non-Newtonian fluids, one of the first analyses of the effects of elastic on such coherent structures was carried out by Kim et al. [12]. In drag-reducing flows, the authors noted that hairpin vortices get larger and weaker due to rotations imposed by the polymer work, which are in opposition to that of the vortices.

Figure 1 shows the iso-contours of the Q-criterion for both Newtonian (left column) and viscoelastic (right column) fluids at four friction Reynolds numbers (Re_τ = 180, 395, 590 and 1000 (being $Re_\tau = u_\tau h/\nu$, where h is the channel half-gap). Using the Q-criterion, the classic hairpin vortices are recovered for the Newtonian fluid and very similar structures are identified for the viscoelastic cases. Moreover, regardless of the fluid, the quantity of hairpin vortices increases with the friction Reynolds number. The same qualitative results are found for the two other classic criteria cited above (Δ and λ_2) and thus such results are not shown.

The elastic effect observed in Fig. 1 is also remarkable. Taking the same friction Reynolds number, the number and the size of hairpin vortices change reasonably. The results show that, for the same value of the Q-criterion (*i.e.*, for the same intensity of rotation) there are less hairpin vortices in the viscoelastic cases than in the Newtonian, which suggests that the intensity of hairpin vortices is reduced, as already observed by Kim et al. [12]. Furthermore, hairpin vortices seem to be bigger in the viscoelastic cases.

On the other hand, from the perspective of the objective versions, classic hairpin vortices are no longer observed, as shown in Fig. 2. In fact, the objective criteria identify swirling-like regions, but their ensemble does not present any well-defined shape.

4.1 *Thickening Buffer Layer in Viscoelastic Flows*

Since we could not identify any well-defined shape with the objective criteria, a wall-normal plane placed in the middle of the channel and measuring the half-gap height is investigated bellow.

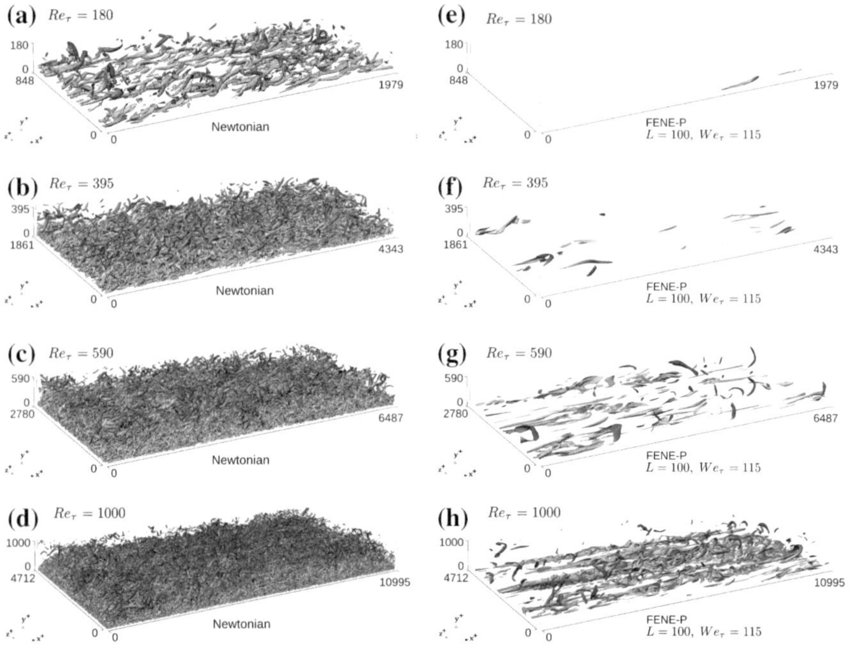

Fig. 1 Iso-contours of Q-criterion ($Q = 1.5$) for Newtonian (*left* column) and viscoelastic (*right* column) fluids at $Re_\tau = 180$, 395, 590 and 1000

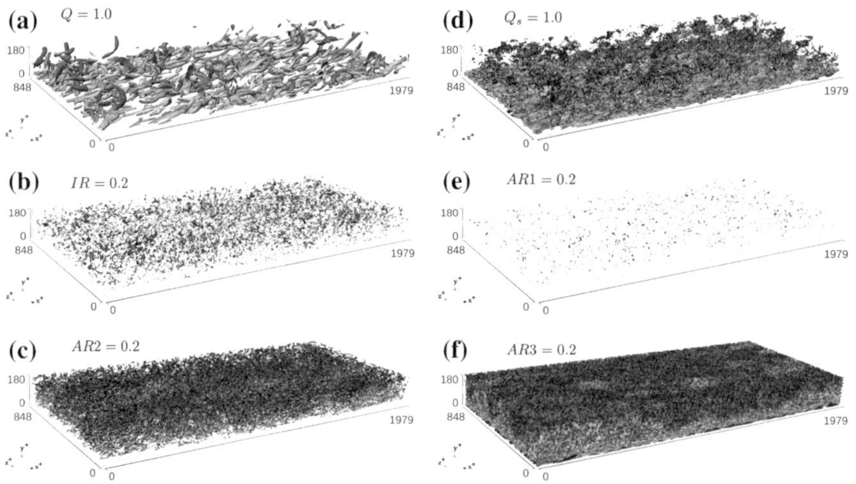

Fig. 2 Iso-contour for non-objective (**a**) and objective (**d**) versions of Q-criterion, and IR (**b**) and AR (**c**, **e** and **f**) criteria for Newtonian fluid at $Re_\tau = 180$

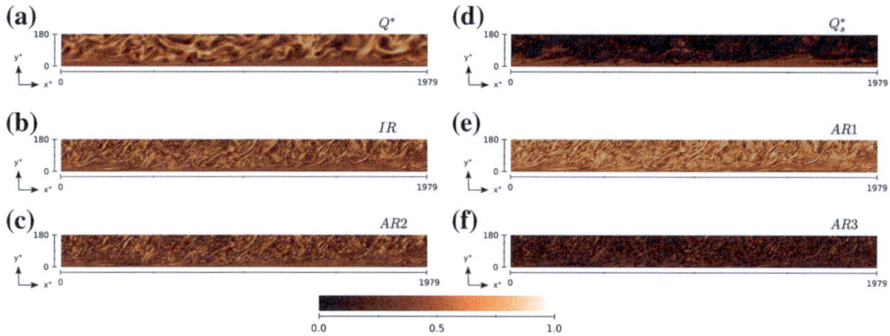

Fig. 3 Contours of non-objective and objective criteria for Newtonian fluid at $Re_\tau = 180$. The xy-plane is located in the middle of the channel

Aiming to better understand the differences observed between objective and non-objective approaches, the criteria are normalized so that the results are always between 0 and 1. The low values (from 0 to 0.5) represent swirling-like regions, whereas the high values (from 0.5 to 1) represent non-swirling-like regions. If the normalized criteria is equal to 0.5, than, in that point, the intensities of extension and rotation are equivalent. Normalized criteria are marked with a superscript asterisk, except for IR and AR, which are already normalized.

The analysis of the contours of the normalized criteria in wall-normal (xy-) planes enables the observation of an interesting behavior that seems to be related to the drag reduction phenomenon, as depicted in Figs. 3 and 4. Firstly, it is important to notice that from the perspective of the classic Q-criterion, the Newtonian flow (Fig. 3a) is globally dominated by regions where the intensity of the rate-of-rotation is similar to the intensity of the rate-of-strain ($Q^* = 0.5$). Nevertheless, there are some swirling-like ($Q^* < 0.5$, or darker) and non-swirling-like ($Q^* > 0.5$, or lighter) regions dispersed in the plane.

On the other hand, when analyzing the same snapshot from the perspective of the normalized objective version of the Q-criterion (Fig. 3d), the flow seems to be swirl dominated ($Q_s^* < 0.5$), except for a thin layer where $Q_s^* \approx 0.5$. However, the other objective criteria (Fig. 3b, c, e, f) present more homogeneous results qualitatively closer to the those of the Q^*-criterion.

At a first impression, the results for the viscoelastic case depicted in Fig. 4 present, in general, more elongated structures when compared to the Newtonian case (Fig. 3).

Interestingly, comparing the Newtonian and the viscoelastic results for the Q_s^*-criterion (respectively, Figs. 3d and 4d), it is noteworthy that the thickness of the layer where $Q_s^* \approx 0.5$ increases in the latter case. We believe that this is due to the thickening of the buffer layer caused by the presence of flexible polymers, leading to drag reduction. Such physical effect has been already observed by many authors by both experiments [13] and numerical simulations [4, 15, 16, 18], and has been predicted by the two most important theories on drag reduction phenomenon (Lumley's viscous theory [14], and Tabor and De Genne's elastic theory [19]).

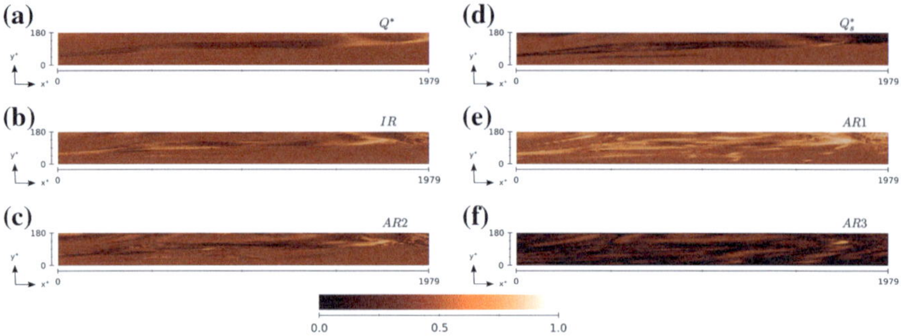

Fig. 4 Contours of non-objective and objective criteria for viscoelastic fluid with $L = 100$ and $We_\tau = 115$ at $Re_\tau = 180$. The xy-plane is located in the middle of the channel

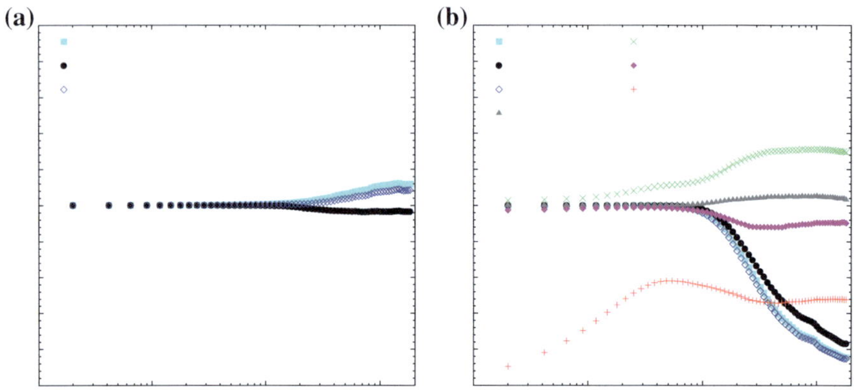

Fig. 5 xz-plane averaged values of non-objective (*left*) and objective (*right*) criteria for Newtonian fluid at $Re_\tau = 180$

The effect that seems to represent the thickening of the buffer layer is even more evident in Figs. 5 and 6, which contain the values of all criteria evaluated in the present work averaged at wall-parallel (xz) planes for Newtonian and viscoelastic fluids, respectively, at $Re_\tau = 180$.

All non-objective criteria depicted in Figs. 5a and 6a present similar behavior. Extensional and swirling motions have the same intensity within the near-wall region and values tend to be slightly extensional near the center of the channel, whereas a very weak rotational tendency is found for the Δ^*-criterion. The major difference is that, for the viscoelastic case, the criteria are equal to 0.5 even near the center of the channel, which is probably related to the thickening of the buffer layer.

Regarding objective criteria in Figs. 5b and 6b, as predicted by Figs. 3 and 4, the IR criterion present a behavior which is similar to those of non-objective criteria. It can be noticed that $AR1 \geq 0.5$ and $AR3 \leq 0.5$ in the whole domain. This fact suggests that there is no perfect extension, where extension acceleration overcomes rotation acceleration in the three directions, nor perfect swirling structure where the opposite happens. What we have, instead, is a situation where at every point of

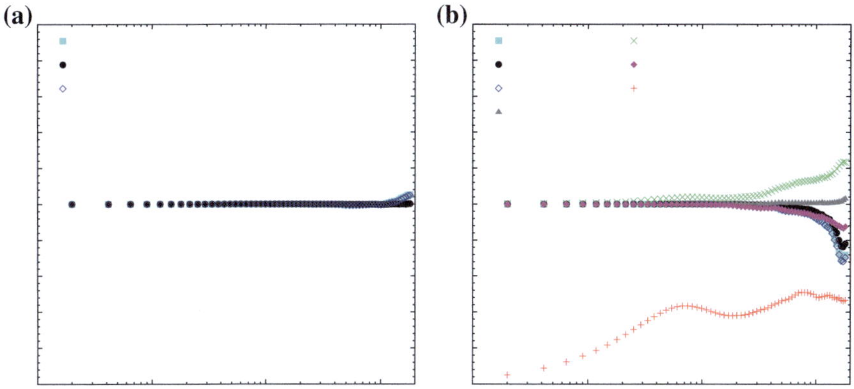

Fig. 6 xz-plane averaged values of non-objective (*left*) and objective (*right*) criteria for viscoelastic fluid with $L = 100$ and $We_\tau = 115$ at $Re_\tau = 180$

the domain there are directions where extension overcomes rotation and vice-versa. The quantity $AR2$ shows that near the wall, the swirling structure has one direction extensional-dominant, another swirling-dominant, and a neutral direction. Near the centerline there are two directions where rotation dominates.

The objective versions of the classic criteria present behaviors which are qualitatively the same. In the near-wall region, they follow the tendency of their non-objective counterpart, with a first layer around 0.5. However, near the center of the channel, these criteria tend to identify swirl-dominated regions, opposing the modest tendency of identifying extensional motion presented by the corresponding classic non-objective criterion (except for Δ^*) in this region. Moreover, because the swirl motion identified by such criteria is reasonably more intense, the deviation gets more evident, enabling an estimation of the thickening, due to the presence of polymers, of the layer related to the buffer layer. Comparing the results of the objective version of the classic criteria in Figs. 5b and 6b, the thickness of the near-wall layer where the value of the criteria is around 0.5 increases from $y^+ \approx 8$ in the Newtonian case to $y^+ \approx 40$ in the viscoelastic case.

It is important to remark that the same analysis have been carried out for $Re_\tau =$ 395, 590 and 1000. Nevertheless, the thickness of the buffer layer seems to depend more on the percentage of drag reduction than on the Reynolds number. The results at higher friction Reynolds are qualitatively similar to the case at $Re_\tau = 180$ differing basically by the possibility to go further on the y^+ scale.

Acknowledgments This work was granted access to the HPC resources of IDRIS under the allocation 2014-i20142b2277 made by GENCI. The authors would also like to express their acknowledgment and gratitude to the Brazilian Scholarship Program *Science Without Borders*, managed by CNPq (National Council for Scientific and Technological Development), for the partial financial support for this research.

References

1. R.J. Adrian, Hairpin vortex organization in wall turbulence. Phys. Fluids **19**(4), 041301 (2007)
2. G. Astarita, Objective and generally applicable criteria for flow classification. J. Non-Newton. Fluid **6**, 69–76 (1979)
3. M.S. Chong, A.E. Perry, B.J. Cantwell, A general classification of three-dimensional flow fields. Phys. Fluids A-Fluid **2**(5), 765–777 (1990)
4. C.D. Dimitropoulos, Y. Dubief, E.S.G. Shaqfeh, P. Moin, S.K. Lele, Direct numerical simulation of polymer-induced drag reduction in turbulent boundary layer flow. Phys. Fluids **17**, 1–4 (2005)
5. R. Drouot, Définition d'un transport associé à un modèle de fluide de deuxième ordre. Comparaison de diverses lois de comportement. C. R. Acad. Sci. A Math. **282**, 923–926 (1976)
6. R. Drouot, M. Lucius, Approximation du second ordre de la loi de comportement des fluides simples. Lois classiques deduites de l'introduction d'un nouveau tenseur objectif. Arch. Mech. **28**(2), 189–198 (1976)
7. F.C. Frank, M.R. Mackley, Localized flow birefringence of polyethylene oxide solutions in a two roll mill. J. Polym. Sci. **14**, 69–76 (1976)
8. G. Haller, An objective definition of a vortex. J. Fluid Mech. **525**, 1–26 (2005)
9. R. Huilgol, Comments on "Objective and generally applicable criteria for flow classification", by G. Astarita. J. Non-Newton. Fluid **7**(1), 91–95 (1980)
10. J.C.R. Hunt, A.A. Wray, P. Moin, Eddies, stream, and convergence zones in turbulent flows, in *Proceedings of Summer Program. Center for Turbulence Research*. Report CTR-S88 (1988), pp. 193–208
11. J. Jeong, F. Hussain, On the identification of a vortex. J. Fluid Mech. **285**, 69–94 (1995)
12. K. Kim, C.F. Li, R. Sureshkumar, L. Balachandar, R.J. Adrian, Effects of polymer stresses on eddy structures in drag-reduced turbulent channel flow. J. Fluid Mech. **584**, 281–299 (2007)
13. T.S. Luchik, W.G. Tiederman, Turbulent structure in low-concentration drag-reducing channel flows. J. Fluid Mech. **190**, 241–263 (1988)
14. J.L. Lumley, Drag reduction by additives. Annu. Rev. Fluid Mech. **11**, 367–384 (1969)
15. T. Min, J.Y. Yoo, H. Choi, Maximum drag reduction in a turbulent channel flow by polymer additives. J. Fluid Mech. **492**, 91–100 (2003)
16. T. Min, J.Y. Yoo, H. Choi, D.D. Joseph, Drag reduction by polymer additives in a turbulent channel flow. J. Fluid Mech. **486**, 213–238 (2003)
17. P.R. Schunk, L.E. Scriven, Constitutive equation for modeling mixed extension and shear in polymer solution processing. J. Rheol. **34**(7), 1085–1119 (1990)
18. R. Sureshkumar, A.N. Beris, R.A. Handler, Direct numerical simulation of the turbulent channel flow of a polymer solution. Phys. Fluids **9**, 743–755 (1997)
19. M. Tabor, P.G. de Gennes, A cascade theory of drag reduction. Europhys. Lett. **7**, 519–522 (1986)
20. M. Tabor, I. Klapper, Stretching and alignment in chaotic and turbulent flows. Chaos Solitons Fractals **4**(6), 1031–1055 (1994)
21. L. Thais, A. Tejada-Martinez, T.B. Gatski, G. Mompean, A massively parallel hybrid scheme for direct numerical simulation of turbulent viscoelastic channel flow. Comput. Fluids **43**, 134–142 (2011)
22. R.L. Thompson, Some perspectives on the dynamic history of a material element. Int. J. Eng. Sci. **46**, 524–549 (2008)
23. R.L. Thompson, P.R.S. Mendes, Persistence of straining and flow classification. Int. J. Eng. Sci. **43**, 79–105 (2005)

Quantification of the Full Dissipation Tensor from an L-Shaped SPIV Experiment in the Near Wall Region

Jean-Marc Foucaut, Christophe Cuvier, Michel Stanislas and William K. George

Abstract An experiment was performed in the LML boundary layer facility to determine all of the derivative moments needed to estimate the dissipation. The Reynolds number was $Re_\theta = 7500$ or $Re_\tau = 2300$. A detailed analysis of the errors in derivative measurements was carried out, as well as applying and using consistency checks derived from the continuity equation and a local homogeneity hypothesis. Local homogeneity estimates of the dissipation are accurate everywhere within a few percent. Both local axisymmetry and local isotropy work almost as well outside of $y^+ = 100$, but only local axisymmetry provides a reasonable estimate close to the wall.

1 Introduction

Stereoscopic PIV is now a recognized method to measure turbulent flow. Many researchers also use this method to compute statistics of the flow such as mean velocity, Reynolds stress tensor, probability density function or spectra [1–3]. SPIV allows the measurement of the three components of the velocity in a plane with an accuracy of about 1–2 % (0.1 pixel). For a turbulent flow, it opens the unique capability of studying the organization of the turbulence. Generally, due to the limited spatial resolution of PIV, the smallest scales of the flow cannot be investigated. One difficulty in the data processing comes from the noise amplification when derivatives have to be computed [4]. However, the velocity gradient is necessary to determine the vorticity, the shear, and specific criteria allowing the detection of vortices such as the Q criterion or the swirling strength which are using tools to study the flow organization. In the present paper, the idea is to propose a method to determine the

J.-M. Foucaut (✉) · C. Cuvier · M. Stanislas
Ecole Centrale de Lille, Laboratoire de Mécanique de Lille (LML), UMR8107 CNRS, Bvd P. Langevin, 59655 Villeneuve d'ascq, France
e-mail: jean-marc.foucaut@ec-lille.fr

W.K. George
Department of Mechanical and Aerospace Engineering, Princeton University, Princeton, NJ 08544, USA
e-mail: georgewilliamk@gmail.com

M. Stanislas et al. (eds.), *Progress in Wall Turbulence 2*,
ERCOFTAC Series 23, DOI 10.1007/978-3-319-20388-1_38

dissipation rate of the turbulent kinetic energy ε as given by Eq. 1, which combines derivatives and statistical computations from a dedicated SPIV experiment.

$$\varepsilon = \nu \left[\left\langle \frac{\partial u_i}{\partial x_j} \frac{\partial u_i}{\partial x_j} \right\rangle + \left\langle \frac{\partial u_i}{\partial x_j} \frac{\partial u_j}{\partial x_i} \right\rangle \right] \tag{1}$$

This dissipation rate is strongly linked to the small scales of turbulence. It is a key parameter of turbulence modelling. To obtain this full dissipation, it is necessary to compute all the components of the instantaneous gradient tensor and to compute the variance of each term together with a few covariances. This corresponds to a total of twelve terms which can be organized in three groups:

$$\begin{aligned}\varepsilon = \nu \Bigg\{ 2 \left[\left\langle \frac{\partial u_1}{\partial x_1} \frac{\partial u_1}{\partial x_1} \right\rangle + \left\langle \frac{\partial u_2}{\partial x_2} \frac{\partial u_2}{\partial x_2} \right\rangle + \left\langle \frac{\partial u_3}{\partial x_3} \frac{\partial u_3}{\partial x_3} \right\rangle \right] + \\ \left[\left\langle \frac{\partial u_1}{\partial x_2} \frac{\partial u_1}{\partial x_2} \right\rangle + \left\langle \frac{\partial u_2}{\partial x_1} \frac{\partial u_2}{\partial x_1} \right\rangle + \left\langle \frac{\partial u_1}{\partial x_3} \frac{\partial u_1}{\partial x_3} \right\rangle + \right. \\ \left. \left\langle \frac{\partial u_3}{\partial x_1} \frac{\partial u_3}{\partial x_1} \right\rangle + \left\langle \frac{\partial u_2}{\partial x_3} \frac{\partial u_2}{\partial x_3} \right\rangle + \left\langle \frac{\partial u_3}{\partial x_2} \frac{\partial u_3}{\partial x_2} \right\rangle \right] + \\ 2 \left[\left\langle \frac{\partial u_1}{\partial x_2} \frac{\partial u_2}{\partial x_1} \right\rangle + \left\langle \frac{\partial u_1}{\partial x_3} \frac{\partial u_3}{\partial x_1} \right\rangle + \left\langle \frac{\partial u_1}{\partial x_3} \frac{\partial u_3}{\partial x_1} \right\rangle \right] \Bigg\}\end{aligned} \tag{2}$$

All of them are difficult to measure accurately. George and Hussain [5] propose to simplify this computation by using different hypotheses in order to reduce the number of terms. These hypotheses are respectively local homogeneity, axisymmetry and isotropy of the small scales. Antonia et al. [6] have studied these different hypotheses on DNS results from a channel flow at low Reynolds number. Balint et al. [7] and Honkan and Andreopoulos [8] published some experimental results of dissipation rate obtained by hot wire anemometry (HWA) in a turbulent boundary layer. A special multiwire probe was specially designed for these measurements. These results are globally of the same order as the DNS but, except for the derivative of the streamwise component, the HWA results are different from the DNS ones. This is why, in the present paper, benefit is taken of Stereoscopic PIV in two normal planes to compute all the derivatives of the three velocity components in a turbulent boundary layer. The limitation of this approach is that PIV resolves a range of scales which is between the size of the field of view for the largest scales and that of the interrogation window for the smallest ones [9]. Emphasis is put particularly here on the derivative filter choice and on the measurement noise management.

2 Experimental Setup

The experiment was performed in the turbulent boundary layer wind tunnel of the Laboratoire de Mécanique de Lille. It is described in detail in Foucaut et al. [10]. The Reynolds number is $Re_\theta = 7500$. The friction velocity is 0.115 m/s (deduced from

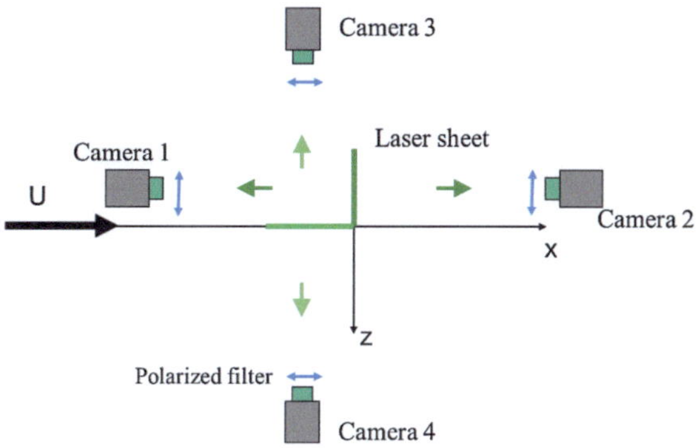

Fig. 1 *Top view* of the experimental setup

the log law with $\kappa = 0.41$) and the boundary layer thickness δ is about 30 cm which give $Re_\tau = \delta^+ = 2300$. Two SPIV planes were used in order to measure the three components of the velocity and the instantaneous derivatives of these components. As shown in Fig. 1, the two planes are normal to the wall and extend in the streamwise and the spanwise directions, respectively. The field size is 8 cm (620 wall units) both in the streamwise and spanwise directions and 12 cm (950 wall units) in the wall normal direction. The final interrogation window size is 24×32 pixel corresponding to 1.4×1.4 mm^2 (12×12 wall units) in both planes. The grid spacing is 0.5 mm (3.8 wall units), which corresponds to a mean overlapping of 66 %. Figure 2 gives an example of the instantaneous field of the streamwise component in both planes. A total of 12000 fields was recorded. At the intersection of the two fields, Foucaut et al. [10] computed an estimation of the measurement uncertainty of the order of 0.1 pixel, which is standard for a good SPIV measurement.

Figure 3 shows the turbulence intensity profiles in the two planes compared with hot wire anemometry (HWA). These results are in good agreement above 20 wall units, except for the spanwise component which presents a difference below 40 wall

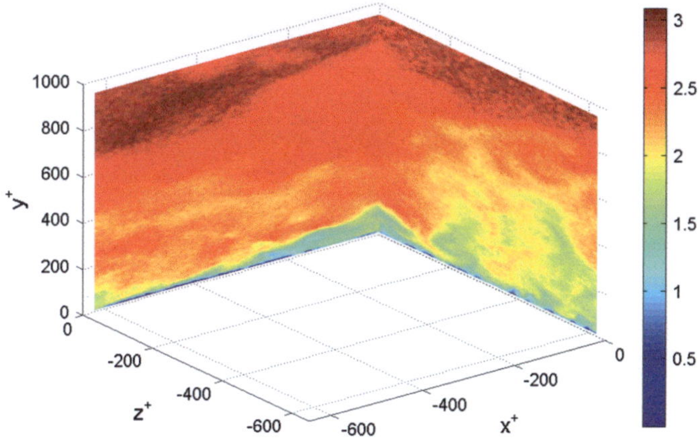

Fig. 2 Example of the instantaneous streamwise component velocity fields

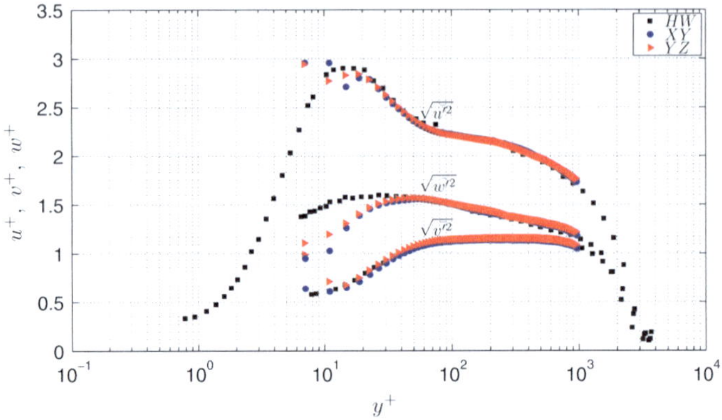

Fig. 3 Turbulent intensity profiles in the two planes compared to HWA by [11]

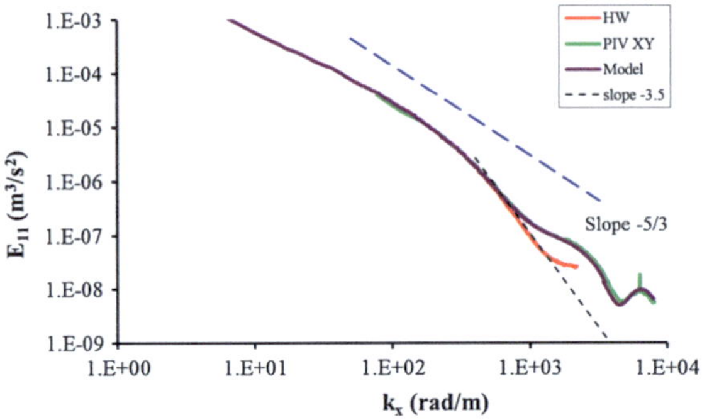

Fig. 4 Comparison at $y^+ = 200$ of the streamwise velocity spectra from PIV with HWA data from [11] using a local Taylor hypothesis

units probably due to the mean velocity gradient at the scale of the X-wire probe. Figure 4 presents the E_{11} spectrum computed from the streamwise SPIV plane at a distance of 200 wall units compared to the HWA spectrum based on a local Taylor hypothesis and the model of Foucaut et al. [9]. The agreement is perfect below k = 900 rad/m. This figure shows also that the spectrum follows a slope of −3.5 for the highest wave numbers. This has no theoretical basis but will be used for the derivative optimization in Sect. 3.

3 Derivative Computation

The derivatives are computed using a second-order centred difference scheme as proposed by Foucaut and Stanislas [4]. As the signal to noise ratio and thus the PIV cut off frequency vary with wall distance, the derivative filter is optimized as a function of this distance. The basic scheme is

$$\frac{\partial u_i}{\partial x_j} = \frac{u_i^{p+n} - u_i^{p-n}}{2n\Delta x_j} + \frac{n^2 \Delta x_j^2}{2} \frac{\partial^2 u_i}{\partial x_j^2} + \sigma_{\frac{\partial u_i}{\partial x_j}}, \tag{3}$$

where Δx_j is the grid spacing along x_j. The first term on the right-hand side corresponds to a second-order derivative scheme. The second term on the right-hand side corresponds to the truncation error. Increasing the value of n decreases the cut off frequency of the scheme. The term $\sigma_{\frac{\partial u_i}{\partial x_j}}$ is the noise error on the derivative proportional to the quadratic sum of the coefficients used to compute the derivative and the random measurement error on the velocity [4]. It is also inversely proportional to the spacing [4]. This term gives an idea of the noise amplification. The higher the n, the lower the noise is. Table 1 gives the values of the cut off frequency and noise level for $1 < n < 4$. For n = 3, the cut off frequency is 925 rad/s, which is of the order of the PIV cut off frequency at $y^+ = 200$ as shown in Fig. 4.

The measurement noise is relatively constant but the signal decreases when the wall distance increases. The cut off frequency of the derivative scheme has also to decrease. For this reason, the derivative is computed from

$$\frac{\partial u_i}{\partial x_j} = \sum_{n=1,4} a_n \frac{u_i^{p+n} - u_i^{p-n}}{2n\Delta x_j}, \tag{4}$$

where the a_n coefficients are optimized as a function of the wall distance to obtain the good cut off frequency. The optimization of this frequency is based on the spectrum model of Foucaut et al. [9] whose signal to noise ratio is one at the cut-off wave number:

$$E_{11PIV} = (E_{11turb} + E_{noise}) \frac{sin(k_c WS/2)}{k_c WS/2} = 2E_{11turb}, \tag{5}$$

where E_{11turb} is the turbulence spectrum measured by HWA, E_{noise} is deduced from the model of [9] plotted in Fig. 3 (see [10]) and WS is the interrogation window size. For the estimation of k_c versus the wall distance, E_{11turb} is approximated by $B(\eta k_c)^{-\alpha}$ where η is the Kolmogorov length scale given by $\eta^+ \approx y^{+1/4}$ in a ZPG boundary layer. In this approximation, B is the energy level at $k = 1/\eta$ of the order of 10^{-10} and α is of the order of 3.5. These two parameters are deduced from a fit on the HWA spectrum as shown in Fig. 4 and are kept constant for all the wall distances.

Table 1 Characteristics of the derivative schemes

No.	k_c (rad/m)	$\sigma_{\frac{\partial u_i}{\partial x_j}}$
1	2800	45.44
2	1350	22.72
3	925	15.15
4	700	11.36

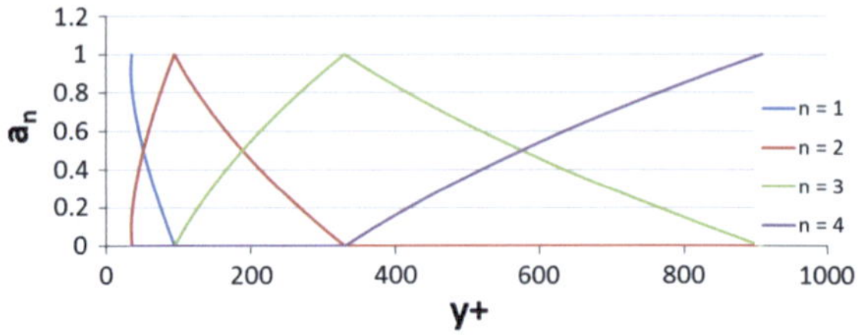

Fig. 5 Evolution of the coefficients a_n versus the wall distance

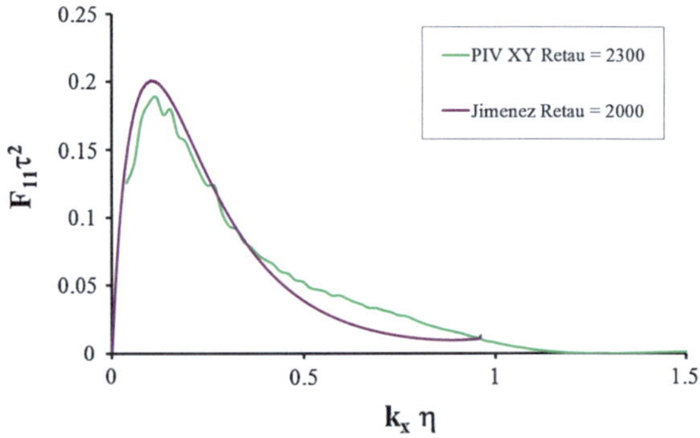

Fig. 6 Spectrum of the term $\left\langle \frac{\partial u_1}{\partial x_1} \frac{\partial u_1}{\partial x_1} \right\rangle$ compared with DNS at 200 wall units

The values of k_c are not very sensitive to the parameter α. A fit with $\alpha = 3$, 3.5 and 4 leads to the same order of k_c. The values of a_n are selected to reproduce the k_c computed from Eq. 5 and are plotted in Fig. 5. The value of k_c decreases monotonously from about 3000 to 800 rad/s when y+ increases from 20 to 900.

Figure 6 shows a comparison of the spectrum of $\frac{\partial u_1}{\partial x_1}$ obtain here by SPIV at $Re_\tau = 2300$ with the DNS of Jimenez [12] at $y^+ = 200$ for a Reynolds number $Re_\tau = 2000$. Both spectra are in relatively good agreement. The PIV slightly underestimates the energy for $k_x\eta$ smaller than 0.4 and overestimates it when $k_x\eta$ is higher, but globally the integral of both give the same value.

As two normal planes were measured here, the term $\left\langle \frac{\partial u_i}{\partial x_2} \frac{\partial u_i}{\partial x_2} \right\rangle$ can be computed in the streamwise or in the spanwise plane or by the product of derivative in each plane. The last possibility gives a variance computed by two different SPIV system. The result is then free from noise as the noises from the two systems are uncorrelated.

$$\left\langle \left.\frac{\partial u_i}{\partial x_2}\right|_{XY} \left.\frac{\partial u_i}{\partial x_2}\right|_{XY} \right\rangle = \left\langle \left.\frac{\partial u_i}{\partial x_2}\right|_{XY} \left.\frac{\partial u_i}{\partial x_2}\right|_{YZ} \right\rangle + \sigma^2_{\left.\frac{\partial u_i}{\partial x_2}\right|_{XY}} \tag{6}$$

By comparison of the three different estimations, the noise can be deduced for each component and each system. As an example, it is found that $\sigma_{\left.\frac{\partial u_1}{\partial x_2}\right|_{XY}} \approx 10\,\text{s}^{-1}$ and $\sigma_{\left.\frac{\partial u_1}{\partial x_2}\right|_{YZ}} \approx 15\,\text{s}^{-1}$ for the present configuration. Following Foucaut and Stanislas [4], from these estimations, the measurement noise of the velocity can be deduced. For each plane and each velocity, it is of the order of 0.1 pixel, which corresponds to the value estimated at the intersection of the two planes [10].

4 Dissipation Rate

4.1 Isotropic Turbulence

The hypothesis of local isotropy allows the reduction of the number of terms down to one:

$$\varepsilon = 15\nu \left\langle \frac{\partial u_1}{\partial x_1} \frac{\partial u_1}{\partial x_1} \right\rangle \tag{7}$$

This is often assumed coupled with a local Taylor hypothesis when a single hot wire is used to estimate the dissipation. It is based on the following relations between the 12 terms of Eq. 2:

$$2\left\langle \frac{\partial u_i}{\partial x_j} \frac{\partial u_i}{\partial x_j} \right\rangle \bigg|_{i=j} = \left\langle \frac{\partial u_i}{\partial x_j} \frac{\partial u_i}{\partial x_j} \right\rangle \bigg|_{i \neq j} = -4\left\langle \frac{\partial u_k}{\partial x_l} \frac{\partial u_l}{\partial x_k} \right\rangle \bigg|_{k \neq l} \tag{8}$$

As can be seen from Fig. 7, which gives the profiles of different terms of Eq. 8, the corresponding relations are well validated for $y^+ > 100$. In Fig. 7 and in the following figure the derivative will be noted $< ui, j\ uk, l >$ to lighten the legends. Below $y^+ = 100$, differences grow when approaching the wall, reaching two and even three orders of magnitude.

4.2 Homogeneous Turbulence

When the hypothesis of local homogeneity is applied, the second term of Eq. 1 is cancelled. This hypothesis allows the ε computation without any crossed term and the dissipation rate is reduced to the sum of 9 derivative terms:

$$\varepsilon = \nu \left\langle \frac{\partial u_i}{\partial x_j} \frac{\partial u_i}{\partial x_j} \right\rangle . \tag{9}$$

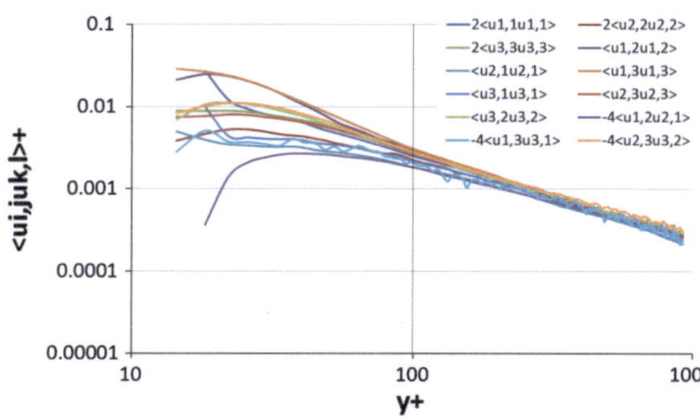

Fig. 7 Terms of ε to evidence their relation in the case of the local isotropy

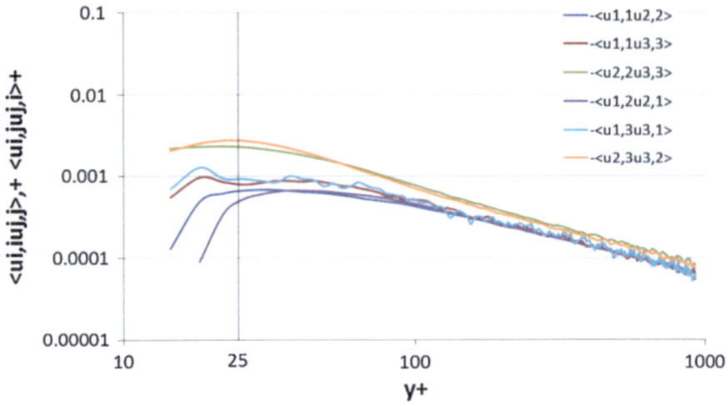

Fig. 8 Covariances of the velocity gradient to evidence their relation in the case of homogeneous turbulence

Coupled with the continuity, this hypothesis is validated if the following relations are verified (see [5]):

$$\left.\left\langle \frac{\partial u_i}{\partial x_j}\frac{\partial u_j}{\partial x_i}\right\rangle\right|_{i\neq j} = \left.\left\langle \frac{\partial u_i}{\partial x_i}\frac{\partial u_j}{\partial x_j}\right\rangle\right|_{i\neq j}. \tag{10}$$

Figure 8 presents the profiles of the different terms of Eq. 10. These relations are well validated except for $i = 1$ and $j = 2$ for which it is true above $y^+ = 25$. As shown in Fig. 3, the accuracy on the velocity decreases below $y^+ = 20$. Due to the use of a centred scheme, it is difficult to conclude because of the uncertainties on the derivative estimation so close to the wall.

4.3 Axisymmetric Turbulence

The hypothesis of axisymmetric turbulence allows computing ε with only four terms [5]:

$$\varepsilon = \nu\left[\left\langle \frac{\partial u_1}{\partial x_1}\frac{\partial u_1}{\partial x_1}\right\rangle + \left\langle \frac{\partial u_1}{\partial x_3}\frac{\partial u_1}{\partial x_3}\right\rangle + \left\langle \frac{\partial u_2}{\partial x_1}\frac{\partial u_2}{\partial x_1}\right\rangle + \left\langle \frac{\partial u_2}{\partial x_3}\frac{\partial u_2}{\partial x_3}\right\rangle\right]. \tag{11}$$

This hypothesis is assumed if the following relations are validated:

$$\begin{gathered} \left\langle \frac{\partial u_2}{\partial x_2}\frac{\partial u_2}{\partial x_2}\right\rangle = \left\langle \frac{\partial u_3}{\partial x_3}\frac{\partial u_3}{\partial x_3}\right\rangle \\ \left\langle \frac{\partial u_1}{\partial x_2}\frac{\partial u_1}{\partial x_2}\right\rangle = \left\langle \frac{\partial u_1}{\partial x_3}\frac{\partial u_1}{\partial x_3}\right\rangle, \left\langle \frac{\partial u_2}{\partial x_1}\frac{\partial u_2}{\partial x_1}\right\rangle = \left\langle \frac{\partial u_3}{\partial x_1}\frac{\partial u_3}{\partial x_1}\right\rangle \\ \left\langle \frac{\partial u_2}{\partial x_3}\frac{\partial u_2}{\partial x_3}\right\rangle = \left\langle \frac{\partial u_3}{\partial x_2}\frac{\partial u_3}{\partial x_2}\right\rangle, \left\langle \frac{\partial u_1}{\partial x_2}\frac{\partial u_2}{\partial x_1}\right\rangle = \left\langle \frac{\partial u_1}{\partial x_3}\frac{\partial u_3}{\partial x_1}\right\rangle. \end{gathered} \tag{12}$$

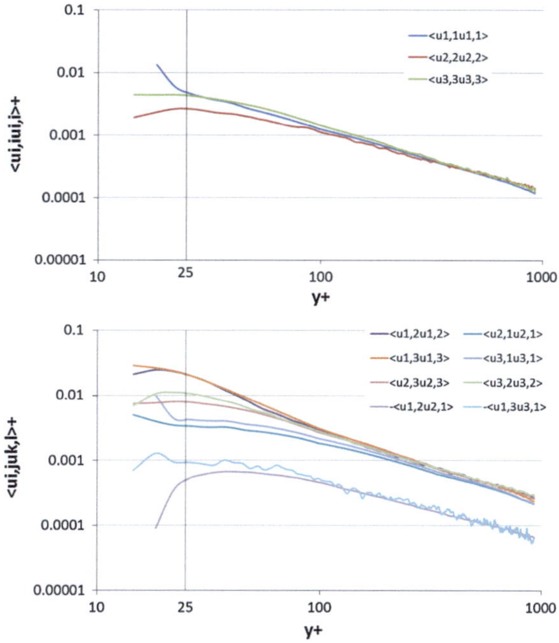

Fig. 9 Terms of ε to evidence their relation in the case of the axisymmetric turbulence

Figure 9 show the profiles of the different terms corresponding to Eq. 12. The corresponding relations are validated for $y^+ > 25$. As for the homogeneity hypothesis, below this value, it is difficult to conclude because of the measurement uncertainties.

4.4 Results

A comparison of the results obtained with the different hypotheses is presented in Fig. 10. They are also compared with the DNS results from Jimenez [12] at a Reynolds number of $Re_\tau = 2000$. The experimental results are in relatively good agreement with the DNS but the slight under estimation is probably due to the filtering of small-scale structures by the SPIV. Figure 11 gives the error on ε computed with each hypothesis using the full dissipation of Eq. 2 as a reference. Apart from some fluctuations due to a lack of convergence of some terms, the dissipation rate computed with the three hypotheses is close above 100 wall units. Below this value the isotropic method underestimates ε significantly. Below 25 wall units it is difficult to conclude on the reliability of the other hypothesis because of the measurement uncertainties. The good agreement between the axisymmetric and full approaches down to the closest wall point can nevertheless be noted.

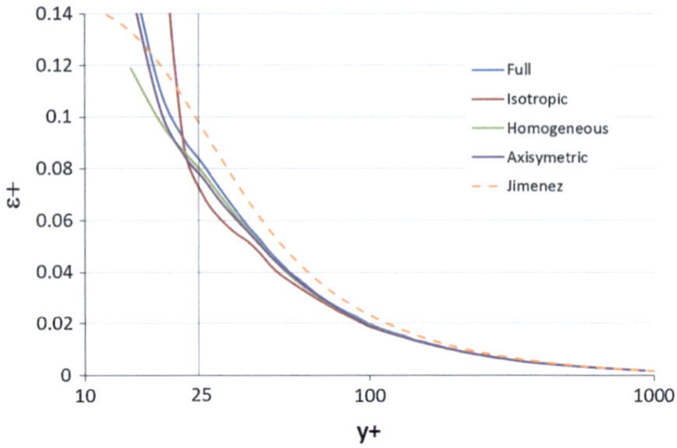

Fig. 10 Comparison of the dissipation rate ε computed with different hypotheses

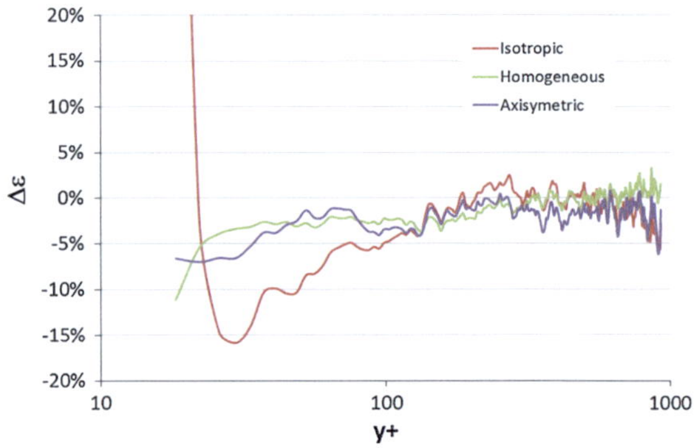

Fig. 11 Errors on the dissipation rate computed with different hypotheses, the full computation is used as a reference

5 Conclusion

For turbulence modelling, it is of prime importance to characterize the dissipation rate. This key parameter is difficult to access in practice as it involves all the terms of the velocity gradient tensor and implies a very good spatial resolution. With this objective, a specific SPIV experiment allowing the derivative computation along the three directions of space was carried out. The measurement noise and the cut off frequency of the derivative schemes were optimized in order to select the best second-order centred scheme depending on the wall distance. The noise characterization was done by a comparison of the variance of derivatives along y obtained from two measurement planes. With the present two planes approach all the terms of the dissipation rate of the turbulent kinetic energy can be computed. As in many cases the full velocity gradient tensor is not available; different hypotheses have been proposed for a long time in the literature to try to assess the dissipation. The comparison of these hypotheses is performed here, thanks to the specificity of the experimental data available. The isotropic turbulence hypothesis is biased below 100 wall units. Both hypotheses of homogeneous and axisymmetric turbulence give results close to the full expression of ε (from Eq. 1) above 25 wall units. At this wall distance, we are very close to the spatial resolution and noise limit of the present experiment, so it is then difficult to conclude about the origin of the differences observed. The experiment could be repeated with a higher spatial resolution in order to obtain more valid points closer to the wall. Nevertheless, in most of the boundary layer, it appears that the axisymmetric hypothesis, which allows us to assess epsilon with a simple SPIV set-up, gives a good estimation of the full turbulent kinetic energy dissipation. Compared to DNS, a small underestimation is observed, which can be attributed to the filtering of the smallest scales. This bias could probably be estimated from a spectral analysis.

Acknowledgments The work was supported through the International Campus on Safety and Inter modality in Transportation (CISIT). S. Coudert is acknowledged for the participation in the experiment.

References

1. R.J. Adrian, C.D. Meinhart, C.D. Tomkins, Vortex organization in the outer layer of the turbulent boundary layer. J. Fluid Mech. **422**, 1–54 (2000)
2. J.-M. Foucaut, S. Coudert, M. Stanislas, J. Delville, Full 3d correlation tensor computed from double field stereoscopic PIV in a high Reynolds number turbulent boundary layer. Exp. Fluids **50**(4, Sp. Iss. SI), 839–846 (2011)
3. S. Herpin, S. Coudert, J.-M. Foucaut, M. Stanislas, Influence of the Reynolds number on the vortical structures in the logarithmic region of turbulent boundary layers. J. Fluid Mech. **716**, 5–50 (2012)
4. J.M. Foucaut, M. Stanislas, Some considerations on the accuracy and frequency response of some derivative filters applied to PIV vector fields. Meas. Sci. Technol. **13**, 1058–1071 (2002)
5. W.K. George, H.J. Hussein, Locally axisymmetric turbulence. J. Fluid Mech. **233**, 1–23 (1991)
6. R.A. Antonia, J. Kim, L.W.B. Browne, Some characteristics of small-scale turbulence in a turbulent duct flow. J. Fluid Mech. **233**, 369–388 (1991)
7. J.L. Balint, J.M. Wallace, P. Vukolavcevic, The velocity and vorticity vector fields of a turbulent boundary layer. Part 2. Statistical properties. J. Fluid Mech. **228**, 53–86 (1991)
8. A. Honkan, Y. Andreopoulos, Vorticity, strain-rate and dissipation characteristics in the near-wall region of turbulent boundary layers. J. Fluid Mech. **350**, 29–96 (1997)
9. J.M. Foucaut, J. Carlier, M. Stanislas, PIV optimization for the study of turbulent flow using spectral analysis. Meas. Sci. Technol. **15**, 1046–1058 (2004)
10. J.M. Foucaut, C. Cuvier, S. Coudert, M. Stanislas, 3d spatial correlation tensor from an l-shaped SPIV experiment in the near wall region, in *In the Present Book* (2014)
11. J. Carlier, M. Stanislas, Experimental study of eddy structures in a turbulent boundary layer using particle image velocimetry. J. Fluid Mech. **535**, 143–188 (2005)
12. S. Hoyas, J. Jiménez, Scaling of the velocity fluctuations in turbulent channels up to $Re_\tau = 2003$. Phys. Fluids **18**(011702), 1–4 (2006)

MIX
Papier aus verantwortungsvollen Quellen
Paper from responsible sources
FSC® C105338

If you have any concerns about our products,
you can contact us on
ProductSafety@springernature.com

In case Publisher is established outside the EU,
the EU authorized representative is:
Springer Nature Customer Service Center GmbH
Europaplatz 3, 69115 Heidelberg, Germany

Printed by Libri Plureos GmbH
in Hamburg, Germany